VELVET ANTS OF NORTH AMERICA

KEVIN A. WILLIAMS,
AARON D. PAN, AND
JOSEPH S. WILSON

Princeton University Press

Princeton and Oxford

To our wonderful families who voluntarily associate with and love us, blemishes, eccentricities, foibles, and all. You are our everything.

Published by Princeton University Press

41 William Street, Princeton, New Jersey 08540

99 Banbury Road, Oxford OX2 6JX

press.princeton.edu

All Rights Reserved

Library of Congress Cataloging-in-Publication Data
Names: Williams, Kevin A., author. | Pan, Aaron D., author. | Wilson, Joseph S., 1980- author.
Title: Velvet ants of North America / Kevin A. Williams, Aaron D. Pan, Joseph S. Wilson.
Other titles: Princeton field guides
Description: Princeton ; Oxford : Princeton University Press, [2024] | Series: Princeton field guides | Includes bibliographical references and index.
Identifiers: LCCN 2023007205 (print) | LCCN 2023007206 (ebook) | ISBN 9780691212043 (paperback) | ISBN 9780691253763 (ebook)
Subjects: LCSH: Mutillidae—North America. | Field guides. | BISAC: NATURE / Animals / Insects & Spiders | SCIENCE / Life Sciences / Ecology
Classification: LCC QL568.M8 W55 2024 (print) | LCC QL568.M8 (ebook) | DDC 595.79—dc23/eng/20230413

LC record available at https://lccn.loc.gov/2023007205
LC ebook record available at https://lccn.loc.gov/2023007206

British Library Cataloging-in-Publication Data is available

Editorial: Robert Kirk and Megan Mendonça
Production Editorial: Mark Bellis
Cover Design: Benjamin Higgins
Production: Steve Sears
Publicity: William Pagdatoon and Caitlyn Robson
Copyeditors: Frances Cooper and Charles J. Hagner
Cover Credit: images by Joseph S. Wilson
Designed by D & N Publishing, Wiltshire, UK

This book has been composed in Garamond Premier Pro (main text) and Proxima Nova (headings and captions)

Printed on acid-free paper. ∞

Printed in China

10 9 8 7 6 5 4 3 2 1

CONTENTS

ACKNOWLEDGMENTS

This field guide would not have been possible without the dedicated researchers and mentors, past and present (including many of our friends), who have assisted in elucidating the amazing world of velvet ants (Mutillidae) to us through field, laboratory, class, and museum work. These include Ernest André, Joseph Bequaert, Hans Bischoff, Clarence Mickel, Osvaldo Casal, Guido Nonveiller, Denis Brothers, Donald Manley, Arkady Lelej, James Pitts, Craig Brabant, Diomedes Quintero, Roberto Cambra, Gabriel Melo, Pedro Bartholomay, David Luz, Guido Pagliano, George Waldren, Justin Schmidt, and Juriya Okayasu. We also pre-emptively acknowledge future students, researchers, and nature lovers. We hope this work, as flawed as it might be, assists you in exploring our incredible world and learn more about the amazing velvet ants. Our apologies for any errors that we may have included, and which you will certainly rectify. Please also consult the formal acknowledgments at the end of the book.

INTRODUCTION

1.1 WHAT IS A VELVET ANT? THE INDESTRUCTIBLE INSECT!

Members of the wasp family Mutillidae are the most interesting and beautiful animals on Earth. These solitary wasps are commonly known as velvet ants (albeit with some controversy, see 1.7). They are parasitoids of immature stages of other insects, especially bees. Female velvet ants are always wingless, while males are usually fully winged and capable of flight. Nearly 5,000 species are known in the world. Velvet ants occur in virtually every habitat type and every continent, except Antarctica. They seem to be especially abundant in tropical forests and woodlands (e.g., miombo and cerrado), grasslands and savannas, and deserts and semi-arid scrublands.

The wingless females loosely resemble large hairy solitary ants, but they are only distant relatives of the true ants (Formicidae). These females are defended by a tremendous suite of natural defenses that leads to another fun nickname: the indestructible insect (Gall et al. 2018). Many other common names used for these insects are also allusions toward their defensive features, which are outlined below.

1.1.1—The colors—Panda ant, *Formiga-feiticeira* (sorceress ant)

Picture yourself as an insectivorous predator, and you see an apparently tasty morsel bumbling along out on the open soil. It constantly taps its antennae on the soil and periodically pokes its head into various divots on the ground. As you approach, you notice a bright contrasting color pattern with long hair-like setae poking up. The colors may be some combination of black, red, orange, yellow, white, or even metallic purple.

These bright colors and distinct patterns possessed by diurnal female velvet ants provide a warning to predators that they are "not to be trifled with" (see 1.3.3). The use of bright color patterns to warn predators away is called aposematism. Diurnal males also have these bold colors, but they just do not possess the impressive defenses to follow through like the females can. More detail will be provided in later sections and chapters about the "fabulous" setae coloration and patterns of female and male velvet ants, but in a sense, many velvet ants want to be noticed in a crowd. If you were to remove the colorful setae from a female velvet ant, you would have an insect that looks very much like a typical true ant, at least superficially. Lots of creatures like to eat ants, such as lizards, birds, and toads, among others. How do you make yourself different enough to not look like a delicious ant? A "fur" coat with style!

A recent paper by Hines et al. (2017) provides some valuable information on what chemicals are contributing to the vivid colors that adorn mutillids. Many different colors in the animal kingdom are due to organic molecules like carotenoids, melanins, pterins, and ommochromes for pigmentation. However, it appears that melanins, particularly pheomelanins for bright colors like red, orange, and yellow, and intermediate concentrations of eumelanins and pheomelanins for black coloration, likely determine velvet ant setal and cuticle coloration. White setal coloration analysis in the velvet ant species used in the study (*Dasymutilla gloriosa*) was interesting in that there was a peak in the 295 nm range of the spectrum wavelength, indicating that an as yet an unknown compound, along with low levels or absence of melanins, may be contributing to the setal coloration of the white setae desert forms. Although not known, it is possible that velvet ants that possess metallic setal and/or cuticle coloration may result from structural coloration, like those seen in jewel scarabs and *Morpho* butterflies, but this needs to be investigated.

1.1.2—The quickness

Against your better judgment, you decide to pursue this colorful oddity and chase it down. Immediately, it starts frantically running away, ducking under every leaf in sight and even trying to dig itself into the soil.

It is hard to characterize the gait of an entire family of insects, but to us, the locomotion of velvet ants resembles that of a sleep-deprived parent who is already 15 minutes late to an important appointment (months in the making), searching for their car keys on a LEGO-strewn surface. Their locomotion kicks up to a whole new level of speedy zaniness when a potential threat (especially in the form of a nature photographer) is observed, with females ducking, weaving, and utilizing all their parkour skills. An important note: a velvet ant wants to go about life as it wants, it does not want to waste energy with a potential predator or threat. Her locomotion and traveling pace provide the wasp with many chances to find resources for her young, food and water for herself, and shelter when necessary. The hurried locomotion helps her cover a lot of ground in these various quests.

Velvet ants also use the environment to their advantage as they race across the landscape to avoid becoming an entrée—a spiny plant like a cactus, euphorb, or yucca makes a wonderful temporary haven to avoid an actively hunting predator. Even if a plant is not armed, dense bunch grasses or leaf litter can provide enough cover for a velvet ant to remain undetected once it has reached the foliage or even provides an unsuspected escape to another section of brush.

For understanding their absolute speed, very few studies exist, although Schmidt & Blum (1977) clocked *Dasymutilla occidentalis* at cursorial speeds of 13.8 ($\pm$ 3) cm/sec or around 0.5 km/hr—not bad for an insect tank if you ask us.

If you look at the locomotion of a true ant (Formicidae) it just feels different from that of the heroes in this book. More regimented. More precise. True ants "stay on the trails," typically following the chemical cues or their sister ants as they forage. Ground spiders often have a very fast but fluid or graceful (excluding the tarantulas—who we consider the mastiffs of the Araneae world) movement across the landscape. Large, robust beetles, please forgive us, often walk in an awkward, plodding manner—this excludes the often erratic carabids (and the tiger beetles are pretty vain in their movements—speeding just ahead of you, waiting till you are near them, and racing off again). The movements of velvet ants are pretty unique, and with some practice you can often spot one just by the gait, even before you see the flash of color that adorns her. For large *Dasymutilla* species, you might even notice them from your car on a dirt road. It should be noted that with generalizations there are always exceptions. While most velvet ant types of locomotion and gaits are as described here, some do not conform (prima donnas that they are), and these will be noted in other portions of this book.

1.1.3—The chemicals

While messing with this creature, you corner it and start to smell something funky.

Here the velvet ant has another trick up its sleeve; even if you grabbed it safely, the first bite, or even whiff, may indicate that the velvet ant may be unpalatable. Secretions from the mandible appear to contain a number of organic compounds, but the most common compound found in a number of species appears to be 4-methyl-3-heptanone. The ways that these mandibular secretions affect predators are not well known, but they appear to serve as an irritant and gustatory and/or olfactory warning signal.

The chemical that is excreted when a velvet ant is held with forceps is pungent, and I would describe it as a sweet-sour-slightly fetid decay smell. No, I have never tried to eat a velvet ant.—AARON

While we usually consider the effectiveness of velvet ant defenses against vertebrate predators, some mandibular secretions seem especially helpful against a completely different potential natural enemy:

true ants in the family Formicidae. Some velvet ant secretions seem to mimic alarm pheromones of various ant genera, such as *Solenopsis*, *Camponotus*, and *Pogonomyrmex* (Fales et al. 1980). Imagine getting swarmed by your enemies, then shouting "watch out behind you," and causing them to scatter in chaos while you make your escape.

Do all velvet ant species produce these stinky mandibular secretions? The answer, like so many answers to questions about velvet ant biology and natural history, is we do not know. It has been observed in a number of western hemispheric genera (*Dasymutilla*, *Pappognatha*, *Timulla*, and *Traumatomutilla*), some of which are not closely related, so it likely is a prevalent characteristic within the family. In addition, whether these secretions serve additional purposes in terms of conspecific pheromones, prey manipulation, and such like, are also not known. More studies need to be done!

1.1.4—The squeak—barking ant, *Perritos de Dios* (God's puppies)

Even if the smell does not convince you to give up the chase, this potential prey has another ploy—the strange fuzzy creature starts "screaming" at you.

Velvet ants do not have vocal cords and would likely leave you wanting a different karaoke partner, but they are able to make noise, and this noise appears to be a great warning system geared toward predators. Velvet ants (both female and male) produce sound via stridulation—when two parts of the body are rubbed together to create a noise. In the case of velvet ants, the specialized organs can be found on two adjacent abdominal segments (tergites). These organs basically consist of a file (a series of cuticular ridges), which is rubbed against a rake on the adjoining tergite. The velvet ant will rub the file and rake together, creating a pulsating, buzzing sound. While there do appear to be differences in the shape of the file and rake across different species, there has not yet been a broad investigation into how these differences affect the sound produced.

In a study by Torrico-Bazoberry & Muñoz (2019), the stridulating sound frequency of some South American velvet ants was likely too high in range to deter lizards, which pick up sounds at lower frequencies. They suggest, and it certainly seems reasonable, that such a frequency range would be more effective against potential rodent predators. They also suggest that stridulation may assist in warning conspecifics of a dangerous predator in the vicinity. In addition to stridulation as a defensive behavior, there is some evidence that velvet ants also stridulate during courtship and mating.

1.1.5—The armor—bone ant

Not being deterred by the color, smell, or cacophony from the little speed demon, you chomp onto the creature—and crack a tooth on its crunchy shell.

There could be multiple explanations for why female velvet ants are wingless (addressed later, in section 1.9.1), but one of the primary benefits of losing wings seems to be the enhanced capacity for strong defensive armor. Experiments by Schmidt & Blum (1977), using a force transducer (a force sensor that measures compression and tension forces), found that velvet ant exoskeletons required 11 times more applied force to be crushed than a honey bee exoskeleton. A less quantitative but more relatable understanding of the strength of a velvet ant's exoskeleton is human versus velvet ant. The customary way to preserve and display insect specimens is to pierce the thorax with a stainless steel pin. While impaling specimens on a steel pin is not difficult for most insects, this endeavor is more frustrating with velvet ants—particularly large species of *Dasymutilla*, *Hoplomutilla*, and *Traumatomutilla*. In some cases, the steel pin is bent when applied forcefully to the velvet ant thorax, and the specimen remains undamaged and the human miffed. Some velvet ants are known to "sting from the grave," as the slipping pin can pierce directly into a collector's thumb rather than the velvet ant's body.

Many a velvet ant has been known to be trodden on by the sole of a shoe or boot by accident and come away from the event mildly inconvenienced but uninjured. If a collector is caught by surprise in seeing a velvet ant, an effective strategy is to step on top of the wasp (as long as your shoes have an adequately strong sole) until a collecting container can be prepared.

The exoskeleton of velvet ants is also particularly protective because the thorax forms a single, fused plate. While some other wasp groups are also wingless (1.9.1), they generally have the thorax divided into multiple flexible plates. While they may not seem dissimilar, this literally creates chinks in the armor. Each joint where segments meet is a point of weakness. In terms of species richness or distribution, none of these other solitary wingless wasp groups is as successful as the velvet ants—could female armor be the innovation that allowed velvet ants to rise above the rest? In addition, velvet ants have their head and abdomen heavily armored as well. The second dorsal segment (tergite) is the largest and has thickened exoskeleton; the apparently weaker apical segments, or the junctures between these segments, can be somewhat withdrawn adjacent to the protection of this largest tergite.

1.1.6—The sting—Cow killer, *Mulas del diablo* (the devil's mules)

Although this morsel has been a challenge, you finally think your efforts are about to pay off. You're exhausted from the chase; your eyes, ears, nose, and jaws are experiencing some combination of confusion and discomfort, but you're not giving up. Within a few seconds, your mouth explodes in excruciating pain. You spit out your newest, still very much alive, nemesis. As she runs off into the distance, you wonder why you didn't pay attention to the warning signs and promise yourself to never mess with one of those beasts again.

Most casual observers of Hymenoptera—whether observing honey bees and bumblebees moving from flower to flower in gardens, navigating picnic diplomacy with yellowjackets (do I really need to shoo the wasp off that small piece of hamburger on the table and risk its wrath?), or gingerly avoiding fire ants while mowing in flip flops (don't judge!)—have a pretty good idea why these creatures need to be treated with respect or, at the very least, not harassed: they can cause painful stings. These stinging insects are the world's first or second most diverse group of insects—there is a cordial disagreement with many beetle specialists on who should receive the gold and silver medals for most species. The sting, the correct term for this weapon as opposed to the colloquial "stinger," was originally used for egg laying (and still serves this purpose for more "primitive" Hymenoptera groups). In the Aculeata, or higher Hymenoptera, this ovipositor has been modified to now deliver venom, either as a defense mechanism, or to assist in prey capture, or to immobilize future meals for the wasp's offspring. Because the sting is derived from an egg-laying device, male aculeates are basically harmless—they lack the requisite hardware.

While some may think of a sting as a hollow needle or spear, when envisioning a female velvet ant's sting it is really more like a very sharp-pointed hose that is incredibly flexible and maneuverable. The sting is coiled inside its metasoma and is typically about the length of its entire abdomen when fully extruded. Therefore, the sting of a large species, like *Dasymutilla occidentalis*, can be nearly one centimeter, or half an inch. In fact, recent research shows that mutillids have the longest sting, compared to their body size, of any stinging bee, wasp, or ant (Sadler et al. 2018). A velvet ant can therefore defend herself from any angle, even from above and behind, as seen in this *Sphaeropthalma unicolor* female.

Luckily for velvet ant collectors, mutillids are often slow to sting in defense. Even more luckily, sting potency in mutillids seems to be proportional to body size, meaning that most velvet ants have a mild sting, and many species have difficulty even breaking the skin. In a recent study (Jensen et al. 2021), the main venom component that appears to be the cause of intense pain in those envenomed by *Dasymutilla* species are cysteine-free peptides. The lethality of velvet ant venom is much reduced compared with social species, like the honey bee or harvester ants

Photo by Alex Wild

Sphaeropthalma unicolor

(Schmidt et al. 2021). The venom appears to send a "let go and leave me alone" message and is not meant to cause irreparable harm, at least to vertebrates.

1.2 WHAT DO VELVET ANTS DO?

While we have some information about the ecology and natural history of North American velvet ants, it is certainly not complete or well known for most species (as you will see in the described species sections), even commonly encountered ones. One motivation for creating this field guide is that we want more people to recognize these amazing beasties and report on exactly what they do in terms of activity periods, behavior, hosts, and predator interactions!

1.2.1 Life history—parasitoidism and hosts

Female velvet ants seek out prey hosts to feed their progeny. Unlike an eagle mother, the velvet ant does not provide food to active hatched young. Instead, female velvet ants are more like preppers, laying an egg on the host so that the larva will have all the food it needs when it hatches. Velvet ants can be described as ectoparasitoids: "ecto" because they are located on the outside of the prey item and "parasitoid" because the velvet ant larva feeds on the living host and will eventually kill it. A regular parasite can be detrimental to a host but does not usually kill it—think ticks, leeches, and tapeworms. In terms of host selection, most velvet ant species prefer the immobile, immature-stage insects in the Holometabola. These Holometabola insects go through complete metamorphosis: egg, larva, pupae, and adult growth stages. Examples include the butterflies and moths, beetles, flies, antlions, and wasps and bees. In fact, many velvet ants specifically target young bees and wasps as hosts. These are good targets because in their immature stages they do not move and come pre-packed with natural defenses against the elements and other parasites. Many wasps, like spider-wasps, do not mind using adult animals that can move as prey, because they use the work-around of paralyzing the host before they lay an egg. Velvet ants do not want to go through this rigmarole of having to attack a potentially dangerous adult host; best to just use a dormant/resting "burrito" for their young. As noted, however, there are always exceptions. There is evidence of a species of *Odontophotopsis* that uses the fertilized eggs of the Arizona Sand Cockroach (*Arenivaga genitalis*) as hosts. Since eggs are immobile, these can be used as well, even those from insect groups who go through incomplete metamorphosis. However, there is also the possibility that this report may be somewhat inaccurate. It is possible that the velvet ants that popped out of the cockroach eggs were actually parasitoids of another parasitoid wasp's pupa that was using the cockroach eggs as its own host.

Studies that have observed female velvet ants looking for hosts note that they actively use their antennae to tap the ground, likely using chemosensory signals to locate their quarry. In some cases, particularly for aggregated nests of bees, there is a guard bee that has to be dealt with. This usually involves trying to intimidate it into retreating or trying to barge past the guard, who is intent on stopping the female velvet ant with its mandibles (the bee often is aware that its sting is ineffective against the hard cuticle of the mutillid). Once the female velvet ant has found an immobile pupa, she will use her sting to probe the potential host, sometimes only superficially but at other times with a deeper pierce. One recent study (Jensen et al. 2021) suggested that the venom may act to arrest further development of the host. The egg is laid on the exterior of the host, and when the velvet ant larva hatches, it feeds on multiple areas of the host.

While the method of providing food to sustain immature velvet ants may be a bit gruesome, adults mainly take a more vegan dietary course (although one study on *Pseudomethoca frigida* noted that a female may feed on prepupae or pupae before laying eggs on other individual pupae). Nectar from both flowers and extrafloral nectaries provides food for velvet ants on the go. If you live where there are a bunch of sunflowers, you can often find female velvet ants at the base of the plant or on the stems in the late morning or early evening, particularly when the plant is "weeping" nectar.

1.2.2 Life history—courtship and mating

Sexual dimorphism is the rule for the family Mutillidae, and in many cases the sexes have little in common morphologically. Males are generally winged, and the females are always wingless. There are a few species where wingless males exist, but this is rare. This extreme sexual dimorphism is interesting because velvet ants, like all Hymenoptera, have a haplodiploidy system of sex determination. What that means is that female wasps, bees, and ants are diploid (they have two complete sets of chromosomes, one from the father and one from the mother). Males, on the other hand, are haploid; they have only one complete set of chromosomes. Generally, if the female wants male offspring, she has the option to lay an egg without fertilization, which results in a haploid egg that will develop into a male. If the female wants female offspring, she can choose to fertilize the egg with sperm she has stored just for this use. This fertilized egg will develop into a female wasp. The ability to choose the sex of her offspring comes in very handy for a generalist parasitoid that attacks a wide range of hosts of different sizes, particularly when it comes to mating strategies.

Two general mating strategies have been described for velvet ants: the first is often referred to as in situ mating and the other is known as phoretic copulation. In situ mating happens "in place," often on the ground where the wingless female spends most of her time. Phoretic copulation on the other hand is when a male physically transports a female by flight and/or foot from their initial site of contact during courtship. The pair may mate on a plant, or mating may take place during flight. Species that use phoretic copulation generally have males that are larger than females. Because of haplodiploid sex determination, velvet ant mothers can choose larger host insects to rear their male offspring and smaller hosts for female offspring.

Winged males and wingless females seem to have different selection pressures, driven by their disparate goals and limitations. As such, diagnostic features used to separate species in males are often different from those that are useful for separating females. Males are shorter lived, and sexual selection seems to be a bigger driver for their diversification. Therefore, males of a species are likely to be differentiated by features of the genitalia or secondary sexual characters on the head, legs, or underside of the body. Females are longer lived, and their main selection pressures seem to involve defense against predators and tools to find or fight their hosts. For example, many diagnostic features can be seen in the sculpture of the pygidium (a tool for digging into soils) or the shape of the head (providing differences in musculature for fighting against host bees and wasps). Females additionally seem to show greater variation in detailed aposematic color patterns (see 1.3.3), which protect against vertebrate predators.

Regarding defense specifically, females and males are likely to encounter different suites of predators based on their main mode of locomotion, with females being mainly on the soil and males spending time primarily in the air. This can lead to a phenomenon called "dual sex-limited mimicry," wherein the different sexes of a single species have divergent color patterns that provide mimetic protection differently to males and females, since each sex has evolved color similarities to different models. For females, mimetic color patterns most often are similar to other velvet ant species, while for males, appearance is more likely to resemble other winged wasp families. The classic example of dual sex-limited mimicry is exhibited by *Dasymutilla gloriosa*. Interestingly, there is a chance that this is not truly based on mimicry, since the female's fluffy white coloration seems to be driven just as much by climatic adaptation to their hot desert habitats as it is by mimicry with other similarly colored velvet ant species.

Since males and females are divergent in their overall body structure, their suites of diagnostic features, and even in their coloration, it can be remarkably difficult to determine which males and females are a match. For this reason, many species are known from only males or only females; figuring out the correct sex associations is a primary goal for velvet ant taxonomists. Sadly, the methods that prove most useful for matching sexes require a great deal of money, time, or luck.

If money is no object and equipment is available, matching identical DNA sequences of male and female specimens is a great way to discover a sex association.

Dasymutilla gloriosa: female, male, *gynandromorph*

If you have spent decades studying velvet ant taxonomy and collected thousands of specimens, you can eventually find parallel diagnostic features and recognize overlapping distributions and activity periods, which can confirm a sex association.

Field observations of mating behavior seem like an obvious way to associate males and females, but courtship and mating in velvet ants often occurs very quickly, and a female needs to mate only once to obtain all the sperm she will need for her entire life. Finding and digging out nests of their hosts could allow you to rear males and females together, but these hosts work hard to hide themselves from velvet ants—what hope do we have to find them consistently?

If your field kit includes a horseshoe, rabbit's foot, and four-leaf clover, however, one remarkably rare natural phenomenon might help you discover a new association. Gynandromorphs are rare mutant specimens that have a body morphology that is partly male and partly female. If the gynandromorph has enough tissue of each sex present, it could lead to a sex association. One fantastic use of this tool was a specimen of *D. gloriosa*, which facilitated the first match for that classic dual sex-limited mimicry example. Only about 10 gynandromorph specimens have been discussed in published literature and, in at least four cases, they provided valuable clues for previously unknown sex associations.

1.2.3 Life history—activity periods

Most velvet ants, at least the colorful ones, have activity periods that are typically considered diurnal, meaning they are active when the sun is shining and there is hay to be made. In most areas of central and western North America, you will not find velvet ants active all day; instead they are mainly found mid-morning and in the last few hours of the day, before sunset—perhaps a little nap is in order during the hours in between? In more humid areas, like in the eastern USA and many tropical climates, velvet ants can be found during much of the day and seem to be most active when it is hottest, in the mid- to late afternoon. Unlike for humans, "dry heat" seems more intolerable to velvet ants than the muggy horrors of Florida swamps and Thai jungles.

Some western species take their aversion to desert sunlight to the extreme by becoming nocturnal. In western states, like California and Arizona, more than half of the known species are night active. Nocturnal velvet ants usually come out after sundown, but some species may be active an hour or two before the sun sets, especially at higher elevations or during cooler seasons.

1.3 WHY SHOULD WE CARE ABOUT VELVET ANTS?

Learning about velvet ants is its own reward ... but it might not always pay the bills. In a few specific ways, the biology of velvet ants makes them good candidates for more widely appealing research questions, or maybe even relevant to applied fields, like conservation or biocontrol of agricultural and medical pests.

TOP: *Chrestomutilla glossinae*: female (left) and male (right)

BOTTOM: *Smicromyrme tekensis*: female (left) and male (right)

1.3.1—Parasite interactions

Velvet ants could be considered pests, because many species "dispatch" the pollinators that help to feed us (poor little baby bees). However, velvet ants can help humans by attacking our enemies. In tropical Africa, multiple velvet ant species, such as *Chrestomutilla glossinae* and *Smicromyrme benefatrix*, are parasites of the tsetse fly (*Glossina morsitans*), which is a vector of the deadly African sleeping sickness parasite (*Trypanosoma brucei*). In Iran, two species, *Smicromyrme nikolskajae* and *S. tekensis*, were recently discovered parasitizing the tephritid fruit fly *Carpomya vesuviana*, a serious pest of *Zizyphus jujuba*, jujube or red date—a tasty and sometimes medicinal fruit product. Since fewer than 10% of velvet ant species have known host records, there are likely many other beneficial species around the world waiting to be recognized as such.

1.3.2—Biogeography

The winglessness of females seems to limit their dispersal speed, so they can assist us in understanding biogeography, particularly historical biogeography. Historical biogeography is the study of where organisms lived in the past and how historical events (both geological and climatological) affected their distributions. Since velvet ant diversity in North America is higher in the deserts than the more temperate areas, and because of their limited dispersal abilities, velvet ants have been able to help us understand how the history of change in the deserts has shaped the diverse desert communities found in North America today. For example, some of our work on nocturnal velvet ants led to a better understanding of how the rise of western mountain ranges led to distinct insect communities in California. Other studies have found that the climatic fluctuations during the last ice age helped shape the genetic diversity in our modern velvet ant communities.

1.3.3—Mimicry studies

Although it may seem unfair, appearances matter in the natural world. They matter in terms of attracting mates, in competition for resources, in warning off possible attacks from potential predators, or to blend in and almost disappear into an environment. Diurnal velvet ants often use bright coloration and contrasting markings to make sure that they are noticed. In contrast, velvet ants that are night denizens are typically brown to pale red in coloration, since working in the dark does not require a vibrant wardrobe. What is particularly interesting about velvet ants and their aposematic (warning) coloration is that different genera and species often have similar color patterns, indicating that mimicry is occurring. To establish a Müllerian mimicry ring, a scientist must demonstrate that multiple well-defended species share similar appearance, that these similarly colored species overlap in distribution, and that the color patterns arose from convergent evolution, rather than shared ancestry.

What is also interesting is that velvet ant diversity and species richness do not necessarily correlate with the number of color syndromes in a particular area. Africa may be home to the largest velvet ant fauna in the world (although more studies need to be done in Asia and South America to determine whether this is the case), but the number of major color syndromes in Africa is far lower than that found in North America. Why? What is the cause of this? Is it due to evolutionary and geologic history differences between the two continents, potential predator diversity differences, the number of different ecoregions/natural communities, geographic barriers, genetic (mutation) restrictions or flexibility within genera/subfamilial taxa, or some other factors? It is likely a combination of multiple causes; this exciting field is open to more exploration.

Through our and other studies, various color syndromes that likely form Müllerian mimicry complexes have been recognized. In some cases, they may be too broad or too narrow, but at the moment, they form a convenient mechanism for sorting velvet ants.

1.3.4—Velvet ants as pets

Since velvet ants are hardy, attractive, and energetic, they can make surprisingly good pets. To keep a female velvet ant as a pet in the cheapest way possible, you will need only a large see-through plastic container with a tight-fitting lid, some sand from outside, and a few rocks or leaves for shelter. You can feed them a small bite of fruit (like half a grape) once or twice a week and they should be fine for a while. You can improve the habitat and maybe extend the life of your pet by giving them a larger terrarium, a heat lamp, some extra humidity for eastern species, and more consistent sources of food and water, like a commercially available food jelly cup or a micro-waterer. Since velvet ants are heavily armored and uninterested in eating other adult insects, multiple specimens (of multiple species) can be kept together, and they can even be kept with other hard-bodied insects, like darkling beetles.

At this point, they have not been bred in captivity, so you will need to capture your pet from the wild. This means that you cannot have a perfect knowledge of the pet's life span, since you will not know how old the velvet ant was when you caught her. A well-cared-for, young velvet ant female of a large-bodied *Dasymutilla* or *Pseudomethoca* species could potentially live for up to two years. Even in captivity, they are constantly on the move and just as entertaining as fish in a tank. You can also show your friends something they have probably never seen before.

It is not certain that they will ever be more than an oddity on the desks of quirky entomologists. If, however, some enterprising pet-breeder finds the secrets to mass production of fresh velvet ant females, and then markets them well, velvet ants could become just as popular in classroom education and the pet trade as walking sticks or hissing cockroaches.

1.4 WHERE DO VELVET ANTS LIVE?

Many insect groups, especially bees and hunting wasps, have distinct genus- or species-level differences in food sources, so researchers learn collecting "hacks" to target these food sources as a step toward collecting the target organism. Squash bees in the genus *Eucera* (*Peponapis*), for example, are specialists that collect pollen only from squash flowers, so researchers looking for these bees can simply look for squash flowers to easily find them. Beetle researchers enter the field armed with dozens of different collecting tools, including various baits, sieves, and even a trusty old crowbar for peeling bark from downed trees. The majority of velvet ants, however, seem to be generalist parasitoids of immature insects, which are widespread in the environment, and the adult food sources seem even more ubiquitous. Even when the targeted hosts are completely unrelated, such as the apparently rare fly parasitoids (1.3.1) versus the typical hunters of baby bees, the velvet ant's target prey are still sought in the soil, and the velvet ant collector's target prey are found crawling above the soil.

Other than the obvious behavioral distinction (and different collecting methods: see 1.5) between nocturnal and diurnal species, all of the dozens of velvet ant species in a given locality seem to be doing more or less the same thing. There is, however, one major distinction in host behavior that seems to impact velvet ant behavior and morphology: ground-nesting versus arboreal-nesting. Many host species nest in twigs, cavities, or mud-nests that are situated above the soil, particularly in trees and sometimes on human habitations. Unlike the majority of velvet ant species, females that home in on arboreal hosts usually have the pygidium undefined (Fig. 6.2.4a,b). Additionally, the foretarsus lacks a "rake" of flattened setae, and the body is usually more slender and sparsely setose. In eastern Asia at least, there seems to be a difference in "walking speed" between these general host-searching cohorts; arboreal velvet ants have a slower gait, likely due to the higher stakes of falling from trees (at least in time wasted). One would imagine that these species would be more frequently encountered above the ground (and they probably do spend the majority of their lives there), but from personal experience, most specimens are still observed crawling on the ground, like their ground-nesting relatives. Furthermore, ground-nesting species are often seen above ground while they seek out nectar or sap for adult feeding.

Velvet ants are often thought to be most prevalent in desert habitats, especially since some of the world's most famous species, like *Dasymutilla gloriosa* (3.1.1.1) or *D. magnifica* (3.3.1.1), are prevalent in those areas. In North America, particularly when you add in the nocturnal species, desert habitats are certainly a haven for velvet ant diversity. Other habitats, however, host even higher diversity around the world, particularly tropical forests and grasslands. In North America, even though the species-level diversity seems lower, diurnal velvet ants appear to be more abundant in grasslands and forests than in deserts.

This book mainly focuses on the diurnal species in the continental USA (many of these species also occur in Canada and Mexico). Political borders are not particularly useful for describing velvet ant distributions, especially in the larger western states. Terrestrial ecoregions are useful, but there are too many categories to list when summarizing the distribution of a species, though we have included the ecoregions on the distribution maps. For this book, we subdivide the USA into three broad categories: western USA, central USA, and eastern USA; then subdivided those regions into two or four regions. These subdivisions are not officially sanctioned, and many other field guides or documents will divide these areas in a different manner. For our current knowledge of velvet ant distribution, however, they seem useful. These regions are often mentioned in the species accounts in chapters 3 and 5, and are especially useful in the keys listed in the appendix (see 9.3).

The eastern USA includes most of the humid states east of the Mississippi River. If you draw an imaginary line from Lake Michigan to Houston, Texas, and then excise some of the plains habitats in Indiana, Kentucky, Tennessee, and Alabama, you have got a good grasp on our concept for the eastern USA. There are 62 velvet ant species, spread across 127 species accounts (chapters 3 and 5). The velvet ant fauna in the eastern USA seems relatively uniform, with gradually greater diversity and abundance the farther south you go. Multiple species do seem to be restricted to southern climes, particularly Florida and the southern extremities of states surrounding the Gulf of Mexico; 10 species seem to be exclusive to that region. Outside this southeastern zone, only two rare species seem to have localized distributions in the northeast (it is not clear whether this is an artifact of their rarity rather than their true distribution), and a few more seem to be relatively cold-adapted and range from the northern Great Plains east to the Great Lakes or New England. Two eastern regions are used in this book: Florida–Gulf has 54 species, and broadly the eastern has 52. Fewer than 10 of the eastern species are known to be nocturnal, and each of these is also included in chapters 3 and 5.

The central USA is loosely defined as areas west of that imaginary Lake Michigan–Houston line and east of the Rocky Mountains, Colorado Plateau, and Chihuahuan Desert. There are 117 diurnal velvet ant species, spread across 197 species accounts (chapters 3 and 5). The two regions that we treat in the central USA are loosely referred to as the Great Plains (from Dallas and Odessa north to Montana, North Dakota, Minnesota, and Wisconsin) and southern Texas (basically the portions of Texas west of Houston, south of Waco, and east of Big Bend). In total, 78 diurnal species (chapters 3 and 5) and an additional 20 nocturnal species (chapter 6) are known from the Great Plains (six species are apparently exclusive to the Great Plains); 103 diurnal species (chapters 3 and 5) and an additional 20 nocturnal species (chapter 6) are known from southern Texas (33 species are apparently exclusive to this region).

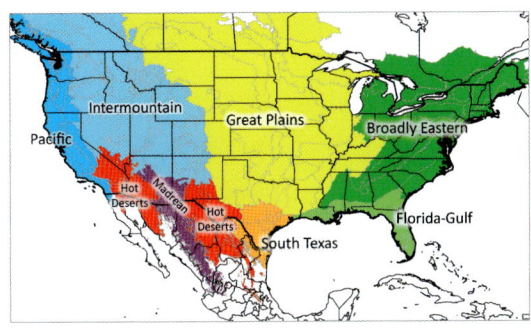

Map of regions used in this book.

The western USA is defined as everything west of and including the Rocky Mountains, Colorado Plateau, and Chihuahuan Desert; there are 137 diurnal species (chapters 3 and 5) and a whopping 159 nocturnal species (chapter 6). The western USA is divided into four regions: Pacific, Intermountain West, North American hot deserts, and Madrean Archipelago (basically mountain island extensions of the Sierra Madre Occidentals and Sierra Madre Orientals)/Arizona mountains. The Pacific region includes everything in California, Oregon, and Washington occurring west of the Sierra Nevada and the accompanying southern California mountain ranges and their eastern slopes. So far, 21 diurnal species and 28 nocturnal species are known from this region (21 species are apparently exclusive to this region). The Intermountain West includes the Rocky Mountains, Colorado Plateau, Great Basin, and northern regions found east of the Sierra Nevada. Although it is the largest western region by area, it includes fewer species than the following regions. So far, 35 diurnal species and 39 nocturnal species are known from the Intermountain West (10 species are apparently exclusive to this region).

The hot deserts, taken as a whole, include the Mojave, Sonoran, and Chihuahuan Deserts. This region, broadly defined, includes 73 diurnal species (chapters 3 and 5) and 108 nocturnal species (chapter 6). The relative total number of species in each desert, however, is comparable to many of the other broadly defined regions, and each desert includes nearly as many or more species than the entire eastern USA: Chihuahuan (99 total species, seven exclusive), Sonoran (99 total species, 26 exclusive), and Mojave (61 total species, six exclusive).

The diversity in hot deserts is impressive, but the true peak of velvet ant diversity is found in the Madrean Archipelago and Arizona mountains. This region is a northern extension of the Sierra Madre in Mexico, and the region includes mid- to high-elevation areas in central and southern Arizona and southeastern New Mexico. This region is bordered by the most diverse of our deserts, the Chihuahuan and Sonoran, to the east and west, and the Rocky Mountains to the north. Although this is the smallest region by area, it has the highest diversity (unless you lump all three hot deserts into one region). So far, 76 diurnal species (chapters 3 and 5) and 73 nocturnal species (chapter 6) are known from this region. Of these, 55 are exclusive to this region, at least in the USA; many of them are more widespread in Mesoamerica.

1.5 HOW CAN I CATCH VELVET ANTS?

1.5.1—Diurnal females

Walk and watch ... To catch a velvet ant, you must think like a velvet ant. These wasps spend their entire lives walking and scanning their habitat in search of rare treasures in the soil; their success is directly correlated with the amount of time spent searching. The velvet ant collector likewise must walk and scan the habitat for rare treasures crawling on the soil, and the longer you search the more you find. As much as we wish there were, there really are not many shortcuts or secret tips that will get you loads of specimens, but we will try to share some helpful information here.

Female mutillids are most commonly seen on the ground, and specifically on open ground. It is unclear whether these wasps are more abundant on open or covered soil, but it is certainly easier to see them when they are not hiding beneath layers of plant foliage. However, vast expanses of perfectly barren soil do not seem optimal for mutillids either. The aim is to find areas that are surrounded by, interspersed with, or in close proximity to vegetation. Therefore, in heavily vegetated regions, you often find most specimens in the least covered microhabitats, like dirt roads, recently eroded hillsides, and river or pond shores. Conversely, in sparsely vegetated areas, the best habitats seem to be those offering the most cover, like the bases of scattered bushes.

The fun part begins after you locate a female velvet ant. Since they can fight back and they run quickly, there is no guarantee of bringing home a velvet ant that you see in the field. The safest and generally most efficient method for capturing a velvet ant requires only a medium-sized, empty plastic tube. It is helpful for this tube to be transparent, so that you can locate and examine the insect during capture. To collect a specimen, you simply need to place the empty tube over the top of the insect, then manipulate the direction of your tube to allow the wasp to climb inside. Once the insect is inside the tube and attempting to scale the inner surface, you can either slide the cap on from below or gently invert the tube so that the mutillid slides to the bottom. It is important to be quick, but not violent, when placing the tube over a specimen. In many instances, the tube's edge will land between the mesosoma and metasoma, impeding the mutillid's ability to crawl inside the tube, and potentially severing the insect in two halves. While using this method, patience and manual dexterity are needed to manipulate the angle of the tube and potentially prod the specimen with the tube's cap. With many specimens it can take three to five tries to get a good angle and catch with the tube, but the insect can typically be herded to an easier location for capture. This method requires minimal gear and can be a safe, rapid method for capturing mutillids.

Surprisingly, an aerial insect net can also be a useful tool for catching flightless velvet ants. When collecting in deep forests, females are most frequently seen crawling on top of leaf litter. This is a tenuous situation, because the velvet ant can quickly hide herself under layers of leaves and eventually dig herself into the soil to avoid capture. The best way to catch a specimen in this circumstance is to grab a handful of leaf litter around the wasp and throw this into the net. Then you can carefully pick out leaves until the wasp is found again, hopefully, inside the net bag. Often the velvet ant gets impatient, and you will see her running up the side of the net. Some caution is useful here, since you can also accidentally throw hidden spiders and scorpions into the net along with your intended quarry, or accidentally pinch the velvet ant between your hand and the plant material long enough to get stung.

Lastly, there are passive methods for collecting diurnal females using pitfall traps. This trap consists of a cup, bowl, or trench dug into the soil, into which insects will accidentally fall. There are risks of capturing small mammals or reptiles, so these should not be used in areas with sensitive or

ABOVE: Three *Dasymutilla* females in tubes.

RIGHT: Kevin collecting in Brazilian Rainforest.

protected terrestrial wildlife. If the traps are placed in an area that can be checked every day, it is fine to leave them dry and dump out the dirt, true ants, and other creatures after you have checked for a velvet ant haul. If you plan to leave the traps for a while, propylene glycol seems to be the best liquid to collect them into. These traps can be left for a few weeks or months, depending on the weather. Anecdotally, these traps seem to be much more effective in dry habitats than humid ones. Perhaps this is due to the liquid of the trap serving as some degree of an attractant to parched creatures. Lastly, their effectiveness can also be improved by adding a barrier that leads into each pit, effectively widening the trapping area without having to dig out a giant crater. While these pitfall traps do effectively capture diurnal females, they are comparatively more effective for catching the nocturnal forms, as discussed later.

1.5.2—Diurnal males

Velvet ant males, like males of many other species, are primarily concerned with only two things—food and sex. Food for velvet ant males usually consists of nectar from flowers or sap from tree wounds. Most other wasps sustain themselves with similar food sources, so if you see a variety of wasps flying around a particular plant, there is a good chance that you will eventually see some velvet ants there too. If you catch it at the right time, Desert Milkweed (*Asclepias erosa*) is really fantas-

tic for wasps (this tube of wasps was collected in just 30 minutes near Barstow, California).

Regarding the sex part, your other best chance to find male velvet ants starts with finding the females. Males are often seen flying in slow meandering sweeps 1 to 2 feet above the soil in areas where females are active. Because of their dark wings and hairy body, even the colorful species look a bit like fuzzy black blurs zooming around. They seem somehow lazier, with smoother arcs to their flight, than flies or hunting wasps.

After you have seen the male, the best chance to catch him is with an aerial insect net. Males stridulate, just like the females, when disturbed. So, to be sure that you have caught a velvet ant, you can herd him into the end of the net and listen for the squeaking. Once you are sure it is a velvet ant, it is generally safe to reach into the net and grab him with your fingers. Males do not sting, remember. Well, that's only mostly true. Male velvet ants have sharp and pointy genitalia that they use to "pseudo-sting," and it

LEFT: Male safely held with fingers.

FAR LEFT: Kevin collecting wasps on Desert Milkweed in California.

Malaise trap with trays in Arizona.

Malaise trap in Brazilian rainforest.

can be quite uncomfortable. For larger *Dasymutilla* species, it can be painful enough for you to drop the specimen, especially if you start second guessing that you might have a scoliid or other hairy wasp instead of a male velvet ant. Another risk factor is that some velvet ant males carry females with them when they mate (phoretic copulation). At least one colleague was stung by a hidden female that was being carried along by a "harmless" male.

If you want some real volume of male velvet ants, though, you will want to get into the malaise trap game. These traps look like a mesh tent with a jar at the top. Depending on the weather, they need to be checked only every few weeks. They function by impeding the random flight paths of wasps in the area and passively catch hundreds of insects. Malaise traps have been a huge boon to Hymenoptera research around the world. In the right habitat with good placement of the trap, it is not impossible to get 100 or more male velvet ants in a few weeks. One downside is that the wingless females are usually too lazy to crawl all the way up to the collecting jar. Adding some trays of soapy water under a malaise trap can marginally increase your odds of getting both sexes.

Good luck trying to use a malaise trap in West Texas. It will not be up for long.
The winds know weakness when they see it.—AARON

1.5.3—Nocturnal females

Even though their color and size are not very impressive, nocturnal female velvet ants are among the most exciting finds because they are hard to come by; they are also the most poorly understood velvet ants in North America from a taxonomic (and likely ecological) perspective. Due to their generally chubbier and fuzzier dull brown appearance, we sometimes refer to them as *ositas de peluche* (Spanish for "teddy bears"), or *ositas* for short. We would not say it is as hard as finding a needle in a haystack ... more like finding a furry pebble on a dirt road. They can be found in similar habitats and caught using similar methods to the diurnal females (1.5.1), but they are usually smaller and duller in color, and you must spot them within a moving 2-foot radius of illumination. Coming back to microhabitats where you have seen diurnal velvet ants earlier in the day is usually a good idea, especially if there are plants that had lots of wasps crawling

Odontophotopsis female

all over them, like sunflowers. We prefer to use a moderately powerful LED flashlight; if the beam is too bright, there can be a bit too much glare off the soil, other insects will fly into the beam and interfere with your search, and the females could stop moving before you notice them. Although holding the light at a lower angle seems to illuminate more ground, it does not help because the shadows of plant material interfere with your ability to see the creatures, so pointing the light beam nearly straight down seems to work best.

Even after years of collecting these, the circumstances rule out something as useful as a search image, and I find nocturnal females by making a conscious effort to examine and loosely identify every creature that moves within my light beam. It seems to go something like this: ant, camel spider, ant, regular spider, ant, ant, scorpion, ant, beetle, ant, ant, ant, spider, ant, beetle, ant, beetle, ant, TOUCHDOWN! Catching five females in 3 hours is a respectable result, and the most I've ever found on a given night is about 20.—KEVIN

After you have spotted a nocturnal female, the fun is just beginning. They will often take similar evasive maneuvers to the diurnal specimens, especially if ground cover is available. They can sting in defense, although, anecdotally, they seem more hesitant to sting and have a less painful sting than most diurnal females. Nocturnal velvet ants, however, have an extra trick—they like to play possum. They can be ridiculously difficult to tell apart from pebbles, especially when it is dark outside. They also seem to be more patient than their diurnal relatives, and we have watched them remain motionless for at least 10 minutes. Once you have secured the specimen in a tube, however, you just may be in possession of a new scientific discovery, since most nocturnal species are known from the male sex only (see chapter 6).

The pitfall traps discussed above (1.5.1) can get you both diurnal and nocturnal species, but they are even more important for nocturnal species, because nocturnal females are so much harder to find in person. More than half of nocturnal female specimens in museums came from pitfall traps, but only about 20% of diurnal females were collected this way. These pitfall traps can be further enhanced for nocturnal velvet ant collecting by staking a cheap solar patio light into the soil near a given trap.

One last point about nocturnal velvet ants is that, other than scattered records of a few *Timulla* and *Photomorphus* species, scarcely any nocturnal velvet ants are known from east of the Mississippi River. Any nocturnal collecting efforts would be better spent in the paradisiacal climes of Yuma or Barstow, rather than the boring old Smoky Mountains or Niagara Falls.

TOP:
Odontophotopsis female hiding below debris.

MIDDLE AND BOTTOM:
Sphaeropthalma females playing dead.

1.5.4—Nocturnal males

Traditional nocturnal insect collecting involves a blacklight (ultraviolet light bulb) or mercury vapor light placed near a white sheet; the insects can be grabbed by hand or with a jar when they land on the sheet. This can work well for velvet ants, but there are a few shortcomings. Velvet ant males are more energetic than most moths and beetles, so they often leave the sheet before they can be grabbed. Additionally, this method requires time and attention that might be better spent searching for the females, as mentioned above. Using a light trap is better, and we have an effective and inexpensive trap design just for you.

The trap only requires a battery-operated camping lantern, four flat plastic sandwich containers, and some soapy water. It is best to choose a lantern with a fluorescent white U-bulb—most modern LED lanterns are apparently worthless for attracting insects (for some reason, most campers nowadays prefer lanterns that do not get swamped with thousands of insects). Sadly, the lanterns that I used to buy for 10 bucks at Walmart now cost between 20 and 50 bucks on eBay (and if this book has any success, they might be even harder to come by). Perhaps some enterprising folks will bring back fluorescent lanterns for catching bugs or find good LED alternatives. Smaller lanterns that require

LEFT: Traditional light-trapping sheet with mercury vapor lamp in Brazil.

BELOW LEFT AND RIGHT: Lantern traps in Arizona.

LEFT: Kevin retrieving lantern trap sample.

MIDDLE: Lantern trap sample from Corn Springs, California.

RIGHT: UV light trap sample from Corn Springs, California.

only four D-cell batteries (instead of six or eight) are less expensive and easier to transport. When selecting plastic-ware for the trap, choose containers with straight edges that fit together snugly.

The plastic trays should be filled with soapy water, and the lantern can be turned on when the sun sets. One gallon of soapy water is enough for about two traps. There is some risk of the lantern being stolen (try to use inexpensive lanterns), so the trap should be placed far away from busy roads or trails. To assuage suspicions of passersby, it can be useful to write some contact information or something scientific sounding on the lantern, like "Insect pest survey."

In the morning, if there are only a few specimens, they can be removed from the soapy water with forceps in the field. Often, however, there can be hundreds of valuable wasp specimens in a trap. In those cases, it is best to pour the specimens through a wire filter and get them into a whirl-pack or jar to sort back in the lab. If you have other entomological interests (or other entomologist friends), the bycatch could be nice, too. We find these traps to be pretty effective for antlions, roaches, and some of the more obscure beetle and true bug groups.

Unlike with blacklight traps, beetles and moths are typically not over-abundant. This is actually a bonus, because moths ruin everything with their horrible scales. When comparing the two pictured light traps from the Sonoran Desert in California, you can see significantly more insect biomass in the blacklight trap, but there are actually just as many velvet ant specimens in the lantern trap (~50).

Malaise traps (1.5.2) can get you both diurnal and nocturnal species, but they are less important for nocturnal specimens, because the light traps discussed above are so effective and cheap to run.

1.5.5—Host nests

Nearly every collecting method has skewed results for one sex or the other. The one method that generally provides equal numbers of males and females involves starting with their food sources. Digging up host nests in the soil can work. This can be done by digging up the soil where ground-nesting bees or wasps are nesting, then sifting the soil to isolate the nest cells. Leaving out trap nests (often used to collect solitary bees) is another method, and requires less effort, but it gets you a smaller range of species, since more than 80% of North American species attack ground-nesting hosts.

The added bonus of collecting mutillids with these methods is that a full scientific paper can come from a single collecting event: one could associate the males with females and identify the host species all in one go. The downside of collecting this way, however, is that you have to wait for each nest cell to yield its contents. This means that after collecting host nests, either in the soil or in nest traps, you must wait to see what emerges from the nest cells (sometimes for many weeks or months). Furthermore, there is no guarantee that the nests you collected have even been parasitized by velvet ants.

1.6 HOW MANY VELVET ANTS ARE OUT THERE?

Our terrestrial planet is home to between 2 million and 30 million extant species (including archaea, animals, bacteria, fungi, plants, etc.), although one study estimates that Earth may actually host 1 trillion species—a bit too far-fetched if you ask us. When examining the animals, which make up the vast majority of the known species on Earth (around 1.5 million species), the insects, with an estimated 900,000–1,000,000 species, make up a large share of the kingdom's diversity.

One of the first questions people ask about velvet ants is: how many species are there? Whether you are asking about the whole world or just one region, the question is surprisingly difficult to answer. In a recently published checklist of the world's velvet ant species (Pagliano et al. 2020), we can find a first guess answer to our question: 4901 recognized species and subspecies are listed. In the three years between the completion of that list and the book you have in your hands being published, however, 65 new species were described. Every publication about velvet ant taxonomy shifts the answer to our question ... and there are piles of papers that still need writing!

Even in the USA, this question cannot be accurately answered, but that does not mean we will not try. Our starting point is the most recent catalog for Hymenoptera of the USA, which was written in 1979; 494 species and subspecies are listed. Multiple types of taxonomic actions since this publication have shifted this count of species.

The most obvious way to change this count is by discovering new species. Since 1979, 52 new species have been described in the USA. We also have a backlog of at least 20 new North American species (mostly in the nocturnal genera of tribe Sphaeropthalmini, see chapter 6). When people read this book and more eyes are directed to velvet ants, even more species are sure to turn up, and that is always an exciting prospect.

As new species increase the total, the USA species count also drops due to the discovery of synonyms. Synonyms are named and previously recognized species that are later determined to be conspecific with another named species. The shorthand term "sink" is a synonym for synonymize. When multiple names are found to belong to the same species, the oldest published name is kept, and the newer names are sunk. Two facets of velvet ant biology make synonyms more common than in other groups.

Firstly, sexual dimorphism led taxonomists to describe males and females of the same species under different names. When males and females are matched up (sex association), they often each have their own name, causing the newer name to become a synonym. In the 1979 catalog, 401 species were known from only males or only females; only 18% were known from both sexes. In the last 40 years, 45 new sex associations were recognized and now 30% of species in the USA are known from both sexes. Of these new sex associations, 18 required synonyms to be sunk; in the other 27 sex associations, the newly discovered male or female did not already have a name.

Mimicry is another leading cause of synonyms. Because velvet ants are under pressure to match their neighbor's color pattern, widespread species look very different across their distribution. Often, each of these color forms was given a different name. In one of Kevin's earlier papers (Williams et al. 2011), we discovered that 14 previously named forms belonged to a single species, *D. bioculata* (examples: 3.2.3.4, 3.2.8.1, 3.3.4.3, 3.6.5.3, 5.2.2.2, 5.2.5.1, 5.3.2.3, 5.6.4.3). Altogether, 65 synonymous color forms have been officially sunk since 1979.

While many formerly recognized species have been sunk in synonymy, sometimes researchers have been overly zealous in that task. Three of North America's velvet ant species have had their status resurrected, increasing the species count. Finally, three species that were previously known only from Mexico have been found in southeastern Arizona since 1979. These new distribution records further increased our number of species.

In the last 40 years, taxonomic acts from 43 publications have altered the answer to our question, "how many velvet ant species live in the USA?" After these changes, 469 species are officially recognized in the published literature: 232 from males only, 99 from females only, and 138 from both sexes. This does not include about 20 that still need to be described. Since many of the species known from just one sex are likely to fall under synonymy, a conservative estimate might tabulate the higher number of male only or female only forms in each genus plus the number of associated species in all the genera: 395. Therefore, while a concrete answer to the question of how many velvet ant species are in the USA is impossible, we suggest a broadly estimated range of 395 to 489 total species.

Strangely, whether looking at the low or high side of our estimate, the species count is lower than the number from 1979. Surely, many authors who described eventual synonyms would be disappointed to see the names they gave to these wasps dropped from usage.

In 2009, I had to sink my first new species, *Lomachaeta garm* from Colombia, only two years after it was described. Although its orange-colored head was apparently unique, we later found transitional forms throughout northern South America that linked it to *L. hyphantria*, originally recognized from Bolivia. After this, I embraced the synonym as an inevitable facet of velvet ant research. In 2011, I named two Argentinean males *Tobantilla drosos* and *T. ephemeros* (meaning "dew" and "temporary," respectively), because I knew they could eventually be matched to previously described females in the area. Because both species of males and three different females overlapped in distribution, it was impossible to know which was which. It was still important, however, to document the existence of these males and describe their structural features. When the correct sex association is finally figured out, it will be easier to publish, whether that project involves me or not.—KEVIN

The drop in species count for North America in the last 40 years might indicate that the global species count could drop, too. We do not think this is likely for a few reasons. Firstly, most of the synonyms (55/83) were made in the genus *Dasymutilla*, basically the largest-bodied, most colorful, and most widely studied velvet ant genus in the world. For small-bodied genera, like *Lomachaeta*, and nocturnal genera, like *Odontophotopsis*, new species outnumbered synonyms. The body size and habits of most velvet ant genera on other continents are less conspicuous than *Dasymutilla*.

Secondly, in preliminary work on velvet ants from other regions, the number of recognized species nearly always rises substantially. In Trinidad, previous studies found 20 species, but unpublished surveys have brought this count up to 35. The Australian genus *Odontomyrme* has 12 named species; so far, we have recognized an additional 35 new species, based on females only, and estimate there will be even more males. In Thailand, the most recent catalog, from 2005, listed 24 species; after six publications in the last 18 years, the number of recognized species has nearly tripled to 68.

We cannot predict the world's total number of species with great confidence, but if the current count rises by more than 550, which is conceivable, there will be more velvet ant species than mammals (~5450 species worldwide).

1.7 HOW ARE VELVET ANTS CLASSIFIED?

Systematics is broadly defined as the classification of biological life. The two primary components of this field are phylogenetics (recognizing relationships between life forms) and taxonomy (naming and categorizing life forms). An oft-overlooked component of this field, however, is diagnosis. In a perfect scenario, reconstructing the phylogeny of organisms and giving them names would be accompanied by tools that aid in telling all these species and categories of life apart from one another. This is not always the case, though. It is not necessarily common for someone with talents for (or interest in) elucidating historical relationships of organisms, or talents for correctly applying their names, to also have talents for communicating their differences to an audience. While the authors of this book are trained specialists in taxonomy (Kevin), phylogenetics (Joe), and evolution (Aaron), the primary goal of this book is to provide a diagnostic guide for the diurnal species of velvet ants in the USA.

It is sometimes possible to recognize a species immediately by one unique feature, but narrowing life forms down into smaller categories is vital for diagnosis. Categories that aid in diagnosis can be loosely treated in three groups: formal taxonomic groups, informal taxonomic groups, and ecological groups.

Formal taxonomic groups are generally taught as a hierarchy in basic biology classes. Many of us remember some variant of this mnemonic: King Phillip cares only for green spinach, to aid in recognizing the seven main categories of formal taxonomic groups: kingdom, phylum, class, order, family, genus, species. Since our school days, an eighth category, domain, was added, leading to the amended mnemonic, "**D**ude! **K**ing **P**hillip **c**ares **o**nly **f**or **g**reen **s**pinach!" These formal taxonomic groups are designed to apply to monophyletic, also known as natural, groups (classifications that include all of the descendants of a single shared common ancestor). The application and establishment of these names are subject to numerous rules established in the International Code on Zoological Nomenclature (ICZN). For many mega-diverse organisms, additional formal categories are useful, especially between the family and genus levels, including subfamilies, tribes, and sometimes even subtribes.

Using the formal taxonomic groups above, all velvet ants are members of the dominion Eukarya, the kingdom Animalia, the phylum Arthropoda, the class Insecta, the order Hymenoptera, and the family Mutillidae. Within the family Mutillidae there are currently 10 subfamily names in use (see 8.1.2, Brothers & Lelej 2017), with many of them subdivided into tribes or even subtribes. *Dasymutilla occidentalis*, for example, is placed into the following formal taxonomic names: subfamily Sphaeropthalminae, tribe Dasymutillini, genus *Dasymutilla*, and species *Dasymutilla occidentalis*.

Informal taxonomic groups are also useful for organizing the world's life forms. Unlike the formal groups above, these are not subject to the rules of the code (ICZN), often making them more flexible and accessible. Like the formal groups, they are based on monophyletic groups. The two main categories of informal taxonomic groups that have been applied to velvet ants are species-groups and species-complexes. Shockingly enough, species-groups are groups of species, usually including anywhere from one to 30 closely related species; they are somewhat analogous to subgenera but are more flexible to apply. Species-complexes are a bit more ... complex. They are rarely used in velvet ants and apply to situations when two or more genetically distinct lineages have been discovered that are indistinguishable using morphology alone.

Unlike the taxonomic groups discussed above, **ecological group names** are not based on monophyly or phylogenetic relatedness. Instead, they are based on traits shared by organisms, such as similar behavior, geographic range, or coloration. Named ecological groups can include anywhere from two to hundreds of closely or distantly related organisms from a single area or from multiple continents. For example, the nocturnal velvet ants (in the broad sense) are linked by their activity period and include hundreds of species from dozens of genera in five different subfamilies, across nearly every continent.

The primary application of ecological groups in velvet ants involves coloration and participation in Müllerian mimicry rings. To establish a **Müllerian mimicry ring**, a scientist must demonstrate that multiple well-defended species share similar appearance, that these similarly colored species overlap in distribution, and that the color patterns arose from convergent evolution, rather than shared ancestry. As you can imagine for a group in which fewer than half of the species are even known from both sexes, this type of analysis is relatively rare in the literature. Furthermore, similarly colored species often live on totally separate continents, so they cannot be placed into a single mimicry ring. Finally, similarly colored males cannot participate in Müllerian mimicry rings, because they are not defended by painful stings.

Because we cannot accurately apply the concept of a mimicry ring to many diagnostically useful color patterns, we use the term **color syndrome** in this book. Color syndrome is interchangeably used to describe the general color pattern of an insect and as a category for the species with said coloration.

Because these color syndromes were initially discovered in the context of mimicry in North American velvet ants, we have tried to retain much of that terminology in this book. The Western mimicry ring established by Wilson et al. (2012) to group females of 18 species in *Dasymutilla* has been succeeded by the Western color syndrome, which includes 34 females and 29 males from four different genera.

We also introduce a narrower ecological group category, **species-cluster** (hereafter simply labeled **cluster**). These are (usually) small collections of species within a single genus and color syndrome that are linked by some feature (or combination of features) of their color, morphology, and/or distribution. Remember, these clusters are not taxonomic groups and have nothing to do with shared ancestry or relatedness (aside from their membership in the same genus). These clusters are named after a prominent **sentinel species**, which represents the group. The *Dasymutilla gloriosa* cluster (3.1.1), for example, includes *D. gloriosa* and two distantly related species within the genus *Dasymutilla*, which express the Desert color syndrome and are further linked by having the entire body covered with dense white setae.

Some final notes about **common names**. The term "velvet ant" is not technically a correct application for a family-level common name, since velvet ants are wasps and not technically a type of ant. "Velvetant" as a single word would follow the modern convention for common names, but it feels a bit ridiculous. Other scientists have proposed calling them velvet wasps or velvet ant wasps, but none of those flows very well. The common name we use might be outdated within a few decades, but they will always be velvet ants in our hearts, and they will stay that way in this book.

There is a modern trend to provide common names for every species, but we do not find this practical. Since the family has nearly 5000 species, and every genus generally fits the name velvet ant, it would quickly become even more confusing than the binomial scientific names. Should we use the name "connected velvet ant" for both *Dasymutilla connectens* and *Pseudomethoca connectens*? Should we use "Seabra's velvet ant" for *Lophomutilla seabrai*, *Lophostigma seabrai*, and *Lynchiatilla seabrai*? If we added qualifiers for each genus, such as "hairy velvet ant" for *Dasymutilla* or "toothed nocturnal velvet ant" for *Odontophotopsis*, then each common name would start with at least three words, then expand further with the specific epithets. Additionally, sexual dimorphism is a big deal in velvet ants; providing a common name that is somewhat descriptive for both sexes would be impossible for many species. The "thistledown velvet ant," *Dasymutilla gloriosa* (3.1.1.1), has perhaps the most descriptive and well-established common name for a velvet ant, but the males do not look anything like thistledown (5.2.7.1, 5.3.3.4). Rather than try to create convoluted, protracted, and misleading common names for every species, we have presented information about the meaning of each name (etymology) and a pronunciation guide. We hope this will make the scientific names themselves more memorable and accessible to the reader.

1.8 HOW TO USE THIS GUIDE

Even with the fun facts and photos of cute wasps, the main reason you likely have this book is to figure out the name for a velvet ant found here in North America. Unless you jump right into the keys in appendix 9.3 (requires a pinned specimen and good microscope), the most daunting task will be finding out which of the 200-ish pages in chapter 3 or 5 includes your target. Finding that page is a multi-step process of elimination.

STEP ONE is the easiest: does it have wings (chapters 4 and 5) or not (chapters 2 and 3)?

STEP TWO is (usually) also easy: which color syndrome?

There are seven color syndromes each for winged and wingless velvet ants. The figures below should help you recognize most color syndromes, but there are also keys to the color syndromes (see 9.3.2.0, 9.3.4.0), which will help you work out the intermediate and complicated forms.

STEP THREE is more challenging: which species-cluster?

Female Color Syndromes   **Male Color Syndromes**

Desert

Western

Texan

Madrean

Ephuta-like

Timulla-like

Eastern

Cryptic

Sphaeropthalma-like

On the first page for each color syndrome you will find a short key to identify the correct cluster, which will send you to the proper page to find your species in question. These clusters are initially organized by genus. Some useful genus-level characters are mentioned in those keys, but chapters 2 (wingless specimen) and 4 (winged specimen) provide more details about confirming the right genus. Within each genus, clusters are generally separated by differences in coloration, body size, or distribution.

On each cluster page you will find photographs, distribution maps, and discussions of key features that can be used to recognize the species. This provides the best hope for a species identification based on photographs or specimens in the field. For each species account in chapters 3 and 5, the following sections are included.

PHOTOGRAPHS For every female, we have provided a dorsal habitus image, and for every male, a lateral habitus image. In addition to these shots, we have provided photos of live specimens, when available. Also, we have tried to add images of the different variations seen in a given form of a species.

SCALE In lieu of scale bars with our figures, we provide two silhouettes showing the minimum and maximum actual body length known for each species. If you have pirated this book, please set your pdf reader to 100% magnification.

MAPS The maps provided for each species include point data for where a given species has been collected and the World Wide Fund for Nature (WWF) Terrestrial Ecoregions (Ricketts et al. 1999) highlighted for each species. For species that have multiple color forms, the pertinent variant of that species has its point data colored blue, and other populations of the species are shown with black dots. The terrestrial ecoregions are filled in only for the color variant pictured on that page. These maps were compiled based on observations from various online collection and observation databases, material examined from previous publications, and specimens that we have studied in museums. You will notice that some maps have hundreds of data points and others have very few. This can therefore serve as a proxy for how common or rare a given species is.

PRONUNCIATION Pronouncing scientific names can be tricky, since the names are derived from multiple different languages, and the primary source for scientific names, Latin, is considered a dead language. Additionally, native speakers of different languages bring their own usages for various letters into their pronunciations.

In California, I grew up pronouncing the carpenter bee genus *Xylocopa* as "zye-loe-koe-puh." It took me weeks to figure out which creature my Brazilian colleagues were discussing when they said "shee-law-cuh-puh." KEVIN

Our rule has always been that if you say all the letters in the right order and act confident about it, you will be fine. But we have provided pronunciation guides for all the diurnal North American species in chapters 3 and 5 to help readers practice with the words and gain some confidence in using the names. In general, we have tried to use unambiguous letter combinations, or short words in common English use, to designate each syllable of a given name, rather than official diacritical marks that one would need to look up. It is important to remember, though, that the pronunciation guides are not official, and none of the book's authors is a Latin scholar. Our goal is not to provide an official scientific pronunciation but, rather, a loose guide to feel out how people from the western USA born in the late 20th century might say the names.

One more caveat: many species are named after other people, and those names have variable pronunciations as well. When a species is named after a person, we have provided only the name itself and a pronunciation for the suffix. James Chester Bequaert, for example, was an American entomologist of Belgian origin; we do not know whether he pronounced his own name "bay-carr," "bee-quart," "bew-kurt," or in another way entirely. Thousands of people who currently possess the surname Bequaert likely use different pronunciations depending on where they grew up or currently reside. We do not feel it is our place to make pronouncements on other people's names.

ETYMOLOGY Etymology is the study of the meanings of words. In recent publications, it has become conventional to provide an etymology section for each new species to explain the reasons why a name was chosen. Older papers discussed the meaning of their species names much less frequently. Using dictionaries of Latin and Ancient Greek, and a bit of sleuthing based on the original publications or particular features of the given insects, we have tried to explain, or at least provide an educated guess, for the origin and meaning of each species name.

FIELD IDENTIFICATION The first few words in this section will tell you how quickly you should give up. In some cases, a simple detail of the color pattern or body shape will give you an easy field identification. In many cases, the diagnostic features for an insect are difficult to see in the field or in photos and will require a few different angles (perhaps even of the underside of the insect) and maybe a hand lens or well-focused shot with a macro lens. In other cases, the differences between species are microscopic features of body sculpture or the presence of various microscopic carinae or tubercles, and a field identification is literally impossible in most parts of the range for a given species. No matter how much experience someone has, you cannot tell whether the mandible has three teeth, or only two, while the creature is running away from you.

After the note on difficulty, a few short sentences explain which characters will be useful for telling the species apart from other species in the cluster. These features are often useful for a lab identification as well, and in most cases they are not repeated under both headings.

LAB IDENTIFICATION Luckily, there are no cryptic species in North America, at least among the diurnal forms (that we know of). Some genera have more known females than males (like *Pseudomethoca*) or more males known than females (like *Timulla*), so the unknown opposite sexes of some species are unlikely to be included in our book; or more likely, they are currently mixed into series along with a related species.

Still, with those caveats, it should be possible to identify every species armed with this book, a well-preserved specimen, and a dissecting microscope of at least moderate quality (up to about 50× magnification should do it). The features listed under lab identification should be consulted alongside the field identification characters. It is also useful to consult the appropriate keys in the appendix (9.3) because they organize the species in a more systematic fashion and direct the user to the most pertinent diagnostic feature in each circumstance.

MALES (chapter 3) or **FEMALES** (chapter 5). A reference to the page with the opposite sex of a given species or color form is provided, followed by a short mention of how similar the sexes are in appearance.

NOTES This is a catch-all area for discussing the other color forms and variation of a given species, discussing any host information that is known, prognosticating about the opposite sex or close relations, and telling anecdotes about a given species.

The bulk of content in this book is devoted to species diagnosis, as laid out above, particularly for the diurnal velvet ants known from the USA. Chapters 2 and 3 cover the wingless specimens (all the females and a few strange males), while chapters 4 and 5 cover the winged specimens (most of the males). Smaller chapters are provided in the later sections of this book to provide background on nocturnal species (chapter 6), Mesoamerican Mutillidae (chapter 7), and a brief overview of velvet ants on other continents (chapter 8).

The final sections of the book include an index, references to scientific literature used in the book, and a checklist of North American velvet ant species. Most importantly, traditional morphology-based keys are provided to separate the species in each color syndrome. These keys rely on technical structural differences and will nearly always require the reader to have a good grasp of insect morphology, a dead specimen on hand, and a relatively high-powered stereo microscope. These keys, when cross-checked with the pertinent species account from chapter 3 or 5, provide the best accuracy for velvet ant identification. To accompany these keys and the rest of the book, there is a glossary of terms and plates that illustrate morphological features.

Before we get much further, a few structural terms that pop up on more or less every page should be mentioned.

Seta—This is the velvet ant (and really all non-mammal organism) equivalent for the term "hair." A creature or structure that seems hairy should be called setose, but we sometimes use the term "hairy" in an informal sense.

Mesosoma/metasoma—In higher wasps, the thorax and first segment of the abdomen are fused together. The traditionally recognizable "wasp waist" is, technically, placed between the first and second segments of the abdomen. Therefore, hymenopterists use the term "mesosoma" for the apparent thorax (really the three thoracic segments and first abdominal segment) and the term "metasoma" for the apparent abdomen (really the second through last true abdominal segments).

Propodeum—Technically, this is the first segment of the abdomen but also the last segment of the mesosoma.

Tergites/sternites—On the metasoma, the dorsal plates are labeled as tergites, and the ventral plates are labeled as sternites. They are usually abbreviated as T1, T2, etc., or S1, S2, etc. The second metasomal tergite (T2) is always the largest and usually carries the most distinct spots and color patterns. The second segment of the metasoma is technically the third segment of the true abdomen.

Pygidium—This is the last tergite of the metasoma and is usually armed with a sculptured pygidial plate.

All right, let's get started on diagnosing these amazing wasps.

1.9 WAIT A MINUTE, IS THIS CREATURE EVEN A VELVET ANT?

1.9.1—Wingless wasps

Other than true ants (Formicidae), velvet ants seem to be the most notable wingless Hymenoptera, but dozens of other wasp families have at least a few wingless species, and many lineages, as a rule, always have wingless females. The ability of powered flight is noticeably advantageous and is one of the key evolutionary innovations that make insects so successful today. Flight allows them to travel greater distances to find new areas to colonize or search for mates and resources. Flight can provide a better means of escape (than crawling) from predators or more powerful rivals, so that a creature can survive another day. In addition, the presence of wings provides options for a number of other novel uses, including signaling conspecifics, scaring or warding off predators (e.g., the flash of color from a *Morpho* butterfly or the large "eyes" on a *Caligo* owl butterfly), or they can be used as camouflage (e.g., forest moths whose wings look like moss or lichen growth on a branch or bark, like *Somera viridifusca* or *Chlorocalliope calliope*).

TOP: *Typhoctes peculiaris* (Chyphotidae)

MIDDLE: *Methocha* sp. (Thynnidae)

BOTTOM: *Gelis acarorum* (Ichneumonidae)

So, one might ask, why would evolution favor a return to winglessness? In nature, just as in economics, it is important to consider both the seen and unseen costs and benefits of various structures and actions; for all their obvious benefits, wings and flight do come with disadvantages. Powered flight is energetically expensive, so flying organisms require large amounts of food, or at least a consistently reliable food source. Wings are also metabolically expensive during development, apparently limiting the amount of food that can be converted directly to body mass for a developing wasp. Also, flying insects can be hindered from taking off due to ambient temperatures being either too cold, so flying cannot commence until air temperatures become more optimal, or too hot, so overheating of the insect becomes a risk. In addition, flying requires a fair amount of motility and flexibility in the thorax so that the wings can move quickly and easily. Having a thin exoskeleton can assist in this endeavor—think about the dainty nature of quick and agile fliers like flies and butterflies. However, this hinders the defensive capacity of the exoskeleton in terms of thickness and weight—an important characteristic that can protect an organism from predation or aggressive stings and bites (see 1.1.5). Finally, though not of direct detriment, wings are unnecessary for use underground and can be easily damaged by digging through soil. Female velvet ants and other wingless wasps generally spend their lives searching for rare underground prey items for their progeny in inhospitable environs, so they have foregone wings and the sky for tank-like armor and a subaerial existence on the ground.

So, that was a fun tangent, but how can you be sure you are holding a velvet ant, instead of some other wingless wasp? In appendix 9.3.1, you will find a loose key to rule out non-velvet ants in North America. The most important feature to look for is the completely fused female mesosoma, which is known only in velvet ants and occurs in most of our species.

Many of these wingless wasps are superficially similar in appearance to velvet ants and are often seen in similar habitats. The New World nocturnal Chyphotidae (genus *Chyphotes*) and Tiphiidae (subfamily Brachycistidinae) are especially similar to nocturnal velvet ants (see 6.1). Many diurnal wingless wasps also resemble velvet ants and may participate in Müllerian mimicry rings with them, especially the diurnal chyphotid wasp *Typhoctes peculiaris*, which fits right in with the Madrean Females Color Syndrome (FCS) (see 3.4) and tiger beetle parasitoids in the thynnid genus *Methocha*, which match the Eastern or *Timulla*-like FCS (3.5, 3.6). In Europe, the wingless ichneumonid parasitic wasps *Gelis acaroroum* and *Gelis formicarium* were originally placed in the velvet ant genus *Mutilla* by Linnaeus, way back in 1758.

Recent phylogenetic studies (Waldren et al. 2023) have found that Myrmosidae (here treated as a subfamily of Mutillidae) belong in their own separate family. Because so little information is available on these wasps, because they look and behave like velvet ants, and because they have a long history of being treated as velvet ants, we included all the North American species in this book as members of Mutillidae.

1.9.2—Mimicking males

In North America, nearly all male velvet ants are fully winged or have small but obvious dark wings. Many other Hymenoptera, however, are relatively hairy and often have similar coloration to mutillids. In appendix 9.3.1, you will find a loose key to rule out non-velvet ant wasps in North America. The most important feature to look for is the presence of a felt line along the side of T2 (and sometimes also S2). Most of our velvet ant species have this felt line, and only one non-mutillid group (Chyphotidae) possesses this feature.

While many of the apparently harmless male velvet ants resemble their own females or females of other species, other male velvet ants have coloration that approximates different families of stinging

Anoplius marginalis (Pompilidae)   *Paratiphia nevadensis* (Tiphiidae)   *Tachytes birkmani* (Crabronidae)

RIGHT: *Triscolia ardens*
(Scoliidae)

FAR RIGHT: *Pepsis* sp.
(Pompilidae)

BELOW LEFT AND RIGHT:
Psorthaspis sp.
(Pompilidae)

wasps. Male velvet ants in the Eastern Males Color Syndrome (MCS) (5.6) are similar to various Pompilidae, like *Anoplius marginalis*. Males in the *Ephuta*-like MCS (5.4), on the other hand, seem to resemble Tiphiidae, like *Paratiphia nevadensis*. Males in the *Timulla*-like MCS (5.5) often resemble various Crabronidae, like *Tachytes birkmani*. There is a lot of overlap in color between all these potential model families; males from each velvet ant color syndrome could be mimicking the stinging females of dozens of species in five or more different wasp families.

Lastly, many "hairy" or colorful wasps seem to generally match female velvet ant color syndromes as well. And the presence of wings does not stop them from joining in the Müllerian mimicry fun. A few western Scoliidae (especially *Triscolia ardens*) are good matches for the Texan FCS (3.3). The word-famous Tarantula Hawks in the genus *Pepsis* could be co-mimics with the Texan FCS velvet ants, too. The spider wasp genus *Psorthaspis* takes this to the extreme, however, including species that are co-mimics with five different mutillid mimicry rings in North America, including the Eastern (see chapter 3.6) and Western (see chapter 3.2) color syndromes.

1.9.3—Distant relations

Wasps are not the only velvet ant look-alikes. Because female velvet ants are abundant and diverse, and especially well protected, they are great models for Batesian mimics of various non-Hymenoptera

A. *Phidippus* sp. (Araneae: Salticidae) from Arizona.

B. *Hemithyrsocera* sp. (Blattodea: Ectobiidae) from Thailand.

C. *Castianeira* sp. (Araneae: Corinnidae) from Texas.

D. *Phidippus* sp. (Araneae: Salticidae) from Utah.

E. *Graptartia granulosa* (Araneae: Corinnidae) from Mozambique.

F. *Peirates arcuatus* (Hemiptera: Reduviidae) from Thailand.

G. *Brachynemurus nebulosus* (Neuroptera: Myrmeleontidae) from Florida.

Photo by Dan Roueche

LEFT: *Pseudoxycheila tarsalis* (Carabidae) from Panama.
BELOW LEFT: *Creophilus* sp. (Staphylinidae) from Thailand.
BELOW RIGHT: *Glenea cantor* (Cerambycidae) from Thailand.
BOTTOM LEFT: *Polynychus tricolor* (Buprestidae) from Thailand.
BOTTOM RIGHT: Undetermined Cleridae from China.

organisms all over the world. Many spiders resemble velvet ants; in North America, these include jumping spiders in the genus *Phidippus* and sac spiders in the genus *Castianeira*. Some especially striking velvet ant–mimicking sac spiders are seen in *Graptartia granulosa* from eastern Africa and *Coenoptychus pulcher* in southern Asia (Nentwig 1985). Recently, in Australia some velvet ant–mimicking silverfish in the genus *Hemitelsella* were discovered (Smith & Mitchell 2021). Various cockroaches, especially immatures of the sun roach genus *Hemithyrsocera* in southern Asia, bear a striking resemblance to local velvet ants and are seen quickly running over leaf litter, just like the mutillids. Around the world, many true bugs, especially Reduviidae, resemble velvet ants, like *Peirates arcuatus* from Thailand. In the eastern USA, especially in Florida, the antlion larva, *Brachynemurus nebulosus* resembles velvet ants in the Eastern FCS and even spends much of its active hunting time on the surface of the soil, rather than underground like its relatives (Brach 1978).

Many beetles are apparent mimics of velvet ants, as well. The tiger beetle genus *Pseudoxycheila* (ex. *P. tarsalis*) is a purported mimic of *Hoplomutilla* (7.4.5.4) in Central and South America. It was originally thought to be a harmless Batesian mimic, but chemical defenses were later recognized in the beetle, suggesting Müllerian mimicry (Schultz & Puchalski 2001). In forested regions of southeastern Asia, many beetle families bear a striking resemblance to the dominant velvet ant color pattern (8.3.7), including Carabidae (*Therastes batesii*), Staphylinidae (*Creophilus sp.*), Cerambycidae (*Glenea cantor*), Buprestidae (*Polyonychus tricolor*), and many species of Cleridae.

The Cleridae are especially convincing mimics of velvet ants. In addition to their general body shape and coloration, they are often "fuzzier" than other beetles, and they even stridulate in a defensive capacity (see 1.1.3). Around the world, numerous clerids resemble local velvet ants, including species in this book. Here in North America, there are clerid matches for every aposematic female color syndrome presented in chapter 3. Moreover, the type species of the family is named *Clerus mutillarius*. If "checkered beetle" were not already a useful and accepted common name for this group, we would suggest they be named "velvet ant beetles."

Figure	Cleridae species	Velvet ant apparent model	Locality
1.9.3a	Enoclerus zacrypticeroides	Dasymutilla pallene (p. 322), Desert FCS	Nayarit, Mexico
1.9.3b	Trichodes bibalteatus	Dasymutilla occidentalis (3.2.8.2), Western FCS	Texas, USA
1.9.3c	Enoclerus zonatus	Dasymutilla klugii (3.3.1.2), Texan FCS	New Mexico, USA
1.9.3d	Enoclerus opifex	Dasymutilla sicheliana (3.4.6.1), Madrean FCS	Sinaloa, Mexico
1.9.3e	Enoclerus quadrisignatus	Dasymutilla quadriguttata (3.5.1.1), Timulla-like FCS	Missouri, USA
1.9.3f	Enoclerus ichneumoneus	Dasymutilla ursus (3.6.5.1), Eastern FCS	Florida, USA
1.9.3g	Enoclerus longipes	Dasymutilla paraparadoxa (7.4.3.1)	Quintana Roo, Mexico
1.9.3h	Enoclerus silbermanni	Dasymutilla araneoides (7.4.4.1)	Costa Rica
1.9.3i	Enoclerus insidiosus	Pappognatha myrmiciformis (8.2.7.1)	Panama
1.9.3j	Enoclerus obliquevittis	Hoplomutilla spinosa (8.3.3.1)	Rio de Janeiro, Brazil
1.9.3k	undetermined	Lobotilla leucospila (8.3.8.1)	Mozambique
1.9.3l	undetermined	Ephutomorpha paradisiaca (8.2.10.1)	Papua New Guinea
1.9.3m	Clerus mutillarius	Ronisia barbara (8.3.10.2)	Western Europe

DAY WALKERS— DIURNAL FEMALE VELVET ANTS—GENERA

2.0 INTRODUCTION AND OVERVIEW

Females of the genera in this chapter and the next are divided into six higher taxa (three subfamilies and four tribes). This chapter begins with an illustrated key to these higher taxa, then each higher taxon has an overview of the included genera (usually fewer than three in each higher taxon). For each higher taxon, examples of species-level diagnostic features are illustrated, which should help with identifications in chapter 3.

Key to higher taxa (subfamilies and tribes)

1. Mesosoma divided, pronotum forming a separate plate;
   eye ovate, often with setae .... 2.6 (Myrmosinae)

2. Mesosoma entire, shape rectangular, pronotum and
   propodeum usually wider than mesonotum; eye
   vertically ovate, clearly faceted, never setose .... 2.5 (Mutillinae: Trogaspidiini: *Timulla*)

3. Mesosoma entire, either ovate, pear-shaped, fiddle-shaped,
   or subrectangular, always wider mesally or anteriorly than
   posteriorly; eye generally circular or transverse, rarely
   with clear facets, never setose .... Sphaeropthalminae

   a. Mesosoma ovate; T1 shape narrowly cylindrical; tergal
      fringes with simple setae only; diurnal species that are
      usually brightly colored .... 2.4 (Ephutini: *Ephuta*)

   b. Mesosoma fiddle-shaped, pinched at propodeal spiracle,
      propodeal sides parallel or widened posteriorly; T1 shape
      sessile tergal fringes not obviously plumose; diurnal species
      that are usually brightly colored with a wide head .... 2.2 (Pseudomethocini)

   c. Mesosoma usually pear-shaped, not pinched at propodeal
      spiracle, propodeal sides usually narrowed posteriorly; tergal
      fringes never plumose; T1 shape petiolate, disciform, or
      subsessile; diurnal species that are usually brightly colored .... 2.1 (Dasymutillini)

   d. Mesosoma usually pear-shaped, not pinched at propodeal
      spiracle, propodeal sides usually narrowed posteriorly
      (Fig. 2.0.1d); tergal fringes often obviously plumose and
      T1 shape variable (Fig. 6.2.3a–f); usually dull-colored
      nocturnal species .... 2.3 (Sphaeropthalmini)

2.1 SPHAEROPTHALMINAE: DASYMUTILLINI: *DASYMUTILLA* AND *LOMACHAETA*

Dasymutilla

ETYMOLOGY From the Latin *dasy* "hairy" and the name *Mutilla,* which is commonly used as a suffix for velvet ant genera. True to the genus name, *Dasymutilla* includes many of the world's hairiest species, although some *Dasymutilla* can be nearly bald. **IDENTIFICATION** These females have the mesosoma entire (subfamily character) and pear-shaped (tribe character), the tergal fringes simple (tribe character), the pygidial plate defined, and the T1 shape petiolate (genus characters). Unlike

Figure	Taxon	Description
2.0.1a	*Myrmosula rutilans*	Mesosoma: pronotum divided
2.0.1b	*Timulla suspensa*	Mesosoma: entire, rectangular
2.0.1c	*Ephuta margueritae*	Mesosoma: entire, ovate
2.0.1d	*Pseudomethoca meritoria*	Mesosoma: entire, fiddle-shaped
2.0.1e	*Dasymutilla birkmani*	Mesosoma: entire, pear-shaped
2.0.2a	*Timulla suspensa*	Eye: ovate, large, faceted
2.0.2b	*Dasymutilla cirrhomeris*	Eye: circular, large, smooth
2.0.2c	*Photomorphus banksi*	Eye: transverse, small, faceted
2.0.3a	*Ephuta margueritae*	Metasoma: T1 shape cylindrical
2.0.3b	*Pseudomethoca meritoria*	Metasoma: T1 shape sessile
2.0.3c	*Dasymutilla montivagoides*	Metasoma: T1 shape petiolate

Figure	Taxon	Description
2.1.1a	Dasymutilla flammifera	Habitus: lateral view
2.1.1b	Lomachaeta beadugrimi	Habitus: lateral view
2.1.2a	Dasymutilla scaevola	Hind femur: truncate
2.1.2b	Dasymutilla gibbosa	Hind femur: rounded
2.1.3a	Dasymutilla nigripes	Vertex: unarmed
2.1.3b	Dasymutilla quadriguttata	Vertex: armed with tubercles
2.1.4a	Dasymutilla nigripes	Mesosoma: no scale
2.1.4b	Dasymutilla ursus	Mesosoma: scutellar scale only
2.1.4c	Dasymutilla bioculata	Mesosoma: scale and accessory carina
2.1.5a	Lomacheta hicksi	Pygidium: undefined
2.1.5b	Dasymutilla ferruginea	Pygidium: microreticulate
2.1.5c	Dasymutilla vestita	Pygidium: rugose
2.1.5d	Dasymutilla bioculata	Pygidium: striate

2.1.1a

2.1.1b

2.1.2a  · 2.1.2b 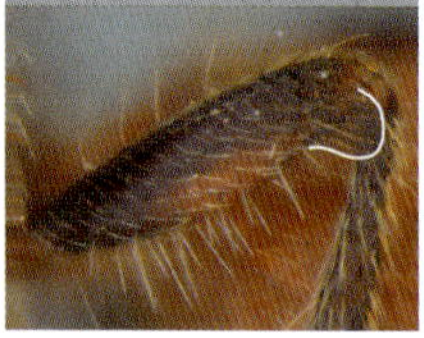 · 2.1.3a 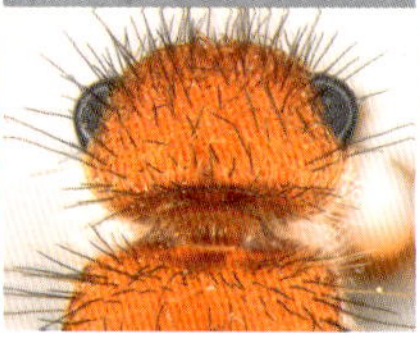 · 2.1.3b

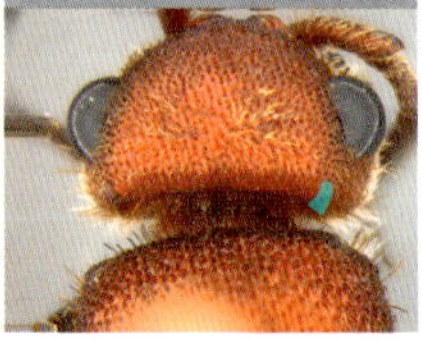

2.1.4a 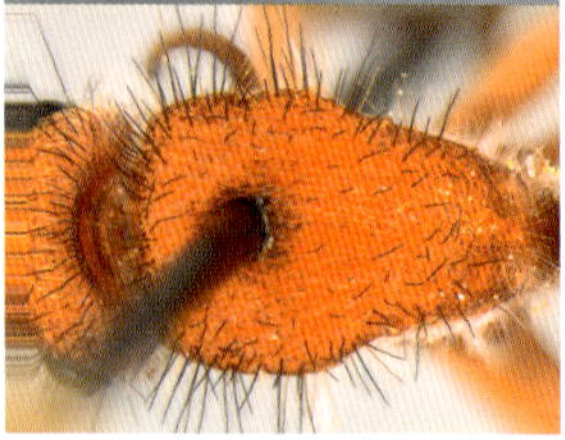 · 2.1.4b  · 2.1.4c

2.1.5a  · 2.1.5b  · 2.1.5c  · 2.1.5d

Lomachaeta, they are usually large, coarsely sculptured, densely setose, and brightly colored insects. **DISTRIBUTION AND DIVERSITY** This genus is diverse across its range from Canada to Panama, though only four species extend into northern Colombia in South America. **NOTES** This is the most frequently collected and examined genus in the USA; *Dasymutilla* are diverse and common in Mesoamerica as well. **KEYS AND CHARACTERS** The species in the USA were recently reviewed by Manley et al. (2020) and there are many useful illustrations of diagnostic features in that paper. The Neotropical species were treated by Manley & Pitts (2007). Many species can be recognized by differences in color and comparative widths and lengths of the head and mesosoma, but a good microscope is needed for some features. The hind femur apex is very useful, but it can be misleading. For the femur to be considered truncate, both the inner and outer lobes of the femur apex should be flat; some individuals belonging to species with a rounded femur apex can have the outer margin somewhat flattened. The presence or absence of tubercles on the head and the types of scutellar scale and carinae surrounding the scale are also useful. These can be difficult to see in overly hairy specimens, so a pin can be used to "shave" the specimen or feel around for the presence of these features. Pygidial sculpture is also important for identification.

Lomachaeta

ETYMOLOGY From the Ancient Greek *loma* "fringe" and *chaeta* "long hair." Males of the type species, *L. hicksi*, have a row of thick black bristles on the T2 posterior margin **IDENTIFICATION** These females have the mesosoma entire (subfamily character) and pear-shaped (tribe character), the tergal fringes simple (tribe character), the pygidium without a pygidial plate, and the T1 shape disciform or subsessile (genus characters). Unlike *Dasymutilla*, they are usually small, faintly sculptured, sparsely setose, and dull-colored insects. **DISTRIBUTION AND DIVERSITY** The genus includes 24 species that range from Canada South to Argentina; 15 of these occur in the USA. **NOTES** These are generally the smallest diurnal velvet ants in the USA. Some species attack arboreal hosts and both sexes can be collected with trap nests. Other species use ground-nesting hosts and are only rarely encountered in pitfall traps or crawling on the dirt. **KEYS AND CHARACTERS** "Abandon all hope, ye who enter here." Williams, Cambra, Bartholomay et al. (2019) has a modern key to the known females, but they are remarkably difficult to separate, even with a well-prepared specimen and a good microscope. Diagnostic features are mentioned for each species, but they are often remarkably difficult to capture in flat images.

2.2 SPHAEROPTHALMINAE: PSEUDOMETHOCINI: *PSEUDOMETHOCA* AND KIN

Invreiella

ETYMOLOGY This genus was named for Italian entomologist Fabio Invrea. **IDENTIFICATION** These females have the mesosoma entire (subfamily character) and fiddle-shaped (tribe character), and the T1 shape sessile (tribe character). This genus has two North American species, which are moderately large (~12 mm) with a strong genal carina with a perpendicular posterior tooth. **DISTRIBUTION AND DIVERSITY** The genus includes 15 Mexican species, two of which extend north into the USA. **NOTES** The genus is widely distributed in higher elevations across Mexico and into Arizona and New Mexico. Males are unknown for all species in this genus. **KEYS AND CHARACTERS** The two species in the USA can be separated by coloration. The recent revision of the genus by Waldren et al. (2020) has well-written and illustrated keys for separating all the species.

Myrmilloides

ETYMOLOGY This genus was named for its similarity to the Old World velvet ant genus *Myrmilla*. **IDENTIFICATION** These females have the mesosoma entire (subfamily character) and fiddle-shaped

Figure	Taxon	Description
2.2.1a	*Invreiella cephalargia*	Habitus: lateral view
2.2.1b	*Myrmilloides grandiceps*	Habitus: lateral view
2.2.1c	*Pseudomethoca paludata*	Habitus: lateral view
2.2.2a	*Myrmilloides grandiceps*	Head: posterior genal and anterior postgenal teeth
2.2.2b	*Pseudomethoca frigida*	Head: small genal tooth
2.2.2c	*Pseudomethoca nephele*	Head: genal and postgenal tooth
2.2.2d	*Pseudomethoca dentigula*	Head: hypostomal tooth only
2.2.2e	*Pseudomethoca sanbornii*	Head: unarmed
2.2.3a	*Pseudomethoca oceola*	Humeral carina: distinct, sharp
2.2.3b	*Pseudomethoca meritoria*	Humeral carina: distinct, blunt
2.2.3c	*Pseudomethoca sanbornii*	Humeral carina: weak, rounded
2.2.4a	*Pseudomethoca sanbornii*	Propodeum: lateral margin simple
2.2.4b	*Pseudomethoca simillima*	Propodeum: lateral margin dentate

2.2.1a

2.2.1b

2.2.1c

2.2.2a

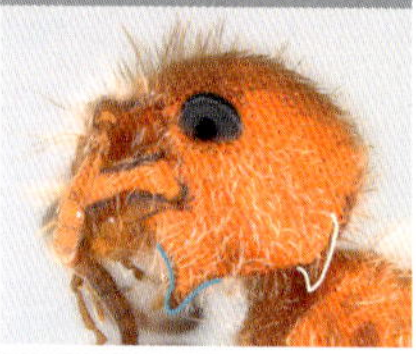

2.2.2b

2.2.2c

2.2.2d

2.2.2e

2.2.3a

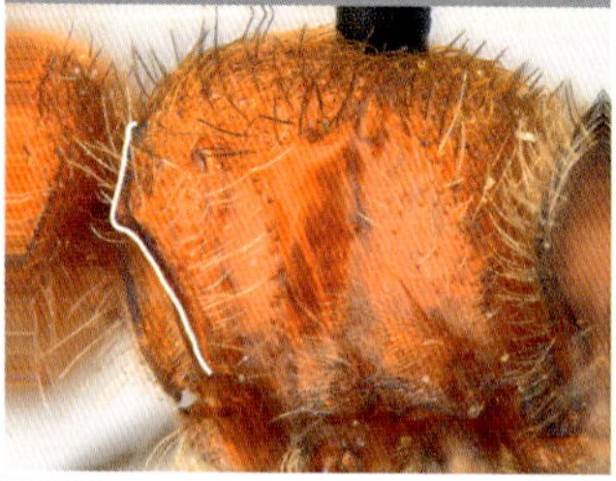

2.2.3b

2.2.3c

2.2.4a

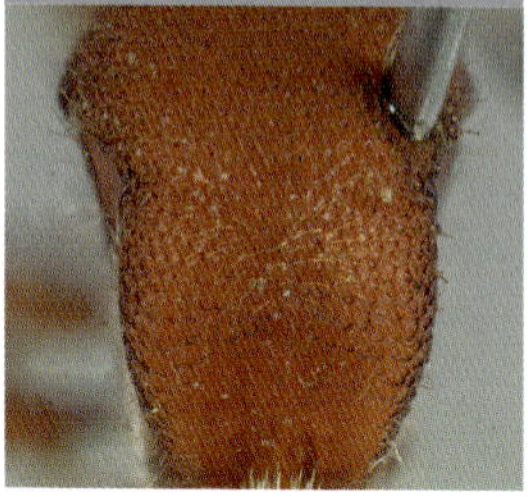

2.2.4b

(tribe character), and the T1 shape sessile (tribe character). This genus has only one species, *M. grandiceps*, and can be recognized by the remarkably large and sharply angled head with a posterior tooth on the gena and a large tooth anteriorly on the postgena. **DISTRIBUTION AND DIVERSITY:** Just one species, found mainly in the central USA. **NOTES** This species is variable in color and somewhat poorly approximates three different color syndromes. See the species pages (3.3.6.1, 3.5.2.1, 3.6.7.1) for more information.

Pseudomethoca

ETYMOLOGY From the Ancient Greek *pseudes* "false" and the genus name *Methocha* (a thynnid wasp that eats tiger beetle larvae). **IDENTIFICATION** These females have the mesosoma entire (subfamily character) and fiddle-shaped (tribe character), and the T1 shape sessile (tribe character). This is the most diverse and variable genus of the Pseudomethocini. Any Pseudomethocini without a genal tooth belong to *Pseudomethoca*, and any small-bodied (less than 6 mm) Pseudomethocini with a genal tooth belong to *Pseudomethoca*. **DISTRIBUTION AND DIVERSITY** There are 40 species in the USA, about 55 Mesoamerican species, and about 40 South American species. **NOTES** Females are often abundant, though not as common as *Dasymutilla*. They are mainly known to attack ground-nesting bees. The large head is theoretically useful to provide a strong bite for removing guard bees from nest entrances. **KEYS AND CHARACTERS** Mickel (1935a) wrote a good key, but this book treats all those species and also includes the more recently discovered ones. Other than body size and coloration, head armature is a useful feature, especially among the small-bodied species. The genal carina is often armed with a tooth, while the postgenal area more rarely has a tooth. Most of the larger-bodied species have the gena simply carinate and the pygidial sculpture helps to differentiate species; these pygidial features are similar to those of Dasymutillini (see section 2.1). These larger species, especially those in the Eastern Females Color Syndrome, are also separated by differences in the humeral carina, which was especially hard to interpret without figures and still takes a bit of practice to recognize. The two most common species in the eastern USA (*Ps. simillima* and *Ps. sanbornii*) are separated by the lateral propodeal margins.

2.3 SPHAEROPTHALMINAE: SPHAEROPTHALMINI: *SPHAEROPTHALMA* AND KIN

This tribe is dominated by nocturnal forms, and only five of the 11 North American genera of Sphaeropthalmini are treated in this chapter. Three of these genera, *Morsyma*, *Protophotopsis*, and *Stethophotopsis*, are represented by a single diurnal species in the USA. *Photomorphus* and *Sphaeropthalma* are large diverse genera with most of their species active at night, but a moderately diverse minority of species in these genera are diurnal, especially in the humid eastern USA. Nocturnal forms of those two genera, and six additional genera, are treated in more detail in chapter 6.

Morsyma

ETYMOLOGY This is a "remix" of the genus name *Myrmosa*. *Myrmosa* was recognized by having a divided mesosoma in the females. The type of this genus and species was a wingless male; the author thought it was a female with a divided mesosoma like that of *Myrmosa*. **IDENTIFICATION** These females have the mesosoma entire (subfamily character) and pear-shaped (tribe character), and the T2 fringe with dense plumose setae (tribe character). The only species in this genus, *M. ashmeadi*, can be recognized by the strong arcuate carina separating the pronotum and mesonotum. Distribution and color are also useful: this coastal California species has orange head and mesosoma, contrasting with black metasoma with a distinct white setal fringe on T2. **DISTRIBUTION AND DIVERSITY** Just one species: *M. ashmeadi*. **NOTES** See the species page (3.5.4.1, 5.7.8.1) for more information.

Photomorphus

ETYMOLOGY Named for being similar to *Photopsis*, an older name for western nocturnal members of *Sphaeropthalma*. **IDENTIFICATION** These females have a unique mesosomal shape that is somewhat intermediate between *Timulla*, *Pseudomethoca*, and other Sphaeropthalmini. The mesosoma is elongate and subrectangular, with the sides pinched at the propodeal spiracle and the propodeum narrower than the mesonotum. They are dull brown in color and generally small-bodied, the gena never has a tooth, the head is never angulate, the mesosoma is always longer than wide, and T2 sometimes has a plumose fringe. **DISTRIBUTION AND DIVERSITY** There are about 10 diurnal species in the central and eastern USA and many more nocturnal ones farther west (6.1.6). **NOTES** These are pretty drab for diurnal mutillids; some eastern species with large eyes seem to be crepuscular or nocturnal. **KEYS AND CHARACTERS** Generally, the diurnal forms occur only in the eastern USA. These species are separated mainly by coloration, mandible structure, and pygidial sculpture. It is currently impossible to differentiate all the nocturnal forms from the central and western USA.

Protophotopsis

ETYMOLOGY The genus name means "early photopsis," reflecting Schuster's (1949) belief that it was loosely related to the nocturnal velvet ants. **IDENTIFICATION** These females have the mesosoma entire (subfamily character) and pear-shaped (tribe character), and the T1 shape broad and sessile (tribe characters). The only species in the USA, *Pr. venenaria*, has a silver or golden head and two pale yellow patches on T3. **DISTRIBUTION AND DIVERSITY** One species lives in the USA (*Protophotopsis venenaria*), one lives in Central America, and two occur in South America. **NOTES** Schuster thought this was related to nocturnal Sphaeropthalmini when he described the genus in 1949. More recently, Brothers & Lelej (2017) recovered it within Dasymutillini in their phylogeny. The lack of axillar armature on the males and presence of a felt line on S2 suggest that it truly belongs with the Sphaeropthalmini, and it is treated as such in this book. To some degree, this also comports with a recent unpublished molecular phylogeny by Waldren (2021) and Waldren et al. (forthcoming). **KEYS AND CHARACTERS** Because there are so few species, distribution is usually diagnostic in *Protophotopsis*.

Sphaeropthalma

ETYMOLOGY From the Ancient Greek *sphaero* "ball" and *ophthalmos* "eye," in reference to the rounded eye shape. Interestingly, there are two *h*'s in *ophthalmos,* but the name was misspelled in the original publication and we are stuck with the single letter *h* -*opthalma* suffix. **IDENTIFICATION** These females have the mesosoma entire (subfamily character) and pear-shaped (tribe character), and the tergal fringes densely plumose (tribe character). This is the most diverse and variable genus of the Sphaeropthalmini, and more detailed information about recognizing the genus is found in section 4. Ruling out the unique coloration and mesosoma shape of other diurnal Sphaeropthalmini in this section can diagnose the diurnal members of this genus treated in this chapter. **DISTRIBUTION AND DIVERSITY** There are two eastern and central diurnal females and about 10 western species, which seem to be mostly nocturnal, but they are brightly colored and sometimes found in daylight. Most species (~95) are dull colored and nocturnal. **NOTES** Eastern and central diurnal species attack mud-nesting wasps and are common inside houses. Western day-active forms are usually found on cool days late in evening and persist in their activity into the night. In some cases, *S. unicolor* individuals were hand collected during the late afternoon and then further individuals were collected in light traps on the same night. **KEYS AND CHARACTERS** The color pattern and T1 shape alone can often work for separating diurnal species. It is important, however, to compare these species with those in the nocturnal chapter since most species in this genus are nocturnal and treated in chapter 6.

Figure	Taxon	Description
2.3.1a	*Morsyma ashmeadi*	Habitus: lateral view
2.3.1b	*Photomorphus impar*	Habitus: lateral view
2.3.1c	*Protophotopsis venenaria*	Habitus: lateral view
2.3.1d	*Sphaeropthalma unicolor*	Habitus: lateral view
2.3.1e	*Sphaeropthalma pensylvanica*	Habitus: lateral view
2.3.1f	*Stethophotopsis maculata*	Habitus: lateral view
2.3.2a	*Protophotopsis venenaria*	Metasoma: T1 shape broad sessile
2.3.2b	*Sphaeropthalma unicolor*	Metasoma: T1 shape short sessile
2.3.2c	*Sphaeropthalma pensylvanica*	Metasoma: T1 shape subsessile
2.3.2d	*Sphaeropthalma marpesia*	Metasoma: T1 shape petiolate

2.3.1a

2.3.1b

2.3.1c

2.3.1d

2.3.1e

2.3.1f

2.3.2a

2.3.2b

2.3.2c

2.3.2d

Figure	Taxon	Description
2.4.1a	*Ephuta margueritae*	Habitus: lateral view
2.4.2a	*Ephuta margueritae*	Frons: large eyes, dense setae
2.4.2b	*Ephuta spinifera*	Frons: small eyes, sparse setae
2.4.3a	*Ephuta scrupea*	Gena: postgenal carina complete
2.4.3b	*Ephuta slossonae*	Gena: postgenal carina incomplete
2.4.4a	*Ephuta floridana*	Hypopygium: basal transverse ridge
2.4.4b	*Ephuta scrupea*	Hypopygium: two basal tubercles
2.4.5a	*Ephuta margueritae*	Pygidium: entirely setose, no plate
2.4.5b	*Ephuta minuta*	Pygidium: plate slender, microreticulate
2.4.5c	*Ephuta slossonae*	Pygidium: plate broad, smooth

2.4.1a

Stethophotopsis

ETYMOLOGY From the Ancient Greek *stetho* "chest" and the formerly recognized genus *Photopsis*. The name refers to the unique mesosternal armature of the male. IDENTIFICATION These females have the mesosoma entire (subfamily character) and pear-shaped (tribe character), and T2 with a plumose fringe (tribe characters). The only species, *S. maculata*, can be recognized by color and the body is entirely orange except for two black patches at the base of T2. DISTRIBUTION AND DIVERSITY Just one species, *Stethophotopsis maculata*, which lives in the Madrean Archipelago of Arizona and northern Mexico. NOTES See the species page (3.4.16.1) for more information.

2.4 SPHAEROPTHALMINAE: EPHUTINI: *EPHUTA*

Ephuta

ETYMOLOGY The genus name was apparently a newly invented word. IDENTIFICATION These females have the mesosoma entire (subfamily character) and ovate (tribe character), the eye vertically ovate (tribe character), and the T1 shape narrow cylindrical (tribe and genus character). Additionally, they lack felt lines, unlike the other Sphaeropthalminae. DISTRIBUTION AND DIVERSITY There are 30 species in the USA and many more in Central and South America. NOTES Females are rarely encountered. Many examined specimens are from pitfall traps. In the USA, they have similar distribution to *Timulla*, with multiple eastern, central, and Arizonan species, but very few in California. Due to the eye shape, this tribe was included in the Mutillinae until a recent phylogenetic study revealed that they belonged in the Sphaeropthalminae. KEYS AND CHARACTERS Schuster (1951) published the most effective key, which works all right but is missing a few species and can be especially difficult to interpret without reference material. The presence or absence of dense, small, colorful setae on the head is useful. Species can have the eye large or small; this can most easily be recognized in a frontal view, by dividing the interocular distance by the eye height. Ratios of 1.2 or lower indicate a large eye, while ratios greater than 1.3 indicate a small eye. One important feature is the postgenal carina, which is found on the underside of the head. *Ephuta* females can therefore be especially annoying to identify because they curl their head downward when they die. It then becomes necessary to rehydrate the specimen and gently pull the head forward to reveal this feature. Many specimens in collections have their heads removed entirely due to accidents during this procedure. Figures are provided for additional characters used by Schuster (1951) and in the species accounts of the next section. The pygidial shape terminology for *Ephuta* is somewhat confusing, since a pygidium that would be considered wide for an *Ephuta* is much narrower than that seen in any species of *Dasymutilla* (see section 2.1) or *Timulla* (see section 2.5).

2.5 MUTILLINAE: TROGASPIDIINI: *TIMULLA*

Timulla

ETYMOLOGY This name is a "remix" of the type genus for velvet ants: *Mutilla*.

This has happened a lot within the family for other velvet ant genera: *Timulla, Atillum,* and *Tallium*. All rearrangements of *Mutilla*. Strangely, *Tumalli* has not been used yet—a shame if you ask us. I've seen the same thing with a handful of genera in the spiny lobster family Palinuridae. The name *Palinurus* was rearranged for the genera *Linuparus, Nupalirus,* and *Panulirus.*—AARON

IDENTIFICATION These females have the mesosoma entire (subfamily character) and rectangular (tribe character), the eye vertically ovate (subfamily character), and the T1 shape sessile (tribe and genus character). **DISTRIBUTION AND DIVERSITY** There are nearly 30 species in the USA, and many more in Central and South America. **NOTES** This is one of the few genera in the USA wherein males typically carry the females during courtship. For this reason, many mating pairs have been collected and the ratio of species known from both sexes is higher than that seen in many other genera. **KEYS AND CHARACTERS** Coloration and mesosoma shape are important for separating females of *Timulla*. Microscopic differences in the scutellar scale, pygidium, and sculpture in various body regions are also useful for identification. Mickel (1937) has a useful key, but many of the characters, especially regarding the mesosoma shape, are difficult to interpret. The first, and often easiest, step is measuring the propodeal width: two species have the propodeum widest, two have the propodeum narrowest, and the remainder have the propodeum and pronotum similar in width. Secondly, the general shape of the humerus (shoulder) must be interpreted. The humerus is considered "angular" when the lateral margins of the pronotum are relatively straight and parallel and form a perpendicular or obtuse angle with the anterior propodeal margin. The humerus is considered "rounded" when the lateral margins of the pronotum are more curved and convergent anteriorly, more evenly rounding into the anterior pronotal margin. Third, the lateral margins of the mesonotum can be constricted or not. When the mesonotum is considered "constricted," it is clearly narrower than the pronotum and propodeum and usually has a distinct in-step posterior to the pronotal spiracle and/or a lateral notch anterior to the propodeal spiracle; Mickel (1937) referred to this state as "sides of dorsum of thorax distinctly emarginate medially." When the mesonotum is "not constricted," it is only scarcely narrower than the pronotum and propodeum and has relatively straight lateral margins without a distinct notch anterior to the propodeal spiracle ("sides of dorsum of thorax not emarginate medially" in Mickel's key). These features are still somewhat subjective and difficult to interpret without practice, but hopefully this phrasing is clearer, especially when accompanied by the figures presented here.

2.6 MYRMOSINAE: *MYRMOSA* AND KIN

Myrmosa

ETYMOLOGY The genus name was apparently a newly invented word based around the Ancient Greek *myrme* "ant." **IDENTIFICATION** These females have the mesosoma divided (subfamily character), the clypeus with a basal longitudinal carina, S1 with a raised longitudinal process, and T2 without yellow spots (genus characters). Species in this genus have simple antennal tubercles without an interantennal prominence and the mandible tridentate. **DISTRIBUTION AND DIVERSITY** Four females are recognized in the USA, of which two are associated with males. This genus also occurs in the Palaearctic Region. **NOTES** This genus and its relatives were sometimes treated as members of the family Tiphiidae or placed into their own family because they have many features that are not found in other velvet ants. Females have a divided mesosoma and usually have ocelli; males have a jugal lobe in the hindwing and lack felt lines. This is the only velvet ant genus to occur in both the New and Old World. **KEYS AND CHARACTERS** Krombein's (1940) key works pretty well, but it is complicated by the inclusion of Old World species. Differences in head shape and microscopic ridges or tubercles on the head and mesosoma are the most useful characters.

Myrmosula

ETYMOLOGY This genus was named for its similarity to *Myrmosa*. **IDENTIFICATION** These females have the mesosoma divided (subfamily character), the clypeus without a basal longitudinal carina, S1 without a raised longitudinal process, and T2 usually with pale yellow spots (genus characters). Species in this genus usually have either an interantennal prominence or the antennal tubercles armed with a tooth or tubercle and the mandible bidentate. **DISTRIBUTION AND DIVERSITY:** There are about 10 species, with most of the diversity in the western USA. **NOTES** Brothers (1978) wrote a good paper on the life cycle and host interactions of *Myrmosula parvula* with halictid bee species

Figure	Taxon	Description
2.5.1a	*Timulla oajaca*	Habitus: dorsal view
2.5.2a	*Timulla ferrugata*	Mesosoma: propodeum widest, humerus rounded, mesonotum weakly constricted, moderate scale
2.5.2b	*Timulla dubitatiformis*	Mesosoma: equally wide, humerus rounded, mesonotum weakly constricted, no scale
2.5.2c	*Timulla dubitata*	Mesosoma: equally wide, humerus rounded, mesonotum constricted, small scale
2.5.2d	*Timulla vagans*	Mesosoma: equally wide, humerus angular, mesonotum constricted, large scale
2.5.2e	*Timulla leona*	Mesosoma: equally wide, humerus angular, mesonotum not constricted, small scale
2.5.2f	*Timulla n. navasota*	Mesosoma: propodeum narrowest, humerus rounded, mesonotum not constricted, no scale
2.5.3a	*Timulla leona*	Pygidium: striate
2.5.3b	*Timulla suspensa*	Pygidium: rugose basally, microreticulate apically

2.5.1a

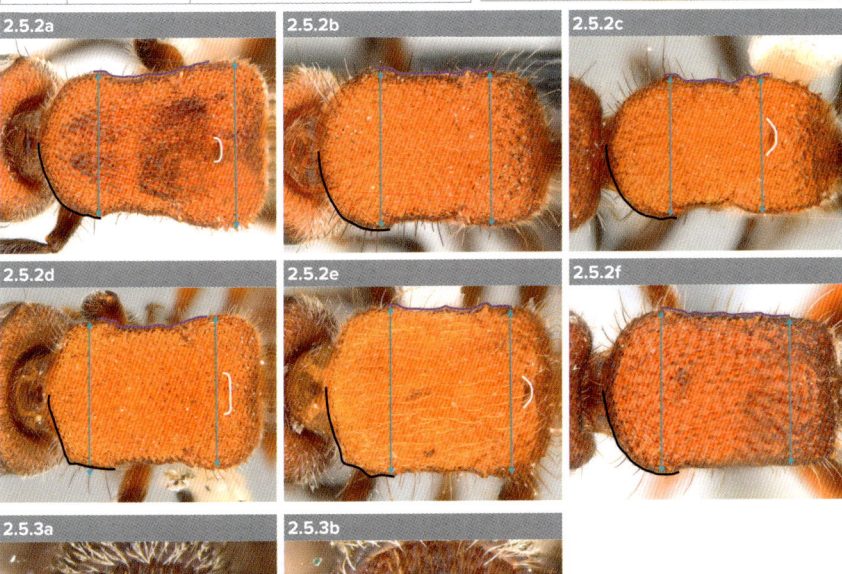

2.5.2a 2.5.2b 2.5.2c
2.5.2d 2.5.2e 2.5.2f
2.5.3a 2.5.3b

Figure	Taxon	Description
2.6.1a	*Myrmosa bradleyi*	Habitus: lateral view
2.6.1b	*Myrmosula pacifica*	Habitus: lateral view
2.6.1c	*Leiomyrmosa spilota*	Habitus: lateral view
2.6.2a	*Myrmosa unicolor*	S1: large ridge; Metacoxa: large tooth
2.6.2b	*Myrmosula exaggerata*	S1: unarmed; etacoxa: small tooth
2.6.3a	*Myrmosa bradleyi*	Frons: ocelli present, vertex quadrate, antennal tubercle unarmed, no interantennal prominence
2.6.3b	*Myrmosa unicolor*	Frons: ocelli present, vertex rounded, antennal tubercle unarmed, no interantennal prominence
2.6.3c	*Leiomyrmosa spilota*	Frons: ocelli absent, mandible tridentate, antennal tubercle unarmed, no interantennal prominence
2.6.3d	*Myrmosula parvula*	Frons: ocelli absent, mandible bidentate, antennal tubercle raised, no interantennal prominence
2.6.3e	*Myrmosula pacifica*	Frons: ocelli absent, mandible bidentate, weak antennal tubercle, interantennal prominence blunt
2.6.3f	*Myrmosula boharti*	Frons: ocelli absent, mandible bidentate, weak antennal tubercle, interantennal prominence sharp
2.6.4a	*Myrmosa unicolor*	Head: clypeus dentate, mandible tridentate
2.6.4b	*Myrmosula pacifica*	Head: mandible with straight lamella, postgena unarmed
2.6.3i	*Myrmosula rutilans*	Head: mandible with concave lamella, postgena dentate

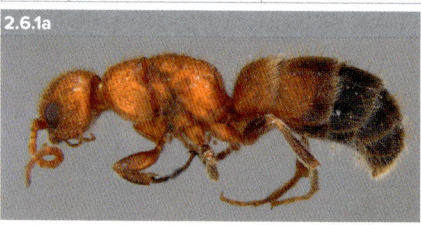

2.6.1a

2.6.2a

2.6.1b

2.6.2b

2.6.1c

Lasioglossum zephyrum, in Kansas. Many of the western specimens examined by Wasbauer were collected on desert mats of *Euphorbia* flowers by R. M. Bohart. At first glance, these wasps can most easily be confused with various *Pseudomethoca* species, often sharing similar coloration with them in a given region. **KEYS AND CHARACTERS** Wasbauer (1973) wrote a good key to separate the females of this genus. They can sometimes be recognized by slight color differences and distribution, but microscopic armature of various structures on the head is usually necessary to confirm a species identification. These particularly include differences in the antennal tubercles and interantennal prominence.

Leiomyrmosa

ETYMOLOGY From the Ancient Greek *leios* "smooth" with the related genus *Myrmosa* as a suffix. The name refers to the overall smooth body with few setae. **IDENTIFICATION** These females have the mesosoma divided (subfamily character), the clypeus without a basal longitudinal carina, S1 without a raised longitudinal process, and T2 usually with pale yellow spots (genus characters). Species in this genus have simple antennal tubercles without an interantennal prominence and the mandible tridentate. **DISTRIBUTION AND DIVERSITY** There is only one species recognized, *Leiomyrmosa spilota*, which occurs in the Sonoran Desert in California (and probably Arizona, see below). **NOTES** This genus was known only from nine specimens of a single species collected from sand dunes near Blythe, California. One dark-bodied individual was recently collected by Bill Warner in Arizona directly opposite of Blythe on the Arizona side of the Colorado River. We cannot be sure yet whether this represents a second species of *Leiomyrmosa* or a color variant of *L. spilota*. The type series and all subsequent specimens were collected only in pitfall traps; it is unlikely that anyone has observed live specimens of these rare wasps in their natural activities.

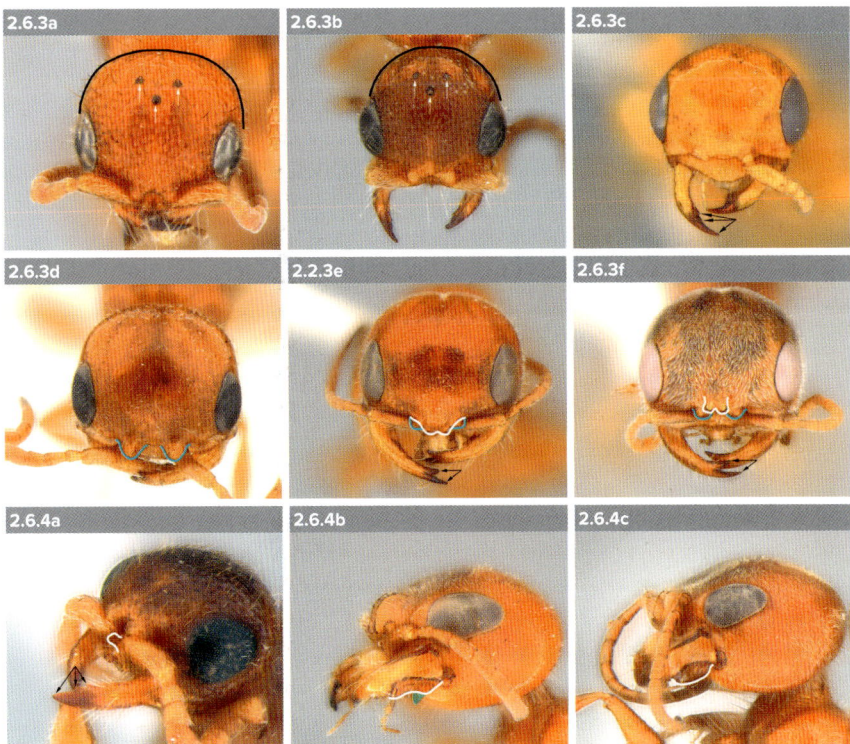

DAY WALKERS—DIURNAL FEMALE VELVET ANTS—COLOR SYNDROMES AND SPECIES

3.0 INTRODUCTION AND OVERVIEW

All female velvet ants are completely wingless. Some males, however, are completely wingless, too, (like *Stethophotopsis maculata*, 3.4.16.1) and others always have tiny wings (like *Myrmilloides grandiceps*, 3.3.6.1). These are treated here in chapter 3 with their females. Some other males, especially *Dasymutilla asopus* and their relatives (see 5.2.5.4), have the wings shortened in some individuals. Since these species also have fully winged populations and because their short wings are dark and obvious, they are included in chapter 5 with the other fully winged males. In rare instances, fully winged males can have their wings mangled by predators or, very rarely, apparently chewed off by themselves on purpose. Those insects are treated in chapter 5, as well.

The key below will help you recognize the Females Color Syndrome (FCS) of a winged velvet ant and get you closer to the target of a species identification.

1. Head and mesosoma cuticle blackened, dorsally with mostly uniformly colored, long, dense setae
    - a. Head, mesosoma, and metasoma dorsal color mostly white .... **3.1 Desert FCS** (p. 51)
    - b. Head, mesosoma, and metasoma dorsal color mostly yellow to red .... **3.2 Western FCS** (p. 62)
    - c. Head and mesosoma color mostly black, metasoma usually reddish .... **3.3 Texan FCS** (p. 82)

2. Mesosoma background color different from metasoma color, or head and/or mesosoma with contrasting pattern
    - a. Head and mesosoma with contrasting patterns; metasoma background color usually reddish or concolorous with mesosoma .... **3.4 Madrean FCS** (p. 91)
    - b. Head and mesosoma reddish-orange with sparse setae (rarely with head blackish); metasoma background color blackish, contrasting with always uniformly reddish-orange mesosoma .... **3.5 *Timulla*-like FCS** (p. 114)

3. Head, mesosomal, and metasomal (at least T2) background cuticle concolorous orange-red or brown, head and mesosoma without distinct setal pattern
    - a. Body color usually reddish-orange, legs usually blackened, metasoma often with distinct black, silvery, and reddish-orange pattern .... **3.6 Eastern FCS** (p. 122)
    - b. Body color dull brown, legs usually light brown, metasoma with faint pattern .... **3.7 Cryptic FCS** (p. 147)

3.1 DESERT FEMALES COLOR SYNDROME

Velvet ants in the Desert Females Color Syndrome (FCS) possess white to pale yellow dorsal setae, which are typically longer than those seen in other color syndromes. Their legs and bodies are generally black, but some species have a reddish underlying cuticle (ex. *Dasymutilla gloriosa* and *D. thetis*). Members of the Desert FCS are concentrated in the hot desert ecoregions of North America, including the Chihuahuan, Sonoran, Mojave, and Baja California Deserts. Members of this syndrome, however, extend into some other regions, particularly the Mediterranean-climate regions of California's Central Valley and some mountainous areas of Arizona and Mexico. In the USA, there are 18 females in three genera with this color syndrome, mainly in the genus *Dasymutilla*.

Many forms could be confused with pale yellow taxa in the Western FCS, but those species are usually included in both sections. None of the species treated here occurs in the Great Plains, so pale yellow individuals from the Great Plains should always be checked with the Western FCS section. There are some Madrean species that could initially seem like they belong here, especially *D. pulchra* (3.4.7.1) and some populations of *D. monticola* (3.4.4.2), but none of the species in the Desert FCS has black setal patches on the mesosoma or pale yellow or orange cuticular spots on T2.

Why are members of this color syndrome clothed in white or very pale setae? One initial popular hypothesis involved purported camouflage—because some species superficially look similar to the hairy fruit of the creosote bush. A recent study, however, indicated that this white coloration evolved in these wasps prior to the arrival of creosote in North America. Instead, the coloration assists in dissipating heat from the female's body. This confers an advantage by extending the activity period during which a female can search for suitable hosts. In the hot deserts, Western FCS species are disadvantaged by necessarily shorter activity periods. It has not yet been established that the non-thistledown desert species also have an advantage in dealing with high temperatures, but abundance in specifically hot regions superficially supports that hypothesis. If white coloration is specifically particularly adaptive to deal with arid, warm environments in southwestern North America, then this likely developed during the Neogene, particularly in the Miocene and/or Pliocene epochs 4–6 million years ago, when widespread desertification occurred in North America.

Even if climatic variables are the main factor driving white color in these wasps, participation in Müllerian mimicry systems also seems to occur. All these insects resemble one another in superficial color, but in numerous cases, minor color features are shared by species within specific regions (and some of these features seem to detract from the thermoregulatory function). For example, there is usually a distinct yellowish tint to the metasomal setae in the *Dasymutilla magna* cluster; if heat management was the sole driving factor in their coloration, why would not the setae be entirely whitish? Furthermore, many species treated here have contrasting black setae, often on the metasoma, which would seem to detract a bit from thermoregulation. Although thermoregulation obviously seems to be an important factor pushing the color pattern of these wasps, it is unlikely to be the sole driver; mimicry, and maybe even camouflage in some cases, may still be at least a secondary factor.

All 18 of these Desert FCS species are known from both sexes. Most of the males (~61%) belong to the analogous Desert Males Color Syndrome (MCS) (5.1), which has a body cuticle that is generally black and the dorsal body setae mostly white to pale yellow, just like these females. dual sex-limited mimicry is common in this color syndrome, and males of some of these species have darker coloration than the females.

This color syndrome extends south into arid regions of Mexico (p. 322). In other arid regions of the world, including North Africa and the deserts of Central Asia (see 8.3.11), and even arid South America, velvet ants also have predominantly whitish setal coloration. In the North American species pictured in this chapter, however, the extent of white setal coverage and the length of these setae, are unmatched, except perhaps by the unusual Australian species, *Ephutomorpha fulvocrinita*.

Why is this lone species from Down Under clothed so similarly to our North American desert friends? Perhaps this species, through a retained mutation or a series of mutations, has come upon the same advantage of dissipating heat in the interior deserts of Australia and being able to become active earlier in the day than other mutillid species.

TOP LEFT: *Dasylabris gussakovski* from Uzbekistan.

TOP RIGHT: *Reedomutilla gayi* from Chile.

LEFT: *Ephutomorpha fulvocrinita* from Australia.

The key below is useful for figuring out which page to visit next.

Genus *Dasymutilla*—T1 shape petiolate, T2 fringe with simple setae

3.1.1 Entire body, including legs, covered with, white setae, widespread .... *D. gloriosa, D. pseudopappus, D. thetis*

3.1.2 Dorsal setae gray to pale yellow, legs often partly gray, Sonoran Desert .... *D. magna, D. connectens, D. eminentia, D. satanas*

3.1.3 Tip of metasoma black or metasoma with black pattern .... *D. nocturna, D. atricauda, D. foxi*

3.1.4 Dorsal setae whitish-gray, legs black, Pacific/Sierra region in California .... *D. sackenii, D. californica, D. coccineohirta, D. aureola*

3.1.5 Dorsal setae whitish, legs mostly black, hot deserts in California .... *D. albiceris, D. imperialis*

Genus *Pseudomethoca*—T1 shape sessile, T2 fringe with simple setae

3.1.6 Dorsal setae uniformly pale grayish yellow .... *Ps. anthracina*

Genus *Sphaeropthalma*—T1 shape petiolate, T2 fringe plumose

3.1.7 Dorsal setae uniformly pale grayish yellow .... *S. edwardsii*

▼ 3.1.1—*Dasymutilla gloriosa* Desert females cluster

Species in this cluster (sometimes referred to as thistledown velvet ants) have the whole body, including the legs and underside, covered with long white setae. Some species with gray legs, like *D. foxi* (3.1.3.3) or *D. eminentia* (3.1.2.3), could be confused with those. They are common and widespread in the southwestern USA, especially in hot arid regions.

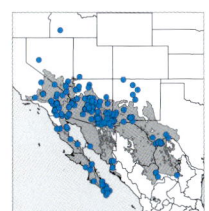

3.1.1.1—*Dasymutilla gloriosa*
DAH-ZEE-MEW-TILL-UH GLOW-REE-OWE-SUH

ETYMOLOGY From the Latin *gloriosa* "glorious," likely referring to the fluffy angel-like appearance of this wasp. **FIELD IDENTIFICATION** Often difficult. The dorsal setae are longer and shaggier than *D. thetis*. Unlike *D. pseudopappus*, the body cuticle is usually reddish and the setae are less regularly oriented; the mesosoma has a partly flattened patch and T2 has a dense patch of especially long setae subbasally. This is the most commonly encountered and most widely distributed Desert FCS species, so, even money, it is likely this one. **LAB IDENTIFICATION** The antennal scrobe has a dorsal carina; the mesosoma has a wide flat scutellar scale with thick carinae anteriorly; the pygidium is flat with the sculpture uniformly rugose or longitudinally rugose. **MALE** (5.2.7.1, 5.3.3.4). Very different from the female; body covered with black and yellow or orange setae. **NOTES** This species was previously thought to be camouflaged with the fruits of creosote bush. Recent studies have found that their white coloration more likely evolved to protect them from high temperatures. The white color allows them to maintain a lower body temperature than orange-colored female congenerics, enabling them to forage for longer periods of time in hot deserts than their relatives. Therefore, this is the most commonly seen species in hot deserts during hot parts of the day or season. This species is most common in North America's hot deserts, but isolated populations were recently discovered in the Great Basin of Nevada and Idaho, too. The male and female of *D. gloriosa* are very different in coloration and are a textbook example of dual sex-limited mimicry. They were not associated until a gynandromorph specimen was discovered (p. 13).

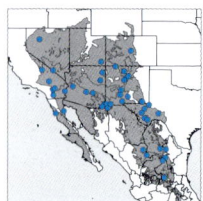

3.1.1.2—*Dasymutilla pseudopappus*
DAH-ZEE-MEW-TILL-UH SOO-DOE-PAPP-USS

ETYMOLOGY From the Ancient Greek words *pseudo* "false" and *pappus* "woolly seed"; the name is a reference to the resemblance of these wasps to creosote fruit. **FIELD IDENTIFICATION** Often difficult. The dorsal setae are longer and shaggier than *D. thetis*. Unlike *D. gloriosa*, the body cuticle is usually blackish; and the setae are more regularly oriented: the mesosoma has erect setae throughout, and T2 has the long setae more uniformly distributed. **LAB IDENTIFICATION** The antennal scrobe has a dorsal carina; the mesosoma has a narrow erect scutellar scale without carinae anteriorly; and the pygidium is somewhat convex with distinct raised striae basally. **MALE** (5.2.7.5, 5.3.3.5). Very different from the female; body covered with black

and yellow, orange, or red setae. **NOTES** While superficially nearly identical to *D. gloriosa*, genetic studies show that these species are only distantly related. *D. pseudopappus* is often found at higher elevations and earlier in the year than *D. gloriosa*.

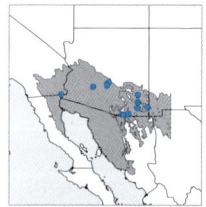

3.1.1.3—*Dasymutilla thetis*
DAH-ZEE-MEW-TILL-UH THAY-TISS

ETYMOLOGY In Greek mythology, Thetis was a goddess of the seas and the mother of Achilles. **FIELD IDENTIFICATION** Relatively easy: this species has shorter and sparser dorsal setae. **LAB IDENTIFICATION** The antennal scrobe lacks a dorsal carina, and the mesosoma lacks a scutellar scale. The mandible is also unique in the genus *Dasymutilla*, having a strong sharp dorsal mandibular carina. **MALE** (5.1.3.3). Similar to female, except with darker cuticle and extensive black setae. **NOTES** *Dasymutilla thetis* is usually smaller than the other thistledown velvet ants and has a narrower distribution, being found in southern Arizona only.

3.1.2—*Dasymutilla magna* Desert females cluster

These species occur mainly in the Sonoran Desert. The dorsal body setae are generally gray or pale yellow, or gray anteriorly and pale yellow apically. Some populations of *D. foxi* (3.1.3.3), treated in the next cluster, have the tergal setae uniformly gray, but they have a wider head than any of the species in this cluster.

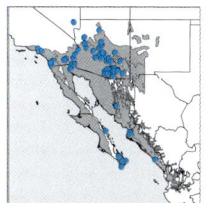

3.1.2.1—*Dasymutilla magna*
DAH-ZEE-MEW-TILL-UH MAGG-NUH

ETYMOLOGY From the Latin *magnus* "big." This species is larger than any of the other desert form species in North America, though some specimens of *D. sackenii* or *D. satanas* can be nearly as large. **FIELD IDENTIFICATION** Usually easy. This large-bodied species has black legs with gray "elbows" and the propodeal setae mostly black. It is larger, more common, and more widespread than the superficially similar *D. connectens*. **LAB IDENTIFICATION** The antennal scrobe has a dorsal carina; the genal carina is sharp and distinct; the mesosoma is longer than wide and has a scutellar scale; the mid and hind femoral apices are rounded; and the S2 sculpture is simply punctate. **MALE** (5.1.1.1). Similar to female. **NOTES** Structurally, this species is similar to *D. magnifica* from the Texan FCS (3.3.1.1).

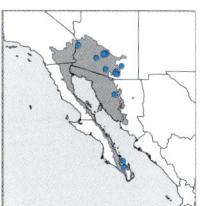

3.1.2.2—*Dasymutilla connectens*
DAH-ZEE-MEW-TILL-UH CUH-NECK-TENS

ETYMOLOGY Derived from the Latin word *conecto* "linked together." It is not clear why that name was chosen for this species. **FIELD IDENTIFICATION** Often easy. This species is similar to *D. magna* but is smaller-bodied. **LAB IDENTIFICATION** The antennal scrobe has a dorsal carina; the genal carina is absent; the mesosoma is longer than wide and has a scutellar scale; the mid and hind femoral apices are truncate; and the S2 sculpture is simply punctate. **MALE** (5.1.1.2). Similar to female. **NOTES** This species is less common and smaller-bodied than *D. magna*. The femora have truncate apices, revealing that *D. connectens* is closely related to *D. nogalensis* from the Texan FCS (3.3.4.5).

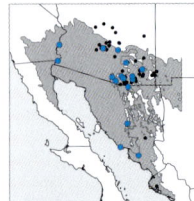

3.1.2.3—*Dasymutilla eminentia*
DAH-ZEE-MEW-TILL-UH EMM-INN-ENN-CHUH

ETYMOLOGY Derived from the Latin word *eminentia* "prominence" or "protuberance," likely referencing the multiple bumps on S2. **FIELD IDENTIFICATION** Often easy. This moderate-sized, compact-bodied species has the leg setae entirely gray. **LAB IDENTIFICATION** The antennal scrobe has a dorsal carina; the genal carina is usually absent; the mesosoma is as wide as long and lacks a scutellar scale; the mid and hind femoral apices are rounded; and the S2 sculpture is scabrous laterally. **MALE** (5.1.1.3). Similar to female. **NOTES** *Dasymutilla*

eminentia is smaller than *D magna* or *D.satanas*. The dorsal setae vary from gray to dark orange, with some populations fitting the Western FCS (3.2.2.2). Based on the wide mesosoma, some populations of *D. foxi* (3.1.3.3) could be confused with this species, but they have a wider head and simply punctate S2 sculpture.

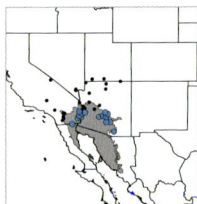

3.1.2.4—*Dasymutilla satanas*
DAH-ZEE-MEW-TILL-UH SAY-TAN-USS

ETYMOLOGY This species was named after Satan, likely in reference to the distribution of this species in the hellish southwestern hot deserts. **FIELD IDENTIFICATION** Usually easy. This large species has the legs entirely black and the propodeal dorsum with setae mostly

pale yellow. **LAB IDENTIFICATION** The antennal scrobe has a dorsal carina; there is no genal carina; the mesosoma is longer than wide and has a scutellar scale; the mid and hind femoral apices are rounded; and the S2 sculpture is simply punctate. **MALE** (5.3.3.3). Much darker coloration than female. **NOTES** In northern and western populations, the dorsal setae are bright orange (see Western FCS, 3.2.5.1). The dorsal setae are shaggier and generally uniformly pale yellow compared with the other species in this cluster.

3.1.3—*Dasymutilla nocturna* Desert females cluster

Unlike the other Desert FCS clusters, the tergites have extensive patches of black setae. The *D. pulchra* cluster (3.4.7.1) from the Madrean FCS could be confused with these species, but they have the mesosoma with a black setal patch. Two of these species are apparently restricted to the Algodones sand dunes in California, but *D. foxi* is relatively widespread in arid regions of Arizona and California.

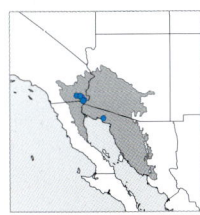

3.1.3.1—*Dasymutilla nocturna*
DAH-ZEE-MEW-TILL-UH KNOCK-TURN-UH

ETYMOLOGY From the Latin *nocturnus* "night active," in reference to the nocturnal behavior in some individuals.
FIELD IDENTIFICATION Easy. This species is often nocturnal; T3–6 have the setae entirely black; the head is narrower than the mesosoma.
LAB IDENTIFICATION The antennal scrobe has a dorsal carina; the head is narrower than the mesosoma; the mesosoma is about as wide as long and has a scutellar scale; and the legs have the setae entirely black. **MALE** (5.1.3.1). Similar to female, with white setae on the apical tergites. **NOTES** This species is most commonly seen in the Algodones sand dunes in California. Like its closest relative, *D. arenivaga* (Western FCS, 3.2.6.1), this species is active both in daylight and after dark.

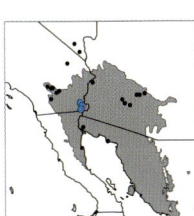

3.1.3.2—*Dasymutilla atricauda*
DAH-ZEE-MEW-TILL-UH AY-TRICK-AWE-DUH

ETYMOLOGY From the Latin words *ater* "black" and *cauda* "tail," in reference to the black tip of the metasoma in females. **FIELD IDENTIFICATION** Usually easy. This species is diurnal; T4–6 have the setae entirely black; the head is as wide as the mesosoma. **LAB IDENTIFICATION** The antennal scrobe lacks a dorsal carina; the head is as wide as the mesosoma; the mesosoma is longer than wide and lacks a scutellar scale; and the legs have the setae entirely black. **MALE** (5.1.3.2). Similar to female, with white setae on the apical tergites. **NOTES** Most populations of *D. atricauda* have orange dorsal setae and are treated in the Western FCS (3.2.6.2). This is a somewhat rare variant of this species; this white desert form occurs only on the Algodones sand dunes in California.

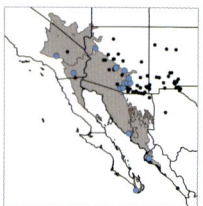

3.1.3.3—*Dasymutilla foxi*
DAH-ZEE-MEW-TILL-UH FOX-EYE

ETYMOLOGY This species was named for the entomologist William J. Fox, who described many velvet ant species in the 1890s. This white variant, found mainly in the Sonoran Desert, used to be named *D. phoenix*, after the city in Arizona. **FIELD IDENTIFICATION** Usually easy. This species is diurnal; there is a mesal patch of black setae on T2–3; and the head is as wide as the mesosoma. **LAB IDENTIFICATION** The antennal scrobe lacks a dorsal carina; the head is as wide as the mesosoma; the mesosoma is as wide as long and lacks a scutellar scale; and the legs have the setae mostly gray. **MALE** (5.5.2.2). Usually with reddish metasomal color. **NOTES** This is one of the most variably colored species in the western USA, being treated in both the Madrean (3.4.6.3) and Western (3.2.7.1) FCS. Unlike most of the other variable species, differently colored individuals are often found together in the same locality. Rarely, this species lacks the black patch on T2–3 and could be confused with *D. eminentia* (3.1.2.3) in the previous cluster, except that species has the head skinnier and the S2 sculpture scabrous laterally.

▼ 3.1.4—*Dasymutilla sackenii* Desert females cluster

These species have white or pale yellow dorsal setae and entirely black legs. These are temperate Pacific or mountain species in California, rarely seen in the true hot deserts. Each of these species varies in color from white to bright red. In all but *D. sackenii*, the Western FCS forms (3.2.4) of these species are more common than their desert forms.

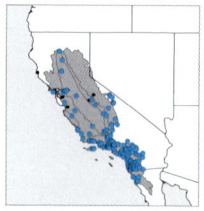

3.1.4.1—*Dasymutilla sackenii*
DAH-ZEE-MEW-TILL-UH SACKEN-EE-EYE

ETYMOLOGY This species was named for the Russian diplomat and entomologist Carl Robert Osten-Sacken, who worked mainly on flies. **FIELD IDENTIFICATION** Often difficult. This species is widespread and has the dorsal setae longer and shaggier than the others in this cluster. Like *D. californica*, the head is slender, but *D. sackenii* individuals are almost always larger in size. **LAB IDENTIFICATION** The antennal scrobe has a dorsal carina; the head is narrower than the mesosoma and is unarmed posteriorly; the mesosoma is longer than wide and has a scutellar scale. **MALE** (5.1.2.1). Similar to female. **NOTES** Some northern populations have orange dorsal setae and fit the Western FCS (3.2.4.4), but unlike the other species in this cluster, this white desert form is more common. This species is recognized as a parasite of the sand wasp *Bembix occidentalis*.

Dasymutilla sackenii

Dasymutilla californica

3.1.4.2—*Dasymutilla californica*
DAH-ZEE-MEW-TILL-UH CAL-IF-OR-NICK-UH

ETYMOLOGY This species was named after the state of California. **FIELD IDENTIFICATION** Often difficult. This species has a slender head like *D. sackenii*, but this desert form of *D. californica* is less common and almost always smaller in

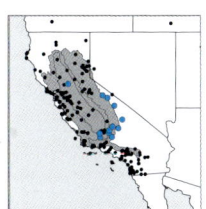

size. **LAB IDENTIFICATION** The antennal scrobe usually lacks a dorsal carina; the head is narrower than the mesosoma and is armed with a posterolateral tubercle; the mesosoma is longer than wide and has a scutellar scale. **MALE** (5.1.2.3). Similar to female. **NOTES** This is a widespread Pacific species, but this white color form is apparently restricted to mountain areas in the southeastern Sierra Nevada, although some pale Central Valley populations fit here. The Western FCS (3.2.4.1) form is more common and widespread.

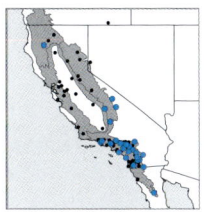

3.1.4.3—*Dasymutilla coccineohirta*
DAH-ZEE-MEW-TILL-UH COX-IN-EE-OH-HURT-UH

ETYMOLOGY From the Latin *coccineus* "red" or "scarlet" and *hirtus* "shaggy" or "hairy," in reference to the shaggy red dorsal setae of many populations. This white desert form used to be named *D. clytemnestra* after the sister of Helen of Troy and wife of Agamemnon from Greek mythology. **FIELD IDENTIFICATION** Often difficult. This species has the mesosoma longer than wide and the head nearly as wide as the mesosoma. **LAB IDENTIFICATION** The antennal scrobe lacks a dorsal carina; the head is about as wide as the mesosoma and is unarmed posteriorly; the mesosoma is longer than wide and has a scutellar scale. **MALE** (5.1.2.4, 5.2.4.2). Similar to female. **NOTES** This white color form is widespread in southern California. Males of *D. coccineohirta* are almost always orange or reddish, even in populations with white females. The Western FCS (3.2.4.2) form is more common and widespread.

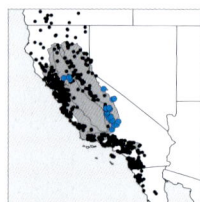

3.1.4.4—*Dasymutilla aureola*

DAH-ZEE-MEW-TILL-UH OUR-EE-OLL-UH

ETYMOLOGY Apparently derived from the Latin word *aureus* "golden," in reference to the golden-yellow dorsal setae of many populations. **FIELD IDENTIFICATION** Often easy. This species has the head much wider than the mesosoma and the mesosoma as wide as long. **LAB IDENTIFICATION** The antennal scrobe usually has a dorsal carina; head is clearly wider than the mesosoma and unarmed posteriorly; mesosoma is as wide as long and lacks a scutellar scale. **MALE** (5.1.2.5). Similar to female. **NOTES** This is a widespread Pacific species, but this Desert FCS form is apparently restricted to the southeastern Sierra Nevada. Some specimens from hot portions of the Central Valley are a pale enough yellow that they could match the desert form. The Western FCS (3.2.1.1) form is more common and widespread.

▼ 3.1.5—*Dasymutilla albiceris* Desert females cluster

Dasymutilla albiceris and *D. imperialis* are similar to the *D. sackenii* cluster, with whitish dorsal setae and black legs, but they are generally found in hotter desert habitats. Unlike the species in the *D. sackenii* cluster above, the two species in this cluster each have the head slender and the mesosoma as wide as long.

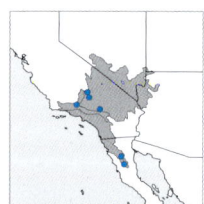

3.1.5.1—*Dasymutilla albiceris*

DAH-ZEE-MEW-TILL-UH AHL-BIH-SEHR-ISS

ETYMOLOGY From the Latin words *albus* "white" and *cerinus* "wax-colored" or "yellowish." **FIELD IDENTIFICATION** Often difficult. This species is superficially similar to *D. sackenii* but has the mesosoma as wide as long and has a narrower geographic distribution. **LAB IDENTIFICATION** Unlike *D. imperialis*, this species lacks a scutellar scale. **MALE** (5.1.2.2). Similar to female. **NOTES** Structurally, this species is similar to *D. vestita* in the Western FCS (3.2.2.1). *Dasymutilla albiceris* is known from a few localities in transition areas between hot deserts and mountain habitats in southern California.

3.1.5.2—*Dasymutilla imperialis*

DAH-ZEE-MEW-TILL-UH IMM-PEER-EE-AHL-ISS

ETYMOLOGY Named for Imperial County, California. **FIELD IDENTIFICATION** Easy. This is the only *Dasymutilla* in the Algodones dunes with white setae on T3–6 and black setae on the legs. **LAB IDENTIFICATION** Unlike *D. albiceris*, this species has a

scutellar scale. **MALE** (5.4.2.2). Different from female; covered entirely with black setae. **NOTES** The male was discovered and named pretty recently (2005); the female was recognized even later (2020). This species is known from very few specimens, all from the Algodones sand dunes in California. It seems to be most active late in the year; most specimens were collected in September or October.

▼ 3.1.6.1—*Pseudomethoca anthracina*

SOO-DOE-METH-OWE-KUH ANN-THRUH-SEE-NUH

ETYMOLOGY From the Ancient Greek *anthrakinos* "coal-black," in reference to the male coloration. **FIELD IDENTIFICATION** Often difficult. Different from all other *Pseudomethoca*, but this species can easily be confused with *D. aureola* (3.1.4.4) except for its sessile T1 shape. **LAB IDENTIFICATION** The head is wider than the mesosoma; the mesosoma is as wide as long; and the T1 shape is sessile. **MALE** (5.4.5.1). Different from female; covered entirely with black setae. **NOTES** This white, desert form of *Ps. antracina* is much rarer than most of the Desert FCS species and can be easily confused with *D. aureola* because of the short mesosoma and wide head. The white color form treated here is only rarely found in the Central Valley of California; most individuals have orange or red dorsal setae and fit the Western FCS (3.2.1.1). There is one undescribed *Pseudomethoca* species from Baja California Sur, Mexico, with similar coloration except that it has a black patch of setae on T2.

ABOVE: New species of *Pseudomethoca* from Baja California.

▼ 3.1.7.1—*Sphaeropthalma edwardsii*

S-FAIR-OPP-THAL-MUH EDWARDS-EE-EYE

ETYMOLOGY This species was named after its collector, Henry Edwards, an English stage actor, writer, and entomologist in the late 19th century. **FIELD IDENTIFICATION** Often difficult. This species overlaps in distribution with the *D. sackenii* cluster (3.1.4), but it has the leg setae mostly whitish-gray and the tergal fringes plumose. It generally has denser, more erect dorsal setae and darker body cuticle than other nocturnal

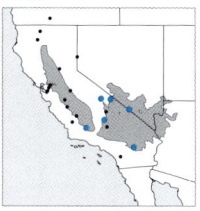

species (p. 283). **LAB IDENTIFICATION** The mandible lacks a dorsal or ventral tooth basally; the mesonotal setae are simple or brachyplumose; the T1 shape is petiolate; the tergal fringes are plumose; and T6 has a defined microreticulate pygidial plate. **MALE** (5.7.12.4). Different from this female form, with the dorsal setae mostly orange. **NOTES** This species is usually nocturnal but is sometimes seen in daylight. Similarly colored nocturnal genera (6.1) usually have the body cuticle pale brown and the dorsal body setae sparser. This color form of *S. edwardsii* overlaps with Desert FCS forms of *D. aureola*, *D. californica*, and *D. coccineohirta* on the eastern slopes of the Sierra Nevada.

3.2 WESTERN FEMALES COLOR SYNDROME

This is the most common color syndrome that all three of us encounter (Kevin in California, Aaron in Texas, and Joe in Utah). The Western Females Color Syndrome (FCS) is almost entirely found west of the Mississippi River valley and extends throughout much of northern and central Mexico (p. 322), but it is not present in Central America. Members are covered in short to long hair that ranges from yellow to orange to cardinal red in coloration. The underlying cuticle and legs are typically black in color. Currently, the Western FCS includes 34 species in four genera.

When the dorsal setae are very pale yellow, these species could be confused with the Desert FCS. A few species treated here are difficult to separate from the Eastern, Madrean, or *Timulla*-like FCS in the central and northwestern USA, especially when the dorsal setae are sparse and the tergites have black, silvery, and orange color patterns. If the head and mesosomal setae lack black or silvery patches, this rules out the Madrean FCS. If the T2 disc has a large yellow-orange cuticular patch mesally, this rules out the *Timulla*-like FCS. In the central USA, if the head and mesosoma setae are moderately dense and erect and T2 has the cuticle largely black anteriorly and laterally, this rules out the Eastern FCS.

Of the 33 species in this FCS, 26 (~79%) are known from both sexes. Most of the males (~88%) belong to the analogous Western Males Color Syndrome (MCS) (5.1), which has the body cuticle mainly black or dark reddish and the dorsal body setae, at least on the mesonotum and T2 disc and/or T3–5, mostly yellow, orange, or red, just like these females. Only a small handful of species in the Western FCS have males with different coloration from the females.

Although this color syndrome is abundant and famous in North America, it is rarely seen on other continents. A few isolated, interesting taxa, like *Dasylabris schultzei* in southwestern Africa (Namibia) and *Quwitilla blattoserica* in Chile, do a pretty good impression of our western Nearctic friends. Additionally, the bizarre Thai species *Cockerellidia sohmi* (8.2.9.1) has long reddish setae and is unlike any other velvet ants in the region.

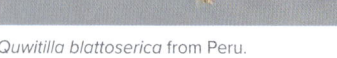

Quwitilla blattoserica from Peru.

Ephutomorpha lutaria from Australia.

The key below is useful for figuring out which page to visit next.

—Genus *Dasymutilla*—T1 shape petiolate; T2 fringe with simple setae

3.2.1 T3–5 setae uniformly yellow, orange, or red, like T2; mesosoma as wide as long; head much wider than mesosoma; Pacific states .... *D. aureola*

3.2.2 T3–5 setae uniformly yellow, orange, or red, like T2; mesosoma as wide as long; head not wider than mesosoma .... *D. vestita, D. eminentia, D. erythrina*

3.2.3 T3–5 setae uniformly yellow, orange, or red, like T2; mesosoma longer than wide; Great Plains and eastern Arizona to Texas ... *D. calorata, D. leda, D. stevensi, D. bioculata, D. nogalensis*

3.2.4 T3–5 setae uniformly yellow, orange, or red, like T2; mesosoma longer than wide; Pacific and northwestern states .... *D. californica, D. coccineohirta, D. sackenii, D. flammifera*

3.2.5 T3–5 setae uniformly yellow, orange, or red, like T2; mesosoma longer than wide; species with especially long and shaggy setae; Mojave and Sonoran Deserts .... *D. satanas, D. arenivaga*

3.2.6 T3–6 or T4–6 setae entirely black; dorsal setae long, dense, and shaggy; Mojave, Sonoran, and Great Basin Deserts .... *D. arenivaga, D. scitula, D. atricauda*

3.2.7 T3–6 with extensive pattern of black or silvery setae; mesosoma as wide as long ..... *D. foxi, D. furina, D. vestita, D. montivagoides*

3.2.8 T3–6 with extensive pattern of yellow, orange, black or silvery setae; mesosoma longer than wide .... *D. bioculata, D. occidentalis, D. radkei, D. campanula, D. californica*

—Genus *Invreiella*—T1 shape sessile, head with genal tooth ventrally, T2 fringe with setae simple

3.2.9 Head massive, armed with large genal tooth .... *I. manleyi*

—Genus *Pseudomethoca*—T1 shape sessile, head unarmed ventrally, T2 fringe with setae simple

3.2.10 Head and mesosoma with dense, mostly erect dorsal setae .... *Ps. anthracina, Ps. flammigera, Ps. aureovestita*

3.2.11 Head and mesosoma with sparse mostly flat dorsal setae .... *Ps. propinqua*

—Genus *Sphaeropthalma*—T1 shape sessile or petiolate, T2 fringe with setae plumose

3.2.12 Dorsal setae uniformly yellow, orange, or red; leg setae usually gray .... *S. unicolor, S. edwardsii*

▼ 3.2.1.1—*Dasymutilla aureola*

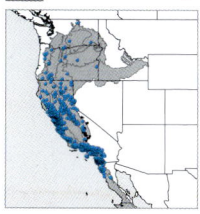

DAH-ZEE-MEW-TILL-UH OUR-EE-OLL-UH

ETYMOLOGY Apparently derived from the Latin word *aureus* "golden," in reference to the golden-yellow dorsal setae of many populations. **FIELD IDENTIFICATION** Often easy. This species has the head much wider than the mesosoma and the mesosoma as wide as long. **LAB IDENTIFICATION** The antennal scrobe usually has a dorsal carina; the head is clearly wider than the mesosoma and unarmed posteriorly; the mesosoma is as wide as long and lacks a scutellar scale. **MALE** (5.2.4.3). Similar to female. **NOTES** This is a widespread Pacific species, which has the dorsal setal color varying from nearly white (see Desert FCS, 3.1.4.4) to bright scarlet red. The brighter red specimens are more commonly seen in southern California, often at somewhat higher elevations, but specimen

age can affect the brightness of the dorsal setae. Based on records from inaturalist.com, this is the second most commonly observed velvet ant species in the world, behind only *D. occidentalis* (3.2.8.2, 3.6.1.1). Males (5.2.4.3), however, are more rarely seen than females and, surprisingly, are encountered less often than males of other Pacific species. This species has been associated with host bees in the genera *Anthophora* and *Melissodes*.

▼ 3.2.2—*Dasymutilla vestita* Western females cluster

Dasymutilla vestita is widespread throughout the central and western USA, while the other two are southwestern species. The tergal setae (at least from T2's posterior half to T5) are uniformly yellow, orange, or red; the mesosoma is about as wide as long (and always lacks a scutellar scale); and the head is not wider than the mesosoma. The mesosomal shape can be difficult to interpret in the field (and sometimes in the lab), so these should be compared with many species in the following clusters.

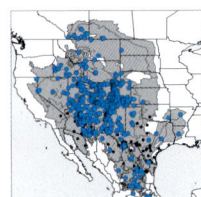

3.2.2.1—*Dasymutilla vestita*
DAH-ZEE-MEW-TILL-UH VESS-TEE-TUH

ETYMOLOGY Apparently from the Latin *vestis* "garment," in reference to the colorful dorsal setae. **FIELD IDENTIFICATION** Often easy. This species is more widespread than the others in this cluster and has the leg setae entirely black. **LAB IDENTIFICATION** The body cuticle is usually entirely black; the antennal scrobe has a dorsal carina; the genal sculpture is less coarse than the vertex sculpture; the vertex is unarmed posteriorly; the mesosoma is as wide as long and lacks a scutellar scale; the hind trochanter is unarmed; S2 has the sculpture simply punctate; and the pygidial sculpture is rugose. **MALE** (5.2.6.1). Similar to female. **NOTES** This is one of the most common species in the western USA, especially in the Intermountain West, and expands east to Arkansas and Louisiana, north into Alberta, Canada, and as far south as Oaxaca, Mexico. In some regions of western Texas and New Mexico, the apical tergites are largely blackened (3.2.7.3), and in southern Texas (and sporadically throughout the southwestern deserts), this species fits the Texan FCS (3.3.2.2).

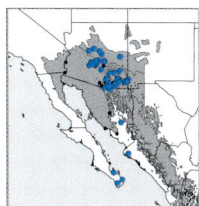

3.2.2.2—*Dasymutilla eminentia*
DAH-ZEE-MEW-TILL-UH EMM-INN-ENN-CHUH

ETYMOLOGY Derived from the Latin word *eminentia* "prominence" or "protuberance," likely referencing the multiple bumps on S2. **FIELD IDENTIFICATION** Often easy. The leg setae are mostly gray and the dorsal setae are usually orange or yellow. **LAB IDENTIFICATION** The body cuticle is variably black to reddish-brown; the antennal scrobe has a dorsal carina; the genal sculpture is nearly as coarse as the vertex sculpture; the vertex is unarmed posteriorly; the mesosoma is as wide as long and lacks a scutellar scale; the hind trochanter has a small tooth; S2 has the sculpture scabrous laterally; and the pygidial sculpture is rugose. **MALE** (5.2.3.2). Similar to female. **NOTES** This color form is more commonly seen at moderately high elevations in Arizona and New Mexico, while specimens from lower elevations in hot deserts more often fit the Desert FCS (3.1.2.3).

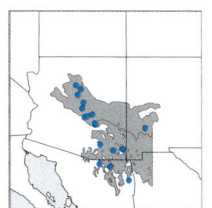

3.2.2.3—*Dasymutilla erythrina*
DAH-ZEE-MEW-TILL-UH AIR-EETH-RYE-NUH

ETYMOLOGY From the Ancient Greek *erythros* "red," in reference to the dorsal setal color. **FIELD IDENTIFICATION** Sometimes easy. The distribution is more restricted, and the dorsal setae are brighter red than other species in this cluster. There are often scattered gray setae on the legs. **LAB IDENTIFICATION** The body cuticle is black laterally and ventrally but mostly pale orange dorsally beneath the red setae; the antennal scrobe has a dorsal carina; the genal sculpture is as coarse as the vertex sculpture; the vertex is unarmed posteriorly; the mesosoma is as wide as long and lacks a scutellar scale; the hind trochanter is unarmed; S2 has the sculpture simply punctate; and the pygidial sculpture is rugose. **MALE** (5.2.5.3). Similar to female. **NOTES** This species is rare in Arizona, but it is apparently the most common species in mountainous areas of Mexico (p. 322). Most of the Mexican populations have gray setae on the legs and a variably large

black patch on the metasoma, but many specimens in Arizona have the leg setae black and at most a tiny black patch on the T2 fringe. Where this species overlaps with *D. vestita* in Arizona, the latter species often has brighter red dorsal setae than other populations (Fig. 3.2.2.1a). *Dasymutilla erythrina*, however, has the dorsal setae even more brilliant and shining, emphasized by the pale orange cuticle beneath these setae.

▼ 3.2.3—*Dasymutilla calorata* Western females cluster

These species are mainly restricted to New Mexico, Texas, and the Great Plains, but *D. stevensi* extends west into Arizona, and a rare variant of *D. nogalensis,* known only from Arizona, is treated here. The tergal setae (at least from T2's posterior half to T5) are uniformly yellow or orange and the mesosoma is longer than wide. The mesosomal shape can be difficult to interpret in the field (and sometimes in the lab), so these should also be compared with other western *Dasymutilla* clusters.

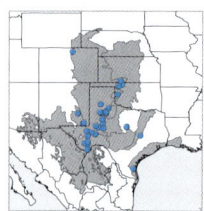

3.2.3.1—*Dasymutilla calorata*
DAH-ZEE-MEW-TILL-UH CAL-OWE-RATT-UH

ETYMOLOGY From the Latin *caloratus* "heated," perhaps in reference to its occurrence in Texas, which has hotter weather than Minnesota, where the author of the species, Clarence Mickel, lived. **FIELD IDENTIFICATION** Often easy. In addition to the large size, there are usually more black setae on the propodeal dorsum and anterior portion of T2 than other species in the cluster. **LAB IDENTIFICATION** The head is narrower than the mesosoma; the antennal scrobe has a distinct dorsal carina; the gena has a weak carina; the mesosoma is longer than wide and has a scutellar scale; the dorsal propodeal setae are usually black; and the pygidial sculpture is striate.

setae are usually black; and the pygidial sculpture is striate. **MALE** (5.2.6.2). Similar to female. **NOTES** Along with *D. klugii* (3.3.1.2) and *D. occidentalis* (3.2.8.2), this is one of the largest species in Texas and the Great Plains. Additionally, the color pattern is somewhat intermediate between those of other large species, and some specimens have the mesosoma more extensively black than normal, making them difficult to separate from *D. klugii* (3.3.1.2). The previously recognized species *D. hispidaria* was recently determined to be an aberrant form of *D. calorata* with the typically black setae turned yellow-orange, likely due to specimen age or some kind of mutation.

Dasymutilla calorata

D. hispidaria

D. klugii

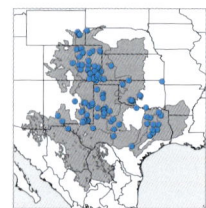

3.2.3.2—*Dasymutilla leda*
DAH-ZEE-MEW-TILL-UH LAY-DUH

ETYMOLOGY In Greek mythology, Leda was an Aetolian princess who became a Spartan queen. **FIELD IDENTIFICATION** Difficult. The head is wider than any other species in this cluster. **LAB IDENTIFICATION** The head is wider than the mesosoma; the antennal scrobe lacks a dorsal carina; the gena lacks a carina; the mesosoma is longer than wide and has a scutellar scale; the dorsal propodeal setae are usually yellow or orange; and the pygidial sculpture is striate. **MALE** Unknown. **NOTES** This species has a massive head, almost like *Pseudomethoca*, but has a distinct scutellar scale and petiolate T1 shape. Structurally, this species

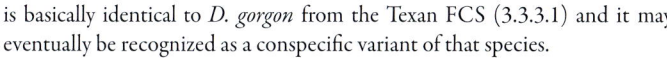

is basically identical to *D. gorgon* from the Texan FCS (3.3.3.1) and it may eventually be recognized as a conspecific variant of that species.

3.2.3.3—*Dasymutilla stevensi*

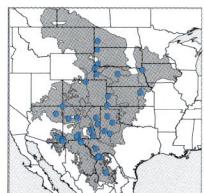

DAH-ZEE-MEW-TILL-UH STEVENS-EYE

ETYMOLOGY This species was named after O. A. Stevens, who collected the type specimen in 1923. **FIELD IDENTIFICATION** Very difficult. The head is wider than most species in this cluster, but not as wide as *D. leda*, and the body size is usually smaller than *D. leda*. The dorsal body setae are usually duller yellow or orange than *D. nogalensis* and the leg setae are usually entirely black. **LAB IDENTIFICATION** The head is scarcely wider than the mesosoma; the antennal scrobe usually lacks a dorsal carina; the gena lacks a carina; the mesosoma is longer than wide and lacks a scutellar scale; the dorsal propodeal setae are usually yellow or orange; and the pygidial sculpture is striate. **MALE** (5.2.6.4). Similar to female. **NOTES** This is the only species of this cluster that lacks a scutellar scale. Structurally this species is similar to *D. nupera* from the Texan FCS (3.2.4.1) and can be dif-

ficult to separate from that species in some individuals with interspersed black and yellow setae on the mesosomal dorsum.

3.2.3.4—*Dasymutilla bioculata*

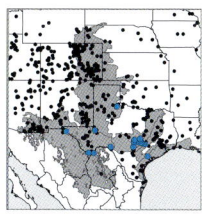

DAH-ZEE-MEW-TILL-UH BYE-OCK-EW-LAW-TUH

ETYMOLOGY From the Latin *bi* "two" and *oculus* "eye," in reference to the two orange spots on T2 in many males. **FIELD IDENTIFICATION** Very difficult. The head is narrower than in most species in the cluster, and T2 has an orange cuticular patch beneath the orange setae. **LAB IDENTIFICATION** The head is narrower than the mesosoma; the antennal scrobe has the dorsal carina weak or absent; the gena lacks a carina; the mesosoma is longer than wide and has a scutellar scale; the dorsal propodeal setae are usually yellow or orange; and the pygidial sculpture is striate. **MALE** (5.2.2.2, 5.2.5.1). Often similar to female. **NOTES** This variant of *D. bioculata* is not commonly encountered. Most *D. bioculata* specimens in the Western FCS have black and/or silvery setae on the apical tergites (3.2.8.1). Various other *D. bioculata* populations fit in the Eastern FCS (3.6.5.3) or Texan FCS (3.3.4.3).

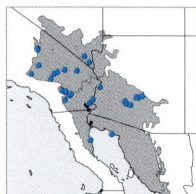

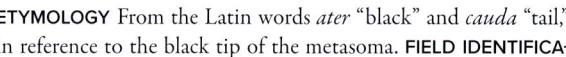

3.2.6.2—*Dasymutilla atricauda*
DAH-ZEE-MEW-TILL-UH AY-TRICK-AWE-DUH

ETYMOLOGY From the Latin words *ater* "black" and *cauda* "tail," in reference to the black tip of the metasoma. **FIELD IDENTIFICA-TION** Usually easy. This species is diurnal; T2–3 have the setae yellow or orange; T4–6 have the setae entirely black; and the head is about as wide as the mesosoma. **LAB IDENTIFICATION** The eye is moderate in size, its diameter generally shorter than F1 + 2 combined; the antennal scrobe lacks a dorsal carina; the head is as wide as the mesosoma; the mesosoma is longer than wide and lacks a scutellar scale; and the pygidial sculpture is usually rugostriate. **MALE** (5.2.7.3). Similar to female, with yellow or orange setae on the apical tergites. **NOTES** In color and distribution, this species seems to follow *D. arenivaga*, even though these species are not closely related. Throughout most of the Mojave and Sonoran Deserts, *D. atricauda* are yellow or orange, but in the Algodones dunes, they have white dorsal setae (3.1.3.2).

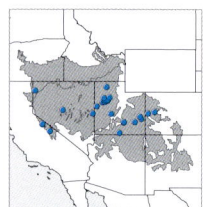

3.2.6.3—*Dasymutilla scitula*
DAH-ZEE-MEW-TILL-UH SCIT-OO-LUH

ETYMOLOGY From the Latin *scitulus* "beautiful" or "elegant." **FIELD IDENTIFICATION** Sometimes easy. Like *D. arenivaga*, T2 has yellow or orange setae and T3–6 have the setae entirely black. This species is diurnal and distribution in the Great Basin is a useful diagnostic feature. This species has denser and shaggier setae than populations of *D. bioculata* in that region and lacks any trace of silvery setae on the tergites or sternites. In transitional areas between the Great Basin and Mojave Desert, a microscope is usually needed to separate it from *D. arenivaga*. **LAB IDENTIFICATION** The eye is moderate in size, its diameter generally shorter than F1 + 2 combined; the antennal scrobe has a distinct dorsal carina; the head is narrower than the mesosoma; the mesosoma is usually longer than wide and has a scutellar scale; and the pygidial sculpture is irregularly rugose. **MALE** (5.2.2.3). Very similar to female. **NOTES** This is one of the most commonly encountered species in the Great Basin, but because people do not collect there very often, it is not especially abundant in museums.

 ### 3.2.7—*Dasymutilla foxi* Western females cluster

These species occur in the southwestern USA, mainly in Arizona and New Mexico. They can be recognized by having variable amounts of black and/or silvery setae on the apical tergites **and** having the mesosoma wider than long. They sometimes overlap with species in the *D. arenivaga* (3.2.6) and *D. bioculata* clusters (3.2.8) and have similar coloration.

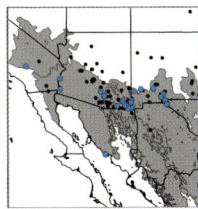

3.2.7.1—*Dasymutilla foxi*
DAH-ZEE-MEW-TILL-UH FOX-EYE

ETYMOLOGY This species was named for the entomologist William J. Fox, who described many velvet ant species in the 1890s. **FIELD IDENTIFICATION** Usually easy. This is the only species in the cluster with bright red dorsal setae. Specimens with duller orange setae can be recognized by the slightly wider head and presence of silver or red setae on the apical tergites posterior to the black setal patch. **LAB IDENTIFICATION** The body cuticle is largely reddish-brown; the mandible is bidentate; the antennal scrobe has the dorsal carina weak or absent; the mesosoma is as wide as long and lacks a scutellar scale; the leg setae are mostly gray; the hind trochanter is unarmed; S2 has the sculpture simply punctate; and the pygidial sculpture is rugose. **MALE** (5.2.3.1, 5.5.2.2). Similar to female. **NOTES** This is one of the most variably colored species in the western USA,

 being treated in both the Madrean (3.4.6.3) and Desert (3.1.3.3) FCS. Unlike most of the other variable species, differently colored individuals are often found together in the same locality. Manley & Taber (1978) discussed a large aggregation of males and females near Tucson in an area with a high population of their hosts in the bee genus *Diadasia*. Many mating pairs were found at the site, and at the

 peak of activity, there seemed to be about five mutillids present per square meter at the site.

3.2.7.2—*Dasymutilla furina*
DAH-ZEE-MEW-TILL-UH FEW-REE-NUH

ETYMOLOGY Perhaps based on the Latin *furia* "madness" or "rage." **FIELD IDENTIFICATION** Often easy. This species and *D. foxi* are the only ones in this cluster known from southeastern Arizona; *D. furina* generally has a narrower head and duller yellow or orange dorsal setae. **LAB IDENTIFICATION** The body cuticle is largely reddish-brown; the mandible is bidentate; the antennal scrobe has a dorsal carina; the mesosoma is as wide as long and lacks a scutellar scale; the leg setae are mostly gray; the hind trochanter has a small tooth; S2 has the sculpture scabrous laterally; and the pygidial sculpture is rugose. **MALE** (5.2.3.3). Similar to female. **NOTES** This species is structurally identical to *D. eminentia* (3.1.2.3, 3.2.2.2), except that it has a large patch of black setae on the metasoma. So far, it has been found only near the city of Douglas, Arizona, where it can be quite abundant.

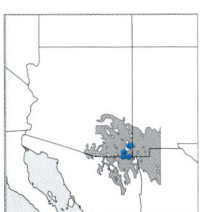

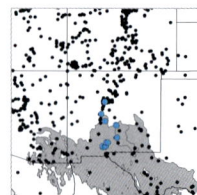

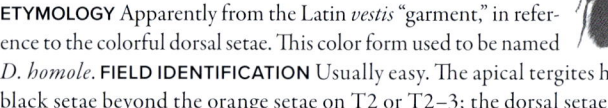

3.2.7.3—*Dasymutilla vestita*
DAH-ZEE-MEW-TILL-UH VESS-TEE-TUH

ETYMOLOGY Apparently from the Latin *vestis* "garment," in reference to the colorful dorsal setae. This color form used to be named *D. homole*. **FIELD IDENTIFICATION** Usually easy. The apical tergites have black setae beyond the orange setae on T2 or T2–3; the dorsal setae are moderately dense and erect; and the T2 cuticle is entirely black. **LAB IDEN-**

TIFICATION The body cuticle is usually entirely black; the mandible is bidentate; the antennal scrobe has a dorsal carina; the mesosoma is as wide as long and lacks a scutellar scale; the leg setae are entirely black; the hind trochanter is unarmed; S2 has the sculpture simply punctate; and the pygidial sculpture is rugose. **MALE** (5.2.6.1). Similar to female. **NOTES** This variant is found mainly in New Mexico, in mountainous areas around the Las Cruces area.

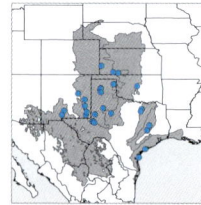

3.2.7.4—*Dasymutilla montivagoides*
DAH-ZEE-MEW-TILL-UH MON-TIVV-AGG-OI-DEEZ

ETYMOLOGY This species was named for its similarity to *Mutilla montivaga*, which is now recognized as a synonym of *Pseudomethoca propinqua* (3.2.11.1). Some populations of this species used to be named *D. nigricauda*, for the mostly black apical tergites. **FIELD IDENTIFICATION** Usually easy. The dorsal setae are sparser and flatter than others in this cluster, and the T2 disc has a large patch of yellow-orange cuticle. **LAB IDENTIFICATION** The body cuticle is mostly black, except for the yellow patch on T2; the mandible is tridentate; the antennal scrobe lacks a dorsal carina; the mesosoma is as wide as long and lacks a scutellar scale; the leg setae are usually entirely black; the hind trochanter is unarmed; S2 has the sculpture simply punctate; and the pygidial sculpture is rugostriate. **MALE** Unknown. **NOTES** Based on the tridentate mandible, this species is closely related to *D. asopus* (3.6.5.2) and *D. waco* (3.3.2.1). Based on coloration, distribution, and similarity to both sexes of *D. asopus*, the male of this species is likely to be *D. hector* (5.7.3.1) or *D. neomexicana* (5.2.5.4) ... or both, if those two males turn out to be conspecific.

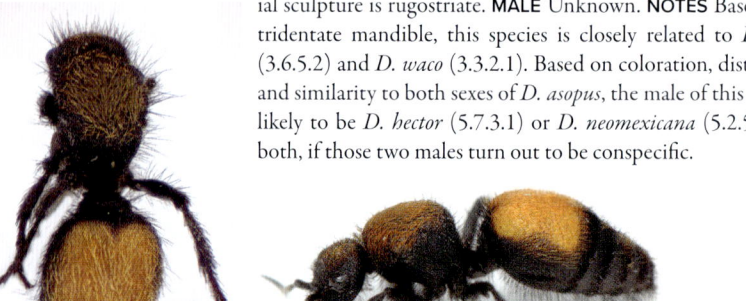

3.2.8—*Dasymutilla bioculata* Western females cluster

These species occur somewhat sporadically in the central and western USA. They can be recognized by having variable amounts of black and/or silvery setae on the apical tergites **and** having the mesosoma longer than wide. They sometimes resemble species in the *D. arenivaga* cluster (3.2.6) but rarely overlap with those species in distribution. They can most easily be confused with the *D. foxi* cluster (3.2.7), except for the mesosoma shape. Many of them could also be confused with the Madrean, Texan, or *Timulla*-like FCS, based on the sparser setae, often darkened head and mesosoma, and contrasting metasomal patterns.

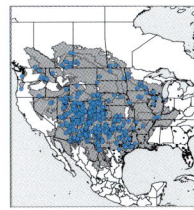

3.2.8.1—*Dasymutilla bioculata*
DAH-ZEE-MEW-TILL-UH BYE-OCK-EW-LAW-TUH

ETYMOLOGY From the Latin *bi* "two" and *oculus* "eye," in reference to the two orange spots on T2 in many males. **FIELD IDENTIFICATION** Sometimes easy. The head is narrower than the mesosoma and rounded posteriorly; and the propodeum has the dorsal setae yellow, orange, or red. Other than these shared characters, each regional color form has different diagnostic features. In most southern areas, the dorsal mesosomal setae are relatively dense and the T3–5 setae are entirely black or black mesally and silver laterally. In northern and eastern areas, the color is similar to *D. campanula* and *D. radkei*, but *D. bioculata* is more widespread and common than those species. **LAB IDENTIFICATION** The body cuticle is variable but often largely orange-brown; the antennal scrobe has the dorsal carina distinct, weak, or absent; the head is narrower than the mesosoma and unarmed posteriorly; the mesosoma is longer than wide and has a scutellar scale with a transverse wavy carina anterior to the scale; and the pygidial sculpture is striate. **MALE** (5.2.2.2, 5.2.5.1). Often similar to female. **NOTES** In the central and western USA, these are the most common forms of this widespread and variable *Dasymutilla* species (3.2.3.4, 3.3.4.3, 3.6.5.3). The most prevalent form in the northern Great Plains and northwestern states has the apical setae of T2–3 mostly black and is somewhat intermediate between the Eastern and Western FCS. The form most commonly seen in Arizona, Nevada, and Utah generally has T3–5 with silver setae laterally and black setae mesally; they are loosely intermediate between the Madrean and Western FCS. Finally, the most common form in New Mexico, Texas, and the southern Great Plains has the setae of T3–5 usually entirely black. They loosely resemble the *D. arenivaga* cluster (3.2.6) but have shorter dorsal setae and occur farther east than any of those species. Each of these color forms can be locally abundant, especially in sandy habitats.

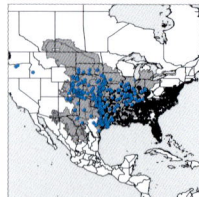

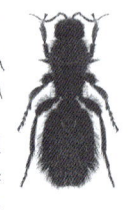

3.2.8.2—*Dasymutilla occidentalis*

DAH-ZEE-MEW-TILL-UH OX-SID-ENT-AHL-ISS

ETYMOLOGY From the Latin *occident* "western." This color form used to be treated as a separate subspecies, *D. o. comanche*. **FIELD IDENTIFICATION** Easy. The propodeal dorsum has the setae black; the T2 disc and T4–6 have the setae mostly yellow or orange; and the T2 fringe and T3

have the setae black. The only species with similar coloration is *D. californica*, but that species is smaller and rarer than *D. occidentalis*, and they overlap in distribution only in Idaho and western Oregon. **LAB IDENTIFICATION** The body cuticle is mostly black or dark reddish-brown without distinctly lighter cuticular patches; the antennal scrobe has a distinct dorsal carina; the head is narrower than the mesosoma and unarmed posteriorly; the mesosoma is longer than wide and has a scutellar scale, usually with carinae anterior and lateral to the scale; and the pygidial sculpture is striate. **MALE** (5.2.1.1). Similar to female. **NOTES** In the eastern USA, this species (3.6.1.1) has brighter red setae that are more "sleek and smooth." Specimens in western areas have the setae longer and shaggier and tending to be orange or pale golden-yellow. Shockingly, some populations of the "Common Eastern Velvet Ant" were found in Idaho and Oregon, which are the only populations that seem to merit the specific epithet, *occidentalis* "western."

3.2.8.3—*Dasymutilla radkei*

DAH-ZEE-MEW-TILL-UH RADKE-EYE

ETYMOLOGY This species was named after William R. Radke, who collected the type specimen in 1996. **FIELD IDENTIFICATION** Often difficult. The head and mesosoma are largely blackened with sparse orange setae, contrasting with the black T2 with a large yellow-orange patch; and the head is narrow and rounded posteriorly. **LAB IDENTIFICATION** The body cuticle is mostly orange-brown with a large yellow-orange patch on T2; the antennal scrobe lacks a dorsal carina; the head is clearly narrower than the mesosoma and is armed with a posterolateral tubercle; the mesosoma is longer than wide and has a small scutellar scale without accessory carinae; and the pygidial sculpture is striate. **MALE** Unknown. **NOTES** This rare species is known from New Mexico and western Texas. They are structurally similar to *D. ursus* (3.6.5.1) and may eventually be recognized as a regional color variant of that species.

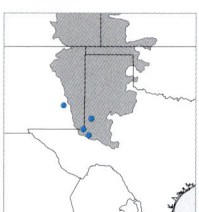

3.2.8.4—*Dasymutilla campanula*

DAH-ZEE-MEW-TILL-UH CAMP-ANN-EWE-LUH

ETYMOLOGY From the Latin *campana* "bell," in reference to the bell-shaped, orange patch on T2. **FIELD IDENTIFICATION** Easy. The head and mesosoma are dull orange-brown, contrasting with the black T2 with a large yellow patch; and the head is sharply angular posterolaterally. **LAB IDENTIFICATION** The body cuticle is mostly orange-brown with a

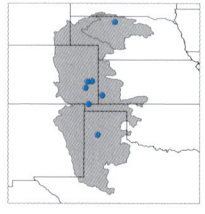

large yellow-orange patch on T2; antennal scrobe lacks a dorsal carina; the head is about as wide as the mesosoma and is armed with a posterolateral tubercle; the mesosoma is longer than wide and has a scutellar scale with carinae anterior and lateral to the scale; and the pygidial sculpture is striate. **MALE** Unknown. **NOTES** This species is structurally similar to *D. parksi* (3.5.1.2) but has very different coloration. The male might be *D. gentilis* (5.6.4.4) or *D. meracula* (5.6.4.5) based on distribution and similarity to *D. quadriguttata*.

ABOVE: *D. campanula*

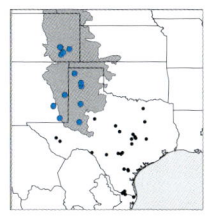

3.2.8.5—*Dasymutilla texanella*
DAH-ZEE-MEW-TILL-UH TEX-UH-NELL-UH

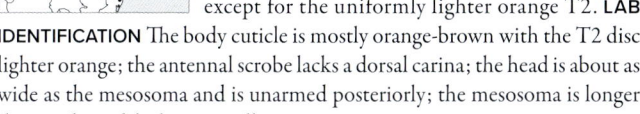

ETYMOLOGY This species was named after the state of Texas. **FIELD IDENTIFICATION** Sometimes difficult. The head is about as wide as the mesosoma; the body is dark orange-brown except for the uniformly lighter orange T2. **LAB IDENTIFICATION** The body cuticle is mostly orange-brown with the T2 disc lighter orange; the antennal scrobe lacks a dorsal carina; the head is about as wide as the mesosoma and is unarmed posteriorly; the mesosoma is longer than wide and lacks a scutellar scale; and the pygidial sculpture is striate. **MALE** Unknown. **NOTES** This species is structurally similar to *D. nigripes*. Specimens from southern Texas have the mesosoma darker and more clearly fit the Texan FCS (3.3.5.1).

ABOVE AND RIGHT:
D. texanella

3.2.8.6—*Dasymutilla californica*
DAH-ZEE-MEW-TILL-UH CAL-IF-OR-NICK-UH

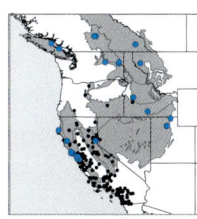

ETYMOLOGY This species was named after the state of California. This variant used to be recognized as a separate subspecies, *D. c. clio*. **FIELD IDENTIFICATION** Usually easy. Small to medium in size; the head is much narrower than the mesosoma; the tergal setae are mostly orange with a patch of black setae on T2 posteromesally and T3 mesally. **LAB IDENTIFICATION** The

body cuticle is mostly black or dark reddish-brown without distinctly lighter cuticular patches; the antennal scrobe has the dorsal carina weak or absent; the head is clearly narrower than the mesosoma and is armed with a posterolateral tubercle; the mesosoma is longer than wide and has a scutellar scale with carinae anterior and lateral to the scale; and the pygidial sculpture is generally rugose. **MALE** (5.2.4.1, 5.4.2.3). Often similar to female, sometimes much darker. **NOTES** This is an uncommon variant of *D. californica* (3.2.4.1) that, in general color pattern, resembles a small-bodied *D. occidentalis* (3.2.8.2). This form seems to occur mainly along the northern Pacific coast in California and in the northeastern extent of the range for this species. In Idaho and eastern Oregon, they actually overlap in distribution with the much larger *D. occidentalis*.

▼ 3.2.9.1—*Invreiella manleyi*

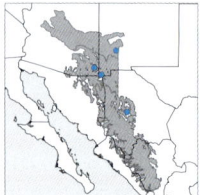

INVREA-EE-ELL-UH MANLEY-EYE

ETYMOLOGY Named after the velvet ant researcher Donald G. Manley. **FIELD IDENTIFICATION** Often difficult. This species could be confused with the *Ps. anthracina* cluster (3.2.10) but has sparser dorsal setae, or confused with the *Ps. propinqua* cluster (3.2.11), but the head is much wider. Neither of those *Pseudomethoca* clusters have a genal tooth like *I. manleyi*. **LAB IDENTIFICATION** The gena is armed with large blunt tooth posteriorly. **MALE** Unknown. **NOTES** This species is similar in coloration to many other *Invreiella* species from Mexico (p. 322), but it is structurally most similar to *I. cephalargia* (3.4.10.1).

Photo by George Waldren

▼ 3.2.10—*Pseudomethoca anthracina* Western females cluster

These species occur in the western USA. They are the only *Pseudomethoca* species with dense and largely erect yellow, orange, or red setae on the dorsum of the head and mesosoma. They can most easily be confused with *Ps. propinqua* (3.2.11.1) or *Dasymutilla aureola* (3.2.1.1).

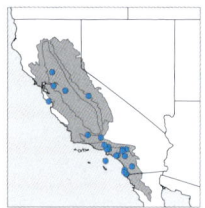

3.2.10.1—*Pseudomethoca anthracina*

SOO-DOE-METH-OWE-KUH ANN-THRUH-SEE-NUH

ETYMOLOGY From the Ancient Greek *anthrakinos* "coal-black," in reference to the male coloration. **FIELD IDENTIFICATION** Sometimes difficult. Can be separated from other *Pseudomethoca* by the uniformly yellow, orange, or red setae on T2–6, but this species can easily be confused with *D. aureola* (3.2.1.1) except for its sessile T1 shape. **LAB IDENTIFICATION** The head is wider than the mesosoma; the dorsal setae are pale yellow, orange, or red; and the setae from T2–6 are uniformly yellow, orange, or red. **MALE** (5.4.5.1). Different from female, covered entirely with black setae. **NOTES** This form of *Ps. antracina* is more common than the Desert FCS (3.1.6.1) form. This species can be easily confused with *D. aureola* because of the short mesosoma and wide head.

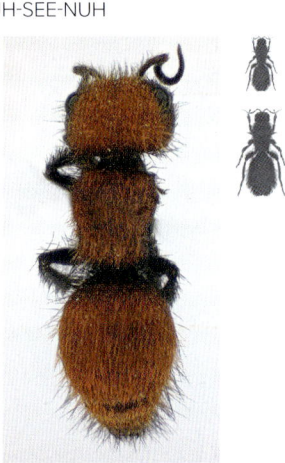

By random chance, back in 2001, this was the first velvet ant that I ever tried to identify using an officially published scientific paper (Hurd's 1949 key to the *Dasymutilla* species of California). At that time, I didn't know how to tell the genera apart, and I wasn't aware that my specimen wasn't in the *Dasymutilla* key, so it didn't go well!—KEVIN

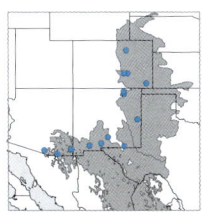

3.2.10.2—*Pseudomethoca aureovestita*
SOO-DOE-METH-OWE-KUH OW-REE-OWE-VESS-TEE-TUH

ETYMOLOGY From the Latin *aureus* "golden" and *vestis* "garment," in reference to the coloration. **FIELD IDENTIFICATION** Often difficult. The dorsal setae are generally dull orange and the apical metasomal patch is composed of interspersed black and silvery setae. **LAB IDENTIFICATION** The head is about as wide as the mesosoma; the dorsal setae are generally

dull orange; and T3–5 have the setae entirely orange or with interspersed black and silvery setae. **MALE** (5.2.8.2). Somewhat similar to female. **NOTES** The type of this species, and that of *Ps. flammigera*, were both collected in the Huachuca Mountains of Arizona. While *Ps. flammigera* seems to be restricted to Arizona, *Ps. aureovestita* occurs farther east, into New Mexico and western Texas.

FAR LEFT: *Pseudomethoca aureovestita* from Arizona

LEFT: *Pseudomethoca aureovestita* from New Mexico

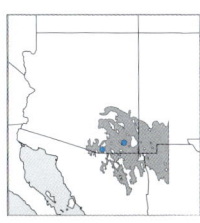

3.2.10.3—*Pseudomethoca flammigera*
SOO-DOE-METH-OWE-KUH FLAMM-IJJ-AIR-UH

ETYMOLOGY Derived from the Latin *flamma* "blaze" or "fire," in reference to the reddish dorsal setae. **FIELD IDENTIFICATION** Often difficult. The dorsal setae are generally red and the apical metasomal setal patch is more distinctly black mesally and silver sublaterally. **LAB IDENTIFICATION** The head is about as wide as the mesosoma; the dorsal setae are generally red; and T3–5 have a mesal patch of black setae with silvery setae sublaterally. **MALE** Unknown. **NOTES** This species is separated from *Ps. aureovestita* only by somewhat subtle and subjective setal color differences. Both species occur together in Arizona and *Ps. flammigera* is apparently restricted to that state. Further studies may eventually reveal that this species is simply a color variant of *Ps. aureovestita*, rather than a separate species.

▼ 3.2.11—*Pseudomethoca propinqua* Western females cluster

These species occur in the central and western USA. They have sparser dorsal setae than the *Ps. anthracina* cluster (3.2.10). They are somewhat intermediate between the Western and Eastern FCS and both of these species are also treated in the *Ps. simillima* cluster (3.6.10.1). Unlike *Invreiella manleyi* (3.2.9.1), the gena is unarmed, and the head is only slightly wider than the mesosoma.

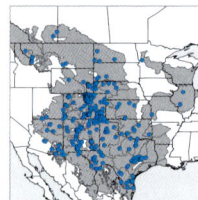

3.2.11.1—*Pseudomethoca propinqua*
SOO-DOE-METH-OWE-KUH PRO-PINK-WUH

ETYMOLOGY From the Latin *propinquus* "near," "neighboring," or "resembling." The author suggested this was similar to the male of *D. vestita*. **FIELD IDENTIFICATION** Often difficult. The fringe of T2 usually has a broad band of black setae and the mesosomal pleura are mostly blackened; this species is more widespread than *Ps. paludata*. **LAB IDENTIFICATION** The humeral carina is weakly defined; the mesosomal pleura are mainly bare and have the cuticle largely blackened; the lateral propodeal margin lacks a weak row of short teeth; the T2 fringe has a broad band of black setae; and the pygidial sculpture is rugose. **MALE** (5.2.8.1). Similar to female,

with tergal setae mostly yellow. **NOTES** This is a common central and western species, whose color pattern is intermediate between the Western and Eastern FCS. This species has been associated with the host bees *Nomia melanderi* Cockerell (Halictidae) and *Melissodes pallidisignata* Cockerell (Apidae).

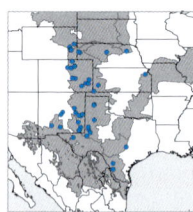

3.2.11.2—*Pseudomethoca paludata*
SOO-DOE-METH-OWE-KUH PAL-LOO-DAH-TUH

ETYMOLOGY This species was apparently named for the paludamentum, a cloak worn by military commanders in Rome. **FIELD IDENTIFICATION** Often difficult. The fringe of T2 has the setae mostly silvery, with only a small posteromesal black patch and the mesosomal pleura are usually entirely orange-brown. **LAB IDENTIFICATION** The humeral carina is strongly defined; the mesosomal pleura are mainly bare and have the cuticle mainly orange; the lateral propodeal margin has a row of short teeth; the T2 fringe has the setae mostly silver, with a small posteromesal black patch; and the pygidial sculpture is rugose. **MALE** (5.6.5.3). Much darker than female. **NOTES** This species is less common than *Ps. propinqua*. Like that species, the coloration is intermediate between the Western and Eastern FCS.

3.2.12—*Sphaeropthalma unicolor* Western females cluster

These species are most often found in the Pacific states (California, Oregon, and Washington). They are the only diurnal *Sphaeropthalma* species with dense uniformly yellow, orange, or reddish dorsal setae. They can be separated from the other Western clusters by having plumose setae on the tergal fringes, but they can sometimes be confused with nocturnal species, which are not treated at the species level in this book (see chapter 6).

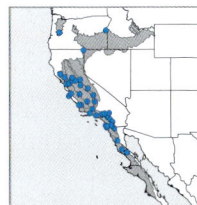

3.2.12.1—*Sphaeropthalma unicolor*
S-FAIR-OPP-THAL-MUH EWE-NICK-UH-LURR

ETYMOLOGY This name is Latin for "one color." Ironically, every individual female has three colors: yellow to red up top, black laterally, and silver hairy legs. Further, the bright dorsal setae vary between populations (like most Pacific species with the Western pattern), ranging from silvery-white to dull yellow-brown to bright scarlet and all the shades in between. **FIELD IDENTIFICATION** Sometimes easy. The sessile T1 shape and relatively wide head are diagnostic, but this species can be confused with *D. aureola* (3.2.1.1), which has a petiolate T1 shape, and *Ps. anthracina* (3.2.10.1), which has the head even wider and angular in shape. Additionally, the legs have gray setae. **LAB IDENTIFICATION** The mandible shape (6.2.2b) combined with the uniformly bright dorsal setae are diagnostic for this species. The mesonotal setae are simple or brachyplumose; the T1 shape is sessile; the tergal fringes are plumose; and T6 has a defined pygidial plate. **MALE** (5.7.12.3). Often similar to female but generally with paler cuticle color. **NOTES** This species is generally considered to be nocturnal in behavior but is frequently seen actively walking during the day, especially on cooler days or at higher elevations.

Photo by Alex Wild

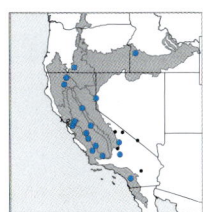

3.2.12.2—*Sphaeropthalma edwardsii*
S-FAIR-OPP-THAL-MUH EDWARDS-EE-EYE

ETYMOLOGY This species was named after its collector, Henry Edwards, an English stage actor, writer, and entomologist in the late 19th century. **FIELD IDENTIFICATION** Often difficult. The petiolate T1 shape and uniformly bright dorsal setae are diagnostic in the genus *Sphaeropthalma*, but this species resembles many members of *Dasymutilla*, except for its plumose tergal fringes and mostly gray leg setae. **LAB IDENTIFICATION** The mandible lacks a dorsal or ventral tooth basally; mesonotal setae are simple or brachyplumose; T1 shape is petiolate; tergal fringes are plumose; T6 has a defined microreticulate pygidial plate. **MALE** (5.2.9.1, 5.7.12.4). Often similar to female. **NOTES** Many *Odontophotopsis* (6.1.5) species in the Pacific region are similar in appearance, but they have dense

flattened plumose setae on the mesonotum (6.2.1a) and their dorsal setae are generally duller orange, never bright red. *Sphaeropthalma marpesia* (3.4.15.1) is similar to this species in structure and color but has a large black setal patch on the T2 disc.

Sphaeropthalma marpesia

3.3 TEXAN FEMALES COLOR SYNDROME

The Texan Females Color Syndrom (FCS) is characterized by individuals having a black head and mesosoma, and the metasoma largely yellow, orange, or red. This color syndrome is particularly common in Texas, New Mexico, and Oklahoma in the USA, and in Chihuahua, Coahuila, and Sonora in Mexico, though some species with this coloration can be found even in southern California, Nevada, or Utah. Currently, this FCS includes 16 species in three genera.

Sometimes aged or damaged Western FCS individuals can resemble these Texan forms, due to setae being worn off the head and mesosoma—so if you are having difficulty placing an individual in this FCS, check out the Western FCS. Additionally, a few Madrean FCS (see 3.4.8.1, 3.4.8.2) species can have the head and mesosoma darker than the reddish-orange metasoma, but these occur only in Arizona or western New Mexico, have silvery setal markings on the head or mesosoma, and usually a contrasting pattern of black, silvery, and reddish color on the metasoma.

Of the 16 species treated here, 10 (~63%) are known from both sexes. All of these known males fit in the analogous Texan MCS (5.3), with the head and mesosoma black and the metasoma mostly yellow, orange, or red.

This FCS does not seem to extend south into the Neotropical realm in Mesoamerica (7.3.1). There do not seem to be species with this exact pattern elsewhere in the world, but many species around the world are predominantly black, with distinct orange or red markings on the metasoma only. These forms are abundant in arid regions in southern South America (8.3.6), and some examples can be found in Australia (such as *Ephutomorpha distinguenda*) and India (such as *Trogaspidia fumipennis*).

Ephutomorpha distinguenda from Australia.

Trogaspidia fumipennis from India.

The key below is useful for figuring out which page to visit next.

—Genus *Dasymutilla*—T1 shape petiolate; T2 fringe with simple setae

3.3.1 Large species (usually more than 12 mm) with mesosoma longer than wide, head narrower than mesosoma, and antennal scrobe carina distinct .... *D. magnifica, D. klugii, D. clotho*

3.3.2 Mesosoma short, as wide as long .... *D. waco, D. vestita*

3.3.3 Head distinctly wider than mesonotum; mesosoma longer than wide .... *D. gorgon*

3.3.4 Head not distinctly wider than long; mesosoma longer than wide; dorsal body setae dense, T3–6 setae generally yellow, orange, or red .... *D. wileyae, D. nupera, D. klugiodes, D. bioculata, D. nogalensis*

3.3.5 Dorsal body setae generally sparse, apical tergites with setae largely black and/or silvery .... *D. uniguttata, D. texanella*

—Genus *Myrmilloides*—T1 shape sessile, head massive with ventral teeth

3.3.6 Body sparsely setose; T2 with obscure silvery setal patches .... *M. grandiceps*

—Genus *Pseudomethoca*—T1 shape sessile, head unarmed ventrally

3.3.7 Body more densely setose; T2 setae mostly yellow or orange, without silvery patches .... *Ps. pigmentata, Ps. brazoria*

 ### 3.3.1—*Dasymutilla magnifica* Texan females cluster

These species are relatively widespread in the southern portions of the central and western USA. These are the largest Texan FCS species. They have the mesosoma longer than wide, the head narrower than the mesosoma, the antennal scrobe carina distinct, the vertex unarmed, the mesosoma with a scutellar scale, the femoral apices rounded, and T2 without distinct orange-yellow cuticular patches. Smaller-bodied individuals could be confused with the *D. nupera* cluster (3.3.4).

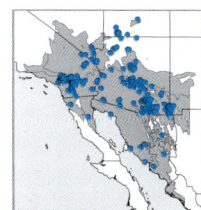

3.3.1.1—*Dasymutilla magnifica*
DAH-ZEE-MEW-TILL-UH MAGG-NIFF-ICK-UH

ETYMOLOGY From the Latin *magnificus* "great," "noble," "eminent," in reference to its large size and bright coloration. **FIELD IDENTIFICATION** Often difficult. Usually, the large body size, long mesosoma, and coloration is diagnostic, but in eastern areas, where this species overlaps with *D. klugii,* the two species are nearly indistinguishable. Smaller-bodied specimens from Arizona can be difficult to separate from *D. nogalensis* (3.3.4.5). **LAB IDENTIFICATION** The antennal scrobe has a distinct dorsal carina; the genal carina and sculpture are coarse; the mesosoma is longer than wide and has a scutellar scale; and the fringes of S2–5 have the setae mostly orange or red. **MALE** (5.3.3.1). Similar to female. **NOTES** This species occurs mainly in the Mojave and Sonoran Deserts, and mountainous areas of Arizona. It generally has shaggier setae than *D. klugii* and *D. clotho* and has the metasomal sternites with mostly reddish setae. This is a species commonly seen in field guides and spectacular to see running across the barren desert floor. Older specimens usually have the reddish metasomal setae turned pale yellow, and sometimes even the black mesosomal setae are lightened, maybe due to some sort of sun-bleaching effect.

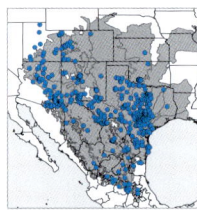

3.3.1.2—*Dasymutilla klugii*

DAH-ZEE-MEW-TILL-UH KLUG-EE-EYE

ETYMOLOGY This species was named after the German entomologist Johann C. F. Klug, who described many Hymenoptera species in the early 1800s. **FIELD IDENTIFICATION** Often difficult. Usually, the large body size, long mesosoma, and coloration are diagnostic, but in western areas, where this species overlaps with *D. magnifica*, they are nearly indistinguishable. Smaller-bodied specimens can be difficult to separate from the *D. nupera* cluster (3.3.4). **LAB IDENTIFICATION** The antennal scrobe has a distinct dorsal carina; the genal carina and sculpture are relatively weak; the mesosoma is longer than wide and has a scutellar scale; and the fringes of S2–5 have the setae mostly black. **MALE** (5.3.1.1). Similar to female. **NOTES** This species is widespread in Arizona, New Mexico, Texas, and the Great Plains. It generally has shorter, flatter setae than *D. magnifica* and the metasomal sternites have mostly black setae. Many populations have the anterior portions of

T2 covered with black setae. This is the most commonly encountered species in the Texan FCS. This species has been recorded attacking cicada killer wasps (*Sphecius grandis*) as the hosts.

3.3.1.3—*Dasymutilla clotho*

DAH-ZEE-MEW-TILL-UH CLOTH-OWE

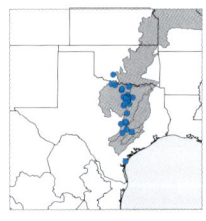

ETYMOLOGY In Greek mythology, Clotho was one of the three Fates. **FIELD IDENTIFICATION** Easy. The color pattern is unique. **LAB IDENTIFICATION** The antennal scrobe has a distinct dorsal carina; the genal carina and sculpture are relatively weak; the mesosoma is longer than wide and has a scutellar scale; and the fringes of S2–5 have the setae mostly black. **MALE** Unknown. **NOTES** This species is a dead ringer for *D. klugii* but has a black band of setae on T2–3. Superficially, it looks like a chimera of *D. klugii* (head and mesosoma) and *D. occidentalis* (metasoma), and the species occurs mainly in eastern Texas, at the margin of the geographic distribution between those species. Further studies could reveal that these three species (and maybe *D. calorata* as well) might all be color variants of a single widespread species.

3.3.2—*Dasymutilla waco* Texan females cluster

These species are found mainly in Texas and the southern Great Plains, but this color form of *D. vestita* has been found as far west as the Mojave Desert in California. These are the only Texan FCS species with the mesosoma as wide as long. In the field, they could be confused with some members of the *D. nupera* cluster (3.3.4).

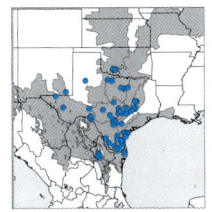

3.3.2.1—*Dasymutilla waco*
DAH-ZEE-MEW-TILL-UH WAY-COE

ETYMOLOGY This species was named for the city of Waco, Texas. **FIELD IDENTIFICATION** Often easy. The short mesosoma and sparse setae, especially above the large orange cuticular patch on T2, are diagnostic. **LAB IDENTIFICATION** The mandible is tridentate; the antennal scrobe lacks a dorsal carina; the vertex is unarmed posteriorly; the mesosoma is as wide as long and lacks a scutellar scale; the mid and hind femoral apices are rounded; and the pygidial sculpture is rugose. **MALE** (5.3.2.1). Similar to female. **NOTES** Based on the tridentate mandible and mesosoma shape, this species is closely related to *D. asopus* (3.6.5.2) and *D. montivagoides* (3.2.7.4).

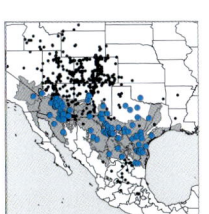

3.3.2.2—*Dasymutilla vestita*
DAH-ZEE-MEW-TILL-UH VESS-TEE-TUH

ETYMOLOGY Apparently from the Latin *vestis* "garment," in reference to the colorful dorsal setae. These females used to be named *D. zelaya*, likely based on an older spelling of the Mexican city of Celaya, Guanajuato. **FIELD IDENTIFICATION** Sometimes difficult. This species has a short mesosoma and denser setae than *D. waco*, but the mesosoma shape can be difficult to recognize in the field. **LAB IDENTIFICATION** The mandible is bidentate; the antennal scrobe has a dorsal carina; the vertex is unarmed posteriorly; the mesosoma is as wide as long and lacks a scutellar scale; the mid and hind femoral apices are rounded; and the pygidial sculpture is rugose. **MALE** (5.3.1.3). Similar to female. **NOTES** This color variant is common in Texas and rarely occurs in desert areas of Arizona, California, and New Mexico. The Western FCS form (3.2.2.1) is more widespread.

3.3.3.1—*Dasymutilla gorgon*

DAH-ZEE-MEW-TILL-UH GORR-GONN

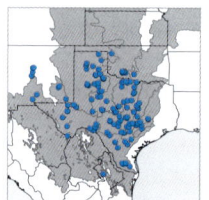

ETYMOLOGY In Greek mythology, gorgons were female monsters that had snakes for hair and could turn people to stone with their gaze; the most famous gorgon was Medusa. **FIELD IDENTIFICATION** Often easy. The head is wider than that of any other species with the Texan FCS. **LAB IDENTIFICATION** The mandible is bidentate; the antennal scrobe lacks a dorsal carina; the vertex is unarmed posteriorly; the mesosoma is slightly longer than wide and has a scutellar scale with some transverse carinae anterior and lateral to the scale; the mid and hind femoral apices are rounded; and the pygidial sculpture is striate. **MALE** (5.3.1.2). Similar to female. **NOTES** This species has a massive head, almost like *Pseudomethoca*, and has a distinct scutellar scale. It seems to be second most commonly encountered Texan FCS species, behind *D. klugii* (3.3.1.2).

3.3.4—*Dasymutilla nupera* Texan females cluster

These species occur mainly in New Mexico, Texas, and the southern Great Plains, but one species occurs in Arizona as well. These are the leftover Texan FCS species with long dense setae. They are generally smaller-bodied than the *D. magnifica* cluster (3.3.1), have a narrower head than *D. gorgon* (3.3.3.1), and a longer mesosoma than the *D. waco* cluster (3.3.2).

3.3.4.1—*Dasymutilla nupera*

DAH-ZEE-MEW-TILL-UH NOO-PAIR-UH

ETYMOLOGY From the Latin *nuperus* "new," "fresh," or "recent." **FIELD IDENTIFICATION** Very difficult. The head is wider than most species in this cluster, but not as wide as that of *D. gorgon*, and the body size is usually smaller than that of *D. gorgon*. **LAB IDENTIFICATION** The mandible is bidentate; the antennal scrobe lacks a dorsal carina; the vertex is unarmed posteriorly; the mesosoma is longer than wide and lacks a scutellar scale; the mid and hind femoral apices are rounded; and the pygidial sculpture is striate. **MALE** Unknown. **NOTES** This is the only species of this cluster that lacks a scutellar scale. Structurally this species is similar to *D. stevensi* from the Western FCS (3.2.3.3).

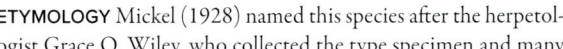

3.3.4.2—*Dasymutilla wileyae*
DAH-ZEE-MEW-TILL-UH WILEY-EE

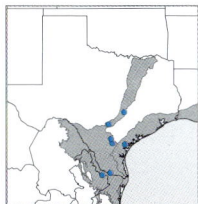

ETYMOLOGY Mickel (1928) named this species after the herpetologist Grace O. Wiley, who collected the type specimen and many other useful velvet ant specimens. **FIELD IDENTIFICATION** Very difficult. The head appears more angular due to the posterolateral tubercles. **LAB IDENTIFICATION** The mandible is bidentate; the antennal scrobe lacks a dorsal carina; the vertex has a linear tubercle posterolaterally; the mesosoma is longer than wide and has a distinct scutellar scale with transverse carinae anterior and lateral to the scale; the mid and hind femoral apices are rounded; and the pygidial sculpture is striate. **MALE** Unknown. **NOTES** This is the only "hairy" Texan FCS species with distinct tubercles on the vertex. Similar coloration and overlapping distribution, coupled with structural similarity with *D. quadriguttata* (3.5.1.1, 3.6.2.4), suggest that the male of this species might be *D. serenitas* (5.3.2.2).

3.3.4.3—*Dasymutilla bioculata*
DAH-ZEE-MEW-TILL-UH BYE-OCK-EW-LAW-TUH

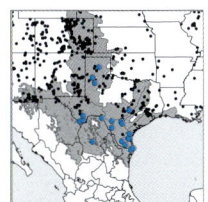

ETYMOLOGY From the Latin *bi* "two" and *oculus* "eye," in reference to the two orange spots on T2 in many males. This form was previously named *D. melanippe*. **FIELD IDENTIFICATION** Very difficult. The head is narrower than in most species in the cluster, and T2 has an orange cuticular patch beneath the orange setae. **LAB IDENTIFICATION** The mandible is bidentate; the antennal scrobe has the dorsal carina weak or absent; the vertex is unarmed posteriorly; the mesosoma is longer than wide and has a distinct scutellar scale with a transverse wavy carina anterior to the scale; the mid and hind femoral apices are rounded; and the pygidial sculpture is striate. **MALE** (5.3.2.3). Similar to female. **NOTES** This is a southern Texan color variant of the most widespread and variable *Dasymutilla* species (3.2.3.4, 3.2.8.1, 3.6.5.3).

3.3.4.4—*Dasymutilla klugiodes*
DAH-ZEE-MEW-TILL-UH KLUG-EE-OWE-DEEZ

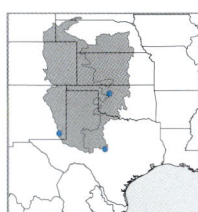

ETYMOLOGY This species was named for its similarity to *D. klugii*. **FIELD IDENTIFICATION** Very difficult. This rare species is very similar to *D. klugii*, except the body is generally smaller and the sternal fringes have the setae mostly orange. **LAB IDENTIFICATION** The mandible is bidentate; the antennal scrobe lacks a dorsal carina; the vertex is unarmed posteriorly; the mesosoma is longer than wide and has a weak scutellar scale; the mid and hind femoral apices are rounded; and the pygidial sculpture is striate. **MALE** Unknown. **NOTES** This species is remarkably rare, originally described only from the type collected in 1930, and it has been subsequently col-

Dasymutilla klugiodes

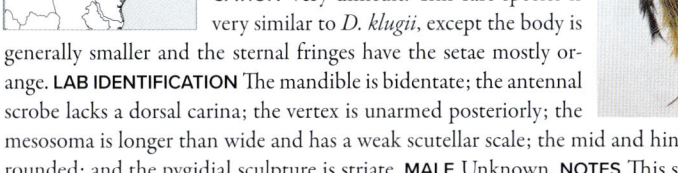

lected only twice (that we know of). Many previous records that were attributed to this species are actually variants of *D. bioculata*.

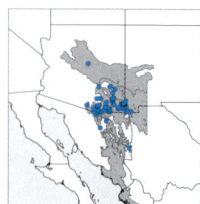

3.3.4.5—*Dasymutilla nogalensis*

DAH-ZEE-MEW-TILL-UH NO-GALL-ENN-SISS

ETYMOLOGY This species was named after the city of Nogales, Arizona. The male used to be named *D. atrifulva* after the black and red coloration. **FIELD IDENTIFICATION** Often difficult. Of the four Texan FCS species in Arizona, this has the widest head and is generally smaller than *D. klugii* or *D. magnifica*. **LAB IDENTIFICATION** The mandible is bidentate; the antennal scrobe has a dorsal carina; the vertex is unarmed posteriorly; the mesosoma is longer than wide and has a distinct scutellar scale with transverse carinae anterior and lateral to the scale; the mid and hind femoral apices are truncate; and the pygidial sculpture is longitudinally rugose. **MALE** (5.3.3.2). Similar to female. **NOTES** This is one of very few Texan FCS species in Arizona; in mountainous areas it seems to be more common than many other species.

 ### 3.3.5—*Dasymutilla texanella* Texan females cluster

These species occur mainly in Texas, but *D. texanella* is sometimes found farther north. They generally have sparser dorsal setae than the other clusters and usually have black or silvery setae on the apical tergites.

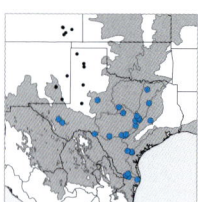

3.3.5.1—*Dasymutilla texanella*

DAH-ZEE-MEW-TILL-UH TEX-UH-NELL-UH

ETYMOLOGY This species was named after the state of Texas. **FIELD IDENTIFICATION** Usually easy. Has sparser setae than the other Texan FCS species, except some *D. waco* specimens, but its mesosoma is longer than that of *D. waco*. **LAB IDENTIFICATION** The mandible is bidentate; the antennal scrobe lacks a dorsal carina; the vertex is unarmed posteriorly; the mesosoma is longer than wide and lacks a scutellar scale; the mid and hind femoral apices are rounded; and the pygidial sculpture is striate. **MALE** Unknown. **NOTES** This species is structurally similar to *D. nigripes* (3.6.5.4), and some northern and western specimens fit more closely with the Western FCS (3.2.8.5) when the mesosoma setae and cuticle are lighter.

ABOVE: *Dasymutilla texanella*

RIGHT: *Dasymutilla uniguttata*

3.3.5.2—*Dasymutilla uniguttata*

DAH-ZEE-MEW-TILL-UH EWE-NIH-GOO-TAH-TUH

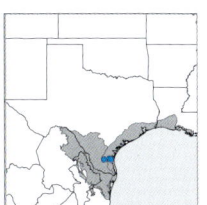

ETYMOLOGY From the Latin *uni* "one" and *gutta* "spot," in reference to the coloration. **FIELD IDENTIFICATION** Easy. The color pattern is unique. **LAB IDENTIFICATION** The mandible is bidentate; the antennal scrobe lacks a dorsal carina; the vertex has a sharp C-shaped tubercle posterolaterally; the mesosoma is longer than wide and has a distinct scutellar scale; the mid and hind femoral apices are rounded; and the pygidial sculpture is striate. **MALE** Unknown. **NOTES** This species is rare in museums, and none of the authors has collected it, but it seems to be relatively abundant in some regions of southern Texas, since the type series included over 100 specimens collected in the 1970s. There is some variation in the size of the yellow patch on T2, and some specimens have the body cuticle almost entirely black.

▼ 3.3.6.1—*Myrmilloides grandiceps*

MEER-MILL-OI-DEEZ GRAN-DISS-EPPS

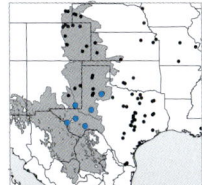

ETYMOLOGY From the Latin *grandis* "large" and *-ceps* "-headed." **FIELD IDENTIFICATION** Easy. The giant head is unique, and the short-winged males are even easier to recognize. The dorsal setae are sparse and T2 has two large silvery setal patches. **LAB IDENTIFICATION** The gena and postgena are each armed with a ventral tooth. **MALE** Similar to female. **NOTES** The males are short winged and generally much rarer than females.

This species is variable in color, also being treated in the Eastern FCS (3.6.7.1) and *Timulla*-like FCS (3.5.2.1). In the field, this color form closely resembles the harvester ant, *Pogonomyrmex rugosus*.

▼ 3.3.7—*Pseudomethoca brazoria* Texan females cluster

These species are found mainly in Texas. They are the only Texan FCS species in the genus *Pseudomethoca* and can be recognized by having the T1 shape sessile. Their generally large heads could lead them to be confused with *D. gorgon* (3.3.3.1), and their generally shorter body shape could lead them to be confused with the *D. waco* cluster (3.3.2).

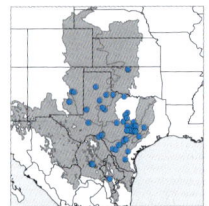

3.3.7.1—*Pseudomethoca brazoria*
SOO-DOE-METH-OWE-KUH BRAW-ZORR-EE-UH

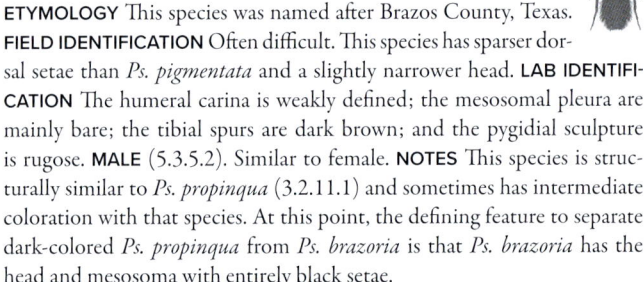

ETYMOLOGY This species was named after Brazos County, Texas. **FIELD IDENTIFICATION** Often difficult. This species has sparser dorsal setae than *Ps. pigmentata* and a slightly narrower head. **LAB IDENTIFICATION** The humeral carina is weakly defined; the mesosomal pleura are mainly bare; the tibial spurs are dark brown; and the pygidial sculpture is rugose. **MALE** (5.3.5.2). Similar to female. **NOTES** This species is structurally similar to *Ps. propinqua* (3.2.11.1) and sometimes has intermediate coloration with that species. At this point, the defining feature to separate dark-colored *Ps. propinqua* from *Ps. brazoria* is that *Ps. brazoria* has the head and mesosoma with entirely black setae.

3.3.7.2—*Pseudomethoca pigmentata*
SOO-DOE-METH-OWE-KUH PIG-MEN-TOT-UH

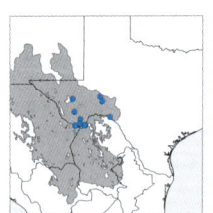

ETYMOLOGY From the Latin *pigmentum* "color" or "paint." **FIELD IDENTIFICATION** Often difficult. This species has denser dorsal setae than *Ps. brazoria* and a slightly wider head. **LAB IDENTIFICATION** The humeral carina is sharply defined; the mesosomal pleura are mainly bare; the tibial spurs are white; and the pygidial sculpture is rugose. **MALE** (5.3.5.1). Similar to female. **NOTES** This is a relatively rare species.

Kevin and I saw a female from a distance in Fort Davis, Texas. Before we got closer, a bird flew down and attacked the velvet ant. When we caught the wasp, her apical tergites were apparently "bitten" off with no apparent negative effects to the bird. Maybe this predator avoided the sting by biting off half of the metasoma, or maybe the bird gave up on eating the rest of the wasp because it fought back.—JOE

3.4 MADREAN FEMALES COLOR SYNDROME

The Madrean Females Color Syndrome (FCS) is more difficult to define than many of the other color syndromes in North America. This FCS is variable both in terms of color pattern and geography. The overall look of species in this color syndrome is based on a red or brown background color, often with a silvery or golden head, and with contrasting whitish, yellow, reddish, and black patterns on the mesosoma and/or metasoma. The Madrean FCS is so named because many species in this color syndrome occur in the Sierra Madre Occidental Mountains in Mexico, and southern Arizona and New Mexico. In the USA, there are 49 species from nine different genera that fit in this color syndrome.

Because this color syndrome is so variable and widespread, many species here can be mistaken for other color syndromes. Females that resemble the Desert FCS belong here only if they have a black setal patch on the mesosoma or yellowish cuticular spots on T2. Females that resemble the Western, Texan, *Timulla*-like, or Eastern FCS belong here only if they have contrasting black or silvery setal or cuticular patches on the head or mesosoma. Females that resemble the Cryptic FCS usually have darker black legs or body patches, or brighter silvery or golden markings. There are still numerous exceptions, so more pertinent diagnostic features and comparisons are provided in the species accounts below. For simplicity, no species in the genus *Timulla* are treated in this color syndrome; they are all placed into either the Eastern or *Timulla*-like FCS.

Of the 49 females in the Madrean FCS, only 25 (52%) are known from both sexes. There is no direct male analog to the Madrean FCS. About 48% of known males fit the *Timulla*-like Males Color Syndrome (MCS) (5.5), 28% fit the *Ephuta*-like MCS (5.4), 28% fit the *Sphaeropthalma*-like MCS (5.7), and a small handful fit in with other color syndromes.

The Madrean FCS extends throughout Mesoamerica, with various species having similar coloration to species in the USA (see 7.3.5, 7.3.6, 7.3.7). Various unrelated species all over the world loosely resemble the species in this FCS (see 8.3.10, 8.3.15, 8.3.16).

The key below is useful for figuring out which page to visit next.

—Genus *Dasymutilla*—T1 shape petiolate; T2 fringe with setae simple; pygidium with defined plate

3.4.1   Head and mesosoma entirely pale yellow-orange, except for a mesal white patch on the propodeum .... *D. heliophila*

3.4.2   Head with long, dense, white setae, contrasting with sparsely setose mesosoma .... *D. dammersi*

3.4.3   Head quadrate with dense bright golden setae, contrasting with reddish-brown mesosoma with sparser silvery markings .... *D. dilucida, D. ferruginea*

3.4.4   Head and mesosoma covered entirely with silver or golden setae .... *D. snoworum, D. monticola, D. birkmani*

3.4.5   Head and propodeum with dense silvery or pale golden setae dorsally, mesosoma usually with black or brown patch, small-bodied species .... *D. birkmani, D. eurynome, D. bonita*

3.4.6   Head and mesosoma cuticle blackish with white to pale golden setal markings; metasoma with red or orange pattern ... *D. sicheliana, D. asteria, D. foxi, D. saetigera, D. toluca*

3.4.7   Head, mesosoma, and metasoma with black and white coloration only .... *D. pulchra, D. arachnoides*

3.4.8   Head and mesosoma with reddish or orange cuticle color .... *D. dionysia, D. cirrhomeris, D. fasciventris*

—Genus *Protophotopsis*—T1 shape sessile; pygidium convex without plate

3.4.9   Tiny species with golden head; T3 with two pale yellow patches .... *Pr. venenaria*

—Genus *Invreiella*—T1 shape sessile; head with large blunt genal tooth

3.4.10 Large species with golden head and gena armed with large tooth .... *I. cephalargia*

—Genus *Pseudomethoca*—T1 shape sessile; head unarmed or with small sharp tooth ventrally

3.4.11 Head covered with dense golden setae, medium to large species
(6–15 mm) .... *Ps. contumax, Ps. praeclara, Ps. connectens, Ps. quadrinotata*

3.4.12 Head covered with dense golden setae, small species
(less than 5 mm) .... *Ps. toumeyi, Ps. bequaerti, Ps. klotsi, Ps. perditrix, Ps. peremptrix*

3.4.13 Head with sparse setae; T2 with two posterior pale yellow
cuticular spots .... *Ps. bethae, Ps. mulaiki*

3.4.14 Head with sparse setae, T2 without pale yellow cuticular spots ... *Ps. donaeanae, Ps. sonorae*

——Genus *Sphaeropthalma*—T1 shape petiolate; T2 fringe with setae plumose

3.4.15 T2 with broad, transverse, black band basally .... *Sp. marpesia*

—Genus *Stethophotopsis*—T1 shape subsessile; T2 fringe with setae plumose

3.4.16 T2 with two basal black patches .... *St. maculata*

—Genus *Ephuta*—T1 shape slender cylindrical

3.4.17 Head setae white or silver .... *E. argenticeps, E. albiceps, E. coloradella*

3.4.18 Head setae golden, from eastern region .... *E. floridana, E. baboquivari*

3.4.19 Head setae golden, from central and
western regions .... *E. tumacacori, E. sudatrix, E. margueritae, E. auricapitis*

—Genus *Myrmosula*—Pronotum divided from mesonotum

3.4.20 Head with dense silver or golden setae; T2 with pale yellow
cuticular patches .... *M. rutilans, M. boharti, M. nasua*

▼ 3.4.1.1—*Dasymutilla heliophila*

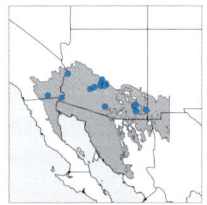

DAH-ZEE-MEW-TILL-UH HEE-LEE-AWE-FILL-UH

ETYMOLOGY From the Ancient Greek *helio-* "relating to the sun" and *-philos* "beloved," in reference to its occurence in hot desert regions. **FIELD IDENTIFICATION** Easy. The color pattern is unique. **LAB IDENTIFICATION** The antennal scrobe has a dorsal carina; the posterolateral angles of the head are rounded; the mesosoma lacks a scutellar scale; the mid and hind femora have their apices truncate; and the pygidial sculpture is microreticulate. **MALE** (5.7.2.1). Darker in color than female. **NOTES** This species is an apparent mimic of harvester ants, especially *Pogonomyrnex*

Photo by Jillian Cowles

maricopa. It is generally rare in collections, even though it has a wide range across the Sonoran Desert in California and Arizona. On an expedition with Justin Schmidt and James Pitts in 2005, Kevin collected nearly 200 diurnal velvet ants near Willcox, Arizona. About 50 of these specimens were *D. heliophila,* which was apparently more than all the previously examined specimens in other museums combined. In an impromptu experiment, Dr. Schmidt kept nearly 100 specimens of various species alive in terraria to measure their longevity. Although most species lived for at least 30 days beyond their collection and some were alive for longer than a year, nearly all the *D. heliophila* died within one week of their capture. Their rarity could be linked with their short adult life span.

▼ 3.4.2.1—*Dasymutilla dammersi*

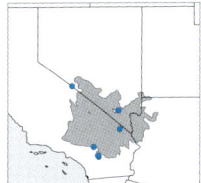

DAH-ZEE-MEW-TILL-UH DAMMERS-EYE

ETYMOLOGY This species was named in honor of Commander Charles M. Dammers, who collected many important velvet ants from southern California in the 1930s. **FIELD IDENTIFICATION** Easy. The color pattern is unique. **LAB IDENTIFICATION** The antennal scrobe has a dorsal carina; the posterolateral angles of the head are rounded; the mesosoma lacks a scutellar scale; the mid and hind femora have their apices truncate; and the pygidial sculpture is microreticulate. **MALE** (5.5.5.2). Somewhat similar to female. **NOTES** The type specimens were collected in Palm Springs, California in the 1930s, and few had been seen since.

On the first day of a weekend expedition with my dad, I found a female of *D. dammersi* near Big Bear Lake, above 5000 feet in elevation. A few more females were collected that day, and a male was nearly caught, but it got away. Since this species is so rare, I decided to spend the remaining two days of our expedition at the site. I ended up finding about 10 females and saw one more male, which escaped. It was frustrating, but at the end of the trip, I was getting ready to head back home, with a plan to return the following year. While my dad was driving up to finish packing up our collecting equipment, I saw a female running across the dirt road. Amazingly, a male was chasing her, and I slammed my net down over him. As my dad drove up, he saw me kneeling in the road, pinching the male in net material with a look of determined concentration. I finally was able to grab the male and get him into a collecting tube and then grabbed up the female After getting back to the lab, I realized that this male had not been previously named.—KEVIN

Because the species had not been found in low desert habitats since its discovery by Dammers, and these specimens were from high elevation, we wondered if the original collecting locality might be wrong, or if the types had actually come from mountains surrounding Palm Springs. Amazingly, an inaturalist.com observation from 2020 was found from low elevation, just outside Palm Springs. Usually, we think species are rare because they have a limited range of usable habitats or live only in distant uncommonly sampled regions. *Dasymutilla dammersi,* however, is found within a few miles of densely populated areas and occurs in both high- and low-elevation habitats. The reason for the rarity of this species remains a mystery. Maybe they have a short life span in the adult stage, meaning that they are only collectible for a few days per year. The apparent sister species, *D. heliophila,* seems to have a shorter life span than other velvet ants.

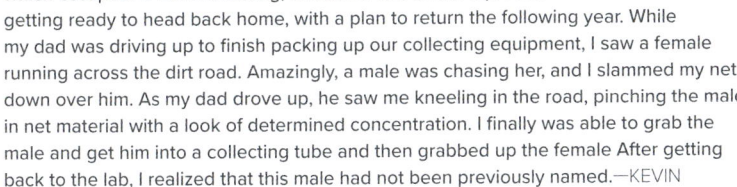

3.4.3—*Dasymutilla dilucida* Madrean females cluster

These species occur mainly in Arizona and New Mexico. They can be recognized by their quadrate and densely golden heads. In that respect, they are similar to *Invreiella cephalargia* (3.4.10.1) and some *Pseudomethoca* clusters (3.4.11, 3.4.12), but their T1 shape is petiolate.

3.4.3.1—*Dasymutilla dilucida*
DAH-ZEE-MEW-TILL-UH DYE-LOO-SIDD-UH

ETYMOLOGY From the Latin *dilucidus* "clear" or "bright," likely in reference to the attractive color pattern. **FIELD IDENTIFICATION** Difficult. They are usually smaller than *D. ferruginea* and have the T2 fringe with a white mesal spot. **LAB IDENTIFICATION** The antennal scrobe lacks a dorsal carina; the posterolateral angles of the head are carinate; the mesosoma lacks a scutellar scale; the mid and hind femora have their apices truncate; and the pygidial sculpture is microreticulate. Unlike *D. ferruginea*, T2 has a white setal spot mesally on the T2 fringe. **MALE** Unknown. **NOTES** This species seems to have a more restricted distribution than *D. ferruginea*. From personal experience, these are some quick and erratically moving velvet ants; the specimen pictured here was difficult to photograph because she just would not stop running. Based on similar size, morphology, and overlapping distributions, the male could be *D. apicalata* (or at least some populations that are currently mixed up with that species; see 5.5.5.1).

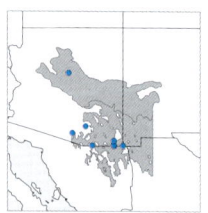

3.4.3.2—*Dasymutilla ferruginea*
DAH-ZEE-MEW-TILL-UH FUR-OO-JINN-AY-UH

ETYMOLOGY From the Latin *ferrugineus* "rusty," likely referencing the reddish-brown base body color. **FIELD IDENTIFICATION** Difficult. They are usually larger than *D. dilucida* and have the T2 fringe entirely black. **LAB IDENTIFICATION** The antennal scrobe lacks a dorsal carina; the posterolateral angles of the head are carinate; the mesosoma lacks a scutellar scale; the mid and hind femora have their apices truncate; and the pygidial sculpture is microreticulate. Unlike *D. dilucida*, the T2 fringe is entirely black. **MALE** Unknown. **NOTES** Males are unrecognized, but *D. apicalata* (5.5.5.1) and *D. sophrona* (5.5.2.3) are good candidates for the sex association based on similar size, morphology, and distribution.

3.4.4—*Dasymutilla snoworum* Madrean females cluster

These species are widespread in the central and western USA. They have the head and mesosoma covered entirely with silver or golden setae. They are similar to the *D. birkmani* cluster (3.4.5) and, depending on the density and intensity of these golden setae, could be confused with some species in the Eastern (3.6.6.1) or *Timulla*-like FCS (3.5.1).

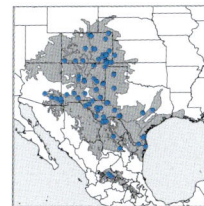

3.4.4.1—*Dasymutilla snoworum*
DAH-ZEE-MEW-TILL-UH SNOW-ORR-UMM

ETYMOLOGY The suffix *-orum* indicates a species named in honor of a family of people. This species was named for the entomologist Francis H. Snow, who established the insect collection at University of Kansas back in 1870, and his son, Frank. **FIELD IDENTIFICATION** Usually easy. The dorsal silvery or golden setae on the head and mesosoma, coupled with the large head, are diagnostic. **LAB IDENTIFICATION** The antennal scrobe lacks a dorsal carina; the posterolateral angles of the head are rounded; the mesosoma lacks a scutellar scale; the mid and hind femora have their apices truncate; and the pygidial sculpture is microreticulate. **MALE** (5.3.4.1). Generally darker than female. **NOTES** This species is widespread in the southwest; specimens from Texas and the Great Plains often have silvery or duller orange setae than the bright glittering

golden color of specimens from Arizona. Females in the Great Plains and Texas can be difficult to separate from *D. scaevola* of the Eastern FCS (3.6.6.1) when the golden setae are dull or worn by age.

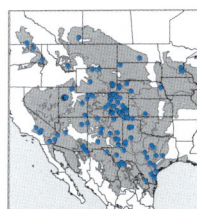

3.4.4.2—*Dasymutilla monticola*
DAH-ZEE-MEW-TILL-UH MON-TICK-OWE-LUH

ETYMOLOGY A Latin term that means "mountain dweller." The females used to be named *D. paenulata* or *D. caneo*. **FIELD IDENTIFICATION** Sometimes easy. The dorsal silvery or golden setae of the head and mesosoma, coupled with the slender head, are diagnostic. Some populations of *D. birkmani* in Texas are nearly identical to this species (see below). **LAB IDENTIFICATION** The antennal scrobe lacks a dorsal carina; the posterolateral angles of the head are rounded; F1 is only scarcely longer than F2; the mesosoma has a distinct scutellar scale with transverse carinae anteriorly and laterally; the mid and hind femora have their apices rounded; the T2 disc has moderately fine punctures and lacks a black and silver target-shaped setal patch mesally; the T2 fringe setae are mostly black, with small silvery patches laterally; and the pygidial sculpture is striate. **MALE** (5.5.3.1, 5.7.2.3). Often similar to female, sometimes much darker in coloration. **NOTES** This is one of the smallest-bodied *Dasymutilla* species.

Specimens from southwestern desert habitats tend to have the head and mesosoma setae silver and the metasoma nearly black, with yellowish spots on T2. Specimens from farther north and east, however, have the head and mesosoma setae duller golden and the metasoma more reddened with a less distinct yellowish patch.

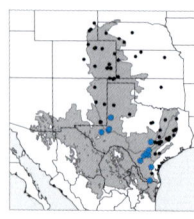

3.4.4.3—*Dasymutilla birkmani*

DAH-ZEE-MEW-TILL-UH BIRKMAN-EYE

ETYMOLOGY This species is named after G. Birkman, who collected many important mutillid specimens in Texas in the late 1800s. **FIELD IDENTIFICATION** Very difficult. This form is nearly identical to *D. monticola*, except the T2 fringe is mostly silvery, with only a small black patch mesally. **LAB IDENTIFICATION** The antennal scrobe lacks a dorsal carina; the posterolateral angles of the head are either rounded or angular; F1 is only scarcely longer than F2; the mesosoma has a distinct scutellar scale with transverse carinae anteriorly and laterally; the mid and hind femora have their apices rounded; the T2 disc has moderately fine punctures and lacks a black and silver target-shaped setal patch mesally; the T2 fringe is mostly silvery,

with only a small black patch mesally; and the pygidial sculpture is striate. **MALE** (5.5.4.2, 5.6.2.1). Usually darker in coloration than female. **NOTES** This rare Texan variant is especially difficult to separate from *D. monticola*.

▼ 3.4.5—*Dasymutilla birkmani* Madrean females cluster

Dasymutilla birkmani is common and widespread in the central USA, while the other species are rare and have more restricted ranges in Arizona or Texas. These are generally small-bodied species that have the head and propodeum covered with dense silvery or pale golden glittering setae. When the setae are duller (often from specimen age or cleanliness), these could be confused with the *D. sicheliana* (3.4.6) or *D. dionysia* (3.4.8) clusters.

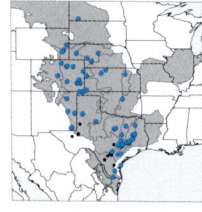

3.4.5.1—*Dasymutilla birkmani*

DAH-ZEE-MEW-TILL-UH BIRKMAN-EYE

ETYMOLOGY This species is named after G. Birkman, who collected many important mutillid specimens in Texas in the late 1800s. **FIELD IDENTIFICATION** Sometimes difficult. The anterior half of the mesosoma has blackish setae, and T2 lacks a black and silver target-shaped spot. **LAB IDENTIFICATION** The antennal scrobe lacks a dorsal carina; the posterolateral angles of the head are either rounded or angular; F1 is only scarcely longer than F2; the mesosoma has a distinct scutellar scale with transverse carinae anteriorly and laterally; the mid and hind femora have their apices rounded; the T2 disc has moderately fine punctures and lacks a black and silver target-shaped setal patch mesally; and the pygidial sculpture is striate. **MALE** (5.5.4.2, 5.6.2.1). Usually darker in coloration than female. **NOTES** This species is widespread in Texas and the Great Plains. In addition to the color variation (see 3.4.4.3), the head shape varies quite a bit, with some larger individuals having the head with a large posterolateral tubercle that makes the head shape more angular.

Dasymutilla birkmani

Dasymutilla eurynome

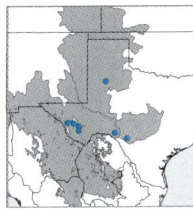

3.4.5.2—*Dasymutilla eurynome*
DAH-ZEE-MEW-TILL-UH EWE-REE-NO-MUH

ETYMOLOGY Many women in Greek mythology were named Eurynome. **FIELD IDENTIFICATION** Sometimes difficult. The mesosomal setae are mostly silver except for a small black patch in the scutellar area, and T2 has a black and silver target-shaped spot. **LAB IDENTIFICATION** The antennal scrobe lacks a dorsal carina; the posterolateral angles of the head are rounded; F1 is much longer than F2; the mesosoma has a distinct scutellar scale with transverse carinae anteriorly and laterally; the mid and hind femora have their apices rounded; the T2 disc has coarse punctures and a black and silver target-shaped setal patch anteromesally; and the pygidial sculpture is striate. **MALE** Unknown. **NOTES** This rare Texan species is known from females only. Based on the long first antennal flagellomere and black and silver target-shaped spot on T2, this species is apparently closely related to *D. bonita* and *D. saetigera* (3.4.6.4).

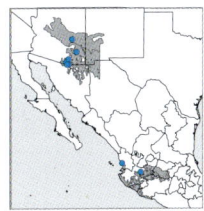

3.4.5.3—*Dasymutilla bonita*
DAH-ZEE-MEW-TILL-UH BOW-NEE-TUH

ETYMOLOGY *Bonita* is Spanish for "pretty," but this species was more likely named for a town in Arizona where the type was collected. **FIELD IDENTIFICATION** Sometimes difficult. The mesosomal setae are mostly silver, except the pronotum has blackish setae, and T2 has the target-shaped spot with the silvery setae inconspicuous. **LAB IDENTIFICATION** The antennal scrobe lacks a dorsal carina; the posterolateral angles of the head are rounded; F1 is much longer than F2; the mesosoma has a distinct scutellar scale with transverse carinae anteriorly and laterally; the mid and hind femora have their apices rounded; the T2 disc has coarse punctures and a black and silver target-shaped setal patch mesally; and the pygidial sculpture is striate. **MALE** (5.5.2.4). Similar to female, with the metasomal setae mostly golden-orange. **NOTES** This species is apparently more common in Mexico, especially in Nayarit and Sinaloa. Within the USA, we have seen only a very few specimens in Arizona. Shockingly, it does not appear to live in Cochise County, which is especially rich in velvet ant diversity. Perhaps it is excluded from that region by competition from other mutillid species.

▼ 3.4.6—*Dasymutilla sicheliana* Madrean females cluster

Most of these species live in Arizona, but this color form of *D. foxi* is widespread in the central and southwestern USA. These species have the head and mesosoma black with contrasting black and white to pale golden setal patterns; the metasoma is predominantly reddish-orange or at least has an orange or reddish patch on the T2 disc. Some of these could be confused with members of the *D. birkmani* (3.4.5) or *D. dionysia* (3.4.7) clusters.

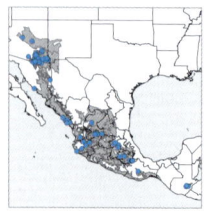

3.4.6.1—*Dasymutilla sicheliana*
DAH-ZEE-MEW-TILL-UH SICHEL-EE-ANN-UH

ETYMOLOGY This species was named for the French entomologist Frédéric J. Sichel. **FIELD IDENTIFICATION** Sometimes difficult. The metasomal cuticle is entirely reddish-brown with sparse black setae and dense patches of whitish setae. **LAB IDENTIFICATION** The antennal scrobe has a dorsal carina; the mesosoma is longer than wide and has a distinct scutellar scale with transverse carinae anteriorly and laterally; the mid and hind femora have their apices rounded; the T2 cuticle is reddish-brown with two white setal patches; and the pygidial sculpture is striate. **MALE** (5.3.4.2, 5.5.2.1). Somewhat similar to female, without

white setal markings on metasoma. **NOTES** This species is widespread in Mexico and common in Arizona.

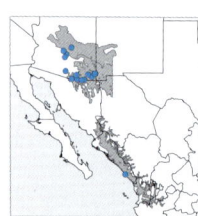

3.4.6.2—*Dasymutilla asteria*
DAH-ZEE-MEW-TILL-UH ASS-TEER-EE-UH

ETYMOLOGY From the Latin *aster* "a star," likely referencing the beautiful color pattern. **FIELD IDENTIFICATION** Sometimes difficult. This species is usually smaller than *D. sicheliana* and has the T2 markings composed of bright red setae, rather than reddish cuticle. **LAB IDENTIFICATION** The antennal scrobe has a dorsal carina; the mesosoma is longer than wide and has a distinct scutellar scale with transverse carinae anteriorly and laterally; the mid and hind femora have their apices rounded; the T2 cuticle is black with two white setal patches and a large anteromesal patch of bright red setae; and the pygidial sculpture is striate. **MALE** Unknown. **NOTES** Using DNA sequences, we have matched some males with females of *D. asteria*, but, so far, we have not found any differences between males of this species and *D. sicheliana* (5.5.2.1). In general, this species is smaller and rarer than *D. sicheliana*. Haddock (1967) noted this species attacking the sand wasp host *Microbembex nigrifrons*.

3.4.6.3—*Dasymutilla foxi*
DAH-ZEE-MEW-TILL-UH FOX-EYE

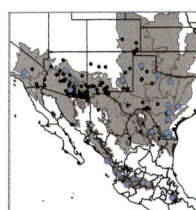

ETYMOLOGY This species was named for the entomologist William J. Fox, who described many velvet ant species in the 1890s. This form used to be named *D. dugesii*, in honor of the Mexican naturalist Alfredo Duges. **FIELD IDENTIFICATION** Usually easy. This is the only species in the cluster known from California, Texas, and the Great Plains. The mesosoma is as wide as long and the T2 disc has a large patch of reddish setae. **LAB IDENTIFICATION** The antennal scrobe lacks a dorsal carina; the mesosoma is as wide as long and lacks a scutellar scale; the mid and hind femora have their apices rounded; the T2 disc has a large patch of reddish setae; and the pygidial sculpture is rugose. **MALE** (5.5.2.2). Similar to female. **NOTES** This color form is apparently widespread throughout the range of *D. foxi*, but regionally it can be less common than the Western FCS (3.2.7.1) or Desert FCS (3.1.3.3) forms.

3.4.6.4—*Dasymutilla saetigera*
DAH-ZEE-MEW-TILL-UH SEE-TIJJ-AIR-UH

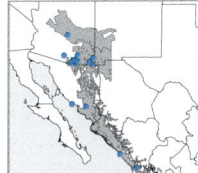

ETYMOLOGY From the Latin *saeta* "bristle of rough hair." **FIELD IDENTIFICATION** Usually easy. The head has the setae mostly black, and T2 has two pale yellow cuticular spots and an anteromesal target-shaped spot of black and silvery setae. **LAB IDENTIFICATION** The antennal scrobe has a dorsal carina; the mesosoma is longer than wide and has a distinct scutellar scale with transverse carinae anteriorly and laterally; the mid and hind femora have their apices rounded; the T2 cuticle is orange-brown with two yellow cuticular spots and an anteromesal target-shaped spot of black and silvery setae; and the pygidial sculpture is striate. **MALE** (5.5.5.5). Similar to female. **NOTES** More than other species in this cluster, *D. saetigera* specimens often have the head and mesosomal cuticle reddened and could be confused with the *D. dionysia* cluster (3.4.7). The mostly black head setae are different from any species in that cluster.

3.4.6.5—*Dasymutilla toluca*
DAH-ZEE-MEW-TILL-UH TOE-LOO-CUH

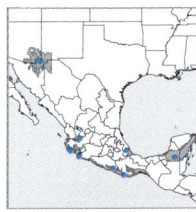

ETYMOLOGY Named for a city in Mexico. **FIELD IDENTIFICATION** Easy. This rare species has a color pattern that is unique in the USA. **LAB IDENTIFICATION** The antennal scrobe has a dorsal carina; the mesosoma is usually as wide as long and lacks a scutellar scale; the mid and hind femora have their apices rounded; the T2 disc has a large patch of orange cuticle that resembles four coalescent spots; and the pygidial sculpture

is rugose. **MALE** Unknown. **NOTES** This species is widespread and variable in Mexico, but in the USA it has apparently been seen only twice, in Cochise County, Arizona.

▼ 3.4.7—*Dasymutilla pulchra* Madrean females cluster

These species are remarkably rare in southern Arizona, but they are more widespread in Mexico and Central America. They have the entire body black and marked with white to pale golden setal patches, and loosely resemble the *D. nocturna* cluster (3.1.3) in the Desert FCS.

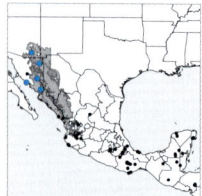

3.4.7.1—*Dasymutilla pulchra*
DAH-ZEE-MEW-TILL-UH PULL-KRUH

ETYMOLOGY From the Latin *pulcher* "beautiful." **FIELD IDENTIFICATION** Easy. The mesosoma is mostly black with a transverse arcuate band of white setae, and T3–6 have the setae entirely whitish. **LAB IDENTIFICATION** The color pattern is diagnostic. **MALE** (5.5.4.4). Similar to female, with a distinct yellow patch on T2. **NOTES** This species superficially resembles the Desert FCS (3.1), except for the partly black mesosomal setae. This is one of the most common and variable species in Mexico and Central America (see 7.4.2.2, 7.4.3.2), but this pattern from arid regions is rare (fewer than 10 specimens examined) and we have seen only one male specimen from the USA (though that record is somewhat dubious, see 5.5.4.4).

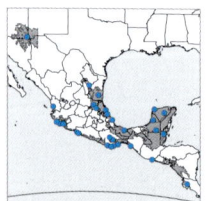

3.4.7.2—*Dasymutilla arachnoides*
DAH-ZEE-MEW-TILL-UH UH-RACK-NOID-EEZ

ETYMOLOGY From the Ancient Greek *arakhne* "spider" with the suffix *-oides* "similar to," referring to the spider-like appearance. **FIELD IDENTIFICATION** Often easy. The mesosoma has a black setal patch surrounded by whitish setae, and T5–6 have the setae mostly black. **LAB IDENTIFICATION** The color pattern is diagnostic. **MALE** Unknown. **NOTES** This species is structurally similar to *D. sicheliana*, but that species has the legs and metasoma largely red. Only two *D. arachnoides* have ever been recorded from Arizona, and we have not been able to study those specimens in person. It is possible that they may just be darkened variants of *D. sicheliana*, rather than true *D. arachnoides*. Another possibility is that *D. arachnoides* and *D. sicheliana* belong to a single widespread and variable species.

3.4.8—*Dasymutilla dionysia* Madrean females cluster

These species live in Arizona, predominantly in mountainous areas. They have the body cuticle predominantly orange or reddish-brown and have contrasting patterns of black and white to golden setal patches. Some of these could be confused with members of the *D. sicheliana* (3.4.6) or *D. birkmani* (3.4.5) clusters.

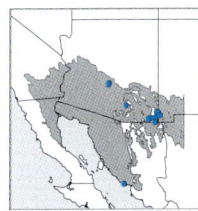

3.4.8.1—*Dasymutilla dionysia*
DAH-ZEE-MEW-TILL-UH DYE-OWE-NYE-ZEE-UH

ETYMOLOGY In Greek mythology, Dionysus was the god of wine and fertility. **FIELD IDENTIFICATION** Usually easy. The T2 disc lacks a distinct pattern of lighter cuticle or setae and is usually lighter in color than the head and mesosoma. **LAB IDENTIFICATION** The mesosoma is longer than wide and lacks a scutellar scale; the T2 disc has large coarse punctures with sparse, erect setae and lacks any distinct setal or cuticular color patches; and the pygidium is mostly microreticulate. **MALE** (5.1.1.4, 5.2.7.4). Very different from female, dorsally with dense gray and/or yellow setae. **NOTES** This species is not especially rare, but none of the authors had collected it prior to 2019.

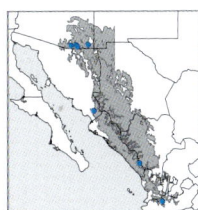

On the western slopes of the Chiricahua Mountains, I finally collected a specimen on a roadside near the grave site of the famous Old West outlaw Johnny Ringo. In the following two years, I collected about 20 more specimens, including the male, in a few sites near Douglas, Arizona.—KEVIN

3.4.8.2—*Dasymutilla cirrhomeris*
DAH-ZEE-MEW-TILL-UH KEER-OWE-MARE-ISS

ETYMOLOGY From the Ancient Greek *kirrhos* "orange" and *meros* "thigh," in reference to the orange legs. **FIELD IDENTIFICATION** Usually easy. The black anteromesal patch on T2 separates it from the other species in this cluster; the pale orange body color separates this species from the *D. sicheliana* cluster, particularly *D. toluca* (3.4.6.5). **LAB IDENTIFICATION** The mesosoma is slightly longer than wide and has a distinct scutellar scale; the T2 disc has moderately coarse punctures with denser setae and has a distinct anteromesal patch of black setae and cuticle; and the pygidium is striate. **MALE** Unknown. **NOTES** This species is widespread in Mexico but rare in Arizona. None of the authors has ever collected this species.

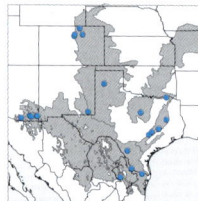

3.4.8.3—*Dasymutilla fasciventris*
DAH-ZEE-MEW-TILL-UH FASS-IVV-ENN-TRISS

ETYMOLOGY From the Latin *fascis* "bundle" and *venter* "belly," referencing the large white pit on S2 of the male. The female used to be named *D. citromaculosa*, after the lemon-colored spots on T2. **FIELD IDENTIFICATION** Easy. This species is longer and narrower than any other *Dasymutilla* species in the USA. The distinct Y-shaped setal mark on the mesosoma and four pale yellow spots on T2 are diagnostic. **LAB IDENTIFICATION** The mesosoma is much longer than wide and lacks a scutellar scale; the T2 disc has moderately coarse punctures and four subcircular spots of pale yellow cuticle; and the pygidium is rugose. **MALE** (5.7.2.2). Similar to female. **NOTES** This is the most widespread and northernmost distributed species in the *D. paradoxa* species-group (see 7.4.3.1, 8.2.7.2). It spans the Neotropical and Nearctic biogeographic realms and has been found in a few widely separated mountainous localities in Arizona.

3.4.9.1—*Protophotopsis venenaria*
PRO-TOE-FOE-TOP-SISS VENN-ENN-ARE-EE-UH

ETYMOLOGY Apparently derived from the Latin *veneno* "potion" or "poison." **FIELD IDENTIFICATION** Often difficult. The yellow spots on T3 are unique, but this species is similar in appearance to small-bodied *Ephuta* (3.4.17, 3.4.19) and *Pseudomethoca* (3.4.12) clusters in the area. **LAB IDENTIFICATION** The head is narrower than the mesosoma; the T1 shape is very broadly sessile; the T2 disc setae are uniformly black; the T2 fringe is composed of white setae; and T6 is convex and without a pygidial plate. **MALE** (5.4.11.1). Darker in coloration than female. **NOTES** This tiny diurnal wasp is rarely encountered in the field. It seems very strange that T3 has pale yellow cuticular markings; nearly all diurnal velvet ants have their dominant metasomal markings placed on the larger T2 disc.

3.4.10.1—*Invreiella cephalargia*
INVREA-EE-ELL-UH SEFF-UH-LARGE-EE-YUH

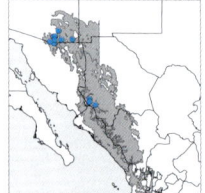

ETYMOLOGY The name *cephalargia* is derived from Latin "large-headed." **FIELD IDENTIFICATION** Easy. This species could be confused with the *Ps. contumax* cluster below but has the T2 disc with a mesal bilobed patch of silver setae mesally, or confused with the *Ps. toumeyi* cluster (3.4.12), but has the body size much larger. **LAB IDENTIFICATION** The gena is armed with large blunt tooth posteriorly.

MALE Unknown. **NOTES** This species was originally placed in the genus *Pseudomethoca* but was transferred to *Invreiella* during the recent revision of that genus (Waldren et al. 2020). The color pattern is different from any of the other 13 species in that genus.

3.4.11—*Pseudomethoca contumax* Madrean females cluster

Most of these species are found in hot regions of California and Arizona, but *Ps. contumax* is widespread in the western and central USA. The head is covered with dense golden setae, and the body size is moderately large (~6–12 mm). Many Madrean FCS species have a large golden head like these. Unlike *Invreiella cephalargia*, the gena is unarmed ventrally; unlike the *D. dilucida* cluster (3.4.3), the T1 shape is sessile; and unlike *Protophotopsis venenaria* (3.4.9.1) and the *Pseudomethoca toumeyi* cluster (3.4.12), the body length is usually greater than 6 mm.

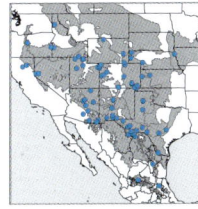

3.4.11.1—*Pseudomethoca contumax*
SOO-DOE-METH-OWE-KUH CONN-TOO-MAX

ETYMOLOGY The name "*contumax*" is derived from Latin "insolent" or "rude." **FIELD IDENTIFICATION** Easy. The bright golden head and unique ⊥⊥-shaped pattern of silver setae on T2 are diagnostic. **LAB IDENTIFICATION** The gena and hypostomal carina are unarmed; the humeral carina is strongly defined; the mesosomal pleura are mainly bare; the lateral propodeal margin has a row of short teeth; the T2 fringe setae are usually entirely black; and the pygidial sculpture is rugose. **MALE** (5.4.8.1). Coloration much darker than female. **NOTES** Females are commonly found crawling on the ground near sunflowers or on sunflower

stems. This is one of the northernmost distributed members of the Madrean FCS but also extends south into many Mexican states. Unlike many other velvet ants, *Ps. contumax* is found from the spring through the late summer months.

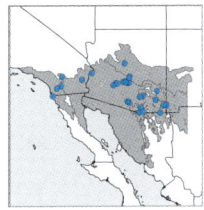

3.4.11.2—*Pseudomethoca praeclara*
SOO-DOE-METH-OWE-KUH PREE-CLAIR-UH

ETYMOLOGY The name *preclara* is derived from Latin, "bright" or "famous." **FIELD IDENTIFICATION** Easy. The bright golden head and T2 disc with the setae mainly pale orange are diagnostic. **LAB IDENTIFICATION** The gena and hypostomal carina are unarmed; the humeral carina is strongly defined; the mesosomal pleura are mainly bare; the lateral propodeal margin has a row of short teeth; the T2 fringe setae are silver laterally; and the pygidial sculpture is rugose. **MALE** (5.4.8.2). Coloration much darker than female. **NOTES** This species is common and abundant in the Sonoran Desert and Madrean Archipelago.

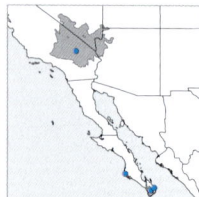

3.4.11.3—*Pseudomethoca connectens*
SOO-DOE-METH-OWE-KUH CUH-NECK-TENZ

ETYMOLOGY The name *connectens* is derived from Latin, "link" or "fasten together." **FIELD IDENTIFICATION** Easy. The bright golden head and unique black setal hourglass shape with the lateral setae silvery make this species quite distinct. **LAB IDENTIFICATION** The gena and hypostomal carina are unarmed; the humeral carina is strongly defined; the mesosomal pleura are mainly bare; the lateral propodeal margin has a row of short teeth; the T2 fringe setae are silver laterally; and the pygidial sculpture is rugose. **MALE** Unknown. **NOTES** This rare species was known only from Baja California (Mexico) until recent collections confirmed that it is also found in the Mojave Desert.

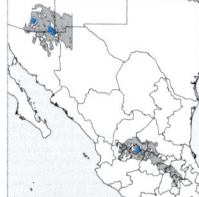

3.4.11.4—*Pseudomethoca quadrinotata*
SOO-DOE-METH-OWE-KUH QUAD-REE-NO-TOTT-UH

ETYMOLOGY Derived from the Latin *quadrum* "four" and *notatus* "marked," in reference to the yellow spots on T2. **FIELD IDENTIFICATION** Easy. The bright golden head and four pale yellow cuticular spots on T2 are diagnostic. **LAB IDENTIFICATION** The gena is unarmed; the hypostoma is armed with a ventral tooth; the T2 fringe setae are silver laterally; and T6 is convex without a pygidial plate. **MALE** Unknown. **NOTES** This species and *Ps. wickhami* (3.5.3.1, 3.6.9.2) are the only members of *Pseudomethoca* in the USA that have the pygidium undefined. This species is generally smaller than the other species in the cluster, but rarely as small as the *Ps. toumeyi* cluster below.

 ### 3.4.12—*Pseudomethoca toumeyi* Madrean females cluster

These species are scattered throughout the western and central USA. The head is covered with dense golden setae, and the body size is very small (less than 5 mm). Many Madrean FCS species have a large golden head, as these do. These species are smaller in size than *Invreiella cephalargia* (3.4.10.1) and the *Ps. toumeyi* cluster (3.4.12); unlike the *D. dilucida* cluster (3.4.3), the T1 shape is sessile; and unlike *Protophotopsis venenaria* (3.4.9.1), T6 has a defined pygidial plate.

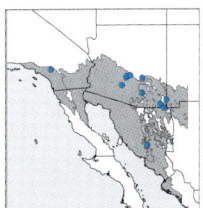

3.4.12.1—*Pseudomethoca toumeyi*
SOO-DOE-METH-OWE-KUH TOUMEY-EYE

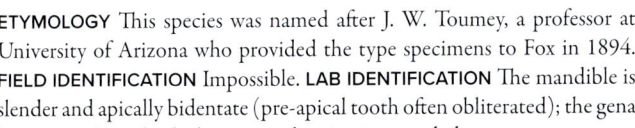

ETYMOLOGY This species was named after J. W. Toumey, a professor at University of Arizona who provided the type specimens to Fox in 1894. **FIELD IDENTIFICATION** Impossible. **LAB IDENTIFICATION** The mandible is slender and apically bidentate (pre-apical tooth often obliterated); the gena has a ventral tooth; the hypostomal carina is expanded to form an apparent tooth; the mesosoma is about as wide as long; the T2 disc has two circular spots of silvery setae; the T2 fringe setae are black; and the pygidium is elongate with faint punctures. **MALE** Coloration much darker than female. **NOTES** This species has been associated with the andrenid host bee, *Perdita portalis* Timberlake (Krombein 1992).

3.4.12.2—*Pseudomethoca bequaerti*
SOO-DOE-METH-OWE-KUH BEQUAERT-EYE

ETYMOLOGY This species was named after J. Bequaert, who collected the type specimen. **FIELD IDENTIFICATION** Often impossible. This is the only species in this cluster in many parts of the northwestern USA, where the color pattern is diagnostic. **LAB IDENTIFICATION** The mandible is apically widened and tridentate; the gena has a ventral tooth; the hypostomal carina is only weakly elevated; the mesosoma is about as wide as long; the T2 disc has two circular spots of silvery setae; the T2 fringe setae are black; and the pygidium is elongate with faint punctures. **MALE** Unknown. **NOTES** This is a widespread species in the western USA. They are relatively common, given their small body size.

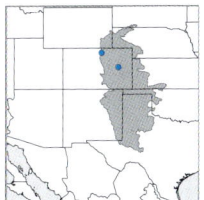

3.4.12.3—*Pseudomethoca klotsi*
SOO-DOE-METH-OWE-KUH KLOTS-EYE

ETYMOLOGY This species was named after A. B. Klots, who collected the type specimen in 1932. **FIELD IDENTIFICATION** Impossible. **LAB IDENTIFICATION** The mandible is slender and bidentate; the gena has a ventral tooth; the hypostomal carina is weakly swollen; the mesosoma is longer than wide; the T2 disc has two circular spots of silvery setae; the T2 fringe setae are black; and the pygidium is elongate and microreticulate. **MALE** Unknown. **NOTES** This rare species is somewhat intermediate between *Ps. frigida* (3.7.2.1) and *Ps. bequaerti* and is also known from the transition zone

3.4.14—*Pseudomethoca donaeanae* **Madrean females cluster**

These species occur mainly in southern Arizona, but *Ps. donaeanae* is more widespread in the southwestern USA. They are medium-sized species with black setae on the vertex and mesonotum (at least mesally), and without distinct yellow spots on T2.

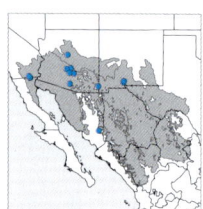

3.4.14.1—*Pseudomethoca donaeanae*
SOO-DOE-METH-OWE-KUH DOE-NYUH-ANN-EE

ETYMOLOGY This species was named for Dona Ana County in New Mexico. **FIELD IDENTIFICATION** Difficult. This species has the head setae silver laterally and black mesally, but they are sparse and inconspicuous. **LAB IDENTIFICATION** The mesosoma has a large erect tubercle near the juncture of the dorsal and posterior propodeal surfaces mesally. **MALE** (5.5.8.1). Often similar to female. **NOTES** This is the only species of *Pseudomethoca* with an apparent scutellar scale, though it is not clear whether this transverse tubercle is structurally analogous with the scutellar scale seen in other velvet ant lineages.

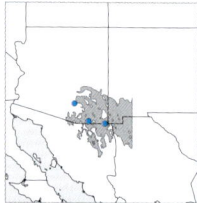

3.4.14.2—*Pseudomethoca sonorae*
SOO-DOE-METH-OWE-KUH SONORA-EE

ETYMOLOGY Kevin named this species after his daughter, Sonora Rose Williams. He collected this new species about two months after she was born. **FIELD IDENTIFICATION** Difficult. This species has the dorsal head setae entirely black, but this can be difficult to see in the field. **LAB IDENTIFICATION** The scutellar area is flat and unarmed. **MALE** (5.4.9.2). Much darker in coloration than female. **NOTES** This species is apparently known only from five specimens: one collected in 2007, one in 2020, and three in 2021. So far, they have been found only in southern Arizona. The females from near Douglas, Arizona, have a darker cuticle than the specimen from Sierra Vista, Arizona. Superficially, this species resembles many species with the Eastern FCS (3.6.10).

 3.4.15.1—*Sphaeropthalma marpesia*
S-FAIR-OPP-THAL-MUH MAR-PEE-ZHEE-UH

ETYMOLOGY In Ancient Greek and Roman mythology, Marpesia was the queen of the Amazons. **FIELD IDENTIFICATION** Easy. The color pattern is diagnostic, especially the black transverse patch on T2. **LAB IDENTIFICATION** The mesonotal setae are simple or brachyplumose; the T1 shape is petiolate; the tergal fringes are plumose; and T6 has a defined microreticulate pygidial plate.

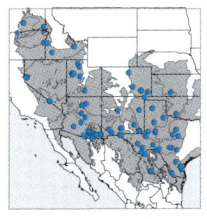

MALE (5.5.10.1). Generally darker with sparser setae than female. **NOTES** This predominantly nocturnal species is widespread in the central and western USA. Like other nocturnal species treated in this chapter (3.2.12), they are most often seen in daylight on colder days or in colder habitats. Various other nocturnal velvet ants have contrasting patterns that resemble the Madrean FCS, and they are rarely encountered in daylight, but they are too poorly understood to treat in this chapter. Additional information about these wasps is provided in chapter 6 (6.1, 6.8).

 ### 3.4.16.1—*Stethophotopsis maculata*
STETH-OWE-FOE-TOP-SISS MACK-EWE-LAW-TUH

ETYMOLOGY From the Latin *macula* "spot," referring to the black patches on T2. **FIELD IDENTIFICATION** Easy. The color pattern is unique, particularly in having the two separated black patches at the base of T2. **LAB IDENTIFI-CATION** The mesonotal setae are simple or brachyplumose; the T1 shape is subsessile; the tergal fringes are plumose; and T6 (or T7 in males) is convex without a pygidial plate. **MALE** Wingless (Fig. 3.4.16.1c), identical to female. **NOTES** Males are wingless but retain their ocelli and tegulae, unlike the females. Based on the bright color pattern and small male ocelli, we presume that this is a diurnal species. This rare species is known from scattered mountainous habitats in southern Arizona and Mexico.

Female

Wingless male

 ### 3.4.17—*Ephuta argenticeps* Madrean females cluster

These three species have disparate distributions: *E. argenticeps* in the Pacific states, *E. auricapitis* in Texas, and *E. coloradella* in the Rocky Mountains. They have the head covered with silver setae, unlike the *E. tumacacori* cluster (3.4.19), which have golden heads. As with all *Ephuta*, these should also be compared with the *E. scrupea* cluster (3.7.6) in the Cryptic FCS.

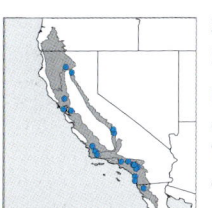 ### 3.4.17.1—*Ephuta argenticeps*
EFF-OO-TUH ARR-JENT-ISS-EPPS

ETYMOLOGY From the Latin *argentum* "silver" and *-ceps* "headed," in reference to the silver head color. **FIELD IDENTIFICATION** Sometimes easy. This is apparently the only silver-headed *Ephuta* species in Pacific regions of California. **LAB IDENTIFICATION** The head is covered with dense silvery setae; the postgenal carina is weakly defined; the T2 disc usually has two whitish setal patches; T3–5 each have a mesal spot of whitish setae; T6 has a pygidial plate defined by lateral carinae; and the hypopygium has a transverse ridge basally. **MALE** (5.4.13.2). Much darker than female. **NOTES** The white spots on the T2 disc and T3–5 are obliterated in many specimens. It is not yet clear whether this is due only to wear and tear on specimens or comes about from intraspecific variation.

3.4.17.2—*Ephuta albiceps*
EFF-OO-TUH ALL-BISS-EPPS

ETYMOLOGY From the Latin *albus* "white" and -*ceps* "headed," in reference to the whitish head color. **FIELD IDENTIFICATION** Sometimes easy. This is apparently the only white-headed species of *Ephuta* in Texas. **LAB IDENTIFICATION** The head is covered with dense whitish setae; the postgenal carina is complete; the T2 disc has two whitish setal patches; T3–5 each have a mesal patch of whitish setae; T6 has a pygidial plate defined by lateral carinae; and the hypopygium is uniformly punctate basally. **MALE** Unknown. **NOTES** This rare species is structurally similar to *E. auricapitis* (3.4.19.3), except for the color of the head setae and a slight variation in the apex of the hypopygium. Eventually, these species might be recognized as conspecific synonyms of one another.

3.4.17.3—*Ephuta coloradella*
EFF-OO-TUH CALL-OWE-RAD-ELL-UH

ETYMOLOGY This species was named after the state of Colorado. **FIELD IDENTIFICATION** Sometimes easy. This is apparently the only silver-headed species of *Ephuta* in Colorado. **LAB IDENTIFICATION** The head is covered with dense silvery setae; the postgenal carina is weakly developed; the T2 disc usually has two whitish setal patches; T3–5 usually have the setae entirely blackish; T6 has a pygidial plate defined by lateral carinae; and the hypopygium has two lateral tubercles basally. **MALE** Unknown. **NOTES** This rare species is apparently known in the literature from only the single holotype specimen. The specimen pictured here matches the original description of *E. coloradella* in most features and was also collected in Colorado, but it has faint white setal patches on

T3–5. It is possible that the type of *E. coloradella* is missing these setae because of wear and tear on the specimen, or that this feature is variable in the species, or that this pictured specimen is actually a separate undescribed species.

3.4.18—*Ephuta floridana* Madrean females cluster

These species occur mainly in the southeastern USA. They have the head covered with golden setae. These setae are often inconspicuous or can be worn off with age, so these should also be compared with the *E. scrupea* cluster (3.7.6) in the Cryptic FCS.

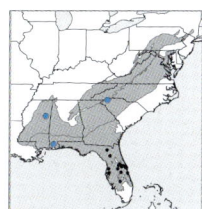

3.4.18.1—*Ephuta floridana dietrichi*
EFF-OO-TUH FLOW-RIDD-ANN-UH DIETRICH-EYE

ETYMOLOGY Named for the state of Florida. This subspecies, *E. f. dietrichi*, was named after H. Dietrich, who collected the type specimen in Mississippi in 1932. **FIELD IDEN-TIFICATION** Difficult. This is the only eastern species with the head setae golden (though they are often inconspicuous) and the T2 disc without any silvery setal patches. **LAB IDENTIFICATION** The vertex is covered with golden setae; the postgenal carina is mostly obliterated; the T2 disc lacks any trace of whitish setal patches; T3–5 usually have the setae whitish mesally; T6 has a pygidial plate defined by lateral carinae; and the hypopygium has a transverse ridge basally. **MALE** (5.7.13.2). Somewhat similar to female, without the golden head setae. **NOTES** There are two subspecies; *E. f. floridana* has sparser, duller golden setae on the head (see Cryptic FCS, 3.7.6.4), and *E. f. dietrichi* has denser, bright golden head setae.

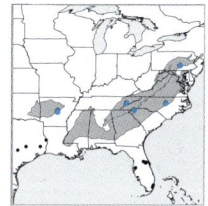

3.4.18.2—*Ephuta margueritae xanthocephala*
EFF-OO-TUH MARGUERITA-EE ZANTH-OWE-SEFF-ALL-UH

ETYMOLOGY Schuster (1951) named this species after his wife, Olga Marguerite Schuster, in appreciation of her constant help during his study of this genus. This subspecies, *E. m. xanthocephala*, was named in reference to the more distinctly yellow-gold head setae. **FIELD IDENTIFICATION** Difficult. This is the only eastern species with the head setae golden (though they are often inconspicuous) and the T2 disc with silvery setal patches. **LAB IDENTIFICATION** The vertex is covered with golden setae; the postgenal carina is complete; the T2 disc has two whitish setal patches; T3–5 usually have the setae whitish mesally; T6 is convex without a pygidial plate; and the hypopygium has two tubercles basally. **MALE** (5.4.13.9). Much darker than female. **NOTES** This species is structurally nearly identical to *E. sudatrix* (3.4.19.4) from Texas and is usually larger in size than *E. floridana*. The subspecies pictured here, *E. m. xanthocephala* has brighter and denser golden setae on the head than the nominal subspecies, which seems to fit the Cryptic FCS (3.7.6.5). Schuster (1951) discussed three males and a female that were reared from Hymenoptera cocoons under stones in Pennsylvania, which were apparently pompilid spider wasp hosts.

▼ 3.4.19—*Ephuta tumacacori* Madrean females cluster

These species occur mainly in Arizona and Texas. They have the head covered with golden setae, unlike the *E. argenticeps* cluster (3.4.17), which have silver heads. As with all *Ephuta* specimens, these should also be compared with the *E. scrupea* cluster (3.7.6) in the Cryptic FCS.

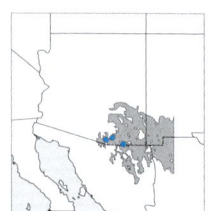

3.4.19.1—*Ephuta tumacacori*
EFF-OO-TUH TOO-MUH-CUH-CORE-EE

ETYMOLOGY This species was named after the Tumacacori Mountains in Arizona. **FIELD IDENTIFICATION** Usually easy. This species is larger-bodied with a wider head than any other *Ephuta* females in Arizona. **LAB IDENTIFICATION** The vertex is covered with dense golden setae; the postgenal carina is complete; the T2 disc has two whitish setal patches; T3–5 usually have the setae whitish sublaterally and distinctly black mesally; T6 is convex without a pygidial plate and is covered by dense, curled setae; and the hypopygium has two flat, smooth ridges basally. **MALE** Unknown. **NOTES** *Ephuta tumacacori* is the largest *Ephuta* female in the USA and has a large quadrate head. This and other Western members of *Ephuta* are rare and hard to catch. They can sometimes be found on open ground or in leaf litter.

3.4.19.2—*Ephuta baboquivari*
EFF-OO-TUH BABA-KEY-VARR-EE

ETYMOLOGY This species was named after the Baboquivari Mountains in Arizona. **FIELD IDENTIFICATION** Impossible. **LAB IDENTIFICATION** The head is covered with dense golden setae; the postgenal carina is complete; the T2 disc apparently lacks whitish setal patches; T3–5 usually have the setae mostly blackish; T6 has a pygidial plate defined by lateral carinae; and the hypopygium has a transverse ridge basally. **MALE** Unknown. **NOTES** In southeastern Arizona, twice as many *Ephuta* males (6 spp.) than females (3 spp.) are recognized. Females that loosely fit the concept used here for *E. baboquivari* are relatively variable in coloration, and some other features, and may eventually be recognized as distinct species.

3.4.19.3—*Ephuta auricapitis*
EFF-OO-TUH OUR-ICK-APP-ITT-ISS

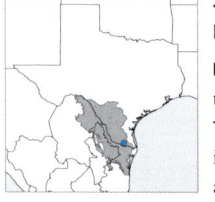

ETYMOLOGY From the Latin *aurum* "gold" and *capitis* "head," in reference to the gold head color. **FIELD IDENTIFICATION** Impossible. **LAB IDENTIFICATION** The head is covered with dense pale golden setae; the postgenal carina is complete; the T2 disc has two whitish setal patches; T3–5 each have a mesal patch of whitish setae; T6 has a pygidial plate defined by lateral carinae; and the hypopygium is uniformly punctate basally. **MALE** Unknown. **NOTES** This rare species is structurally similar to *E. albiceps* (3.4.17.2), except for the color of the head setae and a slight variation in the apex of the hypopygium. Eventually, these species might be recognized as conspecific synonyms of one another.

RIGHT: *Ephuta auricapitis.*
FAR RIGHT: *Ephuta sudatrix.*

3.4.19.4—*Ephuta sudatrix*
EFF-OO-TUH SOO-DAY-TRICKS

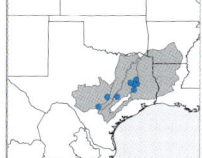

ETYMOLOGY Apparently derived from the Latin *sudatus* "sweat," likely in reference to their occurrence in hot and humid southern Texas. **FIELD IDENTIFICATION** Impossible. **LAB IDENTIFICATION** The vertex is covered with golden setae; the postgenal carina is complete; the T2 disc has two whitish setal patches; T3–5 usually have the setae mostly blackish; T6 is convex without a pygidial plate; and the hypopygium has two tubercles basally. **MALE** Unknown. **NOTES** This species is similar to *E. margueritae* from the previous cluster, but the head setae are usually brighter and denser, the white T2 spots are usually larger, and the T3–5 setae are usually mostly black.

▼ 3.4.20—*Myrmosula rutilans* Madrean females cluster

These species occur mainly in the western USA. They are the only member of Myrmosinae with bright silver or golden setae on the head. These can sometimes be confused with the *Myrmosula parvula* cluster (3.7.9) in the Cryptic FCS if the head setae are dull or rubbed off.

3.4.20.1—*Myrmosula rutilans*
MEER-MOE-SOO-LUH ROO-TILL-AHNZ

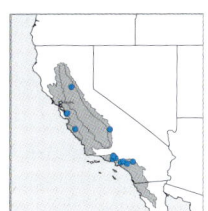 

ETYMOLOGY From the Latin *rutilans* "reddened," likely referring to the background body color. **FIELD IDENTIFICATION** Very difficult. This species is the most common *Myrmosula* species in the Pacific regions and generally has denser, pale golden head setae than the others in that region (3.7.9.3, 3.7.9.4). **LAB IDENTIFICATION** The interantennal prominence has two blunt, ventrally divergent, smooth ridges; the hypostomal carina is edentate; the ventral mandibular lamella is convex; and the pale yellow T2 patches are generally subcircular or subtriangular. **MALE** Unknown. **NOTES** *Myrmosula rutilans* is found in temperate Pacific regions. An apparent fossil of this species was found in the La Brea tar pits. The density and color of the head setae are variable, so this species is also treated in the Cryptic FCS (3.7.9.5).

FAR LEFT: Typical *Dasymutilla quadriguttata*

MIDDLE LEFT: Formerly called *Dasymutilla nitidula*

LEFT: Formerly called *Dasymutilla electra*

FCS forms are also variable in color. The different forms pictured here used to be all treated as different species until a DNA analysis revealed that *D. quadriguttata* was a much more variable species than previously thought.

3.5.1.2—*Dasymutilla parksi*
DAH-ZEE-MEW-TILL-UH PARKS-EYE

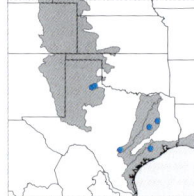

ETYMOLOGY This species was named after H. B. Parks, who collected the type specimen in 1931. **FIELD IDENTIFICATION** Very difficult. The propodeum has an orange setal brush and the head shape is posteriorly flat with the posterolateral angles especially sharp. These features are subtle and difficult to recognize in the field. **LAB IDENTIFICATION** The vertex has the posterolateral tubercle sharply triangular and more widely separated than the epaulets; and the propodeum has a dense brush of orange setae on the posterior surface. **MALE** Unknown. **NOTES** This species is structurally similar to *D. campanula* (3.2.8.4) but has very different coloration. The male might be *D. gentilis* (5.6.4.4) or *D. meracula* (5.6.4.5) based on distribution and similarity to *D. quadriguttata*.

3.5.1.3—*Dasymutilla curticeps*
DAH-ZEE-MEW-TILL-UH CURT-ISS-EPPS

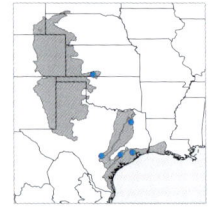

ETYMOLOGY From the Latin *curtus* "short" and -*ceps* "headed," in reference to the truncate vertex. **FIELD IDENTIFICATION** Very difficult. The propodeum has an orange setal brush, and the head shape is posteriorly flat, but the posterolateral angles are not especially sharp. These features are subtle and difficult to recognize in the field. **LAB IDENTIFICATION** The vertex has the posterolateral tubercle linear and about as widely separated as the epaulets; and the propodeum has a dense brush of orange setae on the posterior surface. **MALE** Unknown. **NOTES** This species is very similar to *D. parksi* but has the vertex tubercles with a slightly different shape. The male might be *D. gentilis* (5.6.4.4) or *D. meracula* (5.6.4.5) based on distribution and similarity to *D. quadriguttata*.

3.5.2.1—*Myrmilloides grandiceps*
MEER-MILL-OI-DEEZ GRAN-DISS-EPPS

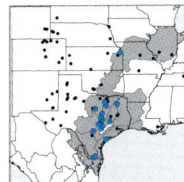

ETYMOLOGY From the Latin *grandis* "large" and *-ceps* "-headed." **FIELD IDENTIFICATION** Easy. The giant head is unique, and the short-winged males are even easier to recognize. They are larger in size than any of the *Pseudomethoca* species that possess teeth on the gena or postgena. **LAB IDENTIFICATION** The gena and postgena are each armed with a ventral tooth. **MALE** Similar to female. **NOTES** The males are short winged and generally more rarely seen than females. This species is variable in color, also being treated in the Eastern FCS (3.6.7.1). Some individuals also have a vague resemblance to the Texan FCS (3.3.6.1), since the head and mesosoma are often darker than the metasoma in western Texas.

3.5.3.1—*Pseudomethoca wickhami*
SOO-DOE-METH-OWE-KUH WICKHAM-EYE

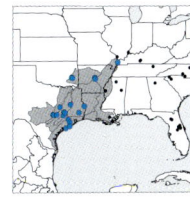

ETYMOLOGY This species was named after H. F. Wickham, who collected the type specimen in the late 1800s. **FIELD IDENTIFICATION** Easy. The color pattern is unique, especially in the widely separated, lateral orange patches on the T2 disc. **LAB IDENTIFICATION** The humeral carina is strongly defined with a sharp tooth-like angle; the mesosomal pleura are mainly bare; the lateral propodeal margin has a row of short teeth; the T2 disc has two medium-sized patches of lighter orange cuticle covered with pale yellow setae at the extreme lateral margins; and T6 is convex without a defined pygidial plate. **MALE** (5.5.7.1). Generally darker than female. **NOTES** *Pseudomethoca wickhami* has many unique features compared with other North American species, especially in the male's mandible shape, undefined pygidium in females, and postgenal armature in both sexes. Many related species, however, occur in Central and South America. This group of species might be eventually recognized as a distinct genus. At the moment, this is the only *Timulla*-like FCS in the genus *Pseudomethoca*, but this form is basically identical to the Eastern FCS form (3.6.9.2), except that the metasoma is darker in populations from the central USA. As more specimens are examined and identified in Texas, it is possible that some populations of other Eastern *Pseudomethoca* might have the metasoma darkened like this. Members of the *Ps. bethae* cluster (3.4.13) are superficially similar to the *Timulla*-like FCS, but they have the T2 disc cuticle largely orange-brown dorsally.

3.5.4.1—*Morsyma ashmeadi*
MORE-SEE-MUH ASHMEAD-EYE

ETYMOLOGY This species was named after the Canadian entomologist William H. Ashmead, who provided the first functional treatment for velvet ant genera at the end of the 19th century. **FIELD IDENTIFICATION** Sometimes easy. The color pattern is unique, especially among diurnal species in the Pacific region, but many nocturnal species are similar in color. **LAB IDENTIFICATION** In both sexes, there is a distinct arcuate carina between the pronotum and mesonotum dorsally; T2 has a distinct posterior fringe of white plumose

setae contrasting with most of the metasomal setae, which are black. **MALE** (5.7.8.1). Often wingless, nearly identical to female. **NOTES** The wingless males are pictured here with the female. Fully winged males (5.7.8.1) are also known, but they seem to occur farther south in the Coastal Range, while these wingless males were mostly collected around the San Francisco Bay.

FAR LEFT: Female

LEFT: Wingless male

 ### 3.5.5.1—*Timulla navasota coahuila*
TIMM-EWE-LUH NAW-VUH-SOE-TUH COE-UH-WEE-LUH

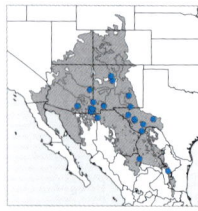

ETYMOLOGY This species was named for the Mexican state of Coahuila. **FIELD IDENTIFICATION** Sometimes easy. This is generally the only nocturnal *Timulla* species in the southwestern hot deserts; the white patches on the T2 disc are broader and more rounded than most other *Timulla*-like FCS *Timulla* species; and the propodeum is narrower than the pronotum and mesonotum. **LAB IDENTIFICATION** See the field identification characters. Additionally, the humerus is rounded; the mesonotum is not constricted; there is no scutellar scale; and the pygidial sculpture has fine, rounded striae. **MALE** (5.3.6.2). Darker than female. **NOTES** This is currently recognized as a subspecies of *T. navasota* (see nominotypical form: 3.6.13.1). This species is typically nocturnal in habit and is apparently the only nocturnal *Timulla* in Arizona.

 ### 3.5.6.1—*Timulla ferrugata*
TIMM-EWE-LUH FAIR-OO-GAH-TUH

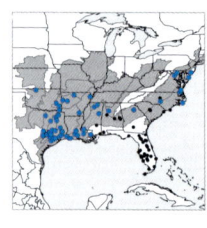

ETYMOLOGY From the Latin *ferrugo* "iron rust," referencing the orange-brown color. **FIELD IDENTIFICATION** Often easy. This is the only *Timulla*-like FCS species with the propodeum wider than the pronotum; additionally, this species usually has the metasoma reddened laterally and black only dorsally. **LAB IDENTIFICATION** The humerus is angular; the mesonotum is constricted; the scutellar scale is moderately wide and posteriorly rounded; the T2 disc has the white setal patches short and linear; and the pygidial sculpture is generally longitudinally rugose. **MALE** (5.6.7.2, 5.7.16.4). Generally darker than female. **NOTES** This species is common in the central and eastern USA, and often fits the Eastern FCS (3.6.14.1). Unlike other *Timulla*-like *Timulla* species, the lateral portions of T2 often have a distinct patch of reddish cuticle; when the metasoma is partly reddened in the other species, that coloration is more diffuse. *Timulla floridensis* (3.6.14.2) is the only other *Timulla* in the USA with the propodeum clearly wider than the pronotum, and it often has the metasoma (and sometimes head) obscurely darkened, so that it resembles the *Timulla*-like FCS. In that species, however, the mesonotum is not constricted and the pygidium is mostly microreticulate with faint rugae.

Photo by Dan Rouethe

Timulla ferrugata

Timulla grotei

▼ 3.5.7.1—*Timulla grotei*

TIMM-EWE-LUH GROTE-EYE

ETYMOLOGY This species was named after Augustus Radcliffe Grote, a British entomol-

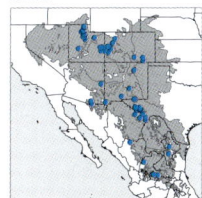

ogist who worked mainly in America. **FIELD IDENTIFICATION** Usually easy. The T2 disc has the setae entirely black, without any white setal stripes basally. **LAB IDENTIFICATION** The humerus is angular; the mesonotum is constricted; the scutellar scale is wide, posteriorly truncate, and has a transverse row of tubercles anterior to the scale; the T2 disc lacks white setal patches basally; and the pygidial sculpture is rugose with the apex microreticulate. **MALE** (5.3.6.1). Generally darker than female. **NOTES** This species seems to occur farther north than most other western *Timulla* and is apparently the only *Timulla* species in central and northern Utah.

▼ 3.5.8.1—*Timulla suspensa*

TIMM-EWE-LUH SUSS-PEN-SUH

ETYMOLOGY Apparently from the Latin *suspensus* "suspended," "anxious," or "doubtful." It is not clear why this name was chosen. **FIELD IDENTIFICATION** Usually easy. The T2 disc has white setal stripes that are continuous from the anterior to posterior margins of T2, but these can often be obliterated or reduced by wear and tear. **LAB IDENTIFICATION** The humerus is angular; the mesonotum is constricted; the scutellar scale is wide, posteriorly truncate, and has a weak transverse furrow anterior to the scale; the T2 disc has long white setal stripes; and the pygidial sculpture is rugose with the apex microreticulate. **MALE** (5.3.6.3, 5.5.14.2). Darker than female. **NOTES** This is the most common form of *T. suspensa*, but some specimens in the Great Plains are treated in the Eastern FCS (3.6.17.2). Like *T. vagans* and *T. grotei*, the head is relatively wide and quadrate in shape.

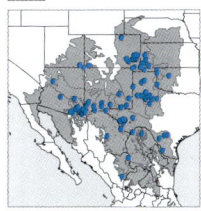

▼ 3.5.9—*Timulla vagans Timulla*-like females cluster

These species occur mainly in the central USA. They have the propodeum and pronotum equally wide and the mesosoma with the humerus (shoulder) obtusely angular or perpendicular. They have short longitudinal patches of whitish setae basally on T2. The metasomal cuticle can vary slightly, and they should be compared with the Eastern FCS *Timulla* clusters (3.6.15, 3.6.17). The mesosomal shape is often difficult to interpret, so they should also be compared with the *T. wilyae* cluster (3.5.10).

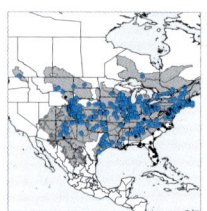

3.5.9.1—*Timulla vagans*
TIMM-EWE-LUH VAY-GUHNS

ETYMOLOGY From the Latin *vagans* "rambling." **FIELD IDEN-TIFICATION** Often difficult. This is the most common and widespread species in the cluster and has a somewhat large quadrate head, but in southern areas, they are difficult to separate from *T. oajaca* and *T. leona*. **LAB IDENTIFICATION** The humerus is angular; the mesonotum is constricted; the scutellar scale is wide, posteriorly truncate, and has a transverse furrow anterior to the scale; the T2 disc has short linear, white setal patches basally; the T2 sculpture is generally fine with the intervals micropunctate and setose; and the pygidial sculpture is rugose with the apex microreticulate. **MALE** (5.5.12.2, 5.6.7.1, 5.7.16.10). Generally darker than female. **NOTES** This is the most common *Timulla* in the USA. Unlike most *Timulla*-like FCS specimens, this species often maintains this coloration even in the eastern USA. The metasomal color is variable, however, and this species is also treated in the Eastern FCS (3.6.17.1). The head is larger and more quadrate in shape than most other *Timulla* species.

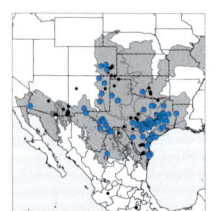

3.5.9.2—*Timulla oajaca*
TIMM-EWE-LUH WAH-HAW-KUH

ETYMOLOGY Named for the Mexican state of Oaxaca. It is confusing that this species does not actually occur in that state. **FIELD IDENTIFICATION** Very difficult. This species has sparser setae and coarser sculpture on T2 than the other species in the cluster, the head is narrower than that of *T. vagans*, and the mesonotal sides are constricted, unlike in *T. leona*. **LAB IDENTIFICATION** The humerus is angular; the mesonotum is constricted; the scutellar scale is wide, posteriorly truncate, and has a weak transverse furrow or row of tubercles anterior to the scale; the T2 disc has short linear, white setal patches basally; the T2 sculpture is coarse with the intervals mostly smooth and bare; and the pygidial sculpture is rugose with the posterior half microreticulate. **MALE** (5.5.14.1). Darker than female. **NOTES** Most specimens of *T. oajaca* fit in the Eastern FCS (3.6.17.3), and most of the *Timulla*-like FCS specimens have the metasoma at least partly reddened.

3.5.9.3—*Timulla leona*
TIMM-EWE-LUH LEE-OWE-NUH

ETYMOLOGY Apparently from the Latin *leonis* "lion" but may be named for the Mexican state of Nuevo Leon, since the type came from Mexico and the author (Blake 1871) named many species after Mexican place names. **FIELD IDENTIFICATION** Difficult. The angular humerus and

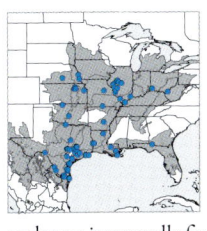

non-constricted mesonotum are diagnostic in the central USA, but these features are difficult to interpret in the field. **LAB IDENTIFICATION** The humerus is angular; the mesonotum is not constricted; the scutellar scale is wide, posteriorly flat or wavy, and has a transverse furrow anterior to the scale; the T2 disc has short linear, white setal patches basally; the T2 sculpture is generally fine with the intervals micropunctate and setose; and the pygidial sculpture is uniformly striate. **MALE** (5.5.14.4). Generally darker than female. **NOTES** Most specimens of *T. leona* fit in the Eastern FCS (3.6.17.4). This is one of the only *Timulla*-like FCS *Timulla* species with the pygidium striate, and the striae in this species are straighter and more clearly complete throughout the entire pygidium than in any other *Timulla* in the USA.

3.5.9.4—*Timulla nicholi*
TIMM-EWE-LUH NICHOL-EYE

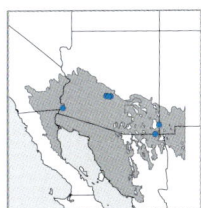

ETYMOLOGY Mickel (1937) named this species in honor of his friend A. A. Nichol, who collected the holotype and one of the paratypes. **FIELD IDENTIFICATION** Sometimes difficult. The metasomal color is generally darker than in the other species known from deserts in Arizona and California. The combination of shortened white setal stripes on T2 coupled with the non-constricted mesonotum are also diagnostic in this region. **LAB IDENTIFICATION** The humerus is angular; the mesonotum is not constricted; the scutellar scale is wide, posteriorly truncate or wavy, and has a weak transverse furrow anterior to the scale; the T2 disc has short linear, white setal patches basally; the T2 sculpture is moderately coarse with the intervals micropunctate and setose mesally, but smooth and bare laterally; and the pygidial sculpture is mostly microreticulate with some irregular rugae basally. **MALE** Unknown. **NOTES** This species has an overlapping range with multiple unassociated *Timulla* males in the southwestern USA. Mickel (1937) thought it was the female of *T. neobule*, but proof for this has yet to be found in the 85 years since.

3.5.10—*Timulla wileyae Timulla*-like females cluster

These species occur mainly in the central USA. They have the propodeum and pronotum equally wide and the mesosoma with the humerus (shoulder) more or less evenly curved. The metasomal cuticle can vary slightly, and they should be compared with the Eastern FCS *Timulla* clusters (3.6.15, 3.6.16). The mesosomal shape is often difficult to interpret, so they should also be compared with the *T. vagans* cluster (3.5.9).

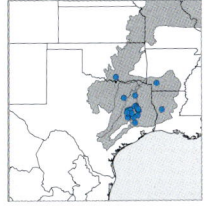

3.5.10.1—*Timulla wileyae*
TIMM-EWE-LUH WILEY-EE

ETYMOLOGY Mickel (1937) named this species after the famous herpetologist Grace O. Wiley, who collected the type specimen and many other useful velvet ant specimens. **FIELD IDENTIFICATION** Difficult. The body size is usually smaller than that of *T. dubitata*, and the mesonotum is not constricted. **LAB IDENTIFICATION** The humerus is rounded; the me-

sonotum is not constricted; the scutellar scale is narrow and rounded posteriorly; the T2 disc has longitudinally ovate, white setal patches; and the pygidial sculpture is rugose with the apex microreticulate. **MALE** Unknown. **NOTES** Based on overlapping distribution and similarity to *T. ornatipennis* (3.6.16), this might be the female of *T. barbata* (5.7.14.1).

Timulla wileyae

Timulla dubitata

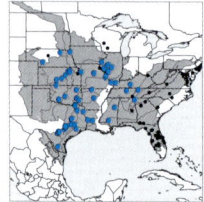

3.5.10.2—*Timulla dubitata*

TIMM-EWE-LUH DOO-BIT-AHT-UH

ETYMOLOGY From the Latin *dubitatus* "doubted." **FIELD IDENTIFICATION** Difficult. The body size is usually larger than *T. wileyae* and the mesonotum is strongly constricted. **LAB IDENTIFICATION** The humerus is rounded; the mesonotum is constricted; the scutellar scale is narrow and rounded posteriorly; the T2 disc has longitudinally ovate, white setal patches; and the pygidial sculpture is rugose with the apex microreticulate. **MALE** (5.6.7.2, 5.7.16.4). Usually darker than female. **NOTES** This species is common throughout the central and eastern USA. This color form is most commonly seen in Texas and the Great Plains, while most other populations fit in the Eastern FCS (3.6.16.1).

3.6 EASTERN FEMALES COLOR SYNDROME

The Eastern Females Color Syndrome (FCS) is represented by females with the mainly reddish or orange cuticle with scant and/or shortened setae. The head and mesosoma are usually uniformly red or orange in color, and the metasoma is mainly red or orange, often with lighter yellowish patches on T2 and black and silvery setal bands. Members of this FCS mostly occur east of the Rocky Mountains and are dominant east of the Mississippi River. If you are on the eastern seaboard, this is the only mutillid color syndrome you are likely to encounter—sorry if you were hoping to spot a Desert or Madrean FCS while beachcombing in Nantucket.

This FCS is more superficially similar (at least in terms of the head, mesonotum, and tergite 1) in coloration and setae density/length to true ants (Formicidae) than many other syndromes. In a paper of ours examining aposematic coloration warnings and potential predators, we suggested that iguanian lizards, which are diurnal, visually stimulated sit-and-wait hunters, are more likely to go after ants and ant look-alikes than other small predators (birds, mammals, other lizard groups, etc.). It is interesting that the vast majority of iguanian diversity and species richness in North America is found in western North America and Mesoamerica, in comparison with eastern USA and Canada, which are dominated by skinks and anoles.

In the USA, there are 45 velvet ant species in six genera that fit in this color syndrome, which can most easily be confused with the *Timulla*-like or Cryptic FCS. Unlike the *Timulla*-like FCS, the metasoma is concolorous with the mesosoma. Many species in the central and eastern USA have the metasomal cuticle color variable, and they are included in both color syndromes. The cryptic and nocturnal species are often more difficult to separate from the Eastern FCS; they are usually small-bodied, have the body color more brown-tinted than red or orange, and have sparse metasomal setae that rarely form a particular pattern. None of the Myrmosinae (3.7.7, 2.7.9) or *Lomachaeta* (3.7.1) species are included in this FCS.

Of the 45 females in the *Timulla*-like FCS, 37 (82%) are known from both sexes. In the USA, the males are distributed in the following color syndromes: *Sphaeropthalma*-like Male Color Syndrome (MCS) (49%), Eastern MCS (43%), *Timulla*-like MCS (19%), Western MCS (14%), and *Ephuta*-like MCS (5%).

The Eastern FCS is rare in northern Mexico and does not seem to occur in Central America (see 7.3), though a few species are intermediate between the Eastern and Madrean FCS (see 7.3.7). This general coloration is uncommon around the world, where they usually seem intermediate with the Madrean FCS (see 7.4.1; 8.3.14) or cryptic forms.

The key below is useful for figuring out which page to visit next.

—Genus *Dasymutilla*—T1 shape petiolate, T6 with defined pygidial plate; T2 fringe with setae simple

3.6.1 Large-bodied with black cuticle and dense setae; without silver setae on metasoma .... *D. occidentalis*

3.6.2 Head armed with posterolateral tubercle; medium-bodied (usually more than 8 mm) with dense propodeal sculpture .... *D. gibbosa, D. quadriguttata, D. angulata, D. rubicunda*

3.6.3 Head angular posterolaterally; small-bodied (usually less than 6 mm) with sparse propodeal sculpture (areolations or scattered tubercles) .... *D. canella, D. macilenta, D. arenneronea*

3.6.4 Head rounded, slender, covered with silver setae .... *D. archboldi*

3.6.5 Head rounded, not especially large or small, covered with orange setae like mesosoma .... *D. ursus, D. asopus, D. bioculata, D. nigripes*

3.6.6 Head rounded, large and quadrate .... *D. scaevola, D. creon*

—Genus *Myrmilloides*—Mesosoma subrectangular; head massive, armed with ventral teeth

3.6.7 T2 with large silvery setal patches .... *M. grandiceps*

—Genus *Pseudomethoca*—Mesosoma fiddle-shaped; T1 shape sessile

3.6.8 Head massive, mesosoma especially short and angular, T1 especially wide .... *Ps. oceola*

3.6.9 T2 with distinct, widely separated lighter orange spots .... *Ps. oculata, Ps. wickhami*

3.6.10 Head not quite as wide, mesosoma usually more rounded, T2 usually without obvious, separated spots .... *Ps. simillima, Ps. meritoria, Ps. sanbornii, Ps. propinqua, Ps. paludata*

—Genus *Sphaeropthalma*—T2 fringe plumose

3.6.11 T1 shape subsessile; pygidium without defined plate .... *S. pensylvanica, S. auripilis, S. contracta*

—Genus *Ephuta*—T1 shape slender cylindrical

3.6.12 Eyes large, mesosoma ovate .... *E. pauxilla, E. sabaliana, E. slossonae*

—Genus *Timulla*—Mesosoma rectangular; T1 shape sessile; head narrow, unarmed ventrally

3.6.13 Propodeum clearly narrower than pronotum .... *T. navasota navasota*

3.6.14 Propodeum clearly wider than pronotum .... *T. floridensis, T. ferrugata*

3.6.15 T1 fringe black, apical tergites often bright orange .... *T. euterpe, T. euphrosyne, T. dubitatiformis*

3.6.16 Humerus rounded .... *T. dubitata, T. barbigera, T. contigua, T. ornatipennis*
3.6.17 Humerus angular .... *T. oajaca, T. vagans, T. suspensa, T. leona, T. tyro*

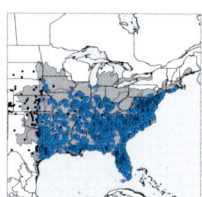

3.6.1.1—*Dasymutilla occidentalis*
DAH-ZEE-MEW-TILL-UH OX-SID-ENT-AHL-ISS

ETYMOLOGY The common name for this species is Common Eastern Velvet Ant, but the specific epithet is Latin for "western." The species was originally named by the Swedish entomologist Caroli Linnaeus in 1758: even the eastern half of North America is farther west than Sweden. **FIELD IDENTIFICATION** Easy. This species is larger-bodied and more densely setose than the other Eastern FCS species. Most importantly, the body lacks silvery setae; all the other Eastern FCS *Dasymutilla* members have the tip of the metasoma with silvery setae. **LAB IDENTIFICATION** The antennal scrobe has a dorsal carina; the vertex is unarmed; the mesosoma is longer than wide and has a distinct scutellar scale; the mid and hind femoral apices are rounded; and the pygidium is striate. **MALE** (5.2.1.1). In a confusing twist, because it resembles the female, the *D. occidentalis* male is treated in the Western, rather than Eastern, MCS (5.6). The Eastern MCS and Eastern FCS patterns are fundamentally different in color but have matching geographic distribution and species composition. **NOTES** This is perhaps the world's most famous velvet ant and is seen in many field guides and textbooks. They were originally thought to be parasitoids of bumblebees, but more recent (and more confident) host studies revealed that they attack the Eastern Cicada-killer wasp (*Sphecius speciosus*) and other large predacious wasps. Populations in the Great Plains and some northwestern areas have the dorsal setae shaggier and yellow to orange in color and are treated in the Western FCS (3.2.8.2).

▼ 3.6.2—*Dasymutilla gibbosa* Eastern females cluster

These species occur mainly in the eastern USA. They have the head armed with a transversely linear or C-shaped tubercle posterolaterally. Unlike the *D. canella* cluster (3.6.3) they have the propodeal sculpture more densely punctate and setose. They can also be easily confused with the *D. ursus* cluster (3.6.4), especially in *D. quadriguttata*, which often has the vertex tubercles low and inconspicuous.

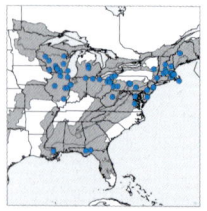

3.6.2.1—*Dasymutilla gibbosa*
DAH-ZEE-MEW-TILL-UH GIBB-OWE-SUH

ETYMOLOGY From the Latin *gibbus* "humped." The species was originally described based on the male only, so this might be a reference to the petiolate T1 shape. **FIELD IDENTIFICATION** Difficult. The head is wider and generally more sharply angular than most specimens of *D. quadriguttata* (but less sharply angular than *D. angulata* or *D. ru-*

bicunda) and is most prevalent in the northern states. Additionally, in *D. gibbosa,* the T2 disc has two large, pale orange patches that are often coalescent mesally; the T2 fringe and T3 are black dorsally and T4–5 have the setae mostly silvery. **LAB IDENTIFICATION** The head is about as wide as the mesosoma; the mandible is straight and bidentate; the vertex has the posterolateral corner angular, with a smooth sublinear tubercle; the mesosoma is longer than wide and has a large scutellar scale with short carinae anterior and lateral to the scale; the propodeum is sparsely setose with moderately dense areolations; and the pygidium is striate. **MALE** (5.4.1.1, 5.5.1.2). Much darker than female. **NOTES** This is the only Eastern FCS *Dasymutilla* species with distinctly white tibial spurs, although some other species can have pale yellow-brown spurs. Members of *Dasymutilla* usually have dark tibial spurs, and members

of their South American sister genus, *Traumatomutilla,* usually have white spurs. Most males of *D. gibbosa* fit in the *Ephuta*-like MCS (5.4.1.1) and look similar to many *Traumatomutilla* species. This species is an important target for molecular phylogenetic studies, but so far, fresh specimens with intact DNA have not been available. This is an uncommon, generally northeastern species, but some populations have been found in the Florida panhandle.

3.6.2.2—*Dasymutilla angulata*
DAH-ZEE-MEW-TILL-UH AYN-GEW-LAW-TUH

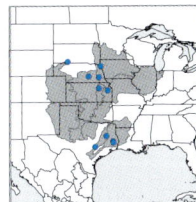

ETYMOLOGY From the Latin *angulatus* "angular," likely referring to the head shape. **FIELD IDENTIFICATION** Difficult. This species has a sharply angular head and usually has the body cuticle duller orange-brown compared with many others in the Eastern FCS that have a more reddish tint to the body color. The T2 disc has two large pale orange patches that are often coalescent mesally; the T2 fringe is entirely black and T3–5 have the setae mostly silvery. **LAB IDENTIFICATION** The head is narrower than the mesosoma; the mandible is straight and bidentate; the vertex has the posterolateral corner sharply angular, with a sharp tubercle; the mesosoma is longer than wide and has a large scutellar scale with carinae anterior and lateral to the scale; the propodeum is sparsely setose with moderately dense areolations; and the pygidium is striate. **MALE** Unknown. **NOTES** This is a rare species known only from the female sex and occurring mainly near the transition areas between the Great Plains and eastern USA.

3.6.2.3—*Dasymutilla rubicunda*
DAH-ZEE-MEW-TILL-UH ROO-BICK-UNN-DUH

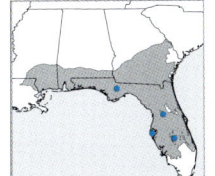

ETYMOLOGY From the Latin *rubicundus* "red" or "ruddy," in reference to the body color. **FIELD IDENTIFICATION** Difficult. This southeastern species is remarkably similar to *D. quadriguttata* but has larger and sharper tubercles on the vertex, and has a brush of thickened dark reddish-brown setae on the propodeum. The tergal fringes have the setae mostly silvery with small mesal black patches. **LAB IDENTIFICATION** The head is nearly as wide as the mesosoma; the mandible is straight and bidentate; the vertex has the posterolateral corner sharply angular, with a large C-shaped tubercle; the mesosoma is longer than wide and has a large scutellar scale with carinae anterior and lateral to the scale; the propodeum has a dense brush of dark reddish-brown setae; and the pygidium is striate. **MALE** Unknown. **NOTES** This southeastern species is very rare and is apparently closely related to *D. campanula* (3.2.8.4) and *D. parksi* (3.5.1.2). Like those species, the male is currently unknown but will probably be difficult to separate from *D. quadriguttata*.

Photo by Efram Goldberg

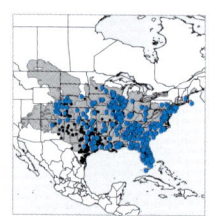

3.6.2.4—*Dasymutilla quadriguttata*
DAH-ZEE-MEW-TILL-UH QUAD-RIGG-OO-TAW-TUH

ETYMOLOGY From the Latin *quadrum* "four" and *gutta* "spot," in reference to the female coloration. **FIELD IDENTIFICATION** Difficult. This species has the head with a more rounded appearance than the others in this cluster, but that makes it more easily confused with other clusters. As the name suggests, there are usually four light orange cuticular spots on T2, but these are often indistinct, and some other species, like *D. asopus* (3.6.5.2), often have similar spots in the eastern USA. The tergal setal markings are quite variable in this species, making identifications based on color challenging. **LAB IDENTIFICATION** The head is usually narrower than the mesosoma; the mandible is straight and bidentate; the vertex has the posterolateral corner somewhat rounded, with a smooth linear tubercle; the mesosoma is longer than wide and has a large scutellar scale with carinae anterior and lateral to the scale; the propodeum is sparsely setose with moderately dense areolations; and the pygidium is striate. **MALE** (5.6.4.1). Much darker than female. **NOTES** The Eastern FCS populations of this species were formerly recognized under

many different names and were only recently synonymized after DNA data became available. Most populations of this species from the central USA fit in the *Timulla*-like FCS (3.5.1.1). This is the most common Eastern FCS species with head tubercles. The head often looks more rounded than other species treated here, but vertex tubercles are always linear and more closely spaced to each other than the distance between the epaulets on the pronotum. The scutellar scale and accessory carinae are distinct, and the propodeum is areolate to punctate with moderate setae.

3.6.3—*Dasymutilla canella* Eastern females cluster

These species occur mainly in the eastern USA, with one in the northeast and two in the southeast. They have the head angular posteriorly, without well-defined linear or C-shaped tubercles. They also have the propodeal sculpture modified, either with broad and shallow areolations or with many scattered tubercles. They can most easily be confused with the *D. gibbosa* cluster (3.6.2). These are generally small and relatively dull-colored members of *Dasymutilla*; they could be confused with the Cryptic FCS.

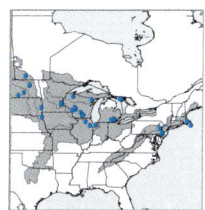

3.6.3.1—*Dasymutilla canella*
DAH-ZEE-MEW-TILL-UH CAN-ELL-UH

ETYMOLOGY Apparently from the Latin *canela* "cinnamon," in reference to the reddish-brown body color. **FIELD IDENTIFICATION** Often difficult. This species is generally smaller-bodied than those in the previous cluster and occurs farther north than the others in this cluster. The T2 disc has two pale orange patches, which are often coalescent mesally; the T2 fringe is entirely black and T3–5 have the setae mostly silvery. **LAB IDENTIFICATION** The head is about as wide as the mesosoma; the mandible is straight

and bidentate; the vertex has the posterolateral corner weakly angulate, with the lateral margins incurving toward this angle; the mesosoma is longer than wide and has a large scutellar scale with carinae anterior and lateral to the scale; the propodeum is sparsely setose with wide complete areolations; and the pygidium is striate. **MALE** (5.6.2.2). Much darker than female. **NOTES** This species is mostly known from the northeastern USA.

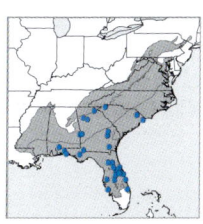

3.6.3.2—*Dasymutilla macilenta*
DAH-ZEE-MEW-TILL-UH MASS-ILL-ENN-TUH

ETYMOLOGY From the Latin *macilentus* "mean," "thin," or "meager." This name was based on the male sex and likely referred to the sparse body setae. The female used to be named *D. chattahoochei*, for the Chattahoochee River. **FIELD IDENTIFICATION** Very difficult. It can be separated from *D. arenneronea* by having the T2 fringe setae mostly black. The head tubercle shape and propodeal sculpture separate these two species from all the other members of the Eastern FCS, but those traits are difficult to recognize in the field. **LAB IDENTIFICATION** The head is about as wide

Photo by Efram Goldber

as the mesosoma; the mandible is straight and bidentate; the vertex has the posterolateral corner distinctly angulate, with the lateral margins basically straight leading to this angle; the mesosoma is longer than wide and has a small scutellar scale without carinae anterior to the scale; the propodeum is sparsely setose with scattered tubercles; and the pygidium is striate. **MALE** (5.6.2.3). Much darker than female. **NOTES** This is a moderately common southeastern species. Like *D. arenerronea*, the head is angular, and the propodeum is mostly bare with scattered tubercles. Until recently, these females were thought to be conspecific (see below).

3.6.3.3—*Dasymutilla arenerronea*
DAH-ZEE-MEW-TILL-UH UH-RENN-ERR-OWE-NEE-UH

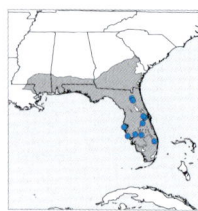

ETYMOLOGY From the Latin *harena* "sand" and *erroneus* "wandering." **FIELD IDENTIFICATION** Very difficult. It can be separated from *D. macilenta* by having the T2 fringe setae mostly silver with only a small black patch mesally. The head tubercle shape and propodeal sculpture separate these two species from all the other members of the Eastern FCS, but those traits are difficult to recognize in the field. **LAB IDENTIFICATION** The head is about as wide as the mesosoma; the mandible is straight and bidentate; the vertex has the posterolateral corner angulate, with the lateral margins incurving toward this angle; the mesosoma is longer than wide and has a small scutellar scale without carinae anterior to the scale; the propodeum is sparsely setose with scattered tubercles; and the pygidium is striate. **MALE** (5.6.2.4). Much darker than female. **NOTES** This species is rarer and apparently has a more restricted distribution than *D. macilenta*. Originally, the female was separated from *D. macilenta* (as *D. chattahoochei*) only by having

Photo by Efram Goldberg

a slightly more rounded head shape. Mickel (1928) could not find any other differences, so he synonymized *D. arenerronea*. Later, Krombein (1954) associated a male with this *D. chattahoochei* complex. In 2009, other females of *D. chattahoochei* were matched with *D. macilenta* (5.6.2.3), leaving two reliably different males associated with the "same" female. A molecular phylogenetic analysis and re-examination of females with those formerly used names showed that they really were different species, so *D. arenerronea* was "resurrected."

3.6.4.1—*Dasymutilla archboldi*
DAH-ZEE-MEW-TILL-UH ARCHBOLD-EYE

ETYMOLOGY This species was named for the Archbold Biological Station in Florida. **FIELD IDENTIFICATION** Often difficult. The small, rounded head covered with silvery setae is diagnostic, but these setae are often indistinct, and other Eastern FCS species sometimes have a few scattered silvery head setae. **LAB IDENTIFICATION** The head is narrower than the mesosoma; the mandible is straight and bidentate; the vertex is rounded and unarmed posteriorly; the mesosoma is longer than wide and has a small scutellar scale without carinae anterior to the scale; the propodeum is sparsely setose with scattered tubercles; and the pygidium is striate. **MALE** (5.5.1.1). With more extensive silver setae than the female. **NOTES** This is a small, rare species restricted to

central Florida sandhills. The mesosoma is similar to *D. arenerronea* and *D. macilenta* in having a small scale without accessory carinae and the propodeum mostly bare with scattered tubercles. The first associated male (5.7.4.3) was much larger than the female and could not be reliably separated from *D. ursus* (it was *D. ursus*). The correct male was matched up using DNA sequences and is small and silvery, just like the female. (5.5.1.1).

Photo by Efram Goldberg

3.6.5—*Dasymutilla ursus* Eastern females cluster

These species are widespread and common in the central and eastern USA. They have the head unarmed posteriorly and generally narrower than the mesosoma (in *D. nigripes* the head is often slightly wider than the mesosoma). They can most easily be confused with the *D. gibbosa* (3.6.2) and *D. scaevola* (3.6.6) clusters. Some of these species loosely resemble species in the *Pseudomethoca simillima* clusters (3.6.10), but these *Dasymutilla* species have the T1 shape petiolate.

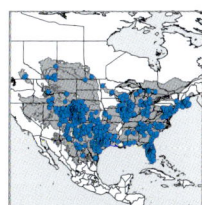

3.6.5.1—*Dasymutilla ursus*
DAH-ZEE-MEW-TILL-UH URR-SUSS

ETYMOLOGY From the Latin *ursus* "male bear," likely based on the hairy body. Until recently, this species was known as *Dasymutilla vesta*. **FIELD IDENTIFICATION** Very difficult. The mesosoma is longer than wide and has a small scutellar scale. The head is generally narrower than *D. nigripes*, and the mesosoma is less robust than *D. bioculata*. In most specimens, the fringes of T2–5 have the setae mostly silvery, with black setae mesally on the T2 fringe. In northern and western areas, the T2 fringe is often entirely black and the T3–5 setae are mostly silvery. The T2 disc often has two lighter orange patches (these are often large and coalescent mesally); it is rarely darkened anteriorly. **LAB IDENTIFICATION** The head is narrower than the mesosoma; the mandible is straight and bidentate; the vertex is rounded and unarmed posteriorly; the meso-

soma is longer than wide and has a small scutellar scale without carinae anterior to the scale; the propodeum is moderately areolate with sparse orange setae; and the pygidium is striate. **MALE** (5.6.3.1, 5.7.4.3). Much darker than female. **NOTES** In raw numbers, this is probably the most common species in the eastern USA. Its comparatively smaller body size and

difficulty in identification make it less commonly represented than species like *D. occidentalis* or *D. quadriguttata* in online resources, like inaturalist.com.

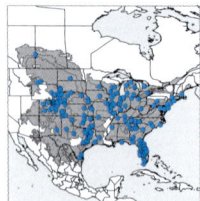

3.6.5.2—*Dasymutilla asopus*
DAH-ZEE-MEW-TILL-UH AY-SOHP-USS

ETYMOLOGY In Greek mythology, Asopus is a name for the gods of rivers. **FIELD IDENTIFICATION** Sometimes difficult. The mesosoma is as wide as long and lacks a scutellar scale. The tergal setae are variable, but the T2 fringe is usually mostly black with a mesal patch of whitish setae. Unlike most specimens of other species in this cluster, T2 often has four separated light orange cuticular spots. **LAB IDENTIFICATION** Unlike other Eastern FCS females, the mandible is tridentate. The head is about as wide as the mesosoma; the vertex is rounded and unarmed posteriorly; the mesosoma is as wide as long and lacks a scutellar scale; the propodeum is moderately areolate with uniformly short setae; and the

pygidium is rugostriate. **MALE** (5.2.2.1, 5.6.4.2, 5.7.4.1). Usually darker than female. **NOTES** Males of this species have more variable coloration than the females and were previously divided into three subspecies, based on color patterns. This species has been associated with bee hosts in the genera *Anthophora*, *Dianthidium*, and *Paranthidium*.

Photo by Efram Goldberg

3.6.5.3—*Dasymutilla bioculata*
DAH-ZEE-MEW-TILL-UH BYE-OCK-EW-LAW-TUH

ETYMOLOGY From the Latin *bi* "two" and *oculus* "eye," in reference to the two orange spots on T2 in many males. **FIELD IDENTIFICA-TION** Very difficult. The mesosoma is longer than wide and has a large scutellar scale with an anterior transverse carina. The head is generally narrower than in *D. nigripes*, and the mesosoma is more robust than in *D. ursus*. The tergal setae are variable, but in southern and western populations, the tergal fringes (at least T2–3) tend to have the setae mainly black. The T2 disc often has apparent lighter orange spots that are widely fused together, with a more or less distinct blackish patch anteromesally. **LAB IDENTIFICA-TION** The head is narrower than the mesosoma; the mandible is straight and bidentate; the vertex is rounded and unarmed posteriorly; the mesosoma is longer than wide and has a large scutellar scale

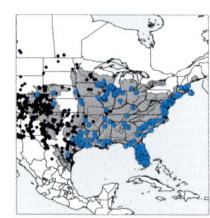

and transverse wavy carina anterior to the scale; the propodeum is moderately areolate with somewhat dense orange setae; and the pygidium is striate. **MALE** (5.2.2.2, 5.6.4.3). Usually darker than female. **NOTES** This is the most widespread and variable *Dasymutilla* species, also being represented in the Texan FCS (3.3.4.3) and Western FCS (3.2.3.4, 3.2.8.1). These Eastern FCS forms used to be called *D. bioculata* in the central USA, *D. lepeletieri* in the eastern USA, and *D. praegrandis* in eastern Texas. This is a sand wasp parasitoid, and throughout its range, *D. bioculata* is commonly encountered in sand dune habitats.

3.6.5.4—*Dasymutilla nigripes*

DAH-ZEE-MEW-TILL-UH NIH-GRIP-EEZ

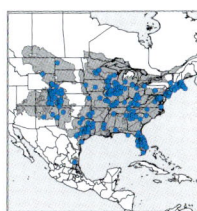

ETYMOLOGY From the Latin *nigra* "black" and *pes* "foot," in reference to the black legs of most populations. **FIELD IDENTIFICATION** Very difficult. The mesosoma is longer than wide and lacks a scutellar scale. The head is generally wider than in *D. ursus*, and the mesosoma is less robust than in *D. bioculata*. In eastern areas, the T2 fringe is usually entirely black and the T3–5 setae are mostly silvery. In western areas, the fringes of T2–4 each have a mesal patch of whitish setae. The T2 disc is usually uniformly reddish-orange (at most slightly darkened anteriorly). **LAB IDENTIFICATION** The head is slightly wider than the mesosoma; the mandible is elongate, curved, and bidentate; the vertex is rounded and unarmed posteriorly; the mesosoma is longer than wide and lacks a scutellar scale; the propodeum is moderately areolate with uniformly short setae; and the pygidium is striate. **MALE** (5.6.3.2, 5.7.4.4). Much darker than female. **NOTES** This species is often nearly as common as *D. ursus* and shares most of its distribution with that species. This species has been recognized as a parasitoid of philanthine wasps in the genera *Cerceris* and *Philanthus*. Specimens from Florida often have the legs reddish, while most northern and central specimens have black legs.

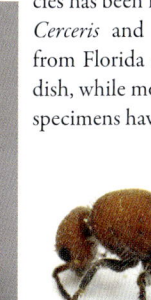

Photo by Efram Goldberg

▼ 3.6.6—*Dasymutilla scaevola* Eastern females cluster

Dasymutilla creon is found in the southern central USA and *D. scaevola* is widespread in the central and eastern USA. They have the head wider than other Eastern *Dasymutilla* clusters and lack tubercles or sharp angles on the posterolateral portion of the head. Some species in the *D. gibbosa* (3.6.2) or *D. ursus* (3.6.5) clusters have the head relatively wide, like these. Because of the wide head, they can loosely resemble some *Pseudomethoca* clusters (3.6.8, 3.6.9, 3.6.10), but these *Dasymutilla* have the T1 shape petiolate.

3.6.6.1—*Dasymutilla scaevola*

DAH-ZEE-MEW-TILL-UH SKEE-VOE-LUH

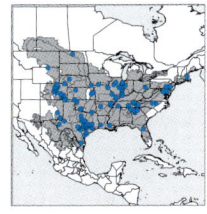

ETYMOLOGY This name was derived from the Latin *scaevus*, "left" or "clumsy." **FIELD IDENTIFICATION** Sometimes difficult. This species is more widespread than *D. creon* and has a large white setal spot on T1. In that respect, and in the head shape, it loosely resembles *S. pensylvanica* (3.6.11.1), but it has the T1 shape more distinctly and narrowly petiolate. The body color is duller orange-brown and the dorsal setae are usually lighter orange and denser than most Eastern FCS *Dasymutilla* species. **LAB IDENTIFICATION** This is the only Eastern FCS species with the mid and hind femoral apices truncate. It lacks a scutellar scale and has the pygidial sculpture microreticulate. **MALE** (5.6.1.1). Much darker than female.

NOTES This species is widespread in the eastern and central USA. It can often be confused with *D. snoworum* in the Madrean FCS (3.4.4.1) depending on whether the head and mesosoma setae are considered pale orange (*D. scaevola*) or golden (*D. snoworum*). Further studies might eventually recognize them as variants of a single species. One gynandromorph specimen was studied from Texas. It has the head of a male, but the rest of the body is female; pretty neat!

FAR LEFT: *Dasymutilla scaevola* female

LEFT: *Dasymutilla scaevola* gynandromorph

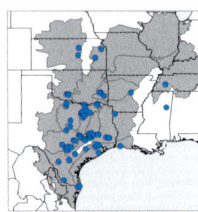

3.6.6.2—*Dasymutilla creon*
DAH-ZEE-MEW-TILL-UH KREE-ONN

ETYMOLOGY This species could have been named after one of many important figures from Greek mythology or history named Creon. **FIELD IDENTIFICATION** Sometimes difficult. The head seems even wider than that of *D. scaevola*, and the T1 fringe has sparse black and white setae. Furthermore, the body sculpture is coarser and the mesosoma has a distinct scutellar scale. **LAB IDENTIFICATION** This is the only *Dasymutilla* species in the USA with the pygidium covered with short setae. Additionally, the mid and hind femoral apices are rounded and the mesosoma has a distinct scutellar scale. **MALE** (5.7.1.1). Nearly identical to female. **NOTES** Of all the Eastern FCS *Dasymutilla* species, males and females of *D. creon* are the most similar in color, except maybe *D. occidentalis* (3.6.1.1, 5.2.1.1). In *Dasymutilla*, dual sex-limited mimicry is considered to be a derived trait, so *D. creon* may be one of the oldest species with the Eastern FCS. Previous authors had correctly guessed this sex association (the female was formerly called *D. bollii*), but the association was not formalized until a mating pair was collected by James Pitts.

 ### 3.6.7.1—*Myrmilloides grandiceps*
MEER-MILL-OI-DEEZ GRAN-DISS-EPPS

ETYMOLOGY From the Latin *grandis* "large" and *-ceps* "-headed." **FIELD IDENTIFICATION** Easy. The giant head is unique, and the short-winged males are even easier to recognize. They are larger in size than any of the *Pseudomethoca* species that possess teeth on the gena or postgena. **LAB IDENTIFICATION** The gena and postgena are each armed with a ventral tooth. **MALE** Similar to female. **NOTES** The males are short winged and generally more rarely seen than females. This species is variable in color and is also treated in the *Timulla*-like FCS (3.5.2.1). Some individuals have an obscure resemblance to the Texan FCS (3.3.6.1), since the head and mesosoma are often darker than the metasoma in specimens from western Texas.

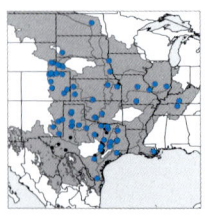

This species is common in the Texas Panhandle and the southern High Plains, and often found in the same communities as abundant populations of harvester ant species (*Pogonomyrmex barbatus* and *Po. rugosus*). After a while, you can easily pick out *Myrmilloides* among a bunch of ants (even from a distance), because they have a distinctive haphazard gait, which is much less regimented than that of the ants. The Texan FCS of this species resembles *Po. rugosus* particularly well, while the Eastern FCS is pretty good at blending in with *Po. barbatus*.

3.6.8.1—*Pseudomethoca oceola*
SOO-DOE-METH-OWE-KUH OWE-SEE-OWE-LUH

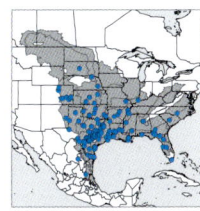

ETYMOLOGY This species was named after Osceola, an influential leader of the Seminole people in Florida. **FIELD IDENTIFICATION** Sometimes difficult. This species is "chunkier" than other *Pseudomethoca*, the mesosoma is more starkly angular, and T1 is especially wide. They are also usually larger in size than other *Pseudomethoca* species. **LAB IDENTIFICATION** The humeral carina is strongly defined with a distinct obtuse angle; the mesosomal pleura are often covered with short dense silvery setae; the lateral propodeal margin has a row of short teeth; T1 is nearly as wide as T2; and the pygidial plate is rugose. **MALE** (5.6.5.1, 5.7.7.2). Usually darker than female. **NOTES** This is a widespread species, being found as far south as Campeche in Mexico. Males seem to be more variable in color than females, fitting in both the Eastern and *Sphaeropthalma*-like MCS. This color variation might be even more extreme if *Ps. flavida* (5.5.8.4) from the *Timulla*-like MCS is eventually recognized as a synonym of *Ps. oceola*.

ABOVE: *Pseudomethoca oceola* from Texas.

LEFT: *Pseudomethoca oceola* from Florida.

3.6.9—*Pseudomethoca oculata* Eastern females cluster

Pseudomethoca oculata occurs only in the southeastern USA, but *Ps. wickhami* is widespread in the central and eastern USA. These species can be recognized by having small, separated bright yellow-orange spots on the T2 disc. They could be confused with some populations of *Ps. oceola* (3.6.8.1), but the shape of the mesosoma and T1 is a bit more slender. They can also be difficult to separate from the *Ps. simillima* cluster (3.6.10). Some Madrean FCS *Pseudomethoca* species, especially *Ps. mulaiki* (3.4.13.2) and *Ps. sonorae* (3.4.14.2), also resemble these species.

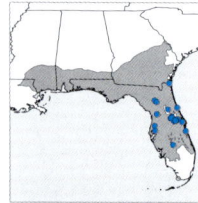

3.6.9.1—*Pseudomethoca oculata*
SOO-DOE-METH-OWE-KUH OCK-EWE-LAW-TUH

ETYMOLOGY From the Latin *oculatus* "having eyes," in reference to the spots on T2. **FIELD IDENTIFICATION** Usually easy. The T2 spots are relatively unique; additionally, the head is much wider than the rounded mesosoma, and the dorsal body setae are mostly blackish. **LAB IDENTIFICATION** The humeral carina is weakly defined; the mesosomal pleura have extensive short, dense silvery setae; the lateral propodeal margin lacks a row of short teeth; the T2 disc has two dorsally situated medium-sized patches of lighter orange cuticle covered with pale yellow setae; and the pygidial plate sculpture has wavy posteriorly divergent striations. **MALE** (5.7.7.1). Somewhat similar to female, without distinct spots or silvery setae on the metasoma. **NOTES** This southeastern species is similar in sculpture and mesosoma shape to *Ps. sanbornii* (3.6.10.3), but it has a much wider head and two distinct orange spots on T2.

Photo by Dan Rouche

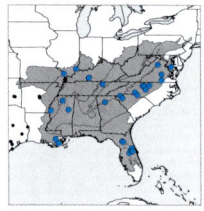

3.6.9.2—*Pseudomethoca wickhami*
SOO-DOE-METH-OWE-KUH WICKHAM-EYE

ETYMOLOGY This species was named after H. F. Wickham, who collected the type specimen in the late 1800s. This Eastern form used to be named *Ps. fattigi* and was the first form to be associated with *Ps. vanduzei*, which was initially named based on the male sex only. **FIELD IDENTIFICATION** Usually easy. The orange T2 spots are very widely separated, and the dorsal setae are mostly bright orange. **LAB IDENTIFICATION** The humeral carina is strongly defined with a sharp tooth-like angle; the mesosomal pleura are mainly bare; the lateral propodeal margin has a row of short teeth; the T2 disc has two

Photo by Efram Goldberg

medium-sized patches of lighter orange cuticle covered with pale yellow setae at the extreme lateral margins; and T6 is convex without a defined pygidial plate. **MALE** (5.5.7.1). Much darker body color than female. **NOTES** *Pseudomethoca wickhami* has many unique features when compared with other North American species, especially in the male mandible shape, undefined pygidium in females, and postgenal armature in both sexes. Many related species, however, occur in Central and South America. This group of species might eventually be recognized as a distinct genus. This species has been recognized as a parasite of the halictic bee *Nomia maneei* Cockerell.

 ### 3.6.10—*Pseudomethoca simillima* Eastern females cluster

These species occur mainly in the central and eastern USA. They can be recognized by having the T2 disc relatively uniform with orange color. They could be confused with some populations of *Ps. oceola* (3.6.8.1), but their mesosoma and T1 shape is a bit more slender. They can also be difficult to separate from the *Ps. oculata* cluster (3.6.9) because they often have large mesally coalescent lighter orange cuticular patches. Sometimes they could be confused with the *Ps. frigida* cluster (3.7.2) in the Cryptic FCS, but Eastern species in that cluster always have a ventral tooth on the gena.

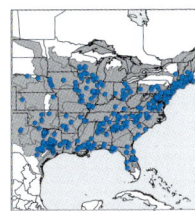

3.6.10.1—*Pseudomethoca simillima*
SOO-DOE-METH-OWE-KUH SEE-MILL-EE-MUH

ETYMOLOGY From the Latin *simillimus* "very similar." The author thought it was similar to eastern populations of *D. bioculata*. **FIELD IDENTIFICATION** Often difficult. This is the commonest species in the cluster. They have the mesosoma somewhat more angular than many others, and T1 and T2 each have a broad posterior band of black setae. **LAB IDENTIFICATION** The humeral carina is strongly defined; the mesosomal pleura are mainly bare; the lateral propodeal margin has a row of short teeth; the T1 and T2 fringes have the setae mostly black; and the pygidial sculpture is striate. **MALE** (5.4.9.1, 5.6.5.2). Much darker in coloration than female. **NOTES** This is a common species in the eastern USA and seems to be active earlier in the year than any other mutillid, based on personal observations by Kevin in Gainesville, Florida. In Sebring, Florida, a bizarrely large aggregation of this species was discovered by Schmidt & Hook (1979). In an open 5 × 10 meter patch of white sand, 263 females were collected in about 1.5 hours. Surprisingly, no males or apparent hosts were found during the event.

Photo by Efram Goldberg

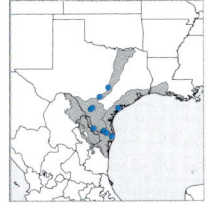

3.6.10.2—*Pseudomethoca meritoria*
SOO-DOE-METH-OWE-KUH MARE-ITT-ORR-EE-UH

ETYMOLOGY From the Latin *meritus* "deserving" or "worthy." **FIELD IDENTIFICATION** Difficult. They have the mesosoma somewhat more angular than many others, and T1 and T2 each have a broad posterior band of black setae. **LAB IDENTIFICATION** The humeral carina is strongly defined; the mesosomal pleura are mainly bare; the lateral propodeal margin has a

row of short teeth; the T1 and T2 fringes have the setae mostly black; and the pygidial sculpture is rugose. **MALE** Unknown. **NOTES** This Texan species is nearly identical to *Ps. simillima*, except that the pygidial sculpture is rugose. It may eventually be revealed to be a synonymous color variant of *Ps. simillima*.

Pseudmethoca meritoria

Pseudmethoca sanbornii

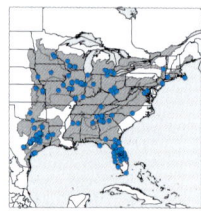

3.6.10.3—*Pseudomethoca sanbornii*
SOO-DOE-METH-OWE-KUH SANBORN-EE-EYE

ETYMOLOGY This species was named after Charles Sanborn, who collected many useful velvet ant specimens from Texas. **FIELD IDENTIFICATION** Often difficult. The mesosoma is more slender and rounded than the others in this cluster, and T1 and T2 have the fringe setae largely silver and only black mesally. **LAB IDENTIFICATION** The humeral carina is weakly defined; the mesosomal pleura are mainly covered with dense, short silvery setae; the lateral propodeal margin lacks a row of short teeth; the T1 and T2 fringes are largely silvery laterally and black only mesally; and the pygidial sculpture is striate or rugose. **MALE** (5.6.5.4, 5.7.7.3). Often similar to female but with silvery setae on the apical tergites. **NOTES** This is a widespread and relatively common species in the eastern and central USA. Their more rounded and smoother mesosoma shape, and comparatively narrow head (for *Pseudomethoca*), make them especially easy to confuse with the genus *Dasymutilla,* but even the slender *Ps. sanbornii* has the T1 shape sessile. This species used to be subdivided into two subspecies, *Ps. s. aeetis* and *Ps. s. sanbornii*, based on variation in the pygidial sculpture. This is the type species of the previously separated genus *Nomiaephagus*, which translates to *Nomia* eater, in reference to the halictid host bees in the genus *Nomia*. If *Ps. frigida* (the type species of *Pseudomethoca*) and their relatives are eventually recognized as a separate genus, the name *Nomiaephagus* could make a comeback.

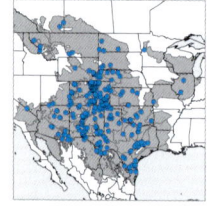

3.6.10.4—*Pseudomethoca propinqua*
SOO-DOE-METH-OWE-KUH PRO-PINK-WUH

ETYMOLOGY From the Latin *propinquus* "near," "neighboring," or "resembling." The author suggested it was similar to the male of *D. vestita*. **FIELD IDENTIFICATION** Often difficult. The fringe of T2 usually has a broad band of black setae, and the mesosomal pleura are mostly blackened; this is the only species in the cluster known to occur west of the Rocky Mountains. **LAB IDENTIFICATION** The humeral carina is weakly defined; the mesosomal pleura are mainly bare and have the cuticle largely blackened; the lateral propodeal margin lacks a weak row of short teeth; the T2 fringe usually has a broad band of black setae; and the pygidial sculpture is rugose. **MALE** (5.2.8.1). Similar to female but with mostly yellow tergal setae. **NOTES** This is widespread in the Great Plains and Intermountain West. The color pattern is intermediate between the Eastern and Western FCS (3.2.11.1).

Pseudmethoca propinqua

Pseudmethoca paludata

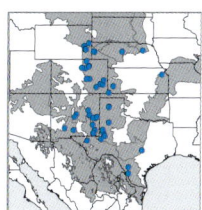

3.6.10.5—*Pseudomethoca paludata*
SOO-DOE-METH-OWE-KUH PAL-LOO-DAH-TUH

ETYMOLOGY This species was apparently named for the paludamen-tum, a cloak worn by military commanders in Rome. **FIELD IDENTIFICA-TION** Often difficult. The fringe of T2 has the setae mostly silvery, with only a small posteromesal black patch, and the meso-somal pleura are usually entirely orange-brown. **LAB IDENTIFICA-TION** The humeral carina is strongly defined; the mesosomal pleura are mainly bare and have the cuticle mainly orange; the lateral propodeal margin has a row of short teeth; the T2 fringe has the setae mostly silver, with a small posteromesal black patch; and the pygidial sculpture is rugose. **MALE** (5.6.5.3). Much darker than fe-male. **NOTES** The color pattern is intermediate between the East-ern and Western FCS (3.2.11.2).

▼ 3.6.11—*Sphaeropthalma pensylvanica* Eastern females cluster

These species are scattered throughout North America. They can be recognized by the plumose tergal fringes and the sessile or subsessile T1 shape. They could easily be confused for many nocturnal species (treated in chapter 6), but they are generally active during the day in the humid eastern or central USA or the cooler northwestern USA.

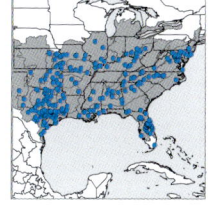

3.6.11.1—*Sphaeropthalma pensylvanica*
S-FAIR-OPP-THAL-MUH PENN-SILL-VANN-ICK-UH

ETYMOLOGY This species was named after the state of Pennsylvania. When the species was described by Lepeletier in 1845, he misspelled the name of the state, and now we are stuck with it. **FIELD IDENTIFICATION** Often easy. The color pattern is unique in the genus *Sphaeropthalma*, espe-cially the small white spot on T1, but the genus could be easily confused with some *Dasymutilla* (3.6.5, 3.6.6). **LAB IDENTIFICATION** The T1 shape is elongate subsessile; the T1 setae are mostly black with a narrow white spot of plumose setae; the T2 fringe is composed of dense, bushy plumose setae; and T6 is convex without a pygidial plate. **MALE** (5.7.11.3). Often sim-

ilar to female but with apical metasomal setae entirely black. **NOTES** This species and *S. auripilis* have the head and mesosoma with coarse sculpture and T1 more or less bell-shaped. This species is common in the central and eastern USA and seems to be mainly diurnal. They are commonly encountered inside houses and are parasites of mud-nesting wasps. A recent paper (Waldren et al. 2020) revealed that this is the only species of the subfamily Sphaeropthalminae known to use phoretic copulation, where the males carry the females during mating. For this reason, males are usually much larger in body size than females.

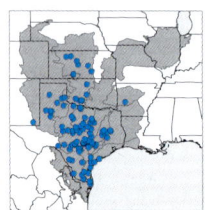

3.6.11.2—*Sphaeropthalma auripilis*

S-FAIR-OPP-THAL-MUH ORR-IPP-EYE-LISS

ETYMOLOGY From the Latin *aurum* "gold" and *pilus* "hair," in reference to the golden setae on the male's metasoma. **FIELD IDENTIFICATION** Sometimes easy. The color pattern is unique among diurnal *Sphaeropthalma* species, especially the wider white band on T1, but it could easily be confused with some *Dasymutilla* (3.6.5, 3.6.6) or nocturnal velvet ants (6.1.8). **LAB IDENTIFICATION** The T1 shape is elongate subsessile; T1 has a broad posterior band of whitish setae; the T2 fringe is composed of dense, bushy plumose setae; and T6 is convex without a pygidial plate. **MALE** (5.7.11.1). Different from female, with the tergal setae mostly golden. **NOTES** This species and *S. pensylvanica* have the head and mesosoma with coarse sculpture and T1 more or less bell-shaped. It is common in Texas and the Great Plains and seems to be active during daylight and after dark.

3.6.11.3—*Sphaeropthalma contracta*

S-FAIR-OPP-THAL-MUH CONN-TRACK-TUH

ETYMOLOGY From the Latin *contractus* "collected" or "accomplished." **FIELD IDENTIFICATION** Often difficult. The color pattern and T1 shape are unique among diurnal *Sphaeropthalma*, but the species could be easily confused with some *Dasymutilla* (3.6.5) and *Pseudomethoca propinqua* (3.6.10.4). **LAB IDENTIFICATION** The T1 shape is short sessile; T1 has sparse setae only; the T2 fringe is composed of plumose setae; and T6 is convex without a pygidial plate. **MALE** (5.7.12.1). Somewhat similar female, with uniformly pale orange setae and cuticle. **NOTES** This is a rare northwestern species with smoother body sculpture than the others in this cluster. The color pattern is similar to many nocturnal species in the western USA (see 6.8).

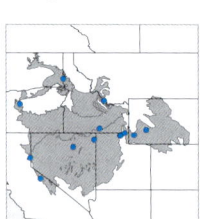

▼ 3.6.12—*Ephuta slossonae* Eastern females cluster

These species occur mainly in the southeastern USA. They are the only *Ephuta* species in the Eastern FCS, and they generally have the cuticle bright reddish-orange and largely blackened legs and antennae. The coloration varies throughout the range of these species, so they should also be compared with the *E. scrupea* cluster (3.7.6), and two of these species are also treated there.

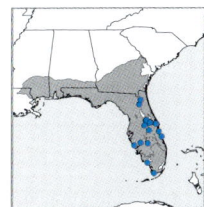

3.6.12.1—*Ephuta slossonae slossonae*
EFF-OO-TUH SLOSSON-EE SLOSSON-EE

ETYMOLOGY This species was named after the American author and entomologist Annie T. Slosson, who collected the type specimen in the 1890s (apparently, the date of collection was not provided on the type specimen). **FIELD IDENTIFICATION** Impossible. **LAB IDENTIFICATION** The vertex has sparse setae only; the postgenal carina is mostly obliterated; the T2 disc lacks any trace of whitish setal patches; the T2 sculpture is very coarse, with the intervals reduced to narrow carinae; T3–5 often have the setae mostly whitish; T6 has a pygidial plate defined by lateral carinae, with the sculpture mostly smooth; and the hypopygium has two small inconspicuous tu-

bercles basally. The eye is very large, and the vertex is relatively short and narrow compared with other *Ephuta*. **MALE** (5.7.13.1). Somewhat similar to female. **NOTES** Schuster (1956) said that females of this species were practically impossible to tell apart from his females of *E. battlei*, but he never got around to describing the females of *E. battlei*. The nominal subspecies, *E. s. slossonae* (Fig. 3.6.12.1a), has the body cuticle more distinctly reddish than the orange-brown *E. s. monochroa* (see Cryptic FCS, 3.7.6.8).

Photo by Efram Goldberg

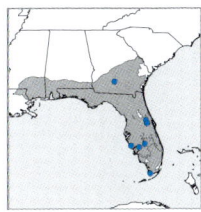

3.6.12.2—*Ephuta sabaliana*
EFF-OO-TUH SAW-BALL-EE-ANN-UH

ETYMOLOGY Named for the Sabalian Zone, a sandy coastal habitat in the southeastern USA. **FIELD IDENTIFICATION** Very difficult. This is apparently the only eastern *Ephuta* species with T3–4 having dense black setal bands and the legs (except the coxae) and antenna (including the scape) cuticle entirely blackened. **LAB IDENTIFICATION** The vertex has sparse setae only; the postgenal carina is mostly obliterated; the T2 disc lacks any trace of whitish setal patches; the T2 sculpture is very coarse, with the intervals reduced to narrow carinae; T3–5 have the setae dense and black, with interspersed golden-orange setae on T5 only; T6 has a pygidial plate defined by lateral carinae, with the sculpture mostly smooth; and the hypopygium has a transverse ridge that is mesally interrupted. **MALE** (5.4.13.12). Much darker than female. **NOTES** This is generally considered to be a rare species, and we have examined only the one female pictured here. Krombein & Evans (1955), however, collected about 100 females in moist muddy salt flats during the spring in Cape Sable, Florida. In general, velvet ants seem to be more frequently active later in the year and when soils are drier. Perhaps some aspect of the host-searching behavior of this species makes them locally abundant in specific habitats only early in the year, which might explain their apparent rarity in collections.

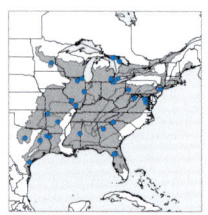

3.6.12.3—*Ephuta puteola*
EFF-OO-TUH PEW-TEE-OWE-LUH

ETYMOLOGY From the Latin *puteolus* "a small well," likely referring to the small dense punctures on the body. **FIELD IDENTIFICATION** Impossible. **LAB IDENTIFICATION** The vertex has sparse setae only; the postgenal carina is mostly obliterated; the T2 disc lacks any trace of whitish setal patches; the T2 sculpture is moderately coarse; T3 has the setae mostly blackish-brown, while T4–5 have the setae largely silvery or golden-orange; T6 has a pygidial plate defined by lateral carinae, with the sculpture mostly smooth; and the hypopygium has a transverse ridge basally. **MALE** Unknown. **NOTES** This species is common (at least for females of *Ephuta*) and widespread in the eastern and central USA. It has been presumed to be the female of *E. pauxilla* (5.4.13.10) based on overlapping distribution, but there is not yet enough evidence for this association. Based on their wide distribution and color variation, the females of multiple species might be currently lumped together under the name *E. puteola*.

▼ 3.6.13.1—*Timulla navasota navasota*

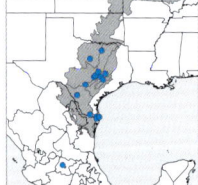

TIMM-EWE-LUH NAW-VUH-SOE-TUH

ETYMOLOGY Named after the city of Navasota, Texas. **FIELD IDENTIFICATION** Sometimes easy. This is generally the only nocturnal *Timulla* species known from southern Texas, and the propodeum is narrower than the pronotum and mesonotum. **LAB IDENTIFICATION** The humerus is rounded; the mesonotum is not constricted; there is no scutellar scale; the T1 fringe is formed of dense white setae; the T2 disc has the setal patches small and subcircular; and the pygidial sculpture is mostly microreticulate. Unlike any other Eastern FCS *Timulla*, the propodeum is narrower than the pronotum and mesonotum. **MALE** (5.5.14.5). Darker than female. **NOTES** In addition to the strange mesosoma shape and nocturnal behavior, this species has a nearly disciform T1 shape that is narrower than other *Timulla*. Because of the duller brownish cuticle color, this species could arguably be placed in the Cryptic FCS, but to make the guide more functional, we have placed all *Timulla* females into the Eastern or *Timulla*-like FCS.

▼ 3.6.14—*Timulla floridensis* Eastern females cluster

These species occur mainly in the eastern USA. They have the propodeum clearly wider than the pronotum. Some species from other clusters can have the propodeum slightly wider than the pronotum. Species in this cluster, however, always have the T1 fringe composed of white setae, ruling out the *T. euterpe* cluster (3.6.15), and always possess a scutellar scale, ruling out the pertinent species in the *T. dubitata* cluster (3.6.16).

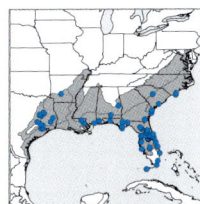

3.6.14.1—*Timulla floridensis*
TIMM-EWE-LUH FLOW-RID-ENN-SISS

ETYMOLOGY This species was named after the state of Florida. **FIELD IDENTIFICATION** Often difficult. This species and *T. ferrugata* have the propodeum clearly wider than the pronotum, but *T. floridensis* is usually smaller in size than *T. ferrugata* and has the mesonotum not constricted; also, the head is often darker than the mesosoma. **LAB IDENTIFICATION** The humerus

is angular; the mesonotum is not constricted; the scutellar scale is usually narrow and posteriorly rounded; the T1 fringe is composed of dense whitish setae; the white setal patches on T2 are short and linear; and the pygidial sculpture is weakly microreticulate with scattered rugae. **MALE** (5.7.15.1). Somewhat similar to female, without black and white metasomal setae. **NOTES** The head and metasomal cuticle are orange-brown but often darker than the pale orange mesosoma, so they could sometimes be confused with the *Timulla*-like FCS, which is very rare in the southeastern USA.

3.6.14.2—*Timulla ferrugata*
TIMM-EWE-LUH FAIR-OO-GAH-TUH

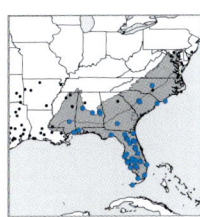

ETYMOLOGY From the Latin *ferrugo* "iron rust," referencing the orange-brown color. **FIELD IDENTIFICATION** Often difficult. This species and *T. floridensis* have the propodeum clearly wider than the pronotum, but *T. ferrugata* is usually larger in size than *T. floridensis* and has the mesonotum constricted; also, the head and mesosoma are always concolorous pale orange. **LAB IDENTIFICATION** The

humerus is angular; the mesonotum is constricted; the scutellar scale is moderately wide and posteriorly rounded; the T1 fringe is composed of dense whitish setae; the white setal patches on T2 are short and linear; and the pygidial sculpture is generally longitudinally rugose. **MALE** (5.7.16.1). Generally darker than female. **NOTES** This species is common in the central and eastern USA and often fits the *Timulla*-like FCS (3.5.10.2).

3.6.15—*Timulla euterpe* Eastern females cluster

Two of these species are located mainly around the Gulf of Mexico, while *T. dubitatiformis* is widespread throughout the whole USA. They are unique in having the T1 fringe with black setae, and two of the species have the apical tergites pale orange and covered with orange setae. Species in the *T. dubitata* cluster can be similar, especially when the T1 setae are sparse (as in *T. contigua*, 3.6.16.3) or rubbed off by wear and tear on a specimen.

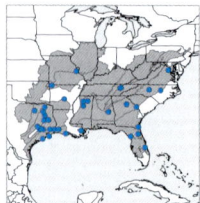

3.6.15.1—*Timulla euterpe*
TIMM-EWE-LUH EH-OO-TAIR-PAY

ETYMOLOGY In Greek mythology, Euterpe was one of the muses and presided over music. **FIELD IDENTIFICATION** Sometimes difficult. The orange apical tergites separate this species from all others except *T. euphrosyne*. Unlike that species, *T. euterpe* is larger-bodied and widespread, has the mesonotum more distinctly constricted (especially near the propodeal spiracle), and has a patch of black setae anteromesally on T2. **LAB IDENTIFICATION** The humerus is angular-rounded; the mesonotum is deeply constricted, especially mesally; there is no scutellar scale; the T1 fringe is formed of dense white setae; the T2 disc lacks white setal patches and has a black setal patch anteromesally; the apical tergites have the setae entirely orange; and the pygidial sculpture is longitudinally rugose and apically microreticulate. **MALE** Unknown. **NOTES** Based on the similarity to *T. barbigera* (3.6.16.2, 5.7.16.2) and overlapping ranges, the male of this species might be *T. compressicornis* (5.7.16.3). Large-bodied individuals of this species are spectacular and easy to recognize, but the smaller individuals are difficult to separate from *T. eyphrosyne*.

3.6.15.2—*Timulla euphrosyne*
TIMM-EWE-LUH EH-OO-FRO-SYE-NAY

ETYMOLOGY In Greek mythology, Euphrosyne was a goddess of good cheer, joy, and mirth. **FIELD IDENTIFICATION** Difficult. This small and rare species is similar to *T. euterpe*, but it has the humerus rounded, and the T2 setae are generally uniformly orange basally. **LAB IDENTIFICATION** The humerus is rounded; the mesonotum is weakly constricted; there is no scutellar scale; the T1 fringe is formed of black setae; the T2 disc lacks whitish setal patches and lacks an anteromesal patch of black setae; the apical tergites have the setae entirely orange; and the pygidial sculpture is mostly microreticulate with faint longitudinal rugae. **MALE** Unknown. **NOTES** This species shares a similar distribution to *T. euterpe*, but it is smaller-bodied and much rarer, being known from only three records (that we know of).

3.6.15.3—*Timulla dubitatiformis*

TIMM-EWE-LUH DOO-BIT-AHT-IF-ORR-MISS

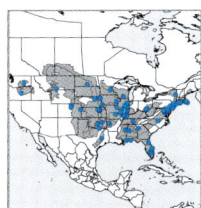

ETYMOLOGY From the Latin suffix *-formis* "shared" or "like," referencing its similarity to *T. dubitata*. **FIELD IDENTIFICATION** Sometimes difficult. The black T1 fringe coupled with black and white apical tergal setae is unique, but this can be hard to recognize in the field, especially when compared with species with sparse white T1 setae (like *T. contigua*, 3.6.16.3) or specimens with the setae worn down by age. **LAB IDENTIFICATION** The humerus is weakly angular; the mesonotum is weakly constricted; the scutellar scale is low and narrow; the T1 fringe is formed of black setae; the T2 disc lacks whitish setal patches; the apical tergites have black and white setal patches; and the pygidial sculpture is mostly microreticulate, with faint rugae. **MALE** Unknown. **NOTES** This widespread species is known from the female only and overlaps in distribution with many unassociated males. It seems likely that this is a cryptic species that will have some of its component populations matched to different species in the future.

 3.6.16—*Timulla dubitata* Eastern females cluster

These species occur mainly in the central and eastern USA. They have the propodeum and pronotum equally wide and the mesosoma with the "round-edged" condition; the humerus (shoulder) is more or less evenly curved. The metasomal cuticle can vary slightly, and they should be compared with the *Timulla*-like FCS *Timulla* clusters (3.5.9, 3.5.10). The mesosomal shape is often difficult to interpret, so they should also be compared with the *T. vagans* cluster (3.6.17).

3.6.16.1—*Timulla dubitata*

TIMM-EWE-LUH DOO-BIT-AHT-UH

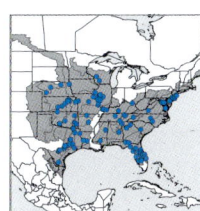

ETYMOLOGY From the Latin *dubitatus* "doubted." **FIELD IDENTIFICATION** Difficult. This species is more widespread, larger-bodied, and has a more deeply constricted mesonotum than most others in this cluster (except *T. barbigera*), and the mesosoma is more slender and rounded than that of *T. barbigera*. **LAB IDENTIFICATION** The humerus is rounded; the mesonotum is constricted; the scutellar scale is narrow and rounded posteriorly; the T1 fringe is formed of dense white setae; the T2 disc has ovate whitish setal patches, which are often sparse or nearly obliterated; and the pygidial sculpture is rugose with the apex microreticulate. **MALE** (5.6.7.2, 5.7.16.4). Usually darker than female. **NOTES** This color form is common in the eastern USA, while many western populations fit the *Timulla*-like FCS (3.5.10.2).

Photo by Efram Goldberg

3.6.16.2—*Timulla barbigera*
TIMM-EWE-LUH BAR-BIJ-AIR-UH

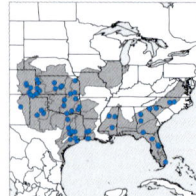

ETYMOLOGY From the Latin *barbiger* "wearing a beard," in reference to the large setal brush on the male scape. **FIELD IDENTIFICATION** Often difficult. The mesosoma is more robust than in the others in this cluster, somewhat similar to that of the *T. vagans* cluster (3.6.17). It is usually larger-bodied than the other species in this cluster. **LAB IDENTIFICATION** The humerus is rounded; the mesonotum is constricted; there is no scutellar scale; the T1 fringe is formed of dense white setae; the T2 disc has the setal patches sparse, nearly obliterated; and the pygidial sculpture is longitudinally rugose. **MALE** (5.7.16.2). Somewhat similar to female. **NOTES** The propodeum of this species is often slightly wider than the pronotum, and the overall appearance is similar to *T. ferrugata* (3.6.14 1). *Timulla barbigera* can be recognized by having the humerus more rounded and the mesosoma without a scutellar scale.

3.6.16.3—*Timulla contigua*
TIMM-EWE-LUH CONN-TIGG-EWE-UH

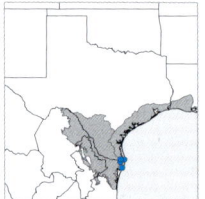

ETYMOLOGY From the Latin *contiguus* "neighboring" or "contiguous." **FIELD IDENTIFICATION** Difficult. The sparse T1 fringe is diagnostic, but it is difficult to recognize in the field. **LAB IDENTIFICATION** The humerus is rounded; the mesonotum is weakly constricted anteriorly; there is no scutellar scale; the T1 fringe is formed of sparse white setae; the T2 disc apparently lacks white setal patches; and the pygidial sculpture is faintly longitudinally rugose anteriorly and microreticulate posteriorly. **MALE** Unknown. **NOTES** We have not seen any specimens of this rare species ourselves, but a photograph was published on bugguide.net. *Timulla contigua* could be easily confused with *T. dubitatiformis* (3.6.15.3), since the setae on the T1 fringe are sparser and fainter than those of other species in this cluster, giving it a black appearance at first glance. In thorax shape and overall appearance, it is almost identical to *T. euphrosyne* (3.6.15.2), but *T. contigua* has the apical tergites black with distinct white setal patches and the T1 fringe has sparse white setae. We don't have a good guess for a male that would match with this species.

3.6.16.4—*Timulla ornatipennis*
TIMM-EWE-LUH ORR-NOT-IPP-ENN-ISS

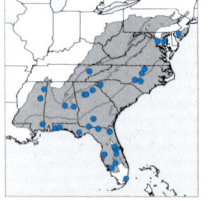

ETYMOLOGY From the Latin *ornatus* "decorated" and *-pennis* "-winged," in reference to the banded wings of the male. **FIELD IDENTIFICATION** Difficult. Unlike the other species in this cluster, the mesonotum is not constricted laterally. **LAB IDENTIFICATION** The humerus is rounded; the mesonotum is not constricted; the scutellar scale is narrow and rounded posteriorly; the T1 fringe is formed of dense white setae; the T2 disc has the setal patches usually obliterated; and the pygidial sculpture is rugose. **MALE** (5.7.14.2). Similar to female. **NOTES** This species is nearly identical structurally to *T. wileyae* from the *Timulla*-like cluster (3.5.10.1).

Photo by Efram Goldberg

*Timulla
ornatapennis*

▼ 3.6.17—*Timulla oajaca* Eastern females cluster

These species range from the central and eastern USA west to Arizona and California. They have the propodeum and pronotum equally wide and the mesosoma with the humerus (shoulder) obtusely angular or perpendicular. The metasomal cuticle can vary slightly, and they should be compared with the *Timulla*-like FCS *Timulla* clusters (3.5.9, 3.5.10). The mesosomal shape is often difficult to interpret, so they should also be compared with the *T. dubitata* (3.6.16) and *T. ferrugata* (3.6.14) clusters.

3.6.17.1—*Timulla oajaca*
TIMM-EWE-LUH WAH-HAW-KUH

ETYMOLOGY Named for the Mexican state of Oaxaca. It is confusing that this species does not actually occur in that state. **FIELD IDENTIFICATION** Sometimes difficult. This color form of *T. oajaca* apparently overlaps only with *T. tyro*. Unlike that species, *T. oajaca* has the body cuticle darker orange-brown and the mesonotum constricted laterally. **LAB IDENTIFICATION** The humerus is angular; the mesonotum is constricted; the scutellar scale is wide, posteriorly truncate, and has a weak transverse furrow or row of tubercles anterior to the scale; the T2 disc has short linear, white setal patches basally; the T2 sculpture is coarse with the intervals mostly smooth and bare; and the pygidial sculpture is rugose with the posterior half microreticulate. **MALE** (5.5.14.1). Darker than female. **NOTES** Counterintuitively, this Eastern FCS form of *T. oajaca* is most prevalent in the western portions of the range for this species, while the *Timulla*-like FCS (3.5.9.2) form is mostly seen in Texas. The dark reddish-brown body color is somewhat reminiscent of the Cryptic FCS. To make the guide more functional, however, we have placed all *Timulla* females into either the Eastern or *Timulla*-like FCS. Very few Eastern FCS species, especially from other genera, are found this far

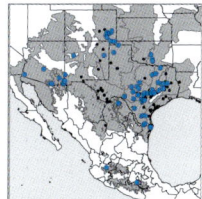

west. Examples like this are one of the primary reasons that we have foregone the use of mimicry rings to categorize species in this book, since co-occurrence with similarly colored model species is a vital component for establishing mimicry. Although the head and mesosoma have sparse setae, and they lack discrete contrasting color patches, this species and *T. tyro* (3.6.17.5) likely participate in the Madrean mimicry ring, at least as imperfect mimics, based on the background body color and contrasting pattern on the metasoma.

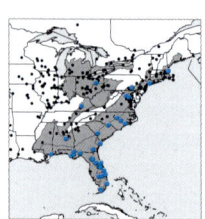

3.6.17.2—*Timulla vagans*

TIMM-EWE-LUH VAY-GUHNS

ETYMOLOGY From the Latin *vagans* "rambling." **FIELD IDENTIFICATION** Difficult. This species and *T. leona* are the only members of this cluster in the eastern USA, but *T. vagans* has a wider head and the mesonotum constricted laterally. **LAB IDENTIFICATION** The humerus is angular; the mesonotum is constricted; the scutellar scale is wide, posteriorly truncate, and has a transverse row of tubercles anterior to the scale; the T2 disc has short linear, white setal patches basally; the T2 sculpture is generally fine with the intervals micropunctate and setose; and the

pygidial sculpture is rugose with the apex microreticulate. **MALE** (5.5.12.2, 5.6.7.1, 5.7.16.10). Generally darker than female. **NOTES** This is the most common *Timulla* in the USA, but this Eastern FCS form is rarer than the *Timulla*-like FCS form (3.5.9.1).

Photo by Dan Roueche

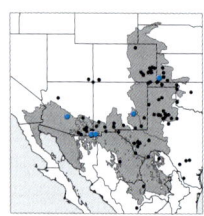

3.6.17.3—*Timulla suspensa*

TIMM-EWE-LUH SUSS-PEN-SUH

ETYMOLOGY Apparently from the Latin *suspensus* "suspended," "anxious," or "doubtful." It is not clear why this name was chosen. **FIELD IDENTIFICATION** Usually easy. The T2 disc has white setal stripes that are continuous from the anterior to posterior margins of T2, but these can often be obliterated or reduced by wear and tear. **LAB IDENTIFICATION** The humerus is angular; the mesonotum is constricted; the scutellar scale is wide, posteriorly truncate, and has a weak transverse furrow anterior to the scale; the T2 disc has long white setal stripes; and the pygidial sculpture is rugose with the apex microreticulate. **MALE** (5.3.6.3, 5.5.14.2). Darker than female. **NOTES** Although this species is widespread in the western and central USA, this Eastern FCS form is only sporadically encountered, while the *Timulla*-like FCS form (3.5.8.1) is more common.

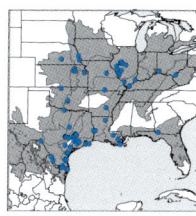

3.6.17.4—*Timulla leona*
TIMM-EWE-LUH LEE-OWE-NUH

ETYMOLOGY Apparently from the Latin *leonis* "lion" but may be named for the Mexican state of Nuevo Leon, since the type came from Mexico and the author (Blake 1871) named many species after Mexican place names. **FIELD IDENTIFICATION** Dif-

ficult. The angular humerus and non-constricted mesonotum are diagnostic in the central USA, but these features are difficult to interpret in the field. **LAB IDENTIFICATION** The humerus is angular; the mesonotum is not constricted; the scutellar scale is wide, posteriorly flat or wavy, and has a transverse furrow anterior to the scale; the T2 disc has short linear, white setal patches basally; the T2 sculpture is generally fine with the intervals micropunctate and setose; and the pygidial sculpture is uniformly striate. **MALE** (5.5.14.4). Generally darker than female. **NOTES** This species and *T. vagans* are the only members of this cluster in the eastern USA, but this species has the pygidium striate and the mesonotum not constricted. This species is structurally similar to *T. tyro* but does not overlap with that species in distribution, usually has the legs blackened, and has the propodeum with more distinct tooth-like tubercles along the posterolateral margin. Evans (1957) discussed this species entering burrows of the sand wasp *Bembix troglodytes*.

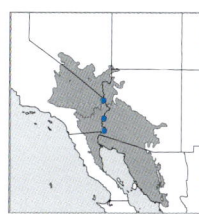

3.6.17.5—*Timulla tyro*
TIMM-EWE-LUH TIE-ROW

ETYMOLOGY From the Latin *tiro* "recruit" or "novice." **FIELD IDENTIFICATION** Sometimes difficult. The coloration and distribution are similar to *T. oajaca*, but *T. tyro* occurs mainly in low-elevation hot deserts and generally has the body

cuticle much paler orange-yellow. **LAB IDENTIFICATION** The humerus is angular; the mesonotum is not constricted; the scutellar scale is wide, posteriorly flat or wavy, and has a transverse furrow anterior to the scale; the T2 disc has short linear, white setal patches basally; the T2 sculpture is generally fine with the intervals micropunctate and setose; and the pygidial sculpture is uniformly striate. **MALE** (5.5.13.2). Generally darker than female. **NOTES** Strangely enough, although this species occurs farther west than any other Timulla, it is placed in the Eastern FCS. Like *T. oajaca*, this species could arguably be placed in the Cryptic FCS, but it is likely to be an imperfect mimic with members of the Madrean mimicry ring. The body color is especially similar to *Dasymutilla heliophila* (3.4.1.1).

3.7 CRYPTIC FEMALES COLOR SYNDROME

The Cryptic Females Color Syndrome (FCS) is represented by females with mainly dull brown cuticle with sparse or shortened setae. The entire body cuticle is usually concolorous, but sometimes the metasoma or legs are obscurely darkened. The metasoma usually lacks distinct spots on T2, and the silvery tergal fringes are duller and do not contrast as much against the brown cuticle as they do against black cuticle in other FCS. Members of this FCS occur throughout North America. There are 43 species in seven genera treated in this FCS, but the majority of nocturnal species (chapter 6) also fit here.

This color syndrome could be most easily confused with the *Timulla*-like or Eastern FCS. Unlike the *Timulla*-like FCS, Cryptic FCS species are diurnal and have uniformly silvery white metasomal setae when the metasomal cuticle is darkened. Unlike the Eastern FCS, they are usually small-bodied, with a body color that is more brown-tinted than red or orange, and have sparse metasomal setae that rarely form a particular pattern. None of the *Dasymutilla* or *Timulla* species are treated in this FCS; those that resemble the Cryptic FCS are likely to be found in the Madrean (3.4.5, 3.4.8) or Eastern (3.6.2–6, 3.6.13–17) FCS.

Of the 43 diurnal females in the Cryptic FCS, 25 (60%) are known from both sexes. In the USA, the males are distributed in the following color syndromes: *Ephuta*-like Male Color Syndrom (MCS) (58%), *Sphaeropthalma*-like MCS (31%), and *Timulla*-like MCS (19%). The vast majority of species with cryptic nocturnal females have males that loosely resemble the *Sphaeropthalma*-like MCS (chapter 6).

The Cryptic FCS occurs around the world, especially in nocturnal and small-bodied species. A few examples can be seen in chapter 8 (see 8.2.3, 8.2.4).

The key below is useful for figuring out which page to visit next.

—Genus *Lomachaeta*—Mesosoma pyriform; T1 shape slender disciform or subsessile; T6 convex without pygidial plate

3.7.1 Body with sparse setae and smooth sculpture ....
L. hicksi, L. warneri, L. beadugrimi, L. powelli, L. osita, L. cirrhomeris, L. vacamuerta, L. argenta

—Genus *Pseudomethoca*—Mesosoma fiddle-shaped; T1 shape sessile; T6 with defined pygidial plate

3.7.2 Head quadrate, wider than
mesosoma .... *Ps. frigida, Ps. torrida, Ps. nephele, Ps. dentigula, Ps. athamas, Ps. nudula*

3.7.3 Head rounded, narrower than mesosoma .... *Ps. gila*

—Genus *Photomorphus*—Mesosoma subrectangular; T1 shape sessile; T6 with defined pygidial plate

3.7.4 T2 fringe setae plumose; mesosomal pluera
mostly smooth .... *Ph. archboldi, Ph. spinci, Ph. auriventris*

3.7.5 T2 fringe setae simple; mesosomal pleura with short,
dense setae .... *Ph. impar, Ph. paulus, Ph. banksi, Ph. alogus*

—Genus *Ephuta*—Mesosoma ovate; T1 shape cylindrical

3.7.6 Eye large and ovate .... *E. spinifera, E. minuta, E. scrupea, E. copano, E. conchate, E. puteola*

—Genus *Myrmosa*—Pronotum separated from mesonotum; body sculpture coarse

3.7.7 T2 without pale yellow spots .... *M. unicolor, M. bradleyi, M. blakei, M. peculiaris*

—Genus *Leiomyrmosa*—Pronotum separated from mesonotum; head narrow; body sculpture mostly smooth

3.7.8 T2 with pale yellow spots .... *L. spilota*

—Genus *Myrmosula*—Pronotum separated from mesonotum; head wide, quadrate; body sculpture mostly smooth

3.7.9 T2 usually with pale yellow
spots .... *M. parvula, M. peregrinatrix, M. pacifica, M. exaggerata, M. rutilans, M. nasuta*

3.7.1—*Lomachaeta hicksi* Cryptic females cluster

These species occur mainly in the southwestern USA, but *L. hicksi* is distributed throughout North America. These are the only *Lomachaeta* females treated in this book.

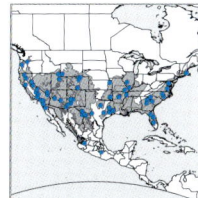

3.7.1.1—*Lomachaeta hicksi*
LOW-MUH-KEE-TUH HICKS-EYE

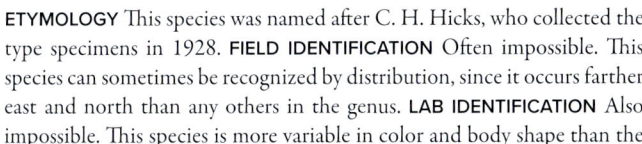

ETYMOLOGY This species was named after C. H. Hicks, who collected the type specimens in 1928. **FIELD IDENTIFICATION** Often impossible. This species can sometimes be recognized by distribution, since it occurs farther east and north than any others in the genus. **LAB IDENTIFICATION** Also impossible. This species is more variable in color and body shape than the other females of *Lomachaeta*. Use the key to species in the appendix (9.3.3.7). The pronotal and propodeal spiracles are usually strongly tuberculate, and the T1 shape is more clearly disciform than other *Lomachaeta* females. **MALE** (5.4.4.3, 5.5.6.1). Generally darker than female. **NOTES** This species is variable in color, from nearly black to pale yellow-brown. This could be the world's most widespread velvet ant species. None of the widespread eastern *Dasymutilla* species expand out to the West Coast or hot deserts, and none of the widespread nocturnal species get out to the East Coast. This species has been reared from nests of the crabronid wasp genera *Solierella* and *Pisonopsis*.

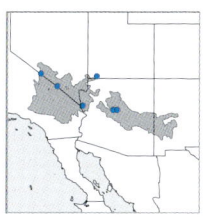

3.7.1.2—*Lomachaeta argenta*
LOW-MUH-KEE-TUH ARE-JENN-TAH-TUH

ETYMOLOGY From the Latin *argenteus* "silvery," in reference to the dorsal mesosomal setae. **FIELD IDENTIFICATION** Impossible. **LAB IDENTIFICATION** Consult the key in the appendix (9.3.3.7). This species has a compact mesosoma and has both the mesonotum and T2 with similar thickened, posteriorly directed, flattened, bristle-like setae. **MALE** Unknown. **NOTES** This species was reared from trap nests with *Solierella* crabronid host wasps. Sadly, no males were reared in these traps.

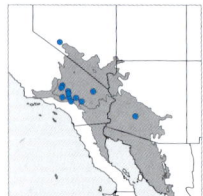

3.7.1.3—*Lomachaeta beadugrimi*
LOW-MUH-KEE-TUH BEE-DOO-GRIMM-EYE

ETYMOLOGY From the Old English *beadugrim* "war-furious," in reference to the blade-like parameres. **FIELD IDENTIFICATION** Impossible. **LAB IDENTIFICATION** Consult the key in the appendix (9.3.3.7). This species and *L. cirrhomeris* have the body cuticle uniformly bright orange, except for the blackened T6. Unlike that species, *L. beadugrimi* has the legs partly blackened and the T3–5 setae interspersed black and silvery. **MALE** (5.5.6.2). Generally darker than female. **NOTES** The sex association was recognized based on a large series of both sexes, which were collected together on spurge flower mats (*Chamaesyce* sp.) by Gordon C. Snelling. Because of their small size, all *Lomachaeta* females are rarely seen in person. One was seen crawling on the soil in an area with numerous *Forelius* ants, which they resemble in coloration and size. It is not clear whether there is some degree of Müllerian mimicry here, or maybe both creatures have cryptic coloration.

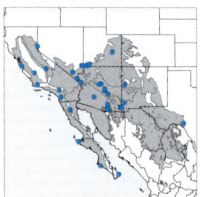

3.7.1.4—*Lomachaeta cirrhomeris*
LOW-MUH-KEE-TUH KEER-OWE-MARE-ISS

ETYMOLOGY From the Ancient Greek *kirrhos* "orange" and *meros* "thigh," in reference to the leg color. **FIELD IDENTIFICATION** Impossible. **LAB IDENTIFICATION** Consult the key in the appendix (9.3.3.7). This species and *L. beadugrimi* have the body cuticle uniformly bright orange, except for the blackened T6. Unlike that species, the legs are usually pale orange, and the setae of T3–5 are entirely black. **MALE** (5.4.3.1). Generally darker than female. **NOTES** Females have a somewhat distinctive color pattern and occur farther north and west than the similarly colored *L. beadugrimi*. They were described along with the males based on material collected in trap nests with their host wasps in the genera *Solierella*, *Pisonopsis*, and *Trypoxylon*.

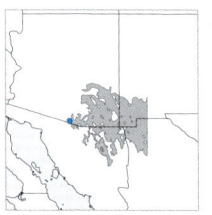

3.7.1.5—*Lomachaeta osita*
LOW-MUH-KEE-TUH OWE-SEE-TUH

ETYMOLOGY From the Spanish *osita* "little bear," based on colloquial references to these small brown wasps as teddy bears by some authors. **FIELD IDENTIFICATION** Impossible. **LAB IDENTIFICATION** Consult the key in the appendix (9.3.3.7). This species has a more elongate mesosoma than other *Lomachaeta* females and does not have parallel bristles on the mesonotum like *L. vacamuerta*. **MALE** Unknown. **NOTES** This rare species is known in museums only from the holotype, which was collected in 2010. Based on its presence in southern Arizona, this might be the female of *L. litosisyra* (5.4.4.1).

3.7.1.6—*Lomachaeta powelli*
LOW-MUH-KEE-TUH POWELL-EYE

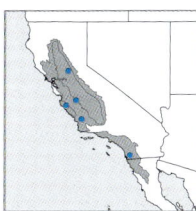

ETYMOLOGY This species was named after J. A. Powell, who collected the type specimens in 1962. **FIELD IDENTIFICATION** Impossible. **LAB IDENTIFICATION** Consult the key in the appendix (9.3.3.7). This dull brown Pacific species has many of the dorsal body setae especially elongate, with many setae longer than the scape. **MALE** (5.5.6.4). Generally darker than female. **NOTES** The sex association was based on both sexes being collected together while they were attacking *Diodontus* wasps near a fire station in the small town of Pozo, California. Unlike the other *Lomachaeta* species with known host records, this species attacks ground-nesting, rather than twig-nesting, hosts.

3.7.1.7—*Lomachaeta vacamuerta*
LOW-MUH-KEE-TUH VAW-CUH-MWER-TUH

ETYMOLOGY The name comes from the Spanish *vaca* "cow" and *muerta* "dead." **FIELD IDENTIFICATION** Impossible. **LAB IDENTIFICATION** Consult the key in the appendix (9.3.3.7). This species has an elongate mesosoma and has both the mesonotum and T2 with similar thickened, posteriorly directed, flattened, bristle-like setae. **MALE** (5.4.4.6). Generally darker than female. **NOTES** Some specimens of this species seem to verge on the Texan FCS, at least as much as is possible for such a tiny wasp. In these specimens, the head and mesosoma are much darker than the orange metasoma.

3.7.1.8—*Lomachaeta warneri*
LOW-MUH-KEE-TUH WARNER-EYE

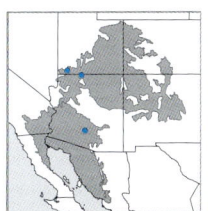

ETYMOLOGY Named in honor of beetle specialist William B. Warner, the collector of the holotype and many other important mutillid specimens. **FIELD IDENTIFICATION** Impossible. **LAB IDENTIFICATION** Consult the key in the appendix (9.3.3.7). This is a pretty unremarkable mutillid and can be identified only by ruling out the other species. The body color is somewhat duller brown than *L. beadugrimi*, and the dorsal body setae are shorter than those of *L. powelli*. **MALE** Unknown. **NOTES** This species was only recently discovered based mainly on specimens collected by William B. Warner.

▼ 3.7.2—*Pseudomethoca frigida* Cryptic females cluster

These species are spread throughout North America. They can be recognized firstly by their small size and cryptic coloration and secondly by their quadrate heads, which are often wider than the mesosoma. They could be confused with members of the *Ps. toumeyi* cluster (3.4.12) from the Madrean FCS depending on how one interprets the brightness and density of the generally pale silvery head setae.

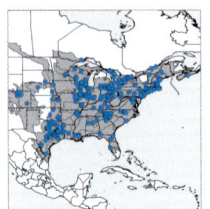

3.7.2.1—*Pseudomethoca frigida*
SOO-DOE-METH-OWE-KUH FRIJJ-IDD-UH

ETYMOLOGY From the Latin *frigidus* "cold," in reference to the cold climate of the type locality in Canada. **FIELD IDENTIFICATION** Sometimes difficult. This species is more common and widespread than other members of this cluster. In the northeastern USA, they are the most commonly seen small velvet ant species. **LAB IDENTIFICATION** The gena has a ventral tooth; the postgena is unarmed; the vertex is sharply angular posterolaterally; the antennal tubercle is unarmed; the mesonotum has a small sharp tooth laterally; the T1 fringe setae are black; and the T2 disc usually has indistinct circular or transverse silvery setal patches. **MALE** (5.4.6.1). Similar to female. **NOTES** This is one of the most commonly encountered velvet ants in the northeastern USA. Brothers (1972) outlined the life cycle of this species in detail. They are parasitoids of the primitively social sweat bee, *Lasioglossum* (*Dialictus*) *zephyrum*. Females had a difficult time entering the nests of their hosts and were often seen wrestling against guard bees. After they had gained access to the nest, however, they seemed to cloak themselves in the scent of their hosts and were able to remain inside the nest for long periods of time, parasitizing many immature bees without being molested. *Pseudomethoca frigida* has also been associated with numerous other halictid bee hosts. This is the type species for the large genus *Pseudomethoca*.

3.7.2.2—*Pseudomethoca torrida*
SOO-DOE-METH-OWE-KUH TORR-IDD-UH

ETYMOLOGY From the Latin *torridus* "dry" or "hot," likely named in contrast to the related *Ps. frigida*, which occurs in cooler habitats farther north. **FIELD IDENTIFICATION** Impossible. **LAB IDENTIFICATION** The gena has a ventral tooth; the postgena is unarmed; the vertex is sharply angular posterolaterally; the antennal tubercle is unarmed; the mesonotum has a small blunt tubercle laterally; the T1 fringe setae are white; and the T2 disc usually has the setae mostly silvery. **MALE** (5.4.6.2). Similar to female. **NOTES** This species was originally described as a southern subspecies of *Ps. frigida*. They have since been erected to be treated as a full species, since there are consistent differences between both sexes, even in areas where they overlap in distribution. This species has been associated with the andrenid host bee *Perdita graenicheri* Timberlake (Krombein 1992).

Photo by Efram Goldberg

3.7.2.3—*Pseudomethoca nephele*
SOO-DOE-METH-OWE-KUH NEFF-ELL-UH

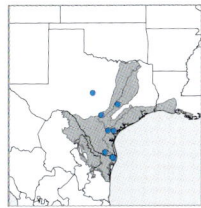

ETYMOLOGY Apparently from the Ancient Greek *nephele* "cloudlike spot," likely in reference to the pale spots on T2. **FIELD IDENTIFICATION** Impossible. **LAB IDENTIFICATION** The gena has a ventral tooth; the postgena has an equally large ventral tooth; the vertex is sharply angular posterolaterally; the antennal tubercle has a blunt tubercle dorsally; the mesonotum has a sharp tubercle laterally; the T1 fringe setae are white; and the T2 disc has two circular patches of silvery setae. **MALE** Unknown. **NOTES** Based on the distribution, head shape, and small body size, this species might eventually be associated with *Ps. gila* (5.4.7.3).

3.7.2.4—*Pseudomethoca dentigula*
SOO-DOE-METH-OWE-KUH DENN-TIGG-EWE-LUH

ETYMOLOGY From the Latin *dentis* "tooth" and *gula* "throat," in reference to the postgenal tooth. **FIELD IDENTIFICATION** Impossible. **LAB IDENTIFICATION** The gena lacks a ventral tooth; the postgena has a ventral tooth; the vertex is slightly angular posterolaterally; the antennal tubercle is unarmed; the mesonotum has a small blunt tubercle laterally; the T1 fringe setae are white; and the T2 disc often has a thick transverse band of silvery setae. **MALE** Unknown. **NOTES** Little is known about the biology of this small, rare species.

3.7.2.5—*Pseudomethoca athamas*
SOO-DOE-METH-OWE-KUH ATH-UH-MUSS

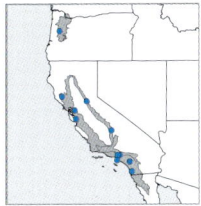

ETYMOLOGY Apparently from the Ancient Greek *a-* "not" and *thama* "crowded," maybe in reference to the sparse setae and punctures on the body. **FIELD IDENTIFICATION** Sometimes easy. This is the only small-bodied, sparsely setose *Pseudomethoca* species in Pacific regions, but they are hard to tell apart from other genera, like *Myrmosula* and *Ephuta*. **LAB IDENTIFICATION** The gena has a ventral tooth; the postgena is unarmed; the vertex is sharply angular posterolaterally; the antennal tubercle is armed with an erect bidentate lamella; the mesonotum has a small sharp tubercle laterally; the T1 fringe setae are white; and the T2 disc usually has two separated circular patches of silvery setae. **MALE** (5.4.7.1). Darker body color than female. **NOTES** The sex association of this species was informally recognized in collections and online but was only recently published officially (Williams 2023). The raised bidentate lamella on the antennal tubercle is unique in the USA, but some apparently new species from Mexico have a similar structure.

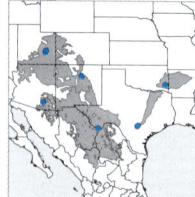

3.7.2.6—*Pseudomethoca nudula*
SOO-DOE-METH-OWE-KUH NUDE-EWE-LUH

ETYMOLOGY From the Latin *nudus* "bare," in reference to the sparse body setae, especially on the head. **FIELD IDENTIFICATION** Impossible. **LAB IDENTIFICATION** The gena and postgena are unarmed; the vertex is sharply angular posterolaterally; the antennal tubercle has a small sharp tooth dorsally; the mesonotum has a small blunt tubercle laterally; the T1 fringe setae are white; and the T2 disc usually has two large patches of silvery setae. **MALE** Unknown. **NOTES** In having the head unarmed, this species is similar to *Ps. gila* but has the head wider and the T2 fringe with black setae (setae white in *Ps. gila*). This is an uncommon but widely distributed species.

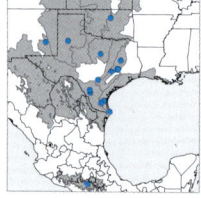

▼ 3.7.3.1—*Pseudomethoca gila*
SOO-DOE-METH-OWE-KUH HEE-LUH

ETYMOLOGY The species was apparently named after a place, but the Gila River, Gila National Forest, and city of Gila Bend are all in Arizona, not Texas. The female used to be named *Ps. ilione*, apparently from the Latin *il-* "not" and Ancient Greek *ion* "violet," maybe in reference to the dull brown body color of females. **FIELD IDENTIFICATION** Difficult. Due to the small size and inconspicuous coloration, this species is hard to tell apart from other *Pseudomethoca* and even other genera, like *Lomachaeta* and various nocturnal velvet ants. They can be diagnosed by their small rounded head; short mesosoma that is constricted at the propodeal spiracles; sessile T1 shape; large silvery setal patches on the T2 disc; and tergal fringes with simple setae only. **LAB IDENTIFICATION** The gena, postgena, antennal tubercle, and mesonotum are unarmed; the vertex is evenly rounded posterolaterally; the T1 and T2 fringe setae are white; and the T2 disc usually has the setae mostly silvery. **MALE** (5.4.7.3). Much darker than female. **NOTES** The sexes were associated based on males and females being found together with their host andrenid bee species, *Pseudopanurgus rugosus* (Robertson).

I was lucky to catch some of these tiny mutillids crawling on open soil in southern Texas, alongside a few *Lomachaeta vacamuerta* (3.7.1.7) females. In the field, they look like tiny, isolated ant (Formicidae) specimens, but they have a slightly different—but hard to describe—gait.—KEVIN

3.7.4—*Photomorphus archboldi* Cryptic females cluster

These species occur in the central or southeastern USA. They can be recognized by the obscurely plumose tergal fringes; on T2–5, these fringes are generally denser than the disc setae of their respective tergites. Except for distribution, they can be confused for many nocturnal species (see chapter 6). They are also difficult to separate from the *Ph. impar* cluster (3.7.5).

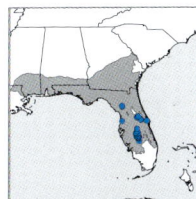

3.7.4.1—*Photomorphus archboldi*

FOE-TOE-MORE-PHUSS ARCHBOLD-EYE

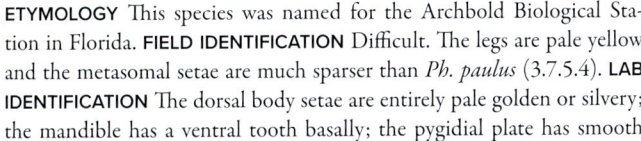

ETYMOLOGY This species was named for the Archbold Biological Station in Florida. **FIELD IDENTIFICATION** Difficult. The legs are pale yellow and the metasomal setae are much sparser than *Ph. paulus* (3.7.5.4). **LAB IDENTIFICATION** The dorsal body setae are entirely pale golden or silvery; the mandible has a ventral tooth basally; the pygidial plate has smooth sculpture. **MALE** (5.7.9.1). Similar to female. **NOTES** The male was discovered before the female. The similar coloration made the sex association pretty obvious.

Photo by Dan Roueche

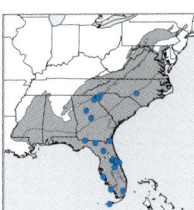

3.7.4.2—*Photomorphus spinci*

FOE-TOE-MORE-PHUSS SPINCE-EYE

ETYMOLOGY It is not clear why this species was named *spinci*, the *-I* suffix indicates it was named after a man, but there is no collector mentioned for the type material. **FIELD IDENTIFICATION** Difficult. The legs are darker than those of *Ph. archboldi*, and the fringes of T3–5 are denser than their respective tergal discs, unlike the *Ph. impar* cluster below. **LAB IDENTIFICATION** The mesonotum, T2 disc, and T3–5 have brown setae; the mandible has a ventral tooth basally; the pygidial plate is mostly smooth or obscurely microreticulate. **MALE** (5.7.9.2). Similar to female. **NOTES** The male was discovered before the female. The similarity to *Ph. archboldi*, coupled with the matching body color, was used to associate the sexes.

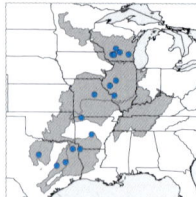

3.7.4.3—*Photomorphus auriventris*
FOE-TOE-MORE-PHUSS ORR-IVV-ENN-TRISS

ETYMOLOGY From the Latin *aurum* "gold" and *venter* "bottom," in reference to the orange-colored apical tergites. **FIELD IDENTIFICATION** Difficult. The fringes of T3–5 are denser than those of their respective tergal discs, unlike the *Ph. impar* cluster below. This species does not overlap in distribution with other species in this cluster. **LAB IDENTIFICATION** The vertex, mesonotum, T2 disc, and T4 have dark brown or reddish setae; the mandible has a ventral tooth basally; the pygidial plate is striate in the basal two-thirds. **MALE** (5.5.9.1). Much darker than female. **NOTES** The overlapping distributions in the northern USA and presence of plumose setae on the tergal fringes were used to associate the sexes. In Texas and the Great Plains this species can be easily confused with many nocturnal species, especially in the genus *Photomorphus* (6.1.6).

▼ 3.7.5—*Photomorphus impar* Cryptic females cluster

These species occur in the central or southeastern USA. They can be recognized by the tergal fringes having simple setae only, and T3–5 having uniformly dense, short setae. They superficially resemble many nocturnal species (chapter 6). They are also difficult to separate from the *Ph. archboldi* cluster (3.7.4), but in that cluster, the fringes of T3–5 are generally much denser than the discal setae of their respective tergites.

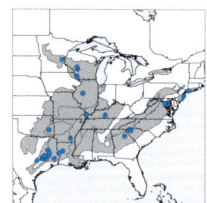

3.7.5.1—*Photomorphus impar*
FOE-TOE-MORE-PHUSS IMM-PARR

ETYMOLOGY Apparently from the Latin *impar* "unequal" or "odd." **FIELD IDENTIFICATION** Impossible. **LAB IDENTIFICATION** The mandible has a strong ventral tooth basally; the legs and apical tergites have the cuticle darker brown than the mesosoma (at least in part); and the T2 disc has appressed dark brown setae. **MALE** (5.7.10.3). Often darker than female. **NOTES** This is the most widespread species in the cluster, being found in the central and eastern USA and overlapping with all of the other species in the cluster.

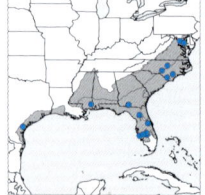

3.7.5.2—*Photomorphus alogus*
FOE-TOE-MORE-PHUSS AHL-OWE-GUSS

ETYMOLOGY Perhaps derived from the Ancient Greek *alogos* "speechless" or "irrational." **FIELD IDENTIFICATION** Impossible. **LAB IDENTIFICATION** The mandible lacks a ventral tooth basally; S1 has a sharp mesal longitudinal ridge; and the apical tergites have the setae mostly silvery. **MALE** (5.7.10.1). Often darker than female. **NOTES** Males of this species are variable in color, with Texan specimens having darker cuticle.

Photomorphus alogus

Photomorphus banksi

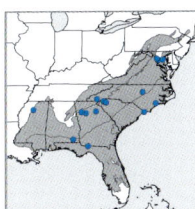

3.7.5.3—*Photomorphus banksi*
FOE-TOE-MORE-PHUSS BANKS-EYE

ETYMOLOGY This species was named after the entomologist Nathan Banks, who collected the type specimen in the early 1900s. **FIELD IDENTIFICATION** Very difficult. This eastern species has T4–5 with the setae and cuticle dark brown, but in the field, this is hard to differentiate from the darkened cuticle with sparser silvery setae seen in *Ph. alogus* and *Ph. impar.* **LAB IDENTIFICATION** The mandible lacks a ventral tooth basally; S1 has a blunt, rounded tubercle; and the T4–5 have the setae mostly dark brown. **MALE** (5.4.10.2, 5.7.10.2). Sometimes similar to female, sometimes entirely blackened. **NOTES** Males of this species are variable in color, with many northern populations having the cuticle entirely black.

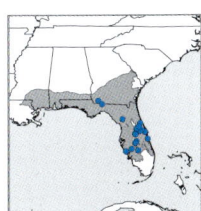

3.7.5.4—*Photomorphus paulus*
FOE-TOE-MORE-PHUSS PAW-LUSS

ETYMOLOGY It is not clear why this species was named *paulus* (originally *paula*); there is no collector mentioned for the type material. **FIELD IDENTIFICATION** Difficult. The legs are pale yellow, and T3–5 have denser setae than those of *Ph. archboldi* (3.7.4.1). **LAB IDENTIFICATION** The mandible has a strong ventral tooth basally; the legs and apical tergites are pale yellow or brown like the mesosoma; and the T2 disc has appressed silvery setae. **MALE** (5.7.9.3). Similar to female, with darker legs. **NOTES** This nocturnal species has larger eyes and lighter coloration than any of the other species in this cluster.

Photo by Efram Goldberg

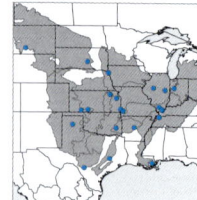

3.7.5.5—*Photomorphus myrmicoides*
FOE-TOE-MORE-PHUSS MEER-MICK-OY-DEEZ

ETYMOLOGY From the Ancient Greek *myrmex* "ant" and *-oides* "resembling," in reference to the similarity in color and size to many ants. **FIELD IDENTIFICATION** Impossible. **LAB IDENTIFICATION** Can be immediately recognized by the uniformly microreticulate pygidial sculpture. Additionally, the mandible lacks a ventral tooth, and T4–5 have the setae dark brown or black. **MALE** Unknown. **NOTES** This is the only unassociated female in the subgenus *Photomorphus* (*Photomorphus*), and it overlaps in distribution with the only unassociated male, *Ph. quintilis* (5.4.10.1). They could be eventually recognized as conspecific forms.

 ### 3.7.6—*Ephuta scrupea* Cryptic females cluster

These species occur mainly in the central and eastern USA. All the cryptically colored species of this genus fit here. They lack dense silver or golden setae on the head and have the body color, including the legs, varying shades of brown. Coloration frequently varies, however, so these should be compared with the *E. floridana* (3.4.18) and *E. slossonae* (3.6.12) clusters.

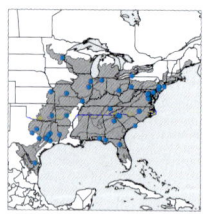

3.7.6.1—*Ephuta scrupea*
EFF-OO-TUH SCREW-PAY-UH

ETYMOLOGY From the Latin *scrupus* "sharp stone," likely in reference to the coarse sculpture of this species. **FIELD IDENTIFICATION** Impossible. **LAB IDENTIFICATION** The vertex has sparse setae only; the postgenal carina is complete; the T2 disc lacks any trace of whitish setal patches; T6 has a pygidial plate defined by lateral carinae, with the sculpture mostly smooth; and the hypopygium has two tubercles basally, but these are sometimes connected to form an apparent transverse ridge. **MALE** (5.4.13.1). Much darker than female. **NOTES** This is the most common and widespread *Ephuta* species in the central and eastern USA. The eye size, setal coloration, and body sculpture are apparently more variable than for other species in this cluster. One male specimen was reared from cocoons of the pompilid spider wasp *Pseudagenia bombycina* according to Schuster (1951).

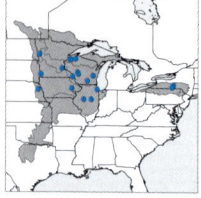

3.7.6.2—*Ephuta conchate*
EFF-OO-TUH CON-KATE-UH

ETYMOLOGY Apparently derived from the Latin *concha* "snail" or "shell," in reference to the coarse body sculpture. **FIELD IDENTIFICATION** Often difficult. This northeastern species has white setal spots on the T2 disc, but these are generally small or often obliterated by specimen wear and tear. **LAB IDENTIFICATION** The vertex is covered with dense dark brown setae; the

postgenal carina is mostly obliterated; the T2 disc has two whitish setal patches; T3–5 usually have the setae mostly orange-brown; T6 has a pygidial plate defined by lateral carinae; and the hypopygium has a two tubercles basally. **MALE** (5.4.13.5). Much darker than female. **NOTES** This northeastern species has the head intervals covered with short, dense setae, similar to the Madrean FCS species (3.4.17–19), except these setae are often dark brown, matching the cuticle of the head.

Ephuta
conchate

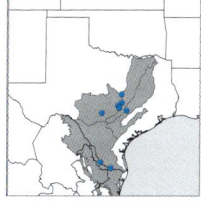

3.7.6.3—*Ephuta copano*
EFF-OO-TUH COE-PAN-OWE

Ephuta copano

ETYMOLOGY This species was named for a ghost town in Texas that was used as a port by smugglers and pirates in the 18th century. **FIELD IDENTIFICATION** Impossible. **LAB IDENTIFICATION** The vertex has sparse setae only; the postgenal carina is mostly obliterated; the T2 disc lacks any trace of whitish setal patches; the T2 sculpture is composed of moderately large subcircular punctures; T3 has the setae interspersed dark brown and golden-orange, while T4–5 have the setae mostly silvery; T6 has a pygidial plate defined by lateral carinae, with the sculpture mostly smooth; and the hypopygium has a transverse ridge basally. **MALE** (5.5.11.1). Darker than the female, at least anteriorly. **NOTES** The female is often larger, with more uniformly bright orange coloration than other *Ephuta* species in Texas.

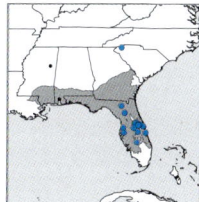

3.7.6.4—*Ephuta floridana floridana*
EFF-OO-TUH FLOW-RIDD-ANN-UH FLOW-RIDD-ANN-UH

ETYMOLOGY Named for the state of Florida. This subspecies *E. f. dietrichi* was named after H. Dietrich, who collected the type specimen in Mississippi in 1932. **FIELD IDENTIFICATION** Difficult. The head setae are golden (but inconspicuous), like *E. margueritae*, and the T2 disc lacks silvery setal patches. **LAB IDENTIFICATION** The vertex is covered with golden setae; the postgenal carina is mostly obliterated; the T2 disc lacks any trace of whitish setal patches; T3–5 usually have the setae whitish mesally; T6 has a pygidial plate defined by lateral carinae; and the hypopygium has a transverse ridge basally. **MALE** (5.7.13.2). Somewhat similar to female, without the golden head setae. **NOTES** There are two subspecies: *E. f. floridana* has sparser, duller golden setae on the head (Fig. 3.7.6.4a), and *E. f. dietrichi* (see Madrean FCS, 3.4.12.1) has denser bright golden head setae.

Photo by Dan Roueche

3.7.6.5—*Ephuta margueritae margueritae*

EFF-OO-TUH MARGUERITA-EE MARGUERITA-EE

ETYMOLOGY Schuster (1951) named this species after his wife, Olga Marguerite Schuster, in appreciation for her constant help during his study of this genus. The subspecies *E. m. xanthocephala* was named in reference to the more distinctly yellow-gold head setae. **FIELD IDENTIFICATION** Difficult. The head setae are golden (but inconspicuous), like *E. floridana*, and the T2 disc has two silvery setal patches. **LAB IDENTIFICATION** The vertex is covered with golden setae; the postgenal carina is complete; the T2 disc has two whitish setal patches; T3–5 usually have the setae whitish mesally; T6 is convex without a pygidial plate; and the hypopygium has two tubercles basally. **MALE** (5.4.13.9). Much darker than female. **NOTES** This species is structurally nearly identical to *E. sudatrix* (3.4.19.4) from Texas and is usually larger in size than *E. floridana*. The nominal subspecies pictured here, has sparser and duller golden setae on the head than *E. m. xanthocephala* in the Madrean FCS (3.4.18.2).

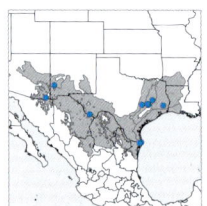

3.7.6.6—*Ephuta minuta*

EFF-OO-TUH MINE-OO-TUH

ETYMOLOGY From the Latin *minutus* "small," in reference to the body size. **FIELD IDENTIFICATION** Impossible. **LAB IDENTIFICATION** The head has sparse setae; the postgenal carina is complete; the T2 disc usually lacks whitish setal patches; the T2 sculpture has moderately small, separated, circular punctures; T3–5 have the setae mostly brown; T6 has a slender pygidial plate with the sculpture densely microreticulate; and the hypopygium has a transverse ridge basally. **MALE** Unknown. **NOTES** This small and rare western species has a unique slender microreticulate pygidium. The subspecies *E. m. modesta*, from Arizona and eastern New Mexico, has denser punctures on the mesosomal pleura and the posterior band of T2 scarcely defined, while the nominal subspecies, *E. m. minuta* from Texas, has sparser punctures on the mesosomal pleura and the posterior band of T2 distinct and silvery.

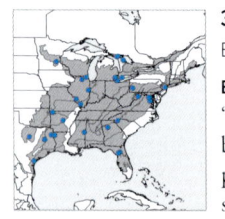

3.7.6.7—*Ephuta puteola*

EFF-OO-TUH PEW-TEE-OWE-LUH

ETYMOLOGY From the Latin word *puteolus* "a small well," likely in reference to the deep body punctures. **FIELD IDENTIFICATION** Impossible. **LAB IDENTIFICATION** The vertex has sparse setae only; the postgenal carina is mostly obliterated; the T2 disc lacks any trace of whitish setal patches; the T2 sculpture is moderately coarse, with longitudinally ovate and nearly contiguous punctures; T3 has the setae mostly blackish-brown, while T4–5 have the setae largely silvery or golden-orange; T6 has a pygidial plate defined by lateral carinae, with the sculpture mostly smooth; and the hypopygium has a transverse ridge basally. **MALE** Unknown.

NOTES This species is common (at least for females of *Ephuta*, which are more rarely seen than most velvet ant genera) and widespread in the eastern and central USA. It has been presumed to be the female of *E. pauxilla* (5.4.13.10) based on overlapping distribution, but there is not yet enough evidence for this association. The sculpture of T2 is coarser than that of the structurally similar species *E. copano* and *E. spinifera*.

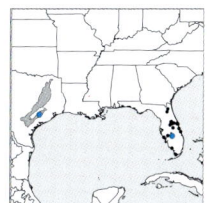

3.7.6.8—*Ephuta slossonae monochroa*
EFF-OO-TUH SLOSSON-EE MAW-NO-CROW-UH

ETYMOLOGY This species was named after the American author and entomologist Annie T. Slosson, who collected the type specimen in the 1890s (apparently, the date of collection was not provided on the type specimen). **FIELD IDENTIFICATION** Impossible. **LAB IDENTIFICATION** The vertex has sparse setae only; the postgenal carina is mostly obliterated; the T2 disc lacks any trace of whitish setal patches; the T2 sculpture is very coarse, with the intervals reduced to narrow carinae; T3–5 usually have the setae mostly whitish; T6 has a pygidial plate defined by lateral carinae, with the sculpture mostly smooth; and the hypopygium has two small inconspicuous tubercles basally. The eye is larger than in other species in this cluster, and the vertex is relatively short and narrow compared with other species of *Ephuta*. **MALE** The male of *E. slossonae slossonae* is known (5.7.13.1) and somewhat similar to the female, but there is some uncertainty about the species-level status for *E. s. monochroa*, and males have not been directly associated with this subspecies. **NOTES** Schuster (1956) said that females of this species were practically impossible to tell apart from his females of *E. battlei*, but he never got around to describing the females of *E. battlei*. The nominal subspecies, *E. s. slossonae* (in the Eastern FCS, 3.6.12.1), has the body cuticle more distinctly reddish than the orange-brown *E. s. monochroa* (Fig. 3.7.6.8a).

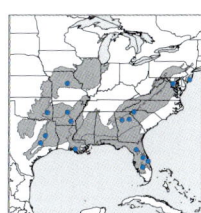

3.7.6.9—*Ephuta spinifera*
EFF-OO-TUH SPINE-IFF-ERR-UH

ETYMOLOGY Derived from the Latin *spina* "spine," in reference to the toothed hind coxa. **FIELD IDENTIFICATION** Impossible. **LAB IDENTIFICATION** The vertex has sparse setae only; the postgenal carina is mostly obliterated; the T2 disc lacks any trace of whitish setal patches; the T2 sculpture is composed of moderate-sized, circular, separated punctures; T3–4 have the setae mostly blackish-brown, while T5 often has the setae partly silvery or golden-orange; T6 has a pygidial plate defined by lateral carinae, with the sculpture mostly smooth; and the hypopygium has a transverse ridge basally. **MALE** (5.4.13.3). Much darker than female. **NOTES** In T2 puncture and other morphological features, this species is similar to *E. copano*, but that species is generally larger in size and has the metasomal cuticle uniformly orange-brown, while *E. spinifera* usually has the cuticle of T3–5 darker brown than that of T2. Krombein & Evans (1955) mentioned that in personal experience, these females often roll up and play dead when disturbed, which seems like a remarkably effective strategy given their resemblance to small pebbles. From our experience, nocturnal female velvet ants utilize this defensive strategy as well (see 1.5.3).

3.7.7—*Myrmosa unicolor* Cryptic females cluster

These species are widely distributed in North America. All females in the genus *Myrmosa* fit in this cluster. They have the clypeus with a basomesal tooth or ridge, the pronotum divided, S1 with a large longitudinal process, and they lack pale yellow spots on T2.

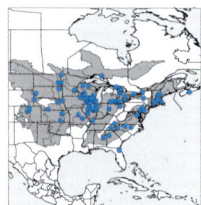

3.7.7.1—*Myrmosa unicolor*
MEER-MOE-SUH EWE-NICK-UH-LURR

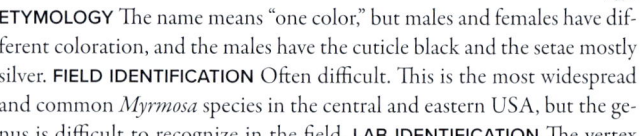

ETYMOLOGY The name means "one color," but males and females have different coloration, and the males have the cuticle black and the setae mostly silver. **FIELD IDENTIFICATION** Often difficult. This is the most widespread and common *Myrmosa* species in the central and eastern USA, but the genus is difficult to recognize in the field. **LAB IDENTIFICATION** The vertex has its lateral margins convergent directly behind the eyes; the antennal scrobe has the dorsal carina weak or absent; the pronotal and propodeal shape are somewhat variable. **MALE** (5.4.14.1). Much

darker than female. **NOTES** *Myrmosa unicolor* is widespread in the central and eastern USA. There are no confirmed published host records, but according to Krombein (1940), this species has been found crawling among nests of the halictid bees *Halictus stultus* and *Halictus* (*Chloralictus*) *pruino-*

sus, and the crabronid wasp *Lindenius errans*. Krombein (1940) also mentioned one specimen with a label saying it was reared from a *Tiphia* cocoon.

Photo by Dan Roueche

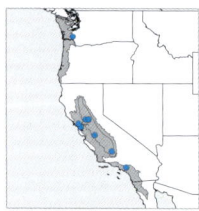

3.7.7.2—*Myrmosa bradleyi*
MEER-MOE-SUH BRADLEY-EYE

ETYMOLOGY This species was named after the entomologist J. C. Bradley, who revised many of the North American velvet ants in 1916. **FIELD IDENTIFICATION** Often difficult. This is apparently the only Myrmosa species in the western USA, but the genus is difficult to recognize in the field. **LAB IDENTIFICATION** The vertex has its lateral margins parallel directly behind the eyes; the antennal scrobe lacks a dorsal carina; the humerus is bluntly rounded; and the propodeum is relatively high with weak teeth posterodorsally and the posterolateral angle bluntly rounded. **MALE** (5.4.15.1). Much darker than female. **NOTES** *Myrmosa bradleyi* is the only known Pacific *Myrmosa* species.

3.7.7.3—*Myrmosa blakei*
MEER-MOE-SUH BLAKE-EYE

ETYMOLOGY This species was apparently named after the American entomologist Charles A. Blake, who described many velvet ant species in the late 1800s. **FIELD IDENTIFICATION** Impossible. **LAB IDENTIFICATION** The vertex has its lateral margins parallel, directly behind the eyes; the antennal

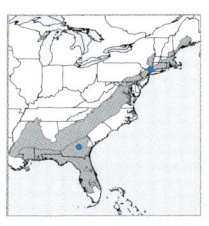

scrobe lacks a dorsal carina; the humerus is prominently dentate; and the propodeum is more highly produced with narrower teeth posterodorsally, and the posterolateral angle somewhat dentate. **MALE** Unknown. **NOTES** The type specimen, collected in New York before 1916, is the only specimen known in the literature, and we have not been able to study that specimen. The specimen pictured in figure 3.7.7.4a is from Georgia and has a wide head and robustly angular mesosoma shape, like that mentioned in the description of *M. blakei*. There is a decent possibility that it is just another specimen of the widespread and variable *M. unicolor* (which also lives in Georgia and New York) or a different new species, rather than true *M. blakei*.

3.7.7.4—*Myrmosa peculiaris*
MEER-MOE-SUH PECK-EWE-LEE-AIR-ISS

ETYMOLOGY From the Latin *peculiaris* "singular" or "unique." **FIELD IDENTIFICATION** Impossible. **LAB IDENTIFICATION** The vertex has its lateral margins parallel directly behind the eyes; the antennal scrobe has a dorsal carina; the humerus is prominently dentate or angular; and the propodeum is more or less flattened, with thick erect teeth posterodorsally. **MALE** Unknown. **NOTES** The type specimen, collected in Kansas in 1923, is the only specimen known in the literature, and we have not been able to study that specimen. The specimen pictured here is also from Kansas and has a weak dorsal carina on the antennal scrobe, but there is a decent possibility that it is just another specimen of the widespread and variable *M. unicolor*, which also lives in Kansas.

3.7.8.1—*Leiomyrmosa spilota*
LAY-OWE-MEER-MOE-SUH SPILL-OWE-TUH

ETYMOLOGY From the Ancient Greek *spilos* "spot," in reference to the pale yellow spots on T2. **FIELD IDENTIFICATION** Difficult. The head is narrower than *Myrmosula*, and the pale yellow spots on T2 are large. **LAB IDENTIFICATION** The head is narrow; the mandible is tridentate; the clypeus is evenly rounded; S1 lacks a longitudinal process; and T2 has two large pale yellow cuticular patches. **MALE** Unknown. **NOTES** This hot desert species (and the whole genus) was previously known only from sandy areas west of Blythe, California. Recently a darker specimen was found in Arizona, but this site is only ~10 miles

RIGHT: Typical *Leiomyrmosa spilota*
FAR RIGHT: Darkened *Leiomyrmosa spilota*

from the Blythe locality. We could not find any structural differences between these new populations and the type series, so we treat them as a regional variant of *L. spilota*. No good male candidate has yet been examined that could match this species. Wasbauer (1973) suggested that the subgenus *Myrmosa* (*Myrmosina*) (5.4.14.2,3) could match these females, but no *Myrmosina* species are known from hot deserts west of Texas.

▼ 3.7.9—*Myrmosula parvula* Cryptic females cluster

These species are widespread in North America. They can be recognized by having the head generally wide and T2 usually with pale yellow cuticular patches. Most species of *Myrmosula* fit here, but these can be easily confused with the *M. rutilans* cluster (3.4.20) in the Madrean FCS, based on one's interpretation of the brightness and density of silvery or golden head setae. In body shape and size, they often resemble *Pseudomethoca* (3.7.2).

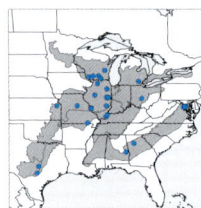

3.7.9.1—*Myrmosula parvula*
MEER-MOE-SOO-LUH PARR-VIEW-LUH

ETYMOLOGY From the Latin *parvus* "little," in reference to the small body size. **FIELD IDENTIFICATION** Often difficult. This is apparently the only *Myrmosula* species in many parts of the central and eastern USA, but it is difficult to separate from *Myrmosa* and other genera, especially in the field. **LAB IDENTIFICATION** The antennal tubercles are separated and edentate, there is no interantennal prominence; the hypostomal carina is dentate; the ventral mandibular lamella is convex; and the pale yellow T2 patches are small or completely obliterated. **MALE** (5.4.16.1). Much darker than female. **NOTES** *Myrmosula parvula* is widespread in the central and eastern USA. Brothers (1978) provided a detailed account of the biology of this species and its interactions with the halictid host bee, *Lasioglossum* (*Dialictus*) *zephyrum* (Smith).

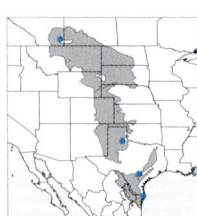

3.7.9.2—*Myrmosula peregrinatrix*
MEER-MOE-SOO-LUH PAIR-UH-GRIN-AH-TRICKS

ETYMOLOGY From the Latin *peregrinus* "exotic," "foreign," or "strange." **FIELD IDENTIFICATION** Impossible. **LAB IDENTIFICATION** The antennal tubercles are separated and have a dorsal tooth-like ridge, there is no interantennal prominence; the hypostomal carina is dentate; the ventral mandibular lamella is convex; and the pale yellow T2 patches are generally small and circular. **MALE** Unknown. **NOTES** *Myrmosula peregrinatrix* is sporadically recorded from Arizona, Texas, and Alberta, Canada. This distribution is strange but may just be an artifact of the general rarity of this species. We have compared specimens from Canada and Texas and found no significant differences.

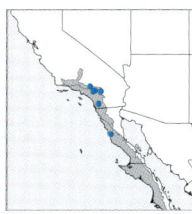

3.7.9.3—*Myrmosula exaggerata*
MEER-MOE-SOO-LUH EGGS-AJJ-URR-AWE-TUH

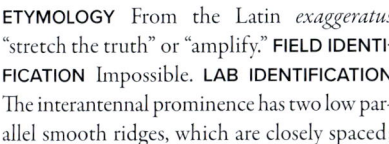

ETYMOLOGY From the Latin *exaggeratus* "stretch the truth" or "amplify." **FIELD IDENTIFICATION** Impossible. **LAB IDENTIFICATION** The interantennal prominence has two low parallel smooth ridges, which are closely spaced; the hypostomal carina is dentate; the ventral mandibular lamella is concave; and the pale yellow T2 patches are generally small and circular. **MALE** Unknown. **NOTES** This species is nearly identical to *M. pacifica*, subtly differing in the length and closeness of the ridges on the interantennal prominence. This species is more common and widespread than *M. pacifica*.

3.7.9.4—*Myrmosula pacifica*
MEER-MOE-SOO-LUH PUH-SIFF-ICK-UH

ETYMOLOGY Named for the Pacific Ocean. **FIELD IDENTIFICATION** Impossible. **LAB IDENTIFICATION** The interantennal prominence has two low parallel smooth ridges, which are widely spaced; the hypostomal carina is dentate; the ventral mandibular lamella is concave; and the pale yellow T2 patches are generally small and circular. **MALE** Unknown. **NOTES** This species is rare and was considered to be endemic to the Antioch sand dunes in California. Because that dune habitat has been greatly degraded, it has been suggested for consideration as a threatened species but is currently not formally listed as such. Wasbauer (1973), however, suggested it could be a synonym of the more widespread *M. exaggerata*, since the variation between populations of *M. exaggerata* was apparently greater than the species-level differences cited by previous authors. Since Wasbauer's treatment of the genus, we have examined additional specimens of *M. pacifica* from Tulare and Monterey Counties in California (some identified by Wasbauer himself in the 1980s). Since the validity of the species is in question, and since it does occur outside the critically damaged Antioch sand dunes, we do not recommend that it be formally treated as a threatened or endangered species—but that's not our decision to make.

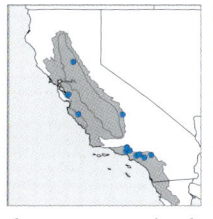

3.7.9.5—*Myrmosula rutilans*
MEER-MOE-SOO-LUH ROO-TILL-AHNZ

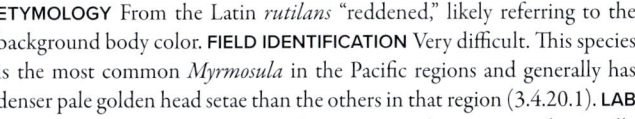

ETYMOLOGY From the Latin *rutilans* "reddened," likely referring to the background body color. **FIELD IDENTIFICATION** Very difficult. This species is the most common *Myrmosula* in the Pacific regions and generally has denser pale golden head setae than the others in that region (3.4.20.1). **LAB IDENTIFICATION** The interantennal prominence has two raised ventrally divergent, smooth ridges; the hypostomal carina is edentate; the ventral mandibular lamella is convex; and the pale yellow T2 patches are generally subcircular or subtriangular. **MALE** Unknown. **NOTES** *Myrmosula rutilans* is found in temperate Pacific regions, like *M. exaggerata* and *M. pacifica*.

An apparent fossil of this species was found in the La Brea tar pits. The density and color of the head setae are variable, so this species is also treated in the Madrean FCS (3.4.20.1).

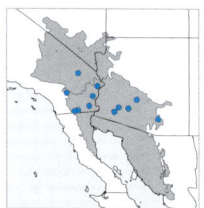

3.7.9.6—*Myrmosula nasuta*
MEER-MOE-SOO-LUH NAY-SOO-TUH

ETYMOLOGY From the Latin *nasutus* "big-nosed," in reference to the raised interantennal prominence. **FIELD IDENTIFICATION** Often difficult. This desert species often has uniquely comma-shaped pale yellow patches on T2 but frequently has the spot round like others in the genus. **LAB IDENTIFICATION** The interantennal prominence is erect and bicarinate, its anterior margin (in lateral view) is convex; the hypostomal carina is dentate; the ventral mandibular lamella is relatively flat; and the pale yellow T2 patches are usually transversely elongate and comma-shaped.

MALE Unknown. **NOTES** This species is widespread in the Mojave and Sonoran Deserts. The density and color of the head setae are variable, so this species is also treated in the Madrean FCS (3.4.20.3).

Photo by Jillian Cowles

DAY FLIERS—DIURNAL MALE VELVET ANTS—GENERA

4.0 INTRODUCTION AND OVERVIEW

Males of the genera in this chapter and the next are divided into six higher taxa (three subfamilies and four tribes). This chapter begins with an illustrated key to these groups, then each higher taxon has an overview of the included genera (usually fewer than three in each higher taxon). For each higher taxon, examples of species-level diagnostic features are illustrated, which should help with identifications in chapter 5.

Key to higher taxa (subfamilies and tribes)

1. Eye with setae, vertically ovate without notch; forewing with veins distinct to outer margin .... **4.6** (Myrmosinae)

2. Eye bare, ovate with notch; forewing with small "empty" stigma, veins not distinct near margin

    a. T1 shape narrowly cylindrical; T2 without felt line .... **4.4** (Sphaeropthalminae: Ephutini: *Ephuta*)

    b. T1 shape subsessile; T2 with lateral felt line .... **4.5** (Mutillinae: Trogaspidiini: *Timulla*)

3. Eye bare, circular without clear notch; forewing with large "full" stigma, veins not distinct near margin, or wings absent or greatly reduced .... Sphaeropthalminae, in part

    a. Axilla armed with posterior tooth, smooth swelling, or truncate lobe (see also 4.1.5); T1 shape usually petiolate or disciform; tergal fringes not obviously plumose; usually bright-colored diurnal species .... **4.1** (Dasymutillini)

    b. Axilla unarmed, at most with weak angle or with defined lateral carina; T1 shape sessile; tergal fringes with simple setae (see 4.2.4); mandible without ventral tooth; mesosternal area without teeth or ridges; felt line present on T2 only; diurnal species .... **4.2** (Pseudomethocini)

    c. Axilla unarmed, at most with weak angle or with small defined lateral carina **and** having at least one of the following: nocturnal behavior with dull brown coloration; T1 shape petiolate; tergal fringes with obvious plumose setae; mandible with ventral tooth; mesosternal area armed with teeth or ridges; felt lines present on both T2 and S2 .... **4.3** (Sphaeropthalmini)

4.1 SPHAEROPTHALMINAE: DASYMUTILLINI: *DASYMUTILLA* AND KIN

Dasymutilla—Cow killers

IDENTIFICATION These males have round, bare eyes (subfamily character), armed axillae (tribe character), and a petiolate T1 shape (genus character). Compared with *Lomachaeta*, they are usually larger, more coarsely sculptured, and more brightly colored insects. **FEMALE** Many species are recognized from both sexes, but the male of the type species, *D. gorgon*, was only recently recognized by

Figure	Taxon	Description
4.0.1a	*Myrmosula rufiventris*	Eye: setose, ovate
4.0.1b	*Timulla neobule*	Eye: bare, notched
4.0.1c	*Dasymutilla asopus*	Eye: bare, circular
4.0.2a	*Myrmosa unicolor*	Wings: veins reaching margin
4.0.2b	*Timulla subhyalina*	Wings: veins not reaching margin, stigma empty
4.0.2c	*Protophotopsis clauseni*	Wings: veins not reaching margin, stigma filled
4.0.3a	*Timulla subhyalina*	Metasoma: T1 shape sessile
4.0.3b	*Ephuta argenticeps*	Metasoma: T1 shape cylindrical
4.0.3c	*Odontophotopsis erebus*	Metasoma: T1 shape elongate, T2 fringe plumose
4.0.4a	*Lomachaeta polemomechana*	Axilla: armed with tooth
4.0.4b	*Pseudomethoca sanbornii*	Axilla: unarmed, weakly angular
4.0.4c	*Photomorphus alogus*	Axilla: unarmed, small lateral carina

4.0.1a

4.0.1b

4.0.1c

4.0.2a

4.0.2b

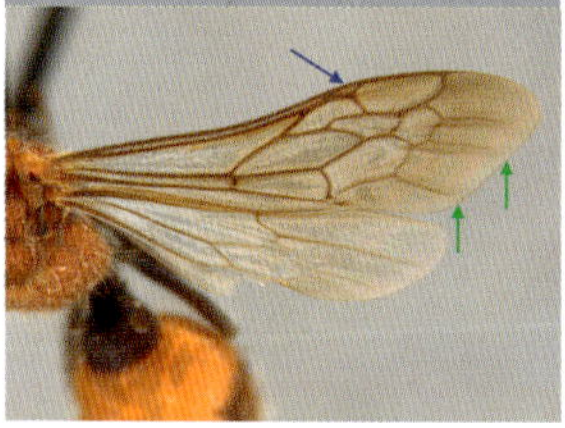

4.0.2c

4.0.3a

4.0.3b

4.0.3c

4.0.4a

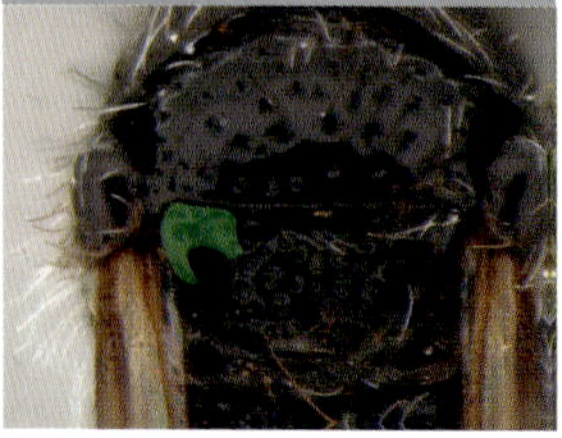

4.0.4b

4.0.4c

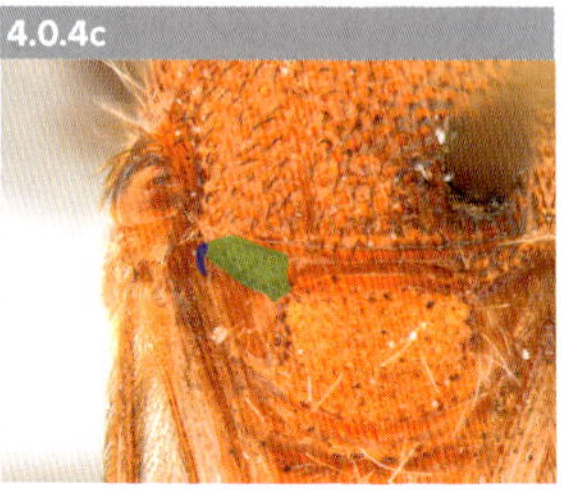

Manley et al. (2020). About 70% of species are known from both sexes. **DISTRIBUTION AND DIVERSITY** This genus is diverse across its range from Canada to Panama, though only four species extend into northern Colombia in South America. **NOTES** *Dasymutilla* is the most readily collected and examined genus in the USA and is often common in Mesoamerica, as well. **KEYS AND CHARACTERS** The species in the USA were recently reviewed by Manley et al. (2020) and there are many useful illustrations of diagnostic features in that paper. The Mesoamerican species were treated by Manley & Pitts (2007). Many can be recognized by color patterns and overall shape, but a good microscope and key are needed for others. The presence or absence of a pit or other modification on S2 is the most important feature for separating similarly colored *Dasymutilla* males; this is rarely visible in photos, though, since it occurs on the underside of the insect. The axillae are also important for identification but often require a microscope to compare; it is often helpful to rub the setae off this area with a pin to see this feature. Other important features include the shape of the head, femur, tibia, and hypopygium. The presence of a fringe on the pygidium is also useful, but it can be misleading because that can vary within a species, and these setae are often rubbed off by natural wear and tear.

Lomachaeta

IDENTIFICATION These males have round bare eyes (subfamily character), armed axillae (tribe character), and a disciform or subsessile T1 shape (genus character). These are small-bodied, smoothly sculptured, and generally black wasps. **FEMALE** The type species, *L. hicksi*, was described from both sexes in the same publication in which the genus was described by Mickel (1936b). About half of the species are known from both sexes. **DISTRIBUTION AND DIVERSITY** The genus includes 24 species, which range from Canada south to Argentina; 15 of these occur in the USA. **NOTES** These are the smallest diurnal velvet ants in the USA. Some species attack arboreal hosts, others use ground-nesting hosts. Males are most frequently collected with malaise traps. **KEYS AND CHARACTERS** For males, genitalia dissections are often needed to make an identification; coloration and mandibular characters are also useful. Williams, Cambra, Bartholomay et al. (2019) has a functional key.

4.2 SPHAEROPTHALMINAE: PSEUDOMETHOCINI: *PSEUDOMETHOCA* AND KIN

Myrmilloides

IDENTIFICATION These males have round bare eyes (subfamily character), sessile T1 shape (tribe character), and short rudimentary wings (genus character). In the only species of this genus, *M. grandiceps*, the head is wider than any *Pseudomethoca* males and is armed below with genal and hypostomal teeth. **FEMALE** Both sexes are known. **DISTRIBUTION AND DIVERSITY** Just one species, *Myrmilloides grandiceps*, which lives mainly in Texas and the Great Plains. **NOTES** See the species entry for more information about males and females of this species (3.3.6.1, 3.5.2.1, 3.6.7.1).

Pseudomethoca

IDENTIFICATION These males have round, bare eyes (subfamily character), sessile T1 shape (tribe character), and fully formed wings (genus character). This is a variable and diverse genus, ranging in size from tiny to large, and in color from dull black and gray to bright red and black. **FEMALE** About half of the species in the USA are recognized from both sexes. **DISTRIBUTION AND DIVERSITY** There are about 40 species in the USA, 40 Mexican and Central American species, and at least 40 more South American species. **NOTES** Males seem to be rarer than females and rarer than males of other genera. **KEYS AND CHARACTERS** Mickel (1924, 1935a) has keys to the species in the USA, but our book works well and includes the newest nomenclature and the handful of species have been described in the last 80 years. They are diagnosed mainly by differences in head shape, tegula structure, sculpture of different body regions, and coloration.

Figure	Taxon	Description
4.1.1a	*Dasymutilla pseudopappus*	Habitus: lateral view
4.1.1b	*Lomachaeta hicksi*	Habitus: lateral view
4.1.2a	*Dasymutilla sicheliana*	Hind tibia: simple, cylindrical
4.1.2b	*Dasymutilla foxi*	Hind tibia: dilated, concave inner face
4.1.2c	*Dasymutilla erythrina*	Hind tibia: apically flared, cylindrical
4.1.3a	*Dasymutilla asopus*	Hind trochanter: unarmed
4.1.3b	*Dasymutilla creon*	Hind trochanter: dentate
4.1.4a	*Dasymutilla nocturna*	Head: ocelli large, vertex mesally swollen
4.1.4b	*Dasymutilla satanas*	Head: ocelli small, vertex rounded
4.1.5a	*Dasymutilla apicalata*	Axilla: large dorsally smooth tooth
4.1.5b	*Dasymutilla scaevola*	Axilla: moderate dorsally punctate tooth
4.1.5c	*Dasymutilla foxi*	Axilla: short dorsally punctate tooth
4.1.5d	*Dasymutilla nigripes*	Axilla: dorsally smooth short lobe
4.1.5e	*Dasymutilla ursus*	Axilla: truncate lobe Mesonotum: swollen posterolaterally
4.1.5f	*Dasymutilla quadriguttata*	Axilla: truncate lobe Mesonotum: not swollen posterolaterally
4.1.6a	*Dasymutilla sicheliana*	S2: mesal seta-filled pit
4.1.6b	*Dasymutilla asopus*	S2: mesal longitudinal "mohawk"
4.1.6c	*Dasymutilla ursus*	S2: unarmed, simply punctate
4.1.6d	*Dasymutilla nogalensis*	S2: mesal patch of microsetae
4.1.7a	*Dasymutilla gloriosa*	Hypopygium: unarmed
4.1.7b	*Dasymutilla vestita*	Hypopygium: posterolateral teeth

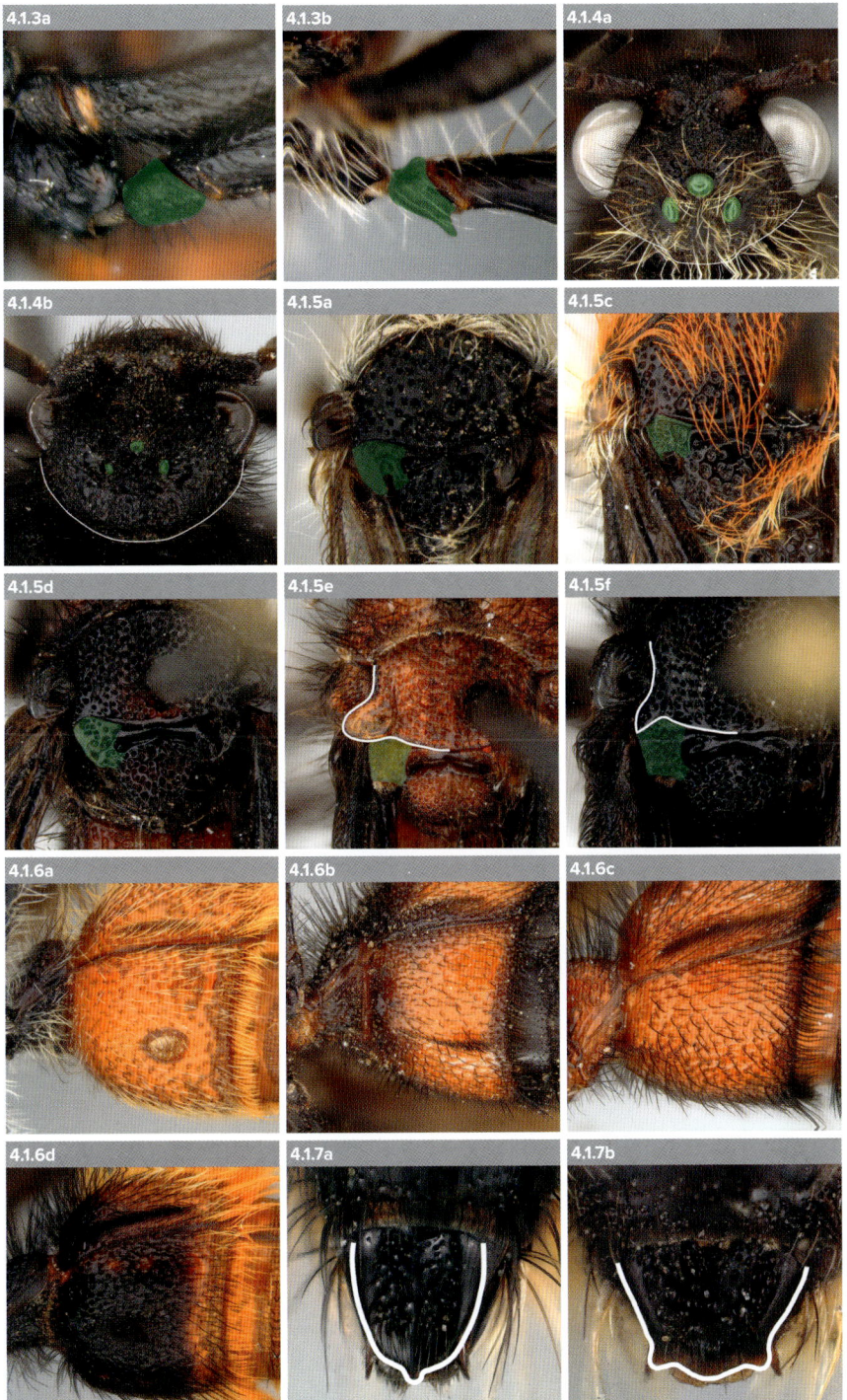

Figure	Taxon	Description
4.2.1a	*Myrmilloides grandiceps*	Habitus: lateral view
4.2.1b	*Pseudomethoca oceola*	Habitus: lateral view
4.2.2a	*Pseudomethoca frigida*	Head: quadrate with sharp posterolateral angles
4.2.2b	*Pseudomethoca simillima*	Head: rounded posteriorly
4.2.3a	*Pseudomethoca sanbornii*	Tegula: evenly curved
4.2.3b	*Pseudomethoca simillima*	Tegula: simply angled
4.2.3c	*Pseudomethoca flavida*	Tegula: angled with flat rim
4.2.4a	*Pseudomethoca frigida*	Tergal fringes: setae simple
4.2.4b	*Pseudomethoca praeclara*	Tergal fringes: bristle-like setae
4.2.5a	*Pseudomethoca gila*	Hypopygium: uniformly sparse setae
4.2.5b	*Pseudomethoca toumeyi*	Hypopygium: lateral tufts of setae

4.2.1a

4.2.1b

4.2.2a

4.2.2b

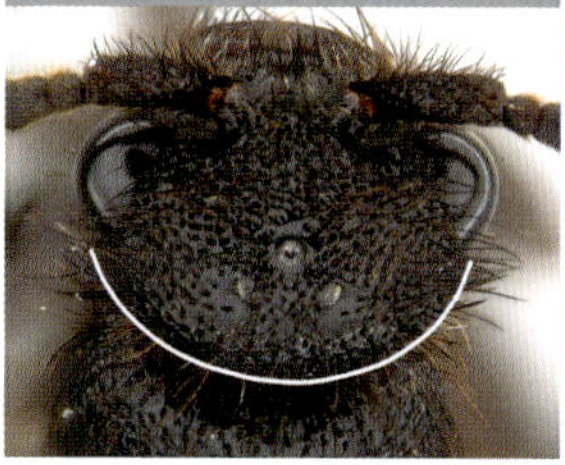

4.2.3a

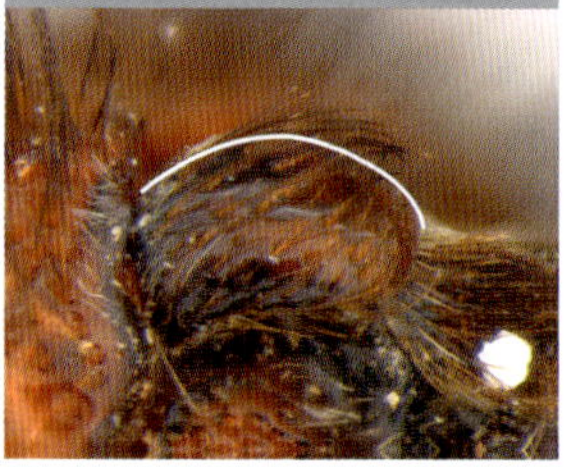

4.2.3b

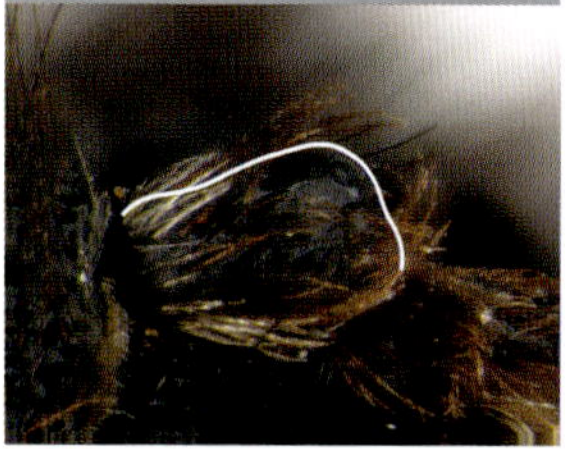

4.2.3c

4.2.4a

4.2.4b

4.2.5a

4.2.5b

4.3 SPHAEROPTHALMINAE: SPHAEROPTHALMINI: *SPHAEROPTHALMA* AND KIN

This tribe is dominated by nocturnal forms, and only five of the 11 North American genera of Sphaeropthalmini are treated in this chapter. Three of these genera, *Morsyma*, *Protophotopsis*, and *Stethophotopsis*, are represented by a single diurnal species in the USA. *Photomorphus* and *Sphaeropthalma* are large genera with most of their species active at night, but a moderately diverse minority of species in these genera are diurnal, especially in the humid eastern USA. Nocturnal forms of these two genera, and six additional genera, are treated in more detail in chapter 6.

Morsyma

IDENTIFICATION These males have round, bare eyes (subfamily character), and T2 with a plumose fringe and the mandible with a ventral tooth (tribe character). The only species in this genus, *M. ashmeadi*, can be separated from other genera by coloration and distribution. It is a small-bodied species from coastal areas of California, with an orange head and mesosoma, and dark black metasoma with a striking band of white plumose setae on T4. It is known in both wingless and fully winged forms. **FEMALE** Mickel recognized the female (in museum collections), but did not publish the association. Quintero & Cambra (2001b) described a female of *Morsyma*. Manley & Pitts (2002) described *Caenotilla* as the suspected female. Pitts (2007a) later formally sank *Caenotilla* under *Morsyma* and discovered that some males were fully winged. **DISTRIBUTION AND DIVERSITY** Just one species, *Morsyma ashmeadi*, which lives in coastal areas of California. **NOTES** Wingless males of *Morsyma ashmeadi* are treated in chapter 3 (3.5.4.1), while the winged forms fit in the *Sphaeropthalma*-like MCS in chapter 5 (5.7.8.1).

Photomorphus

IDENTIFICATION These males have round, bare eyes (subfamily character) and the mesosternal area armed with complexes of ridges (tribe character). This is the only predominantly diurnal genus in the eastern USA with the mesosternal area armed. Other members of this genus, and other similar genera, are discussed in the nocturnal section (chapter 6). **FEMALE** Females are known for most of the diurnal species. Very few of the nocturnal western species have formally published sex associations. **DISTRIBUTION AND DIVERSITY** There are about 10 diurnal species in central and eastern USA; more nocturnal ones live out West. **NOTES** These are usually pretty drab for diurnal mutillids; some eastern species seem to be crepuscular or facultatively nocturnal. **KEYS AND CHARACTERS** Males are sorted by characters of the mandible, ocelli size, mesosternal processes, and coloration. The diurnal males can be identified with the key by Schuster (1958).

Protophotopsis

IDENTIFICATION These males have round, bare eyes (subfamily character) and felt lines present on T2 and S2 (tribe character). This genus has a unique broad and subpetiolate T1 shape. It is the only member of Sphaeropthalmini in North America with an entirely black body, unarmed mesosternal area, and dense, coarse body sculpture. **FEMALES** Schuster (1947) based this genus on males only; Cambra & Quintero (1997) described one new species based on both sexes, and associated males with females for the three Neotropical species. **DISTRIBUTION AND DIVERSITY** One species lives in the USA (*Protophotopsis venenaria*), one lives in Central America, and two occur in South America. **NOTES** Various authors have disagreed about the tribal placement of this genus. Most recently, Brothers & Lelej (2017) placed the genus in the Dasymutillini, but we treat this as a member of the Sphaeropthalmini because of the unarmed axilla and presence of a felt line on S2. **KEYS AND CHARACTERS** Cambra & Quintero (1997) provided a functional key to the species. Because the species are rare and widely separated, distribution alone is usually diagnostic.

Figure	Taxon	Description
4.3.1a	*Morsyma ashmeadi*	Habitus: lateral view
4.3.1b	*Photomorphus alogus*	Habitus: lateral view
4.3.1c	*Protophotopsis venenaria*	Habitus: lateral view
4.3.1d	*Sphaeropthalma pensylvanica*	Habitus: lateral view
4.3.1e	*Sphaeropthalma unicolor*	Habitus: lateral view
4.3.1f	*Stethophotopsis maculata*	Habitus: lateral view

Sphaeropthalma

IDENTIFICATION These males have round, bare eyes (subfamily character) and the tergites usually with distinctly plumose fringes (tribe character). This genus is diverse and variable and serves somewhat as a dumping ground for nocturnal velvet ants that lack mesosternal armature. They can be identified only by ruling out the various other genera (see chapter 6). **FEMALE** Most of the diurnal and brightly colored species are known from both sexes. Less than 20% of the total species have formally published sex associations. **DISTRIBUTION AND DIVERSITY** Most species in this genus are nocturnal, but three eastern and central species are abundant and diurnal. There are also about 10 western species that are brightly colored and sometimes encountered in daylight, although they are more commonly collected at night. Only some of these are treated in chapter 5. **NOTES** The eastern and central diurnal species attack mud-nesting wasps and are common inside houses. The western diurnal forms are usually found on cool days late in the evening and are more colorful than their nocturnal congeners. **KEYS AND CHARACTERS** Differences in coloration are often adequate to separate many of the day-active species, but check chapter 6 for this genus. The best key to separate males of this genus was written by Schuster (1958), but it includes many outdated names and is missing many more recently described species.

Stethophotopsis

IDENTIFICATION These males have round, bare eyes (subfamily character) and T2 with a plumose fringe (tribe character). The only species in this genus, *S. maculata*, can be separated from other genera by coloration and distribution. It is a wingless orange species from the Madrean Archipelago with striking black patches at the base of T4. **FEMALE** The genus was first known from the male only. Pitts recognized and described the female (and another similar female that actually belongs to *Acanthophotopsis*) a few years later based on similar color and distribution. **DISTRIBUTION AND DIVERSITY** Just one species, *Stethophotopsis maculata*, which lives in the Madrean Archipelago. **NOTES** See the species account (3.4.16.1) for more information.

4.4 SPHAEROPTHALMINAE: EPHUTINI: *EPHUTA*

Ephuta

IDENTIFICATION These males have notched-ovate, bare eyes (subfamily character) and a slender cylindrical T1 shape (tribe character). This is the only genus in the tribe that occurs in North America. **FEMALE** Although relatively few species have males associated with females, the genus-level association was made obvious by the similar T1 shape. Schuster (1951) recognized the female of the type species, *E. scrupea*. In the USA, more males are known than females. **DISTRIBUTION AND DIVERSITY** There are 30 species in the USA and many more in Central and South America. **NOTES** *Ephuta* males are common in malaise traps and sometimes seen on leaves or flowers. This genus has a similar distribution to *Timulla*, with many eastern, central, and Arizonan species but very few in California. Due to the eye shape, this tribe was included in the Mutillinae until a recent phylogenetic study revealed that they belong in the Sphaeropthalminae. **KEYS AND CHARACTERS** Schuster (1951) wrote a key that works well, but it can be difficult to use because of reliance on comparative measurements of various structures. Schuster's key also suffers from reliance on color of the hypopygium, which is scored in a relatively subjective manner and is known to vary within a species. We have attempted to use this feature as a last resort after the reader has checked other more definitive features. Aside from coloration and sculpture of the tegula, most of the diagnostic features for species are found on the head. Mandible shape, ocelli size, and comparative size and shape of the subantennal basin and clypeal basin are the most important. In figures 4.2.2a–e, the subantennal basin is highlighted in green, and the clypeal basin is highlighted in blue.

4.5 MUTILLINAE: TROGASPIDIINI: *TIMULLA*

Timulla

IDENTIFICATION These males have notched-ovate, bare eyes (subfamily character) and sessile T1 shape (tribe character). This is the only genus in this tribe that occurs in the New World. **FEMALE** Because males carry females during courtship, a higher number of sex assocations were recognized earlier in this genus than the others. **DISTRIBUTION AND DIVERSITY** There are nearly 30 species in the USA, and many more in Central and South America. **NOTES** This is one of the few genera in the USA wherein males typically carry the females during courtship. For this reason, many mating pairs

Figure	Taxon	Description
4.4.1a	*Ephuta grisea*	Habitus: lateral view
4.4.2a	*Ephuta krombeini*	Frons: contorted mandible, subantennal basin forming tooth
4.4.2b	*Ephuta slossonae*	Frons: broad mandible, subantennal basin much shorter than clypeal basin
4.4.2c	*Ephuta sabaliana*	Frons: slender mandible, subantennal basin much shorter than clypeal basin
4.4.2d	*Ephuta scrupea*	Frons: slender mandible, subantennal basin as long as clypeal basin
4.4.2e	*Ephuta cephalotes*	Frons: subantennal basin as long as clypeal basin, vertex conical
4.4.3a	*Ephuta margueritae*	Tegula: mostly smooth
4.4.3b	*Ephuta pauxilla*	Tegula: coarsely rugose
4.4.3c	*Ephuta tegulicia*	Tegula: with sharp longitudinal carina

4.4.1a

4.4.2a

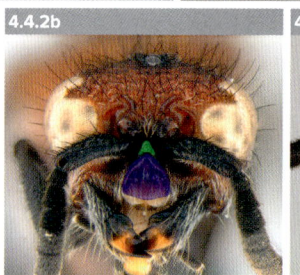

4.4.2b

4.4.2c

4.4.2d

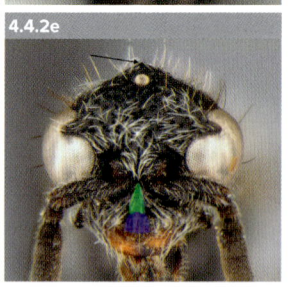

4.4.2e

4.4.3a

4.4.3b

4.4.3c

Figure	Taxon	Description
4.5.1a	*Timulla dubitata*	Habitus: lateral view
4.5.2a	*Timulla vagans*	Head: ocelli small; scape with brush
4.5.2b	*Timulla navasota coahuila*	Head: ocelli large; scape simple
4.5.3a	*Timulla subhyalina*	Mesosternal area: unarmed
4.5.3b	*Timulla vagans*	Mesosternal area: armed with transverse tubercle
4.5.4a	*Timulla dubitata*	Sternites: tubercles on S5–8
4.5.4b	*Timulla vagans*	Sternites: tubercles on S6–8 only
4.5.5a	*Timulla tyro*	T7: apex emarginate, transverse arcuate carina
4.5.5b	*Timulla navasota coahuila*	T7: apex rounded; raised "fin-like" carina
4.5.5c	*Timulla ferrugata*	T7: apex rounded; without distinct carina
4.5.5d	*Timulla vagans*	T7: apex rounded; Y-shaped carina

have been collected, and the ratio of species known from both sexes is higher than that seen in other genera. Some names treated as synonymous in the next chapter, particularly the subspecies designations, are based on soon-to-be published results by George Waldren. **KEYS AND CHARACTERS** Mickel's key (1937) is still useful, but some of the names are outdated. Microscopic differences in the mandible, clypeus, antenna, metasomal sternites, and pygidium (T7) are useful. The pygidial morphology presents some especially useful features; most of the species in the USA have a neat little Y-shaped carina (Fig. 4.5.5d).

4.6 MYRMOSINAE: *MYRMOSA* AND KIN

Myrmosa

IDENTIFICATION These males have ovate, hairy eyes (subfamily character) and S1 with a distinct basal tooth (genus character). They also show distinct constriction between each tergite. The two subgenera can be separated by the presence or absence of a tooth on S4. Members of *Myrmosa* (*Myrmosa*) have a tooth on S2, while *Myrmosa* (*Myrmosina*) species lack a tooth on S4. **FEMALE** Two of the North American species are recognized from both males and females. Both of the associated females belong to the subgenus *Myrmosa* (*Myrmosa*), so it is not clear yet whether there are significant differences between females of the subgenera. Unlocking the secrets of female morphology would be a huge step in determining whether these subgenera should be sunk or remain separated, or whether they even support raising *Myrmosa* (*Myrmosina*) to full genus status. **DISTRIBUTION AND DIVERSITY** Four males are recognized in the USA, two are associated with females. This genus also occurs in the Old World, in Europe and Asia. **NOTES** Strangely, when these insects die, they often raise their metasoma straight up, making them stand out among unsorted wasp specimens. Members of *Myrmosa* in North America are divided in two separate subgenera depending on the base of S2: *Myrmosa* (*Myrmosa*) species have a tooth there, while members of *Myrmosa* (*Myrmosina*) lack this tooth. **KEYS AND CHARACTERS** Krombein's (1940) key works well. New World males of *Myrmosa* (*Myrmosa*) are separated mainly by distribution and setal coloration, while *Myrmosa* (*Myrmosina*) species are separated by the size of their ocelli.

Myrmosula

IDENTIFICATION These males have ovate, hairy eyes (subfamily character) and S1 without a basal tooth (genus character). They also have the tergites more smoothly rounded into one another. **FEMALE** Many females are described, but only one formal sex association is recognized: *M. parvula*. **DISTRIBUTION AND DIVERSITY** Two males are recognized, and the genus is known only from Canada, Mexico, and the USA. **NOTES** Males of this genus could provide a good opportunity for a student project. Piles of undetermined specimens are available in numerous entomological collections. **KEYS AND CHARACTERS** The two known males are separated by color differences: *M. rufiventris* has the metasoma reddish, while *M. parvula* has the metasoma blackish.

4.6.1a

4.6.1b

Figure	Taxon	Description
4.6.1a	*Myrmosa bradleyi*	Habitus: lateral view
4.6.1b	*Myrmosula rufiventris*	Habitus: lateral view
4.6.2a	*Myrmosa unicolor*	Sternites: S1 and S2 with tooth
4.6.2b	*Myrmosa nocturna*	Sternites: S1 with tooth, S2 unarmed
4.6.2c	*Myrmosula rufiventris*	Sternites: S1 and S2 unarmed
4.6.3a	*Myrmosa unicolor*	Metasoma: tergites constricted
4.6.3b	*Myrmosula rufiventris*	Metasoma: tergites not constricted

4.6.2a

4.6.2b

4.6.2c

4.6.3a

4.6.3b

DAY FLIERS—DIURNAL MALE VELVET ANTS—COLOR SYNDROMES AND SPECIES

5.0 INTRODUCTION AND OVERVIEW

Nearly all male velvet ants are fully winged. The few completely wingless species (like *Stethophotopsis maculata*, 3.4.16.1) or species that always have tiny imperceptible wings (like *Myrmilloides grandiceps*, 3.3.6.1) are not treated here in chapter 5 but are only included in chapter 3 with their females. Some species, especially *Dasymutilla asopus* and their relatives (see 5.2.5.4), have the wings drastically shortened in some individuals. Since these species also have fully winged populations, and because their short wings are dark and obvious on the body, they are included here in chapter 5. In rare instances, fully winged males can have their wings mangled by predators or, very rarely, apparently chewed off by themselves on purpose. These insects are treated here in chapter 5, as well.

The key below will help you recognize the color syndrome of a winged velvet ant and get you closer to the target of a species identification.

1. Body cuticle mostly black, mesoscutum with dense white, yellow, orange, or red setae

    a. Head, mesosoma, and metasoma dorsal color mostly white....................**5.1 Desert MCS** (p. 180)

    b. Head, mesosoma, and metasoma dorsal color mostly yellow,
        orange, or red .... **5.2 Western MCS** (p. 186)

2. Head and mesosoma cuticle blackened, covered with black setae

    a. Dominant dorsal metasomal color mostly yellow, orange,
        or red, at least from T2 apex to T6 .... **5.3 Texan MCS** (p. 201)

    b. Metasoma partly reddish-orange, dominant color of
        apical tergites mostly black .... **5.6 Eastern MCS** (p. 252)

    c. Metasomal cuticle and setae entirely black .... **5.4 *Ephuta*-like MCS, in part** (p. 211)

3. Head and mesosoma cuticle blackened, largely covered with silvery or golden setae

    a. Metasomal cuticle black or dark reddish-brown,
        covered with black and silvery setae .... **5.4 *Ephuta*-like MCS, in part** (p. 211)

    b. Metasomal color (at least patch on T2) largely
        yellow, orange, or red .... **5.5 *Timulla*-like MCS** (p. 231)

4. Head and/or mesosoma with cuticle largely
    reddish-orange or brown .... **5.7 *Sphaeropthalma*-like MCS** (p. 262)

5.1 DESERT MALES COLOR SYNDROME

The Desert Males Color Syndrome (MCS) is a direct analog to the Desert Females Color Syndrome (FCS) (3.1). The males are recognized by the dark body cuticle and whitish dorsal setae, especially on the mesoscutum. A few of the species have the metasomal setae more yellow-tinted than those of the head and mesosoma. This color syndrome is predominantly found in arid regions of the southwestern USA. There are 12 species in the genus *Dasymutilla* with the Desert MCS, no other genera are known to fit in this color syndrome.

This color syndrome could be most easily confused with pale yellow species in the Western MCS (5.2.6.7,8). Some species with the *Ephuta*-like MCS (5.4) are similar to these, but the dorsal setae are sparser and the mesoscutum has the setae mostly black. Some species in the *Timulla*-like MCS (5.5.3.1) have the vertex and mesosomal dorsum with silvery setae, but their metasoma is partly reddened with sparse black and silvery setae.

All of the 12 males in the Cryptic FCS are known from both sexes. All but one of these is associated with a Desert FCS species, and some populations of the Madrean FCS species *Dasymutilla dionysia*. The Desert MCS has a few Mexican species but is rarely encountered around the world.

—Genus *Dasymutilla*—Eye circular, T1 shape petiolate, sculpture and setae dense

5.1.1 Dorsal setae of head and mesosoma grayish, tergal setae
usually pale yellow, Sonoran Desert .... *D. magna, D. connectens, D. eminentia, D. dionysia*

5.1.2 Pale dorsal setae often gray or yellow-tinted, Pacific and
adjacent areas .... *D. sackenii, D. albiceris, D. californica, D. coccineohirta, D. aureola*

5.1.3 Pale dorsal setae usually stark white, Sonoran Desert
and Arizona .... *D. nocturna, D. atricauda, D. thetis*

 5.1.1—*Dasymutilla magna* Desert males cluster

These species occur mainly in the Sonoran Desert in Arizona and California. The pale dorsal setae of the head and mesosoma are usually gray-tinted, while the metasomal setae tend to be yellow-tinted. Some pale yellow Western MCS males from desert habitats (5.2.7) are similar to this cluster.

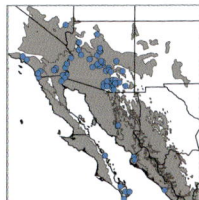

5.1.1.1—*Dasymutilla magna*

DAH-ZEE-MEW-TILL-UH MAGG-NUH

ETYMOLOGY From the Latin *magnus* "big." This species is larger than most other Desert form species. **FIELD IDENTIFICATION** Usually easy. This is a large-bodied species; the legs have the setae black; and the T2 disc setae are gray or yellowish only posteriorly. **LAB IDENTIFICATION** S2 is evenly convex with a distinct seta-filled pit mesally and is simply punctate laterally; and the hind trochanter is unarmed. **FEMALE** (3.1.2.1). Similar to male. **NOTES** This is the largest species in the cluster. Some populations have been found near San Diego and Los Angeles. They also extend south to Baja California Sur and Sinaloa in Mexico, where they seem to lose the yellowish tint on the apical tergites.

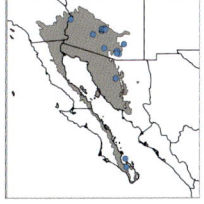

5.1.1.2—*Dasymutilla connectens*

DAH-ZEE-MEW-TILL-UH CUH-NECK-TENS

ETYMOLOGY Derived from the Latin word *conecto* "linked together." It is not clear why that name was chosen for this species. The male used to be named *D. ocydrome*. **FIELD IDENTIFICATION** Usually easy. This is a medium-sized species; the leg setae are mostly black; and the T2 disc setae (and often cuticle) are mostly yellowish. This species is rarer than *D. magna* and *D. eminentia*. **LAB IDENTIFICATION** S2 is mesally flattened with a patch of short, dense setae and the sculpture is punctate laterally; and the hind trochanter is unarmed. **FEMALE** (3.1.2.2). Similar to male. **NOTES** This is an easily recognized, though rarely collected, species; none of the authors has

collected either sex in the field (but one must have goals!). In both sexes, *D. connectens* is nearly identical in morphology to *D. nogalensis* (5.3.3.2), typically with the Texan MCS, but they have drastically different coloration. Until recently, the male was called *D. ocydrome* and that species was thought to vary in dorsal setal color from gray to bright orange. The type of *D. ocydrome* has the same coloration as *D. connectens* and is from Phoenix, where *D. connectens* occurs (but *D. nogalensis* does not). The darker orange-colored specimens of *D. ocydrome* are now recognized as a variant of *D. nogalensis* (5.2.5.2).

5.1.1.3—*Dasymutilla eminentia*
DAH-ZEE-MEW-TILL-UH EMM-INN-ENN-CHUH

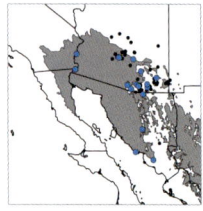

ETYMOLOGY Derived from the Latin word *eminentia* "prominence" or "protuberance," likely referencing the multiple bumps on S2. **FIELD IDENTIFICATION** Usually easy. This is a medium-sized species; the legs have mostly gray setae; the overall body shape is thicker than those of the others. **LAB IDENTIFICATION** S2 is mesally concave and scabrous laterally; and the hind trochanter is armed with a tooth. **FEMALE** (3.1.2.3). Similar to male. **NOTES** This is a variable species, which often has orange dorsal setae, placing some populations in the Western MCS (5.2.3.2).

5.1.1.4—*Dasymutilla dionysia*
DAH-ZEE-MEW-TILL-UH DYE-OWE-NYE-ZEE-UH

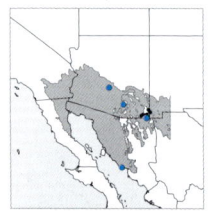

ETYMOLOGY In Greek mythology, Dionysus was the god of wine and fertility. The male used to be named *D. glycera*. **FIELD IDENTIFICATION** Often difficult. This is a small-sized species; the leg setae are mostly black; and the T2 disc setae are yellowish only posteriorly. This species is rarer than *D. magna* and *D. eminentia*. **LAB IDENTIFICATION** S2 is evenly convex

without a seta-filled pit and is simply punctate laterally; and the hind trochanter is unarmed. **FEMALE** (3.4.8.1). Different from male, with short setae and a contrasting pattern of orange, silvery, and black setae. **NOTES** The sex association was only recently discovered, and this color form has been collected alongside males of the more commonly encountered Western MCS (5.2.7.4) forms.

5.1.2—*Dasymutilla sackenii* Desert males cluster

These species occur mainly in the southern mountain ranges, the Central Valley, and adjacent Mojave Desert areas in California. The pale dorsal setae of the body are generally whitish, often slightly tinted pale yellow or gray. Some isolated populations of *Dasymutilla magna* (5.1.1.1) are known from southern Pacific areas in California. Many of these species are also treated in the Western MCS (5.2.4).

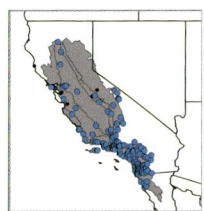

5.1.2.1—*Dasymutilla sackenii*
DAH-ZEE-MEW-TILL-UH SACKEN-EE-EYE

ETYMOLOGY This species was named for the Russian diplomat and entomologist Carl Robert Osten-Sacken, who worked mainly on flies. **FIELD IDENTIFICATION** Often difficult. This is the most common and widespread species in the cluster. The T2 disc setae are mostly black; and S2 has a mesal seta-filled pit. **LAB IDENTIFICATION** The head is nar-

row with a subtle posteromesal expansion; the axilla is truncate posteriorly; S2 has a large circular seta-filled pit; and the hypopygium is clearly longer than wide. **FEMALE** (3.1.4.1). Similar to male. **NOTES** Unlike the other species in this cluster, the Desert MCS form of *D. sackenii* is more common and widespread than the Western MCS form (5.2.4.6). It is generally the largest-bodied and most commonly encountered species in this cluster.

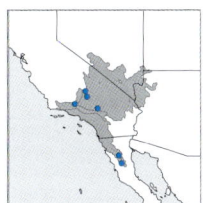

5.1.2.2—*Dasymutilla albiceris*
DAH-ZEE-MEW-TILL-UH AHL-BIH-SEHR-ISS

ETYMOLOGY From the Latin words *albus* "white" and *cerinus* "wax-colored" or "yellowish." **FIELD IDENTIFICATION** Relatively easy. This species is rare and has a restricted distribution. It has a thick body shape, and the T2 disc setae are black.

LAB IDENTIFICATION The head is narrow; the axilla is dentate posteriorly; S2 lacks a seta-filled pit; and the hypopygium is wider than long. **FEMALE** (3.1.5.1). Similar to male. **NOTES** This species is similar to the more widespread *D. vestita* (5.2.6.1), which has the hypopygium dentate. Although *D. albiceris* has the hypopygium unarmed, its shape is short and similar to that of *D. vestita*.

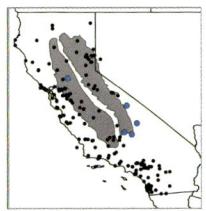

5.1.2.3—*Dasymutilla californica*
DAH-ZEE-MEW-TILL-UH CAL-IF-OR-NICK-UH

ETYMOLOGY This species was named after the state of California. The male used to be named *D. adbita*. **FIELD IDENTIFICATION** Difficult. The head is narrow, S2 lacks a seta-filled pit, and the T2 disc setae are usually entirely black. **LAB IDENTIFICATION** The head is narrow with a posteromesal expansion; the axilla is truncate posteriorly; S2 lacks a seta-filled pit and is evenly convex; the hypopygium is clearly longer than wide. **FEMALE** (3.1.4.2). Similar to male. **NOTES** This white form is rare and mostly seen around Sequoia National Park and the eastern Sierra Nevadas.

The more common Western MCS forms (5.2.4.1) are abundant throughout the Pacific and northwestern USA.

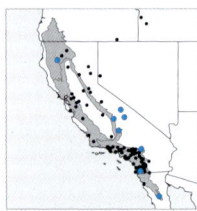

5.1.2.4—*Dasymutilla coccineohirta*
DAH-ZEE-MEW-TILL-UH COX-IN-EE-OH-HURT-UH

ETYMOLOGY From the Latin *coccineus* "red" or "scarlet" and *hirtus* "shaggy" or "hairy," in reference to the shaggy red dorsal setae of many populations. **FIELD IDENTIFICATION** Difficult. The head is narrow, S2 lacks a seta-filled pit, and the T2 disc often has at least half of the setae white. **LAB IDENTIFICATION** The head is narrow without a posteromesal expansion; the axilla is

truncate posteriorly; S2 lacks a seta-filled pit and is mesally flattened; and the hypopygium is clearly longer than wide. **FEMALE** (3.1.4.3). Similar to male. **NOTES** Unlike *D. aureola* and *D. californica*, Desert MCS males of *D. coccineohirta* are less common than Desert FCS females. White female *D. coccineohirta* were formerly called *D. clytemnestra* and are widespread in southern California. Males associated with those females, however, usually have the yellow to red setae typical for this species (5.2.4.2). Desert MCS males of *D. coccineohirta* are only rarely encountered in hotter portions of the California Central Valley or eastern slopes of the Sierra Nevada.

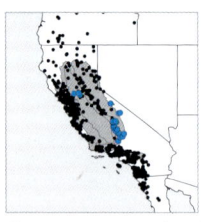

5.1.2.5—*Dasymutilla aureola*
DAH-ZEE-MEW-TILL-UH OUR-EE-OLL-UH

ETYMOLOGY Apparently derived from the Latin word *aureus* "golden," in reference to the golden-yellow dorsal setae of many populations. **FIELD IDENTIFICATION** Often difficult. The head is wider than in other species in this cluster, S2 has a seta-filled pit, and the T2 disc usually has the setae extensively white. **LAB IDENTIFICATION** The head is nearly as wide as the mesosoma; the axilla is dentate posteriorly; S2 has a small mesal seta-filled pit; and the hypopygium is longer than wide. **FEMALE** (3.1.4.4). Similar to male. **NOTES** This white form is rare, mostly seen in the southern Sierra Nevada and eastern slopes of the Sierra Nevada. The more common Western MCS forms (5.2.5.3) are abundant throughout the Pacific and northwestern USA.

▼ 5.1.3—*Dasymutilla nocturna* Desert males cluster

These species occur in Arizona and the Algodones dunes area in California. The pale dorsal setae are generally uniformly white. Some pale yellow Western MCS males from desert habitats (5.2.7) are similar to this cluster.

5.1.3.1—*Dasymutilla nocturna*

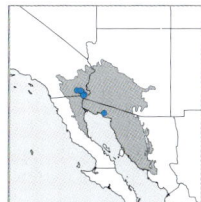

DAH-ZEE-MEW-TILL-UH KNOCK-TURN-UH

ETYMOLOGY From the Latin *nocturnus* "night active," in reference to the nocturnal behavior in some individuals. The male used to be called *D. subhyalina* for the lightened wings. **FIELD IDENTIFICA-TION** Usually easy. This is the only nocturnal *Dasymutilla* in the Algodones dunes; the

eyes are large and the wings are pale brown. **LAB IDEN-TIFICATION** The eyes and ocelli are distinctly enlarged; the axilla is truncate posteriorly; the T2 disc setae are mostly black; and the pygidium is mostly smooth with a posterior setal fringe. **FEMALE** (3.1.3.1). Similar to male, except the apical tergites have black setae. **NOTES** Like its relative *D. arenivaga* (5.2.7.2), this species is active both in daylight and after dark.

5.1.3.2—*Dasymutilla atricauda*

DAH-ZEE-MEW-TILL-UH AY-TRICK-AWE-DUH

ETYMOLOGY From the Latin words *ater* "black" and *cauda* "tail," in reference to the black tip of the metasoma in females. **FIELD IDENTIFICATION** Usually easy. In the Algodones dunes, this species has smaller eyes and darker wings than *D. noctur-na*. **LAB IDENTIFICATION** The

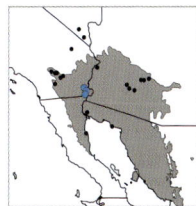

eyes and ocelli are not especially large; the axilla is weakly rounded posteriorly; the T2 disc setae are mostly black; and the pygidium is mostly smooth and has a posterior setal fringe. **FEMALE** (3.1.3.2). Similar to male, except the apical tergites have black setae.

NOTES This is a variable species, which often has brighter orange dorsal setae, placing some populations in the Western MCS (5.2.7.3). This Desert MCS form is apparently restricted to the Algodones dunes in California.

5.1.3.3—*Dasymutilla thetis*

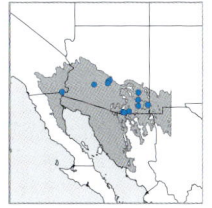

DAH-ZEE-MEW-TILL-UH THAY-TISS

ETYMOLOGY In Greek mythology, Thetis was a goddess of the seas and the mother of Achilles. The male used to be referred to as *D. candida*. **FIELD IDENTIFICATION** Usually easy. This is the only *Dasymutilla* male in Arizona with the pale dorsal setae entirely white. **LAB IDENTIFICATION** The eyes and ocelli are not especially large; the axilla is dentate; the T2 disc setae are mostly black; and the pygidium is marked with irregular striae and lacks an apical setal fringe. **FEMALE** (3.1.1.3). Similar to male but without any black setae. **NOTES** The sex association was

discovered using DNA sequence comparisons. Unlike the other thistledown velvet ants (5.2.7.1), the male and female both have white dorsal setae.

5.2 WESTERN MALES COLOR SYNDROME

The Western Males Color Syndrome (MCS) is a direct analog to the Western Females Color Syndrome (FCS) (3.2). The males are recognized by the dark body cuticle and yellow, orange, or red dorsal setae, especially on the mesoscutum and T2 or T3–6. This color syndrome is found throughout the central and western USA, and also includes the most common eastern species: *D. occidentalis*. There are 29 species in four genera with the Western MCS.

This color syndrome could most easily be confused with partly yellow species in the Desert MCS (5.1.1). A few species can have the mesosomal dorsum orange-red (5.2.8.2, 5.2.10.1), similar to the *Sphaeropthalma*-like MCS, but the orange cuticle is obscured by dense yellowish setae.

Females are recognized for 26 of the 29 (89%) Western MCS males. Most of them (81%) fit the Western FCS, but a few are included in the Eastern (23%), Desert (11%), or Madrean (7%) FCS.

The Western MCS has a few Mexican species but is rarely encountered around the world. The most notable exception would be *Quwitilla blattoserica* in Peru.

Quwitilla blattoserica from Peru

—Genus *Dasymutilla*—Eye circular, T1 shape petiolate

5.2.1 Tergal setae mostly orange or red with black band on T4–5 .... *D. occidentalis*

5.2.2 T2 setae or cuticle orange, T3–7 setae and cuticle black .... *D. asopus, D. bioculata, D. scitula*

5.2.3 Legs with extensive gray setae; setae from T2 fringe to T7 entirely yellow, orange, or red .... *D. foxi, D. eminentia, D. furina*

5.2.4 Setae from T2 fringe to T7 entirely yellow, orange, or red; species mainly from Pacific areas .... *D. californica, D. coccineohirta, D. aureola, D. testaceiventris, D. flammifera, D. sackenii*

5.2.5 Setae from T2 fringe to T7 entirely yellow, orange, or red; T2 cuticle largely orange or red beneath yellow, orange, or red setae .... *D. bioculata, D. nogalensis, D. erythrina, D. neomexicana*

5.2.6 Setae from T2 fringe to T7 entirely yellow, orange, or red; T2 cuticle generally blackish; species from western mountains, Great Plains, and Texas .... *D. vestita, D. calorata, D. myrice, D. stevensi*

5.2.7 Setae from T2 fringe to T7 entirely yellow, orange, or red; T2 cuticle generally blackish; species from southwestern desert areas .... *D. gloriosa, D. arenivaga, D. atricauda, D. dionysia, D. pseudopappus*

—Genus *Pseudomethoca*—Eye circular, T1 shape sessile

5.2.8 Mesonotum and T3–6 with yellow or orange setae .... *Ps. propinqua, Ps. aureovestita*

—Genus *Sphaeropthalma*—Eye circular, tergal fringes plumose

5.2.9 Entire body covered with somewhat dense erect orange setae .... *S. edwardsii*

—Genus *Timulla*—Eye ovate with notch, T1 shape cylindrical

5.2.10 Vertex and mesonotum with orange cuticle and dense orange setae .... *T. barbigera*

▼ 5.2.1.1—*Dasymutilla occidentalis* Western males cluster

DAH-ZEE-MEW-TILL-UH OX-SID-DENT-AHL-ISS

ETYMOLOGY From the Latin *occident* "western." The species was originally named by the Swedish entomologist Caroli Linnaeus in 1758—because even the eastern half of North America is west of Sweden. **FIELD IDENTIFICATION** Usually easy. The apical tergites have mostly yellow, orange, or red setae with a pre-apical black setal band on T4–5. In the eastern USA, this is the only *Dasymutilla* species with reddish setae on the apical tergites. **LAB IDENTIFICATION** The axilla is truncate; S2 has a large circular seta-filled pit mesally; and T7 has an erect fringe of setae posteriorly. **FEMALE** (3.2.8.2, 3.6.1.1). Similar to male. **NOTES** Populations in the eastern USA usually have bright red setae, while western forms are generally duller yellow or orange. Some populations, especially in Florida, have the pre-apical black band obliterated, so the setae from T3–7 are entirely orange or red. These could be confused with *D. calorata*, but that species occurs only in the Great Plains and Texas and has an ovate seta-filled pit on S2 (it is generally circular in *D. occidentalis*).

Typical *Dasymutilla occidentalis* with black setal band on T4–5.

Photo by Efram Goldberg

Atypical *D. occidentalis* from Florida without black setal band on T4–5.

▼ 5.2.2—*Dasymutilla asopus* Western males cluster

These species have T2 partly covered with yellowish setae and T3–7 covered entirely with black setae. One species (*D. scitula*) occurs in the Great Basin, one (*D. asopus*) occurs in the Great Plains, and one (*D. bioculata*) is widespread in central, northern, and western North America and sometimes even in Florida.

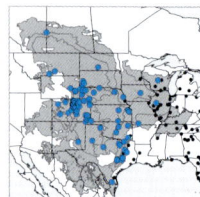

5.2.2.1—*Dasymutilla asopus*

DAH-ZEE-MEW-TILL-UH AY-SOHP-USS

ETYMOLOGY In Greek mythology, Asopus is the name of a god of rivers. This male color form represents the oldest of many names that have been applied to this variable species. **FIELD IDENTIFICATION** Usually easy. The body is thicker than the others in this cluster and S2 has a longitudinal setal-ridge mesally. **LAB IDENTIFICATION** The axilla is dentate, and S2 has a longitudinal row of stiff setae mesally. **FEMALE** (3.6.5.2). The female fits the Eastern FCS. **NOTES** This is a widespread and variable species with males having the Sphaeropthalma-like (5.7.4.1) and Eastern (5.6.4.2) Color Syndromes. Surprisingly, all the females (3.6.5.2) have the Eastern FCS. This and related species often have wings that are greatly reduced in size.

5.2.2.2—*Dasymutilla bioculata*

DAH-ZEE-MEW-TILL-UH BYE-OCK-EW-LAW-TUH

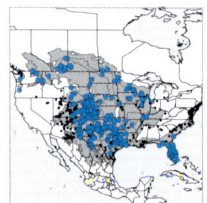

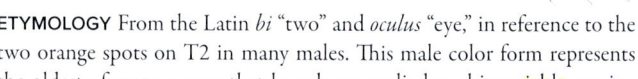

ETYMOLOGY From the Latin *bi* "two" and *oculus* "eye," in reference to the two orange spots on T2 in many males. This male color form represents the oldest of many names that have been applied to this variable species. **FIELD IDENTIFICATION** Usually easy. In Florida, eastern Texas, and far northwestern North America this is the only species in the cluster. The setae are shorter and smoother than those of *D. scitula* and the body is more slender than that of *D. asopus*. **LAB IDENTIFICATION** The axilla is truncate; S2 is variable, with or without a small ovate seta-filled pit mesally; and T7 has an erect posterior setal fringe. **FEMALE** Females in the central and northern USA (3.2.8.1) are loosely similar to the male but usually have some silvery tergal setae; many other females have the Eastern FCS (3.6.5.3). **NOTES** *Dasymutilla bioculata* is one of the most widespread and variable species in North America, and many of the color variants used to be recognized as distinct species. This variant of *D. bioculata* was found mating with three different "species" of females— great evidence for combining some names.

ABOVE RIGHT: *Dasymutilla bioculata* from Florida.

RIGHT: *Dasymutilla bioculata* from Oklahoma.

Dasymutilla bioculata from Nebraska.

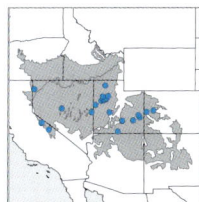

5.2.2.3—*Dasymutilla scitula*
DAH-ZEE-MEW-TILL-UH SKIT-OO-LUH

ETYMOLOGY From the Latin *scitulus* "beautiful" or "elegant." **FIELD IDEN-TIFICATION** Easy. Has longer, denser setae than others in the cluster and lives only in the Intermountain West. **LAB IDENTIFICATION** The axilla is truncate; S2 has a large circular seta-filled pit mesally; and T7 lacks an erect setal fringe posteriorly. **FEMALE** (3.2.6.3). Nearly identical to the male except for the wings. **NOTES** This species is uncommon in museums but can be locally abundant in Nevada and Utah. The overlapping distribution and matching color patterns made the sex association pretty obvious.

5.2.3—*Dasymutilla foxi* Western males cluster

These are Western MCS species with extensive gray setae on the legs and with T3–7 having uniformly orange or red dorsal setae. They are found in Arizona, New Mexico, and western Texas.

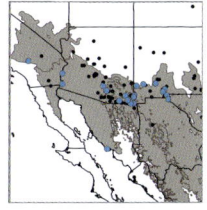

5.2.3.1—*Dasymutilla foxi*
DAH-ZEE-MEW-TILL-UH FOX-EYE

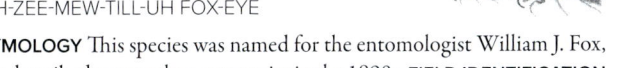

ETYMOLOGY This species was named for the entomologist William J. Fox, who described many velvet ant species in the 1890s. **FIELD IDENTIFICATION** Usually easy. The T2 disc setae and cuticle are largely orange or red, and the dorsal setae are often bright red. **LAB IDENTIFICATION** S2 is simply punctate laterally; the hind trochanter is unarmed; and the hind tibia is apically dilated

and flattened with a concave inner surface. **FEMALE** (3.1.3.3, 3.2.7.1, 3.4.6.3). Often similar to male. **NOTES** This Western MCS form is usually bright scarlet but sometimes more orange-tinted. This is a variably colored species that often expresses the *Timulla*-like MCS (5.5.2.2).

5.2.3.2—*Dasymutilla eminentia*
DAH-ZEE-MEW-TILL-UH EMM-INN-ENN-CHUH

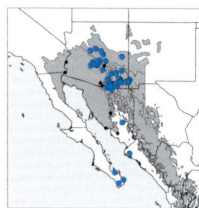

ETYMOLOGY Derived from the Latin word *eminentia* "prominence" or "protuberance," likely referencing the multiple bumps on S2. **FIELD IDENTIFICATION** Moderately easy. The T2 disc cuticle and setae are mostly black, and the dorsal body setae are usually orange. **LAB IDENTIFICATION** S2 is mesally concave and scabrous laterally; the hind trochanter is armed with a tooth; and the hind tibia is cylindrical. **FEMALE** (3.1.2.3). Similar to male. **NOTES** This is a variable species that often has paler gray and yellow dorsal setae, placing some populations in the Desert MCS (5.1.1.3).

5.2.3.3—*Dasymutilla furina*
DAH-ZEE-MEW-TILL-UH FEW-REE-NUH

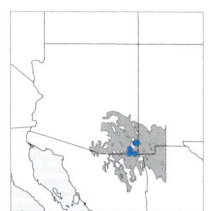

ETYMOLOGY Perhaps based on the Latin *furia* "madness" or "rage." **FIELD IDENTIFICATION** Usually easy. The T2 disc often has the cuticle orange and the setae mainly black; the dorsal body setae are usually yellow or orange in coloration. This species has a narrow geographic range. **LAB IDENTIFICATION** S2 is mesally concave and scabrous laterally; the hind trochanter is armed with a tooth; and the hind tibia is cylindrical. **FEMALE** (3.2.7.2). Similar to male. **NOTES** This species is common around the city of Douglas, Arizona, but is absent from other areas. It is nearly identical to *D. eminentia*, except for the orange cuticular color. It may eventually be recognized as a conspecific variant of that more widespread species.

▼ 5.2.4—*Dasymutilla californica* Western males cluster

These species have the dorsal metasomal setae, at least from the T2 fringe to T7, entirely yellow, orange, or red in coloration. They are common and widespread in the Pacific states (California, Oregon, and Washington), and sporadically found in Idaho, Nevada, and Utah. They are generally absent from the hot desert regions of California and Nevada. Species from the following clusters, especially *D. bioculata* and *D. vestita*, often overlap in distribution with these species and the following clusters should be considered.

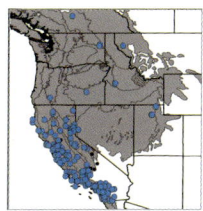

5.2.4.1—*Dasymutilla californica*
DAH-ZEE-MEW-TILL-UH CAL-IF-OR-NICK-UH

ETYMOLOGY This species was named after the state of California. The male used to be named *D. adbita*. **FIELD IDENTIFICATION** Very difficult. Common moderately small species. T2 has the cuticle entirely black, the T2 disc setae are mainly black; S2 is mesally convex and lacks a mesal pit; and the sternal fringes are mainly black. **LAB IDENTIFICATION** The head is narrow and posteromesally expanded; the axilla is truncate posteriorly; S2 is convex without a pit; S7 is elongate with the apex smooth; and T7 lacks a posterior setal fringe. **FEMALE** (3.2.4.1, 3.2.8.6). Similar to male. **NOTES** Together with *D. coccineohirta*, this is the most commonly encountered Pacific *Dasymutilla* male. This color form is more common and widespread than the Desert MCS forms (5.1.2.3), being found as far east as Utah.

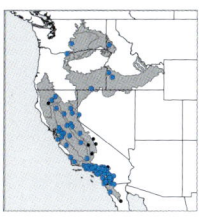

5.2.4.2—*Dasymutilla coccineohirta*
DAH-ZEE-MEW-TILL-UH COX-IN-EE-OH-HURT-UH

ETYMOLOGY From the Latin *coccineus* "red" or "scarlet" and *hirtus* "shaggy" or "hairy," in reference to the shaggy red dorsal setae of many populations. **FIELD IDENTIFICATION** Very difficult—but you like a challenge. Common moderately small species. T2 has the cuticle entirely black, the T2 disc setae are usually largely yellow, orange, or red; S2 is mesally flattened without a pit; and the sternal fringes are mostly black. **LAB IDENTIFICATION** The head is narrow but not posteromesally expanded; the axilla is truncate posteriorly; S2 is mesally flattened without a pit; S7 is elongate with the apex smooth; and T7 lacks a posterior setal fringe. **FEMALE** (3.1.4.3, 3.2.4.2). Usually similar to male, but in some cases, females

have much lighter dorsal setae than males. **NOTES** This is a common species in the Pacific states. Western MCS forms are more common than the Desert MCS form (5.1.2.4), and males from regions dominated by Desert FCS females frequently have bright red dorsal setae. In southern California, this species shows dual sex-limited mimicry, but in northern populations, the males and females are similar in appearance.

5.2.4.3—*Dasymutilla aureola*
DAH-ZEE-MEW-TILL-UH OUR-EE-OLL-UH

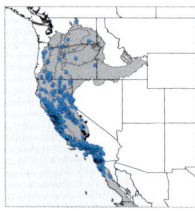

ETYMOLOGY Apparently derived from the Latin word *aureus* "golden," in reference to the golden-yellow dorsal setae of many populations. **FIELD IDENTIFICATION** Usually difficult. Common moderately small species. The head is wider than others in this cluster; T2 often has the cuticle orange, and the T2 disc setae are often mostly yellow, orange, or red; S2 has a small seta-filled pit mesally; and the sternal fringes are usually mostly yellow, orange, or red. **LAB IDENTIFICATION** The head is wide; the axilla is dentate posteriorly; S2 has a small mesal seta-filled pit;

S7 is elongate with the apex smooth; and T7 lacks a posterior setal fringe. **FEMALE** (3.2.1.1). Similar to male. **NOTES** Males are rarer than their conspecific females and also rarer than males of other Pacific *Dasymutilla* species.

Most males that I've seen were crawling on the ground, rather than flying. Perhaps their large head makes flight more difficult; this behavior would make males less likely to be collected in malaise traps or with insect nets.—KEVIN

5.2.4.4—*Dasymutilla testaceiventris*
DAH-ZEE-MEW-TILL-UH TEST-ACE-IVV-ENT-RISS

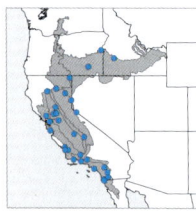

ETYMOLOGY From the Latin *testaceus* "brick" and *venter* "belly," perhaps in reference to the sternal setae usually being reddish. **FIELD IDENTIFICATION** Usually difficult. Rare, moderately small species. T2 often has the cuticle orange; the T2 disc setae are largely yellow, orange, or red; S2 is mesally flattened, without a pit; and the sternal fringes are usually mostly yellow, orange,

or red. **LAB IDENTIFICATION** The hypopygium is unique: apically truncate and punctate throughout. The head is narrow; the axilla is truncate posteriorly; S2 is mesally flattened, without a pit; and T7 lacks a posterior fringe. **FEMALE** Unknown. **NOTES** This is the last remaining unassociated *Dasymutilla* species in the Pacific region. Previous authors thought it might eventually be matched to *D. flammifera*, but the real male of that species was recently associated. Superficially, this male looks like an intermediate between *D. aureola* and *D. coccineohirta*. So might the true *D. testaceiventris* female be hidden among females of one of those species in a museum collection somewhere waiting to be discovered.

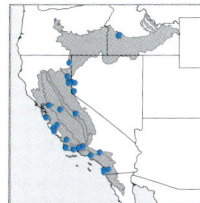

5.2.4.5—*Dasymutilla flammifera*

DAH-ZEE-MEW-TILL-UH FLAMM-IH-FAIR-UH

ETYMOLOGY Derived from the Latin *flamma* "blaze" or "fire," in reference to the long, bright red dorsal setae. **FIELD IDENTIFICATION** Usually difficult. Rare, moderately large species. T2 has the cuticle entirely black; the T2 disc setae are mostly yellow, orange, or red; S2 is mesally convex, without a pit; and the sternal fringes are mostly black. **LAB IDENTIFICATION**

The head is narrow; the axilla is truncate posteriorly; S2 is convex, without a pit; S7 is elongate with the apex smooth; and T7 has a distinct setal fringe posteriorly. **FEMALE** (3.2.4.3). Similar to male. **NOTES** Until recently, this species was known only from the bright red females found in southern California's coastal mountains. Since then, males have been recognized and both sexes were found in the Great Basin and adjacent areas in California, Idaho, and Nevada. Northern populations used to be named *D. dorippa* and have the dorsal setae duller orange.

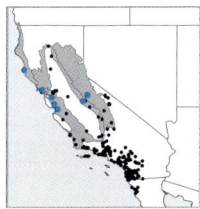

5.2.4.6—*Dasymutilla sackenii*

DAH-ZEE-MEW-TILL-UH SACKEN-EE-EYE

ETYMOLOGY This species was named for the Russian diplomat and entomologist Carl Robert Osten-Sacken, who worked mainly on flies. **FIELD IDENTIFICATION** Often difficult. Rare, moderately large species. T2 has the cuticle entirely black; the T2 disc setae are mostly black; S2 has a large circular seta-filled pit mesally; and the sternal fringes are mostly black. **LAB**

IDENTIFICATION The head is narrow; the axilla is truncate posteriorly; S2 has a large circular seta-filled pit mesally; S7 is elongate with the apex smooth; and T7 lacks a posterior setal fringe. **FEMALE** (3.2.4.4). Similar to male. **NOTES** Like the orange females of *D. sackenii*, Western MCS males are rarer than the Desert MCS form (5.1.2.1) and mostly known from northern coastal areas.

▼ 5.2.5—*Dasymutilla bioculata* Western males cluster

These species occur throughout the western USA but are rarer in the Pacific states and absent from California's Central Valley and coastal areas. They usually have the cuticle of T2 orange beneath the yellow, orange, or red setae. In many cases, the orange patch is missing or totally blocked by the dorsal setae, so these should be cross-referenced with the following Western MCS clusters (5.2.6, 5.2.7), whenever possible.

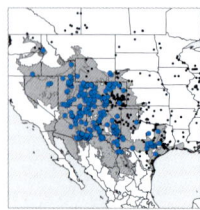

5.2.5.1—*Dasymutilla bioculata*
DAH-ZEE-MEW-TILL-UH BYE-OCK-EW-LAW-TUH

ETYMOLOGY From the Latin *bi* "two" and *oculus* "eye," in reference to the two orange spots on T2 in many males. **FIELD IDENTIFICATION** Often difficult. The dorsal setae are usually dense and orange; and the cuticle of T3–7 is black. This is the most commonly encountered species in this cluster. **LAB IDENTIFICATION** The axilla is truncate; the hind tibia is cylindrical; S2 is

variable, usually without a seta-filled pit mesally; and T7 has an erect posterior setal fringe. T2 always has a yellow-orange cuticular patch beneath the orange or red setae. **FEMALE** (3.2.3.4, 3.2.8.1). Generally similar to male but often with black or silver setae on the apical tergites. **NOTES** *Dasymutilla bioculata* is one of the most widespread and variable species in North America, and many of the color variants used to be recognized as distinct species. This species is most commonly encountered in sandy habitats, where they attack sand wasps (Crabronidae: Bembicinae: Bembicini).

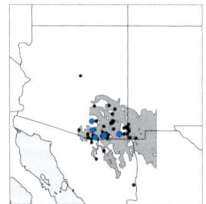

5.2.5.2—*Dasymutilla nogalensis*
DAH-ZEE-MEW-TILL-UH NO-GALL-ENN-SISS

ETYMOLOGY Named after the city of Nogales, Arizona. **FIELD IDENTIFICATION** Often difficult. The dorsal setae are orange; the apical tergites often have the cuticle orange beneath the orange setae. This is a rare species from Arizona. **LAB IDENTIFICATION** The axilla is truncate; S2 is mesally flattened, with a patch of short, dense setae; and the hind tibia is cylindrical. **FEMALE** (3.2.3.5, 3.3.4.5). Usually darker than this male form. **NOTES** Until recently, this male form was treated under the name *D. ocydrome*. That species was separated from *D. nogalensis* males (5.3.3.2) by the yellow or orange dorsal setae of the head and mesosoma, while *D. nogalensis* males were thought to have black head and mesosoma setae in both sexes. Typical *D. ocydrome* specimens (including the holotype) have paler grayish yellow dorsal setae and occur only in localities

that overlap with *D. connectens* (3.1.2.2), including Phoenix, Arizona, and that species was recently synonymized under *D. connectens*. The brighter orange males formerly called *D. ocydrome* were "orphaned" until the recent discovery of rare, aberrant *D. nogalensis* females with orange setae on the head and mesosoma (3.2.3.5). No structural differences have been found between *D. connectens* and either color form of *D. nogalensis*, and the two species may eventually be recognized as synonyms. Until additional data, perhaps including DNA sequences, are available, the species can be separated by the setal color and distribution. *Dasymutilla nogalensis* occurs in mid- to high-elevation mountainous regions of Arizona and always have the metasomal setae dark orange or red (and usually has the head and mesosoma with black setae), while *D. connectens* lives in lowland hot areas of the Sonoran Desert and has the dorsal setae gray or pale yellow.

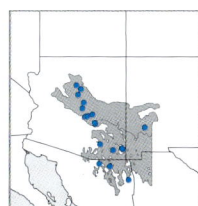

5.2.5.3—*Dasymutilla erythrina*
DAH-ZEE-MEW-TILL-UH AIR-EETH-RYE-NUH

ETYMOLOGY From the Ancient Greek *erythros* "red," in reference to the dorsal setal color. **FIELD IDENTIFICATION** Often easy. The dorsal setae are brighter red than the other species in this cluster. T2 has the setae and cuticle mostly red, even basally. Found only in mountainous areas of Arizona.

LAB IDENTIFICATION The axilla is dentate; S2 has a large flat circular seta-filled pit mesally; and the hind tibia is dilated with the apex flared. **FEMALE** (3.2.2.3). Similar to male. **NOTES** This species is rare in Arizona but is apparently the most common species in mountainous areas of Mexico (p. 322). Most of the Mexican populations have gray setae on the legs, but all the specimens we have seen in Arizona have the leg setae black. Superficially, this species could be easily confused with *D. vestita* (5.2.6.1) in Arizona, and a microscope is often useful.

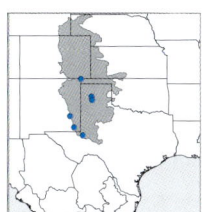

5.2.5.4—*Dasymutilla neomexicana*
DAH-ZEE-MEW-TILL-UH NEE-OH-MECK-SICK-ANN-UH

ETYMOLOGY This species was named after the state of New Mexico. **FIELD IDENTIFICATION** Usually easy. The dorsal setae are shorter and sparser than the others; T2 usually has the pale yellow cuticular patch covering almost the entire T2 disc. **LAB IDENTIFICATION** The axilla is dentate; S2 has a longitudinal setal row mesally; and the hind tibia is cylindrical. **FEMALE** Unknown; could be *D. montivagoides*. **NOTES** This species is nearly identical to *D. hector* (5.7.3.1) and can be separated from that species only by the black mesosoma cuticle color (orange in *D. hector*). Both of these species are structurally identical to *D. asopus* and might eventually be recognized as color variants of that widespread species. Like *D. asopus*, many males have wings that are greatly reduced in size, like the individual pictured in figure 5.2.5.4a.

- -
This is one of the more commonly encountered *Dasymutilla* males in the Texas Panhandle. The tiny wings are adorable!—AARON
- -

▼ 5.2.6—*Dasymutilla vestita* Western males cluster

Most of these species are restricted to the Great Plains, New Mexico, and Texas, but *D. vestita* is widespread, being found virtually throughout the central and western USA, and *D. stevensi* extends west into Arizona. These species have the dorsal metasomal setae, at least from T2 fringe to T7, entirely yellow, orange, or red and the metasomal cuticle entirely blackish. Some variants of *D. occidentalis* (5.2.1) can be confused for these species. The other Western clusters (5.2.4, 5.2.5, 5.2.7) should also be consulted, when possible.

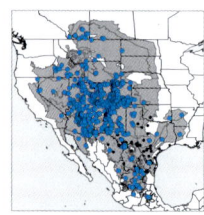

5.2.6.1—*Dasymutilla vestita*
DAH-ZEE-MEW-TILL-UH VESS-TEE-TUH

ETYMOLOGY Apparently from the Latin *vestis* "garment," in reference to the colorful dorsal setae. **FIELD IDENTIFICATION** Usually difficult. Common and widespread. The body shape is shorter and thicker than most others; the T2 disc setae are usually extensively yellow, orange, or red. **LAB IDENTIFICATION** Immediately recognizable by the posterolaterally dentate hypopygium. The tegula is mostly bare, and the axilla is dentate posteriorly. **FEMALE** (3.2.2.1, 3.2.7.3). Similar to male. **NOTES** This is one of the most widespread and common species in the central and western USA and even expands north into Alberta and Saskatchewan, Canada, and south to Oaxaca, Mexico. They vary in dorsal setal color from yellow to bright red, with the brightest red specimens often being found at higher elevations. In some hotter regions of Texas and Arizona they match the Texan Color Syndrome (5.3.1.3).

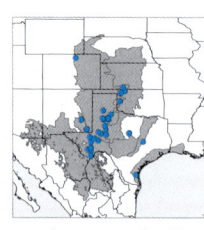

5.2.6.2—*Dasymutilla calorata*
DAH-ZEE-MEW-TILL-UH CAL-OWE-RATT-UH

ETYMOLOGY From the Latin *caloratus* "heated," perhaps in reference to its occurrence in Texas, which has hotter weather than Minnesota, where the author of the species, Clarence Mickel, lived. **FIELD IDENTIFICATION** Often difficult. Large-sized species; the T2 disc setae are mostly black; and S2 has an ovate seta-filled pit mesally. **LAB IDENTIFICATION** The tegula is mostly bare; the axilla is truncate posteriorly; S2 has a distinct ovate seta-filled pit mesally; and T7 has an erect setal fringe posteriorly. **FEMALE** (3.2.3.1). Similar to male. **NOTES** This species can be confused with some specimens of *D. occidentalis* but has the seta-filled pit on S2 ovate in shape, rather than circular, and is known only from the central USA. It is structurally similar to *D. klugii* from the Texan MCS (5.3.1.1).

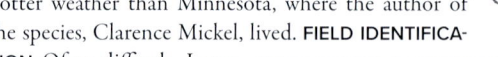

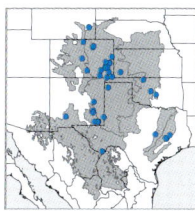

5.2.6.3—*Dasymutilla myrice*
DAH-ZEE-MEW-TILL-UH MY-REECE

ETYMOLOGY Possibly derived from the Ancient Greek *myrike* "tamarisk."
FIELD IDENTIFICATION Often difficult. Medium-sized species; the T2 disc
setae are black; and S2 lacks a pit. **LAB IDENTIFICATION** The tegula is mostly bare; the axilla is truncate posteriorly; S2 is mesally flattened without a
pit; and T7 has an erect setal fringe posteriorly. **FEMALE** Unknown. **NOTES**
Based on distribution and similar body size, this might be the male of *D. leda* (3.2.3.2). *Dasymutilla*
leda and *D. myrice* are both structurally similar to the
respective sexes of *D. gorgon* (3.3.3.1, 5.3.1.2).

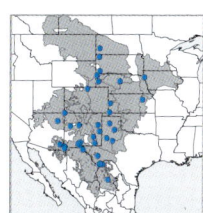

5.2.6.4—*Dasymutilla stevensi*
DAH-ZEE-MEW-TILL-UH STEVENS-EYE

ETYMOLOGY This species was named after O. A. Stevens, who collected the
type specimen in 1923. **FIELD IDENTIFICATION** Often difficult. Usually
small and slender-bodied species; the T2 disc setae are largely orange. **LAB
IDENTIFICATION** The tegula is usually entirely covered with sparse setae; the
axilla is weakly rounded or subdentate posteriorly; S2 is mesally flattened
without a pit; and T7 has an erect setal fringe
posteriorly. **FEMALE** (3.2.3.3). Similar to male.
NOTES Other than the widespread *D. vestita*,
this is the only species of the cluster to expand
west into eastern Arizona, where it could be
confused with *D. dionysia* or *D. atricauda*,
which are treated on the following pages.

 ## 5.2.7—*Dasymutilla gloriosa* Western males cluster

Many of these species are restricted to hot desert habitats in Arizona, California, and
Nevada, although *D. gloriosa* and *D. pseudopappus* are more widespread, extending
east into western Texas and north into Idaho. These species have the dorsal metasomal
setae, at least from T2 fringe to T7, entirely yellow, orange, or red and the metasomal
cuticle entirely blackish. The previous Western clusters should be consulted (5.2.4,
5.2.5, 5.2.6), when possible.

5.2.7.1—*Dasymutilla gloriosa*
DAH-ZEE-MEW-TILL-UH GLOW-REE-OWE-SUH

ETYMOLOGY From the Latin *gloriosa* "glorious," likely referring to the fluffy angel-like appearance
of females. **FIELD IDENTIFICATION** Very difficult. Often the commonest species in hot deserts; the

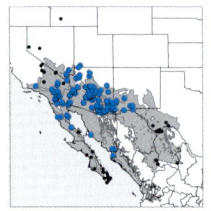

LEFT: *Dasymutilla gloriosa* males

RIGHT: *Dasymutilla gloriosa* gynandromorph

coloration is variable, with the dorsal setae usually dull yellow or orange; S2 has an ovate seta-filled pit. **LAB IDENTIFICATION** The eyes and ocelli are not enlarged; the axilla is truncate posteriorly; S2 has an ovate seta-filled pit mesally; and T7 usually lacks a setal fringe posteriorly. **FEMALE** (3.1.1.1). Very different from male, covered entirely with long white setae. **NOTES** This species is the poster child for dual sex-limited mimicry in velvet ants. The sex association was recognized only after the discovery of a gynandromorph specimen, which had the metasoma mostly composed of male tissue on the right side and female on the left side.

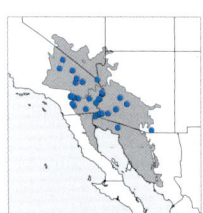

5.2.7.2—*Dasymutilla arenivaga*
DAH-ZEE-MEW-TILL-UH UH-RENN-IVV-AH-GUH

ETYMOLOGY From the Latin *arena* "sand" and *vagus* "wandering," in reference to the desert habitat. The male used to be named *D. megalophthalma* for the large eyes. **FIELD IDENTIFICATION** Often easy. This is the only Western MCS *Dasymutilla* species active at night. Diurnal specimens are harder to identify. The eyes and ocelli are enlarged; the dorsal body setae are pale yellow or orange; the T2 disc setae are usually mostly black; and S2 lacks a pit. **LAB IDEN-**

TIFICATION The eyes and ocelli are enlarged; the antennal scrobe with a distinct dorsal carina; the axilla is truncate posteriorly; S2 lacks a pit; and T7 usually has an erect setal fringe posteriorly. **FEMALE** (3.2.5.2, 3.2.6.1). Similar to male, except metasomal apex is usually black. **NOTES** Structurally, this species is very similar to *D. nocturna* from the Desert MCS (5.1.3.1), except it is more widespread than that species. Both of these species are facultatively nocturnal, being active during the day and at night.

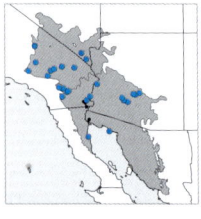

5.2.7.3—*Dasymutilla atricauda*
DAH-ZEE-MEW-TILL-UH AY-TRICK-AWE-DUH

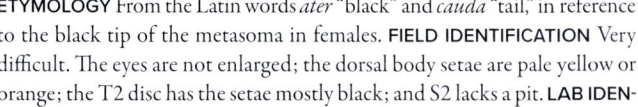

ETYMOLOGY From the Latin words *ater* "black" and *cauda* "tail," in reference to the black tip of the metasoma in females. **FIELD IDENTIFICATION** Very difficult. The eyes are not enlarged; the dorsal body setae are pale yellow or orange; the T2 disc has the setae mostly black; and S2 lacks a pit. **LAB IDEN-TIFICATION** The eyes and ocelli are not enlarged; the antennal scrobe lacks a dorsal carina; the axilla is weakly dentate posteriorly; S2 lacks a pit; and T7 has an erect setal fringe posteriorly. **FEMALE** (3.2.6.2). Similar to male, except apical tergites with black setae. **NOTES** This Western MCS form is more common and widespread than the Desert MCS form (5.1.3.2) from the Algodones dunes in California. Together with *D. arenivaga* and *D. gloriosa*, these males are often found feeding on Desert Milkweed (*Asclepias erosa*) in the Mojave Desert.

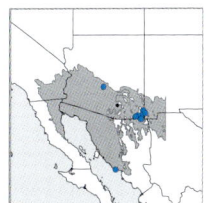

5.2.7.4—*Dasymutilla dionysia*
DAH-ZEE-MEW-TILL-UH DYE-OWE-NYE-ZEE-UH

ETYMOLOGY In Greek mythology, Dionysus was the god of wine and fertility. The male used to be named *D. glycera*. **FIELD IDENTIFICATION** Difficult. The dorsal setae are pale yellow or orange; the T2 disc setae are usually largely yellow; and S2 lacks a pit. **LAB IDENTIFICATION** The eyes and ocelli are not enlarged; the antennal scrobe has a low dorsal carina; the axilla is weakly dentate posteriorly; S2 lacks a pit; and T7 has an erect setal fringe posteriorly. **FEMALE** (3.4.8.1). Different from male, with short setae and a contrasting pattern of orange, silvery, and black setae. **NOTES** The sex association was only recently discovered, and while sexual dimorphism in color patterns is common in velvet ants, this is the first known example of a Western MCS male with a Madrean FCS female.

While collecting in the San Bernardino National Wildlife Refuge (with a permit) in 2020, I found a few females of this species and got excited, since the sex association for this species was a huge mystery, and I didn't even have a good guess for the male sex. One of the females was running with her metasoma raised in the air, and I saw a male "dive-bomb" the ground and start chasing her. Luckily, I was able to catch this male with a net, and then later the female with a plastic tube. I was totally shocked to see this combination of male and female color patterns.—KEVIN

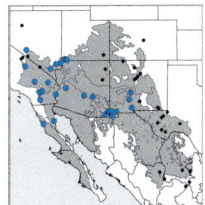

5.2.7.5—*Dasymutilla pseudopappus*
DAH-ZEE-MEW-TILL-UH SOO-DOE-PAPP-USS

ETYMOLOGY From the Ancient Greek words *pseudo* "false" and *pappus* "woolly seed"; the name is a reference to the resemblance of these wasps to creosote fruit. Until recently, the male used to be called *D. phaon*. **FIELD IDENTIFICATION** Sometimes easy. The dorsal setae are often bright scarlet red, and S2 lacks a pit. **LAB IDENTIFICATION** The eyes and ocelli are not enlarged; the antennal scrobe has a distinct dorsal carina; the axilla is truncate posteriorly; S2 lacks a pit; and T7 has an erect setal fringe posteriorly. **FEMALE** (3.1.1.2). Very different from male; covered entirely with long white setae. **NOTES** The sex association for *D. pseudopappus* was only recently discovered. Previously, *D. phaon* was recognized as the most widespread species known only from

males, and *D. pseudopappus* was the most widespread unassociated female. Both species are structurally similar to *D. flammifera* (5.2.4.5). We are confident that *D. phaon* should be sunk into *D. pseudopappus*.

Dasymutilla pseudopappus

 ### 5.2.8—*Pseudomethoca propinqua* Western males cluster

These are the only *Pseudomethoca* with the Western Color Syndrome. Sometimes they could be confused for *Dasymutilla* in the field, except members that genus have the T1 shape petiolate.

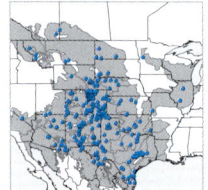

5.2.8.1—*Pseudomethoca propinqua*
SOO-DOH-METH-OWE-KUH PRO-PINK-WUH

ETYMOLOGY From the Latin *propinquus* "near," "neighboring," or "resembling." The author suggested this was similar to the male of *D. vestita*. **FIELD IDENTIFICATION** Usually easy. The dorsal mesosomal cuticle is orange. **LAB IDENTIFICATION** The head is rounded and much narrower than the mesosoma; the tibial spurs are black or white. **FEMALE** (3.2.11.1, 3.6.10.4). Similar to male, except apical tergites with black and silvery setae. **NOTES** This species is commonly encountered throughout the western and central USA.

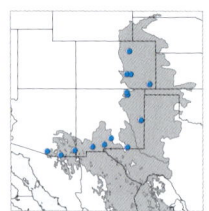

5.2.8.2—*Pseudomethoca aureovestita*
SOO-DOH-METH-OWE-KUH OUR-EE-OH-VEST-EE-TUH

ETYMOLOGY From the Latin *aureus* "golden" and *vestis* "garment," in reference to the coloration. **FIELD IDENTIFICATION** Usually easy. The dorsal mesosomal cuticle is black beneath the orange setae. **LAB IDENTIFICATION** The head is somewhat quadrate and nearly as wide as the mesosoma; the tibial spurs are white. **FEMALE** (3.2.10.2). Similar to male, except apical tergites with black and silvery setae. **NOTES** This is an uncommon species.

▼ 5.2.9.1—*Sphaeropthalma edwardsii*

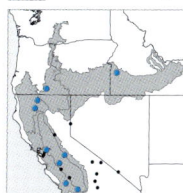

S-FAIR-OPP-THALL-MUH EDWARDS-EE-EYE

ETYMOLOGY This species was named after its collector, Henry Edwards, an English stage actor, writer, and entomologist in the late 19th century. **FIELD IDENTIFICATION** Usually easy. Usually nocturnal. None of the other *Sphaeropthalma* have dark cuticle with dense reddish setae and dark brown wings. **LAB IDENTIFICATION** The tergal fringes are plumose; the body cuticle is blackish; the wings are dark brown; S2 lacks a felt line; and the dorsal body setae are orange or red. **FEMALE** (3.2.12.2). Similar to male. **NOTES** This species could be confused for some Western MCS *Dasymutilla* (5.2.4) or for *Sphaeropthalma marpesia* individuals with the *Timulla*-like MCS (5.5.10.1). A few other nocturnal species, such as *S. mendica*, have the cuticle dark, but their dorsal body setae are sparser and generally white.

▼ 5.2.10.1—*Timulla barbigera*

TIM-EWE-LUH BAR-BIJJ-AIR-UH

ETYMOLOGY From the Latin *barba* "beard" and *-gera* "having," likely in reference to the setal brush on the male's scape. This color form used to be named *T. b. rohweri* after the entomologist S. A. Rohwer. **FIELD IDENTIFICATION** Easy. This is the only *Timulla* species with dense orange dorsal setae on the head and mesosoma. **LAB IDENTIFICATION** The scape has a sparse brush of whitish setae; the mandible is unarmed ventrally; the mesosternal area is armed with a smooth tubercle; the forewing is uniformly dark brown; T7 has a Y-shaped carina. **FEMALE** (3.6.16.2). Has sparser setae and a black and whitish tergal setal pattern. **NOTES** This form is less common than other *T. barbigera* specimens (5.7.16.2), which have mostly black setae and fit the *Sphaeropthalma*-like Color Syndrome.

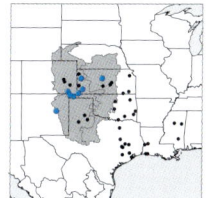

5.3 TEXAN MALES COLOR SYNDROME

The Texan Males Color Syndrome (MCS) is a direct analog to the Texan Females Color Syndrome (FCS) (3.3). The males are recognized by the entirely blackened head and mesosoma and predominantly orange or red metasoma. This color syndrome is found throughout the central and western USA and also includes the most common eastern species: *D. occidentalis*. There are 19 species in four genera with the Texan MCS.

This color syndrome could most easily be confused with the *Timulla*-like and Eastern MCS. Unlike the *Timulla*-like MCS, the head and mesosoma have the setae entirely (or almost entirely)

ABOVE: *Nemka philippa* from India.

TOP RIGHT: *Trogaspidia fervida* from Indonesia.

RIGHT: *Mimecomutilla aurinigra* from Democratic Republic of Congo.

black. Unlike the Eastern MCS, the metasoma is mostly orange, at least from the T2 fringe to T6. A few Eastern species (5.6.1.1, 5.6.7) have the metasomal cuticle largely orange, but their apical tergites have mostly black setae and they live farther northeast than most species in the Texan MCS. Some specimens of *Lomachaeta* (5.5.6) and *Ephuta* (5.5.11) in the *Timulla*-like MCS have the head and mesosoma mostly black, but, for simplicity, no members of those genera are treated here in the Texan MCS.

Females are recognized for 17 of the 19 (89%) Texan MCS males. More than half of them (53%) fit the Texan FCS, but a few are included in the *Timulla*-like (17%), Desert (17%), or Madrean (11%) FCS.

Many species in the Texan MCS also occur in Mexico, especially in the Chihuahuan Desert (7.3.1). This color syndrome is not especially common around the world, but various species have loosely similar coloration, such as *Nemka philippa* from India, *Trogaspidia fervida* from Indonesia, and *Mimecomutilla aurinigra* from the Democratic Republic of the Congo.

—Genus *Dasymutilla*—Eye circular, T1 shape petiolate, sculpture and setae dense

5.3.1 Tergal cuticle entirely black, species from Great Plains
and Texas .... *D. klugii, D. gorgon, D. vestita*

5.3.2 T2 disc with orange cuticular patch beneath orange
setae, species from Texas and Great Plains .... *D. waco, D. serenitas, D. bioculata*

5.3.3 T2 cuticle without distinct orange patch, usually entirely black,
but sometimes largely reddish-orange beneath the red setae,
species from Western states and deserts .... *D. magnifica, D. nogalensis, D. satanas,*
D. gloriosa, D. pseudopappus

5.3.4 Cuticle and setae of T2–6 orange or reddish .... *D. snoworum, D. sicheliana*

—Genus *Pseudomethoca*—Eye circular, T1 shape sessile

5.3.5 T2–7 with dense yellow or orange setae .... *Ps. pigmentata, Ps. brazoria*

—Genus *Timulla*—Eye ovate with notch, T1 shape cylindrical

5.3.6 T2–7 with cuticle and setae orange or golden .... *T. grotei, T. navasota coahuila, T. suspensa*

—Genus *Myrmosula*—Eye ovate with setae

5.3.7 Metasoma entirely orange with sparse black setae .... *M. rufiventris*

5.3.1—*Dasymutilla klugii* Texan males cluster

These species usually occur in New Mexico, Texas, and the Great Plains, but *D. klugii* is more widespread in the western USA. They have the metasomal cuticle entirely black. For hotter deserts in western Texas and New Mexico, compare these species with the *D. magnfica* cluster (5.3.3).

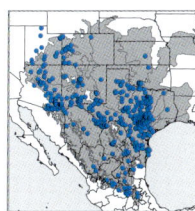

5.3.1.1—*Dasymutilla klugii*
DAH-ZEE-MEW-TILL-UH KLUG-EE-EYE

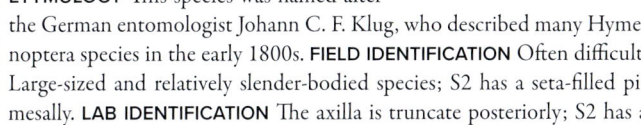

ETYMOLOGY This species was named after the German entomologist Johann C. F. Klug, who described many Hymenoptera species in the early 1800s. **FIELD IDENTIFICATION** Often difficult. Large-sized and relatively slender-bodied species; S2 has a seta-filled pit mesally. **LAB IDENTIFICATION** The axilla is truncate posteriorly; S2 has a distinct, ovate seta-filled pit mesally; the fringes of S2–6 have the setae mostly black mesally; and T7 has an erect setal fringe posteriorly. **FEMALE** (3.3.1.2). Similar to male. **NOTES** This is the largest Texan MCS species in Texas and the Great Plains. Farther west, they can often be confused for *D. magnifica* (5.3.3.1), which has the sternal fringes mostly covered with reddish setae.

Sternal fringes black mesally

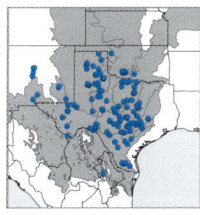

5.3.1.2—*Dasymutilla gorgon*
DAH-ZEE-MEW-TILL-UH GORR-GONN

ETYMOLOGY In Greek mythology, gorgons were female monsters that had snakes for hair and could turn people to stone with their gaze; the most famous gorgon was Medusa. **FIELD IDENTIFICATION** Difficult. Medium-sized, relatively slender-bodied species; S2 lacks a pit. **LAB IDENTIFICATION** The axilla is truncate posteriorly; S2 lacks a seta-filled pit; the fringes of S2–6 have the setae mostly black mesally; and T7 has an erect setal fringe posteriorly. **FEMALE** (3.3.3.1). Similar to male but with larger head. **NOTES** Even though this is the type species for the genus *Dasymutilla*, the male was only recently recognized (Manley et al., 2020).

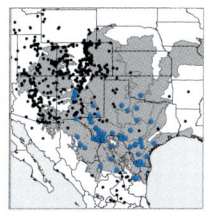

5.3.1.3—*Dasymutilla vestita*
DAH-ZEE-MEW-TILL-UH VESS-TEE-TUH

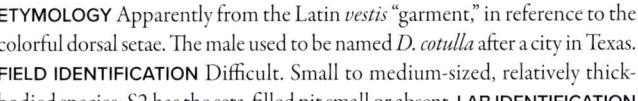

ETYMOLOGY Apparently from the Latin *vestis* "garment," in reference to the colorful dorsal setae. The male used to be named *D. cotulla* after a city in Texas. **FIELD IDENTIFICATION** Difficult. Small to medium-sized, relatively thick-bodied species; S2 has the seta-filled pit small or absent. **LAB IDENTIFICATION** Immediately recognizable by the posterolaterally dentate hypopygium. The

axilla is dentate posteriorly; and S2 has the seta-filled pit small or absent. **FEMALE** (3.3.2.2). Similar to male. **NOTES** Western MCS forms of *D. vestita* (5.2.6.1) usually have a distinct seta-filled pit on S2, but these Texan forms often have the pit reduced or absent.

▼ 5.3.2—*Dasymutilla waco* Texan males cluster

These species occur mainly in Texas. They have the metasomal cuticle mostly black, except T2 has the cuticle. Many species from hotter deserts in the *D. magnfica* cluster (5.3.3) also have T2 partly red or orange. The orange cuticular patch on T2 is often difficult to see beneath the orange setae, so these should be cross-referenced with the previous cluster.

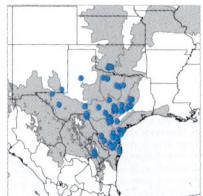

5.3.2.1—*Dasymutilla waco*
DAH-ZEE-MEW-TILL-UH WAY-COE

ETYMOLOGY This species was named for the city of Waco, Texas. **FIELD IDENTIFICATION** Usually easy. The dorsal setae are sparse and the orange cuticular patch covers most of the T2 disc. **LAB IDENTIFICATION** The axilla is dentate posteriorly; S2 has a longitudinal row of setae mesally. **FEMALE** (3.3.2.1). Similar to male. **NOTES** Overall, this species has shorter, sparser setae and more squat body proportions than similarly colored species.

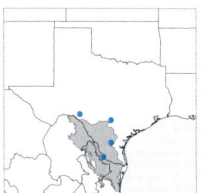

5.3.2.2—*Dasymutilla serenitas*
DAH-ZEE-MEW-TILL-UH SEHR-ENN-EE-TUSS

ETYMOLOGY From the Latin *serenitas* "clearness" or "calmness"; it is not clear why this name was chosen. **FIELD IDENTIFICATION** Difficult. This species is densely setose with a small orange patch on T2, like *D. bioculata*. **LAB IDENTIFICATION** The axilla is truncate posteriorly; S2 has a seta-filled pit mesally; and T7 lacks a posterior setal fringe. **FEMALE** Unknown. **NOTES** This species is similar to Texan forms of *D. bioculata*. It is rarer and apparently restricted to south Texas, T7 lacks a posterior setal fringe, and the clypeus also differs (see key in appendix, 9.3.5.3). In the clypeus and genitalia, this species is structurally similar to *D. quadriguttata*

and its allies. Based on similar coloration, overlapping distribution, and similarity to *D. quadriguttata*, the best candidate for the female sex is *D. wileyae* (3.3.4.2), but there is currently not enough sufficient evidence to associate these species.

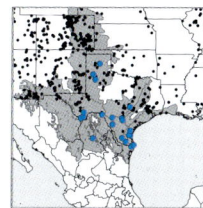

5.3.2.3—*Dasymutilla bioculata*
DAH-ZEE-MEW-TILL-UH BYE-OCK-EW-LAW-TUH

ETYMOLOGY From the Latin *bi* "two" and *oculus* "eye," in reference to the two orange spots on T2 in many males. This color form used to be named *D. chiron*. **FIELD IDENTIFICATION** Difficult. This species is densely setose with a small orange patch on T2, like *D. serenitas*. **LAB IDENTIFICATION** The axilla is truncate posteriorly; S2 with or without a seta-filled pit mesally; and T7 has an erect setal fringe posteriorly. **FEMALE** (3.3.4.3). Similar to male. **NOTES** *Dasymutilla bioculata* is apparently the most widespread and variable species in North America (5.2.2.2, 5.2.5.1, 5.6.4.3).

5.3.3—*Dasymutilla magnifica* Texan males cluster

These species have the metasomal cuticle (at least T2 basally and most of the sternites) blackish and the tergites (at least from the T2 fringe to T7) yellow, orange, or red. They are found mainly in the western hot deserts in Arizona and California, but some of them extend east into Texas and can be confused with the *D. klugii* cluster (5.3.1). Many of them have the metasomal cuticle partly reddened, and they should be compared with the *D. waco* (5.3.2) and *D. snoworum* (5.3.4) clusters.

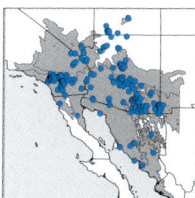

5.3.3.1—*Dasymutilla magnifica*
DAH-ZEE-MEW-TILL-UH MAGG-NIFF-ICK-UH

ETYMOLOGY From the Latin *magnificus* "great," "noble," or "eminent," in reference to its size and bright color. **FIELD IDENTIFICATION** Often difficult. Large-bodied species; the metasomal setae are often bright red; the T2 disc setae are largely orange or red; and S2 has a large seta-filled pit mesally. **LAB IDENTIFICATION** The axilla is truncate posteriorly; S2 has a distinct ovate seta-filled pit mesally; the fringes of S2–6 have the setae largely reddish; and T7 has an erect setal fringe posteriorly. **FEMALE** (3.3.1.1). Similar to male. **NOTES** This is the largest and most commonly

Sternal fringes
entirely reddish

encountered Texan MCS species in the Mojave and Sonoran Deserts. They can often be found feeding on the nectar of desert trees like mesquite (*Prosopis* spp.), Palo Verde (*Parkinsonia florida*), and Desert Ironwood (*Olneya tesota*).

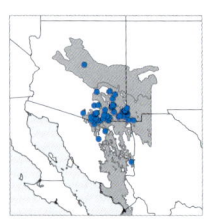

5.3.3.2—*Dasymutilla nogalensis*
DAH-ZEE-MEW-TILL-UH NO-GALL-ENN-SISS

ETYMOLOGY This species was named after the city of Nogales, Arizona. The male used to be named *D. atrifulva* after the black and red coloration. **FIELD IDENTIFICATION** Often difficult. Small to medium-bodied species; the metasomal setae are often bright red; the T2 disc setae and cuticle are often mostly red or orange; and S2 has a mesal patch of short, dense setae. **LAB IDENTIFICATION** Axilla truncate posteriorly; S2 mesally flattened with a patch of short, dense setae; fringes of S2–6 with setae largely reddish; T7 without posterior fringe. **FEMALE** (3.3.4.5).

Similar to male. **NOTES** In both males and females, *D. nogalensis* is nearly identical in morphology to *D. connectens* (3.1.2.2, 5.1.1.2) with the Desert MCS, but they have drastically different coloration. A few populations have the head and mesosoma with orange setae dorsally (5.2.5.2), more closely fitting the Western MCS.

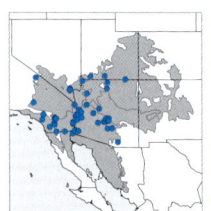

5.3.3.3—*Dasymutilla satanas*
DAH-ZEE-MEW-TILL-UH SAY-TAN-USS

ETYMOLOGY This species was named after Satan, likely in reference to the distribution of this species in the "hellish" southwestern hot deserts (although we are quite fond of them). **FIELD IDENTIFICATION** Often difficult. Large-bodied species; the metasomal setae are pale orange; the T2 disc setae are largely orange; and S2 has a large seta-filled pit mesally. **LAB IDENTIFICATION** The axilla is truncate posteriorly; S2 has a distinct ovate seta-filled pit mesally; the fringes of S2–6 have the setae mostly black mesally; and T7 lacks a setal fringe posteriorly. **FEMALE** (3.1.2.4, 3.2.5.1). Very different from male, dorsally with long, pale yellow or bright orange setae. **NOTES** This species is similar to *D. gloriosa*, but it has some subtle head shape differences (see key in the appendix: 9.3.5.3). Furthermore, in areas where these species overlap, *D. gloriosa* males usually have the head and mesosoma with yellow or orange dorsal setae.

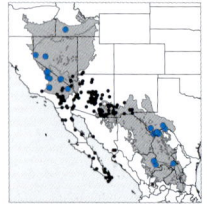

5.3.3.4—*Dasymutilla gloriosa*
DAH-ZEE-MEW-TILL-UH GLOW-REE-OWE-SUH

ETYMOLOGY From the Latin *gloriosa* "glorious," likely referring to the fluffy angel-like appearance of this wasp. This Texan MCS form male used to be called *D. chisos* after the Chisos Mountains in Big Bend National Park, Texas. **FIELD IDENTIFICATION** Often difficult. Small to medium-bodied species; the metasomal setae are usually orange; the T2 disc setae are mostly black;

and S2 has a large seta-filled pit mesally. **LAB IDENTIFICATION** The axilla is truncate posteriorly; S2 has a distinct, ovate, seta-filled pit mesally; the fringes of S2–6 have the setae mostly black mesally; and T7 lacks an erect setal fringe posteriorly. **FEMALE** (3.1.1.1). Very different from male, and covered with long white setae. **NOTES** This species is most prevalent in hot desert habitats but occurs sporadically in sandy low-elevation habitats in the Great Basin in Idaho and Nevada. Although this color form is most prevalent in the Chihuahuan Desert in Texas, the Texan MCS form also occurs in those previously mentioned Great Basin habitats. Like *D. pseudopappus*, males from hot deserts in Arizona and California are more likely to fit the Western MCS (5.2.7.1).

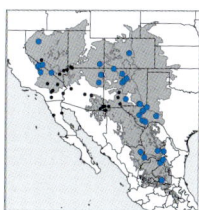

5.3.3.5—*Dasymutilla pseudopappus*
DAH-ZEE-MEW-TILL-UH SOO-DOE-PAPP-USS

ETYMOLOGY From the Ancient Greek *pseudo* "false" and *pappus* "woolly seed"; the name is a reference to the resemblance of these wasps to creosote fruit. Until recently, the male used to be called *D. phaon*. **FIELD IDENTIFICATION** Often difficult. This is a medium- to large-bodied species; the metasomal setae are often bright red; T2 has the setae mostly red; S2 lacks a seta-filled pit. **LAB IDENTIFICATION** The axilla is truncate posteriorly; S2 is convex without a pit; the fringes of S2–6 have the setae largely reddish; and T7 has a fringe of erect setae posteriorly. **FEMALE** (3.1.1.2). Very different from male, and covered with long white setae. **NOTES** While this species occurs throughout desert environments, it is also found at high elevations in Arizona, Colorado, and New Mexico. Like *D. gloriosa*, males from hot deserts in Arizona and California are more likely to fit the Western MCS (5.2.7.5).

5.3.4—*Dasymutilla snoworum* Texan males cluster

These species are mainly found in Arizona and New Mexico. They have the cuticle and setae of T2–6 entirely orange or red. Sometimes they have subtle grayish setae on the head and mesosoma, so they should be cross-checked against some of the *Dasymutilla* clusters in the *Timulla-like* MCS (5.5.2, 5.5.5).

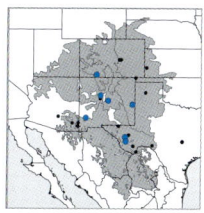

5.3.4.1—*Dasymutilla snoworum*
DAH-ZEE-MEW-TILL-UH SNOW-ORR-UMM

ETYMOLOGY The suffix *-orum* indicates a species named in honor of a family of people. This species was named for the entomologist Francis H. Snow, who established the insect collection at University of Kansas back in 1870, and his son, Frank. **FIELD IDENTIFICATION** Often difficult. The tergal setae are paler orange than *D. sicheliana* and S2 lacks a seta-filled pit. **LAB**

IDENTIFICATION The axilla is dentate posteriorly; the apices of the mid and hind femora are truncate; and S2 lacks a pit. **FEMALE** (3.4.4.1). Different from male; the head and mesosoma have dense golden setae, and the metasoma has a contrasting pattern of black and golden setae. **NOTES** The male of this species has been found mainly in New Mexico, but females are widespread from Arizona to southern Texas. Perhaps some of the *Timulla*-like MCS males (5.5.5.3, 5.5.5.4) are conspecific with this species and simply color variants of *D. snoworum*.

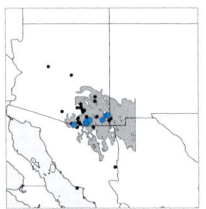

5.3.4.2—*Dasymutilla sicheliana*
DAH-ZEE-MEW-TILL-UH SICHEL-EE-ANN-UH

ETYMOLOGY This species was named for the French entomologist, Frédéric J. Sichel, who worked mainly in the 1800s. **FIELD IDENTIFICATION** Often difficult. The tergal setae are brighter red than *D. snoworum* and S2 has a seta-filled pit mesally. **LAB IDENTIFICATION** The axilla is truncate posteriorly; the apices of the mid and hind femora are rounded; and S2 has a seta-filled pit mesally. **FEMALE** (3.4.6.1). Somewhat similar to male. **NOTES** This is an uncommon male color variant that is usually found in far southeastern Arizona. Most males of *D. sicheliana* have noticeable gray or silver setae on the head and mesosoma, matching the *Timulla*-like MCS (5.5.2.1).

5.3.5—*Pseudomethoca pigmentata* Texan males cluster

These are the only *Pseudomethoca* with the Texan Color Syndrome. Sometimes they could be confused for *Dasymutilla* in the field.

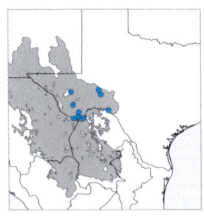

5.3.5.1—*Pseudomethoca pigmentata*
SOO-DOE-METH-OWE-KUH PIG-MEN-TOT-UH

ETYMOLOGY From the Latin *pigmentum* "color" or "paint." **FIELD IDENTIFICATION** Often difficult. The dorsal setae are denser and shaggier than in *Ps. brazoria*. **LAB IDENTIFICATION** The head is somewhat quadrate and nearly as wide as the mesosoma; and the tibial spurs are usually lighter than the tibia. **FEMALE** (3.3.7.2). Similar to male. **NOTES** Males of this species are rare in collections. This species is similar to *Ps. aureovestita* from the Western MCS (5.2.8.2).

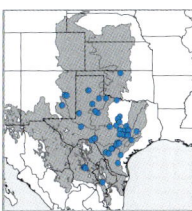

5.3.5.2—*Pseudomethoca brazoria*
SOO-DOE-METH-OWE-KUH BRAW-ZORR-EE-UH

ETYMOLOGY This species was named after Brazos County, Texas. **FIELD IDENTIFICATION** Often difficult. The dorsal body setae are sparser and flat-

ter than in *Ps. pigmentata*. **LAB IDENTIFICATION** The head is rounded and much narrower than the mesosoma; and the tibial spurs are black like the tibia. **FEMALE** (3.3.7.1). Similar to male. **NOTES** Males of this species are rare in collections. This species is structurally similar to *Ps. propinqua* from the Western MCS (5.2.8.1).

▼ 5.3.6—*Timulla grotei* Texan males cluster

These are the only *Timulla* species with the Texan MCS. *Timulla* males with the *Timulla*-like MCS (5.5.12, 5.5.13, 5.5.14) often have the silvery mesosomal setae sparse, and they can easily be confused with these species. Similarly, the Eastern MCS *Timulla* males (5.6.7) can be similar to these but have mostly black setae on the tergites.

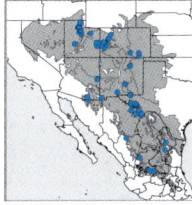

5.3.6.1—*Timulla grotei*
TIMM-EWE-LUH GROTE-EYE

ETYMOLOGY When Blake described this species in 1871, he named this species after his friend A. R. Grote. **FIELD IDENTIFICATION** Often difficult. This diurnal species has a dense white setal brush on the scape. **LAB IDENTIFICATION** The ocelli are small to moderate; the scape has a dense white setal brush; the mandible has a ventral tooth basally; and T7 has a Y-shaped carina. **FEMALE** (3.5.7.1). Not similar to male, the female matches the *Timulla*-like FCS. **NOTES** This is apparently the only dark-bodied *Timulla* male in the northwestern USA, and the thick white setal brush on the scape is usually obvious (at least in the lab). It can be confused with some populations of *T. vagans* (5.6.7.1) and is apparently a close relative of that species and *T. suspensa*.

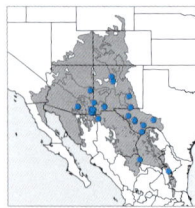

5.3.6.2—*Timulla navasota coahuila*
TIMM-EWE-LUH NAW-VUH-SOE-TUH COE-UH-WEE-LUH

ETYMOLOGY This subspecies was named for the Mexican state of Coahuila. **FIELD IDENTIFICATION** Sometimes easy. This is the only nocturnal *Timulla* in Arizona and New Mexico. **LAB IDENTIFICATION** The ocelli are large; the scape lacks a setal brush; the mandible is unarmed ventrally; and T7 has an erect longitudinal ridge. **FEMALE** (3.5.5.1). Not similar to male, the female matches the *Timulla*-like FCS. **NOTES** This is a subspecies of *T. navasota* (see nominotypical subspecies: 5.5.14.5), but the females have numerous structural differences from that species, so it may eventually be recognized as a full species.

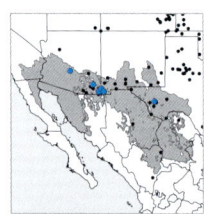

5.3.6.3—*Timulla suspensa*
TIMM-EWE-LUH SUSS-PEN-SUH

ETYMOLOGY Apparently from the Latin *suspensus* "suspended," "anxious," or "doubtful." It is not clear why this name was chosen. **FIELD IDENTIFICATION** Difficult. This is a diurnal species; the scape has a sparse white setal brush. **LAB IDENTIFICATION** The ocelli are small to moderate; the scape has a sparse, white setal brush; the mandible has a ventral tooth basally; and T7 has a

Y-shaped carina. **FEMALE** (3.5.8.1). Not similar to male, the female matches the *Timulla*-like FCS. **NOTES** Males of this species are difficult to identify, even with a microscope, and it is often hard to tell whether the scape is sufficiently dilated and the white setae are dense enough to count as a "fringe." It is also hard to be sure that you are not just handling a *T. grotei* specimen with the scape's fringe rubbed off, which can happen due to specimen age or interactions with predators.

 ### 5.3.7.1—*Myrmosula rufiventris*
MEER-MOE-SOO-LUH ROOF-IVV-ENN-TRISS

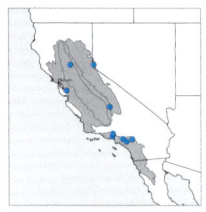

ETYMOLOGY From the Latin *rufus* "red" and *venter* "underside," likely in reference to the reddish metasomal color. **FIELD IDENTIFICATION** Often difficult. The color pattern is unique for this genus, but all small-bodied velvet ants are hard to identify on the wing, even to genus. **LAB IDENTIFICATION** The eye is ovate with setae, and S1 lacks a mesal tooth or ridge. **FEMALE** Unknown, none of the females in the genus looks similar to this male. **NOTES** Males that fit this coloration have been found all over the western

USA, overlapping in distribution with at least five *Myrmosula* females with the Cryptic FCS (3.7.9) or Madrean FCS (3.4.20). A closer examination of these males will likely reveal that multiple species are currently lumped together as *M. rufiventris*, but after the group is revised, only one of them will get to keep the name. The true *M. rufiventris* is from Nevada; it has a beak-like clypeus and a quadrate third submarginal cell of the forewing. Sometimes they have sparse silvery setae on the head and mesosoma, and resemble the *Timulla*-like MCS.

5.4 *EPHUTA*-LIKE MALES COLOR SYNDROME

This color syndrome is named for the predominant coloration of males in the genus *Ephuta* throughout the New World. The males are recognized by having the entire body cuticle black (or dark reddish or brown) and the body setae entirely black or with relatively sparse interspersed black and silvery setae. Sometimes the legs or tegula are orange, but the metasomal cuticle is dark like the mesosoma. This color syndrome is found throughout North America. There are 48 species in eight genera with the *Ephuta*-like Males Color Syndrome (MCS).

Based on the combination of black and whitish coloration, these species could be confused with the Desert MCS (5.1), but species with the *Ephuta*-like MCS are usually smaller in size with sparser setae and almost always have black setae on the mesoscutum. Some specimens with orange legs (5.4.3.1, 5.4.12.1) could be confused with small-bodied members of the *Timulla*-like MCS (5.5.6, 5.5.11), but species in that ring always have T2 at least partly orange or red.

Females are recognized for only 27 of the 48 (56%) *Ephuta*-like MCS males. There is no clear female analog to the *Ephuta*-like MCS. About half of the females (56%) fit the Cryptic Females Color Syndrome (FCS) and others fit the Madrean (26%), Eastern (11%), Western (7%) or Desert (7%) FCS.

This is perhaps the most commonly encountered color syndrome for male velvet ants around the world and is known from nearly every realm. Various examples can be seen in the later chapters of this book, including most of the males of Mesoamerican genera (see 7.2) and many others (7.5.2.1, 7.5.3.1, 8.2.4.2, 8.2.4.3, 8.2.5.2, 8.2.7.1, 8.2.9.1, 8.3.1.1, 8.3.3.1, 8.3.5.1, 8.3.11.1, 8.3.12.1).

—Genus *Dasymutilla*—Eye circular, T1 shape petiolate, sculpture and setae dense

5.4.1 Body setae largely gray, species from eastern USA .... *D. gibbosa*

5.4.2 Body setae entirely black, species from Texas and
California .... *D. nigra, D. imperialis, D. californica*

—Genus *Lomachaeta*—Eye circular, T1 shape narrowly subsessile or disciform, sculpture and setae sparse

5.4.3 Legs pale orange .... *L. cirrhomeris*

5.4.4 Legs blackish .. *L. litosisyra, L. ilex, L. hicksi, L. polemomechana, L. snellingella, L. vacamuerta*

—Genus *Pseudomethoca*—Eye circular, T1 shape sessile

5.4.5 Body setae entirely black .... *Ps. anthracina, Ps. nigricula*

5.4.6 Small species with angular head .... *Ps. frigida, Ps. torrida*

5.4.7 Small species with rounded head .... *Ps. athamas, Ps. toumeyi, Ps. gila*

5.4.8 Medium species with white parallel tergal bristles *Ps. contumax, Ps. praeclara, Ps. carbonaria*

5.4.9 Medium species with simple tergal setae .... *Ps. simillima, Ps. sonorae*

—Genus *Photomorphus*—Eye circular, mesosternal area armed with ridges; T1 shape narrow petiolate

5.4.10 Body entirely black with sparse silvery setae .... *Ph. quintilis, Ph. banksi*

—Genus *Protophotopsis*—Eye circular, mesosternal area unarmed, T1 shape broadly petiolate

5.4.11 Body entirely black with sparse silvery setae .... *Pr. venenaria*

—Genus *Ephuta*—Eye ovate with notch, T1 shape cylindrical

5.4.12 Legs orange .... *E. rufisquamis*

5.4.13 Legs black .... *E. scrupea, E. argenticeps, E. battlei, E. cephalotes, E. conchate, E. ecarinata, E. eurygnathus, E. grisea, E. margueritae, E. pauxilla, E. psephenophila, E. sabaliana, E. spinifera, E. tegulicia, E. krombeini,*

—Genus *Myrmosa*—Eye ovate with setae; tergal junctions constricted

5.4.14 Body entirely black with mostly silvery setae .... *M. unicolor, M. nocturna, M. texana*

5.4.15 Body entirely black with mostly black setae .... *M. bradleyi*

—Genus *Myrmosula*—Eye ovate with setae; tergal junctions not constricted

5.4.16 Body entirely black with sparse silvery and dark brown setae .... *M. parvula*

 ### 5.4.1.1—*Dasymutilla gibbosa*

DAH-ZEE-MEW-TILL-UH GIBB-OWE-SUH

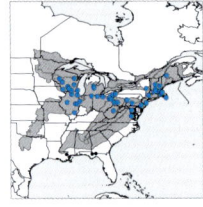

ETYMOLOGY From the Latin *gibbus* "humped," likely in reference to the petiolate T1 shape. **FIELD IDENTIFICATION** Sometimes difficult. No other *Dasymutilla* species in the northeastern USA has this coloration, but species from *Pseudomethoca* (5.4.9.1) and *Photomorphus* (5.4.10) could be confused for this one. **LAB IDENTIFICATION** The axilla is truncate posteriorly; the tibial spurs are white; and S2 lacks a seta-filled pit. **FEMALE** (3.6.2.1). Different from male; lighter brownish body color. **NOTES** Superficially, this species looks more like *Traumatomutilla* (7.2.2) than *Dasymutilla*, and even has the tibial spurs white like most *Traumatomutilla*. Although this species is relatively abundant and common, fresh material has never been available for DNA analysis, and it has not been included in any phylogenetic studies that could help us understand its relations. There is a rarer *Timulla*-like MCS form (5.5.1.2) of this species in the southern portions of its range.

 ### 5.4.2—*Dasymutilla nigra Ephuta*-like males cluster

These are the only *Dasymutilla* with the setae and cuticle entirely black. They could sometimes be confused for similarly colored *Pseudomethoca* (5.4.5.1) or *Myrmosa* (5.4.15.1). Each of these species has a relatively restricted and disjunct distribution.

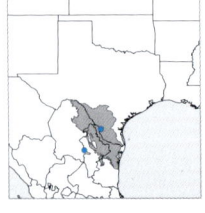

5.4.2.1—*Dasymutilla nigra*

DAH-ZEE-MEW-TILL-UH KNEE-GRAH

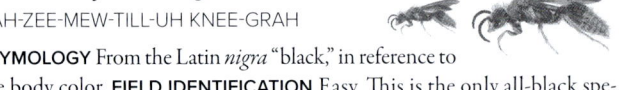

ETYMOLOGY From the Latin *nigra* "black," in reference to the body color. **FIELD IDENTIFICATION** Easy. This is the only all-black species in southern Texas. **LAB IDENTIFICATION** S2 has an ovate seta-filled pit mesally. **FEMALE** Unknown. **NOTES** This species is structurally similar to *D. quadriguttata* (5.6.4.1). The female is unknown but could be *D. uniguttata*

(3.3.5.2) since that species is covered mostly with black setae and also occurs in southern Texas. Some *D. quadriguttata* females from southern Texas (3.5.1.1), however, also have extensive black setae, so *D. nigra* might just be yet another conspecific color variant of that widespread and variable species.

5.4.2.2—*Dasymutilla imperialis*
DAH-ZEE-MEW-TILL-UH IMM-PEER-EE-AHL-ISS

ETYMOLOGY Named for Imperial County, California. **FIELD IDENTIFICATION** Easy. This is the only all-black mutillid in the Algodones dunes in California. **LAB IDENTIFICATION** S2 has a large circular seta-filled pit mesally. **FEMALE** (3.1.5.2). Very different from male, with white dorsal setae. **NOTES** This species is known from very few specimens, all from the Algodones sand dunes in California. It seems to be most active late in the year; most specimens were collected in September or October.

5.4.2.3—*Dasymutilla californica*
DAH-ZEE-MEW-TILL-UH CAL-IF-OR-NICK-UH

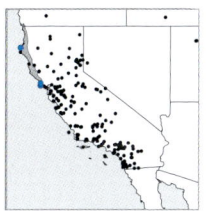

ETYMOLOGY This species was named after the state of California. **FIELD IDENTIFICATION** Usually easy. This is the only all-black *Dasymutilla* in coastal California, but it does overlap with the similarly colored *Pseudomethoca anthracina*. **LAB IDENTIFICATION** S2 lacks a seta-filled pit; the T1 shape is petiolate. **FEMALE** (3.2.8.6). Very different from male, with largely orange dorsal setae. **NOTES** This rare dark variant has been found around the northern California Pacific Coast. The specimen pictured here is somewhat intermediate between the Western and *Ephuta*-like Color Syndromes, but at least one specimen has been seen with the mesosomal setae entirely black like the metasoma.

▼ 5.4.3.1—*Lomachaeta cirrhomeris*

LOW-MUH-KEE-TUH KEER-OWE-MARE-ISS

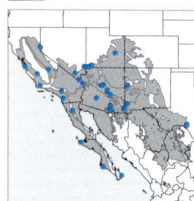

ETYMOLOGY From the Ancient Greek *kirrhos* "orange" and *meros* "thigh," in reference to the leg color. **FIELD IDENTIFICATION** Often difficult. This color pattern is unique in the genus *Lomachaeta*, but *Ephuta rufisquamis* (5.4.12.1) is similar in size, distribution, and coloration. **LAB IDENTIFICATION** The eyes are circular; the mandible has a ventral tooth basally; the T1 shape is disciform; and T2 has a fringe of thick black bristles. **FEMALE** (3.7.1.4). Lighter body color than male. **NOTES** The sex association was discovered by using trap nests (1.5.5), because males and females emerged from the same parasitized hosts. This species has been reared from nests of the crabronid genera *Solierella*, *Pisonopsis*, and *Trypoxylon*.

▼ 5.4.4—*Lomachaeta litosisyra Ephuta*-like males cluster

These are *Lomachaeta* species with the body cuticle, including the legs, entirely black; at most, the tegula or tarsi can be yellow-brown or orange.

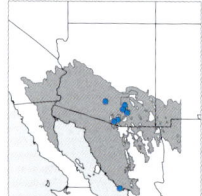

5.4.4.1—*Lomachaeta litosisyra*

LOW-MUH-KEE-TUH LEE-TOW-SISS-EE-RUH

ETYMOLOGY From the Ancient Greek *litos* "simple" and *sisyra* "garment," in reference to the dull gray body setae. **FIELD IDENTIFICATION** Impossible. **LAB IDENTIFICATION** The mandible is unarmed ventrally; the T2 fringe has simple setae only; the paramere is elongate and evenly curved with a long setal tuft apically. **FEMALE** Unknown. **NOTES** Based on their overlapping distributions, *L. osita* (3.7.1.5) might be the female of this species.

5.4.4.2—*Lomachaeta ilex*

LOW-MUH-KEE-TUH EE-LEX

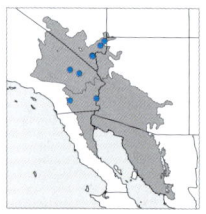

ETYMOLOGY Named after the daughter, Holly, of one of the authors (James Pitts), using the genus name for that plant. **FIELD IDENTIFICATION** Impossible. **LAB IDENTIFICATION** The mandible is unarmed ventrally; the T2 fringe has simple setae only; the paramere is laterally flattened with a ventral fringe of long setae. **FEMALE** Unknown. **NOTES** Based on their overlapping distributions, *L. warneri* (3.7.1.8) might be the female of this species.

5.4.4.3—*Lomachaeta hicksi*
LOW-MUH-KEE-TUH HICKS-EYE

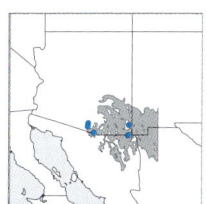

ETYMOLOGY This species was named after C. H. Hicks, who collected the type specimen in 1928. **FIELD IDENTIFICATION** Usually impossible, but this is apparently the only *Lomachaeta* species in the northern, central, and eastern USA. **LAB IDENTIFICATION** The mandible has a ventral tooth basally; the T2 fringe has long black bristles; the paramere is acuminate with mostly short setae. **FEMALE** (3.7.1.1). Usually with lighter body color than male. **NOTES** This is the most widespread and variable *Lomachaeta* species, with some populations having the *Timulla*-like MCS (5.5.6.1). The sex association was discovered because males and females emerged from the same parasitized hosts. This species has been reared from nests of the crabronid genera *Solierella* and *Pisonopsis*.

5.4.4.4—*Lomachaeta polemomechana*
LOW-MUH-KEE-TUH POE-LEMM-OWE-MECK-ANN-UH

ETYMOLOGY From the Ancient Greek *polemikos* "warlike" and *mechanos* "machine." **FIELD IDENTIFICATION** Impossible. **LAB IDENTIFICATION** The mandible is unarmed ventrally; the T2 fringe has simple setae only; the paramere is acuminate without any long setae. **FEMALE** Unknown. **NOTES** This is a rare species that has apparently only ever been collected in malaise traps.

5.4.4.5—*Lomachaeta snellingella*
LOW-MUH-KEE-TUH SNELLING-ELL-UH

ETYMOLOGY Kevin intended to name this species after the Hymenoptera researcher Roy Snelling, who collected the type specimen, but he messed up the suffix, so this species name translates to "little Snelling *Lomachaeta*" rather than "Snelling's *Lomachaeta*." Woops! **FIELD IDENTIFICATION** Impossible. **LAB IDENTIFICATION** The mandible is unarmed ventrally; the T2 fringe has simple setae only; the paramere is dorsoventrally flattened with a rounded apex. **FEMALE** Unknown. **NOTES** This species is structurally similar to *L. beadugrimi* from the *Timulla*-like MCS (5.5.6.2).

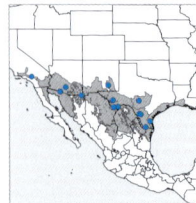

5.4.4.6—*Lomachaeta vacamuerta*

LOW-MUH-KEE-TUH VAW-CUH-MWER-TUH

ETYMOLOGY The name comes from the Spanish *vaca* "cow" and *muerta* "dead." **FIELD IDENTIFICATION** Impossible. **LAB IDENTIFICATION** The mandible is unarmed ventrally; the T2 fringe has simple setae only; the paramere is relatively straight with a setal tuft apically. **FEMALE** (3.7.1.7). Lighter body color than male. **NOTES** Other than *L. hicksi*, this is the only *Lomachaeta* species in southern Texas, and the sex association was based on overlapping distribution in that region.

*While collecting velvet ants in Texas, my father, Ed Williams, tried to communicate to a Spanish-speaking waitress that we were searching for "cow killers," but he mistranslated this to **vaca muerte** "dead cow."*—KEVIN

 ## 5.4.5—*Pseudomethoca anthracina Ephuta*-like males cluster

These are the only *Pseudomethoca* with the color of the setae and cuticle entirely black. They could sometimes be confused for similarly colored *Dasymutilla* (5.4.2.3) or *Myrmosa* (5.4.15.1). One of these species lives in California and the other in Arizona.

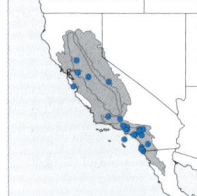

5.4.5.1—*Pseudomethoca anthracina*

SOO-DOE-METH-OWE-KUH ANN-THRUH-SEE-NUH

ETYMOLOGY From the Ancient Greek *anthrakinos* "coal-black," in reference to the male coloration. **FIELD IDENTIFICATION** Usually easy. This is the only all-black *Pseudomethoca* in California but could be confused with some populations of *Dasymutilla californica* (5.4.2.3). **LAB IDENTIFICATION** The lateral propodeal face is irregularly areolate. **FEMALE** (3.1.6.1, 3.2.10.1).

Dorsally with white, yellow, orange, or red setae. **NOTES** Females are more commonly encountered than males. Even given its rarity, this is the most commonly encountered all-black mutillid in California. This color pattern is shared by *Myrmosa bradleyi*, some populations of *D. californica*, and the rare sapygid wasp *Fedtschenkia anthracina*.

5.4.5.2—*Pseudomethoca nigricula*

SOO-DOE-METH-OWE-KUH NIGG-RICK-EWE-LUH

ETYMOLOGY Derived from the Latin *nigra* "black," in reference to the coloration. **FIELD IDENTIFICATION** Easy. This is the only all-black *Pseudomethoca* in Arizona. **LAB IDENTIFICATION** The lateral propodeal face is transversely striate. **FEMALE** Unknown. **NOTES** This is a rare species. The female might be *Ps. quadrinotata* (3.4.11.4), based on similar size, rarity, and distribution, but they are drastically different in color. Structurally, this species is similar to the even rarer *Ps. ajattara* within the Timulla-like MCS (5.5.8.2), which is also known from males only.

Pseudomethoca
nigricula

 ## 5.4.6—*Pseudomethoca frigida Ephuta*-like cluster

These species occur in the central and eastern USA. They are small-bodied and have the head sharply angular posterolaterally when viewed from above. They could most easily be confused with the *Ps. athamas* cluster (5.4.7) or other small-bodied, similarly colored genera, like *Protophotopsis* (5.4.11.1) and *Lomachaeta* (5.4.4).

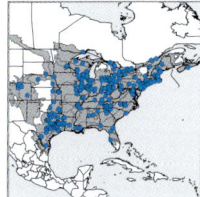

5.4.6.1—*Pseudomethoca frigida*
SOO-DOE-METH-OWE-KUH FRIJJ-IDD-UH

ETYMOLOGY From the Latin *frigidus* "cold," in reference to the cold climate of the type locality in Canada. **FIELD IDENTIFICATION** Often easy. Except where they overlap with *Ps. torrida*, this species is easy to recognize based on the head shape and body size. **LAB IDENTIFICATION** The head is angular posterolaterally; the T2 punctures are moderately dense; and the metasoma (including T7) is usually entirely black. **FEMALE** (3.7.2.1). Lighter in color than male. **NOTES** Brothers (1972) described the life cycle of this species in great detail. One interesting note was that during mating the males gripped the female around the "waist" using their mandibles. This is different from other genera, like *Dasymutilla*, which grasp the female using their legs only, or *Timulla* and *Ephuta*, which grasp the female's "neck" using their mandibles.

Head angular posteriorly

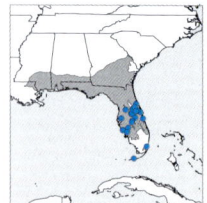

5.4.6.2—*Pseudomethoca torrida*

SOO-DOE-METH-OWE-KUH TORR-IDD-UH

ETYMOLOGY From the Latin *torridus* "dry" or "hot," likely named in contrast to the related *Ps. frigida*, which occurs in cooler habitats farther north. **FIELD IDENTIFICATION** Very difficult. Has a more restricted distribution

and is often smaller-bodied than *Ps. frigida*. **LAB IDENTIFICATION** The head is angular posterolaterally; the T2 punctures are sparse; and the metasoma has T7 lighter reddish-brown than the other tergites. **FEMALE** (3.7.2.2). Lighter in color than male. **NOTES** This rare southeastern species was originally treated as a subspecies of *Ps. frigida* but was raised to full species level subsequently.

▼ 5.4.7—*Pseudomethoca athamas Ephuta*-like males cluster

These are generally southern and western species, ranging from Texas to California. They are recognized by the small body size and heads that are not angular posterolaterally, though they often have a broad and quadrate shape. In the Madrean and Cryptic Female Color Syndromes, there are six small-bodied species that have their male sex unknown. Those males will likely fit in with this cluster when they are eventually discovered. Species in this cluster could be confused with the *Ps. frigida* cluster (5.4.6) or other small-bodied, similarly colored genera, like *Protophotopsis* (5.4.11.1) and *Lomachaeta* (5.4.4).

5.4.7.1—*Pseudomethoca athamas*

SOO-DOE-METH-OWE-KUH ATH-UH-MUSS

ETYMOLOGY Apparently from the Ancient Greek *a-* "not" and *thama* "crowded," maybe in reference to the sparse setae and punctures of the body. **FIELD IDENTIFICATION** Often difficult. This is the only species in the cluster from Pacific regions in California but is difficult to separate from other genera. **LAB IDENTIFICATION** The head is broad and quadrate; the

mandible is broadly tridentate; the clypeus lacks a pre-apical transverse furrow; the forewing has a large stigma that is much longer than the marginal cell; and the hypopygium has uniformly sparse setae. **FEMALE** (3.7.2.5). Lighter in color than male. **NOTES** Based on similar body size, head shape, and distribution, the sex association was pretty obvious, but it was only recently formalized (Williams 2023).

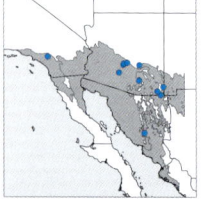

5.4.7.2—*Pseudomethoca toumeyi*

SOO-DOE-METH-OWE-KUH TOUMEY-EYE

ETYMOLOGY This species was named after J. W. Toumey, a professor at the University of Arizona, who provided the type specimens to Fox in 1894. **FIELD IDENTIFICATION** Impossible. **LAB IDENTIFICATION** The head is broad and quadrate; the mandible is broadly tridentate; the clypeus lacks a pre-apical transverse furrow; the forewing has the marginal cell as long as the stigma;

Pseudomethoca toumeyi genitalia

and the hypopygium has a dense tuft of long white setae laterally. Furthermore, the genitalic paramere is broadly paddle-shaped. **FEMALE** (3.4.12.1). More colorful than male. **NOTES** The genitalia of this species are strange and diagnostic. Furthermore, the lateral setal tufts on the hypopygium are different from other described males. We have seen a few specimens with these tufts but different genitalia from *Ps. toumeyi*; these may eventually be associated with other small *Pseudomethoca* species, like *Ps. bequaerti* (3.4.12.2).

5.4.7.3—*Pseudomethoca gila*

SOO-DOE-METH-OWE-KUH HEE-LUH

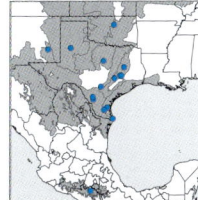

ETYMOLOGY The species was apparently named after a place, but the Gila River, Gila National Forest, and city of Gila Bend are all in Arizona, not Texas. **FIELD IDENTIFICATION** Impossible. **LAB IDENTIFICATION** This name serves as somewhat of a catch-all for unassociated small *Pseudomethoca* males in Texas. In most specimens, the head is broad and quadrate; the mandible is broadly tridentate; the clypeus lacks a pre-apical transverse furrow; the forewing has the marginal cell as long as or longer than the stigma; and the hypopygium has uniformly sparse setae. **FEMALE** (3.7.3.1). Lighter in color than male. **NOTES** This species is structurally variable and is included twice in the keys in the appendix for this reason (see 9.3.5.4). *Pseudomethoca ilione* was sunk under this species after the sexes were associated by Manley & Neff (1989). Many other small-bodied unassociated females of *Pseudomethoca* are known from Texas, such as *Ps. dentigula* (3.7.2.4), *Ps. nephele* (3.7.2.3), *Ps. nudula* (3.7.2.6), and *Ps. scaevolella* (3.4.12.4). Their males, when eventually discovered, will likely be found mixed in with series of *Ps. gila*.

▼ 5.4.8—*Pseudomethoca contumax Ephuta*-like males cluster

These species occur mainly in the central and western USA. They are recognized by the fringes of T2–4 having parallel, thickened, white bristles, but some *Pseudomethoca* with other color syndromes have similar bristles (such as *Ps. donaeanae*, 5.5.8.1). They can be difficult to separate from other *Ephuta*-like MCS *Pseudomethoca* clusters.

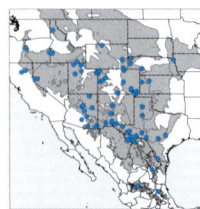

5.4.8.1—*Pseudomethoca contumax*
SOO-DOE-METH-OWE-KUH CONN-TOO-MAX

ETYMOLOGY The name *contumax* is derived from Latin, "insolent" or "rude." The males used to be named *Ps. manca* or *Ps. albicoma*. **FIELD IDENTIFICATION** Sometimes difficult. Except where they overlap with *Ps. praeclara* or *Ps. carbonaria*, the body color and tergal bristles are diagnostic. **LAB IDENTIFICATION** The antennal tubercles are smooth; the tegula

is flat and relatively smooth; the wings are generally clear; and the genitalic parameres are parallel. **FEMALE** (3.4.11.1). More colorful than male. **NOTES** This is a widespread species. Until recently, the males were treated as two separate species (*Ps. manca* and *Ps. albicoma*), which were separated by minor differences in body punctation.

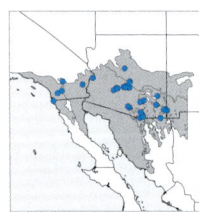

5.4.8.2—*Pseudomethoca praeclara*
SOO-DOE-METH-OWE-KUH PREE-CLAIR-UH

ETYMOLOGY The name *preclara*, is derived from Latin, "bright" or "famous." The male used to be named *Ps. aegeon*. **FIELD IDENTIFICATION** Often difficult. Except where they overlap with *Ps. contumax*, the body color and tergal bristles are diagnostic. **LAB IDENTIFICATION** The antennal tubercles have dense, small punctures; the tegula is flat and relatively smooth; the wings are generally clear; and the genitalic parameres are divergent posteriorly. **FEMALE** (3.4.11.2). More colorful than male. **NOTES** This species is apparently restricted to the Sonoran Desert and Madrean Archipelago, unlike the similarly colored, widespread *Ps. contumax*.

5.4.8.3—*Pseudomethoca carbonaria*
SOO-DOE-METH-OWE-KUH CAR-BOW-NARR-EE-UH

ETYMOLOGY Presumably from the Latin *carbonis* "charcoal," in reference to the blackish body and wing color. **FIELD IDENTIFICATION** Often difficult. The wings are darker than those of other species in this cluster but not necessarily darker than the wings of other *Pseudomethoca* species. **LAB IDENTIFICATION** The antennal tubercles are mostly smooth; the tegula is raised and punctate;

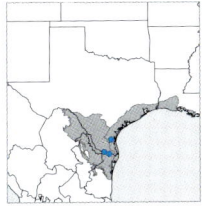

the wings are generally dark brown; and the genitalic parameres are parallel. **FEMALE** Unknown. **NOTES** This rare species is known only from southern Texas.

 ## 5.4.9—*Pseudomethoca simillima Ephuta*-like males cluster

These species are similar in appearance but do not overlap in distribution. *Ps. simillima* is widespread in the central and eastern USA, and *Ps. sonorae* is known only from Cochise County, Arizona. These species are moderate in size, have the head rounded posteriorly, and have the tergal fringes with simple setae only. They can be difficult to separate from other *Ephuta*-like MCS *Pseudomethoca* clusters.

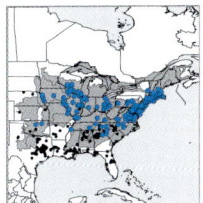

5.4.9.1—*Pseudomethoca simillima*
SOO-DOE-METH-OWE-KUH SEE-MILL-EE-MUH

ETYMOLOGY From the Latin *simillimus* "very similar." The author thought it was similar to eastern populations of *D. bioculata*. This male variant used to be named *Ps. geryon*. **FIELD IDENTIFICATION** Often difficult. Does not overlap in distribution with *Ps. sonorae*, but the characters that separate this species from the *Ps. frigida* (5.4.6) and *Ps. contumax* (5.4.8) clusters are difficult to recognize in the field. **LAB IDENTIFICATION** The head is rounded posteriorly; and the tergal setae are simple and predominantly whitish. **FEMALE** (3.6.10.1). Different from male, with paler body color. **NOTES** The Eastern MCS form (5.6.5.2) of this species seems to be more common and widespread.

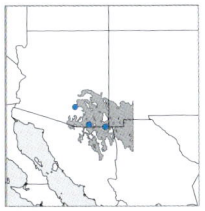

5.4.9.2—*Pseudomethoca sonorae*
SOO-DOE-METH-OWE-KUH SONORA-EE

ETYMOLOGY This species is named after Kevin's daughter, Sonora Rose Williams. **FIELD IDENTIFICATION** Sometimes difficult. Does not overlap in distribution with *Ps. simillima*, but the characters that separate this species from the *Ps. athamas* (5.4.7) and *Ps. contumax* (5.4.8) clusters are difficult to recognize in the field. **LAB IDENTIFICATION** The head is rounded posteriorly; and the tergal setae are simple and predominantly black dorsally. **FEMALE** (3.4.14.2). Different from male, with paler body color. **NOTES** This species was only recently discovered and named. The moderately large body and extent of black setae compared with other *Pseudomethoca* in this area provided evidence to match the male and female.

5.4.10—*Photomorphus quintilis Ephuta*-like males cluster

These are the only *Photomorphus* species in the USA with the body cuticle entirely black. The genus is recognized by having a complex of transverse ridges on the mesosternal area. They could easily be confused for similarly colored species in other genera, especially *Dasymutilla gibbosa* (5.4.1.1) or various *Pseudomethoca* species (5.4.6, 5.4.9.1).

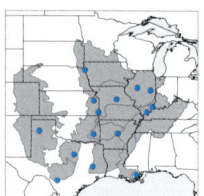

5.4.10.1—*Photomorphus quintilis*
FOE-TOE-MORE-PHUSS KWIN-TILL-ISS

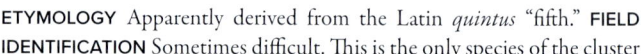

ETYMOLOGY Apparently derived from the Latin *quintus* "fifth." **FIELD IDENTIFICATION** Sometimes difficult. This is the only species of the cluster in the central USA, but the genus is difficult to recognize in the field. **LAB IDENTIFICATION** The mesosternal area is armed with a single broad complex of transverse carinae. **FEMALE** Unknown. **NOTES** Based on overlapping distribution and placement in the subgenus *Photomorphus* (*Photomorphus*), the female might be *Ph. myrmicoides* (3.7.5.5). It is the westernmost species known in the subgenus.

5.4.10.2—*Photomorphus banksi*
FOE-TOE-MORE-PHUSS BANKS-EYE

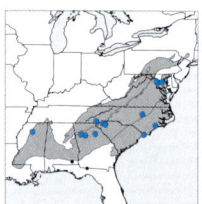

ETYMOLOGY This species was named after the entomologist Nathan Banks, who collected the type in the early 1900s. **FIELD IDENTIFICATION** Often difficult. This is the only species of the cluster in the eastern USA, but the genus is difficult to recognize in the field. **LAB IDENTIFICATION** The mesosternal area is armed with two complexes of transverse carinae; the posterolateral complex is slender and spine-like, unlike the broader, shorter anterior complex. **FEMALE** (3.7.5.3). Lighter in color than male. **NOTES** This color variant seems to occur farther north than the *Sphaeropthalma*-like MCS (5.7.10.2) form.

 5.4.11.1—*Protophotopsis venenaria*

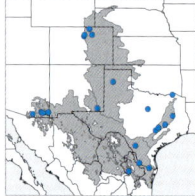

PRO-TOE-FOE-TOP-SISS VENN-ENN-ARE-EE-UH

ETYMOLOGY Apparently derived from the Latin *veneno* "potion" or "poison." **FIELD IDENTIFICATION** It is nearly impossible to recognize this genus in the field. **LAB IDENTIFICATION** The body sculpture is dense and coarse; the axilla is unarmed; the mesosternal area is unarmed; the T1 shape is broadly petiolate; and T2 and S2 each have a felt line. **FEMALE** (3.4.9.1). More colorful than male. **NOTES** This is the only Nearctic species in the genus and one of very few diurnal species with a felt line on S2. A few nocturnal species (6.1.8), like *Sphaeropthalma mendica*, can be black like this, but they have plumose setae on the tergal fringes.

Protophotopsis venenaria

RIGHT: *Ephuta rufisquamis*

▼ 5.4.12.1—*Ephuta rufisquamis*

EFF-OO-TUH ROOF-ISS-KWAH-MISS

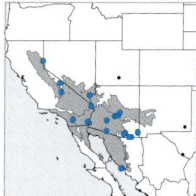

ETYMOLOGY From the Latin *rufus* "red" and *squama* "scale," in reference to the reddish tegula color. **FIELD IDENTIFICATION** Often difficult. This color pattern is unique in the genus *Ephuta*, but *Lomachaeta cirrhomeris* (5.4.3.1) is similar in size, distribution, and coloration. **LAB IDENTIFICATION** The subantennal basin is equal in length to the clypeal basin, and the tegula is evenly rounded with weak sculpture. **FEMALE** Unknown. **NOTES** This is the only *Ephuta* in the USA with reddish legs and antennae, and a black body. This species seems to be most prevalent in hot desert habitats. Some previous authors thought the female might be *E. tumacacori* (3.4.19.1), but that species generally has a larger body size than *E. rufisquamis*.

▼ 5.4.13—*Ephuta scrupea Ephuta*-like males cluster

Most males of *Ephuta* fit here, and this cluster occurs throughout North America. They have the body and legs entirely black or dark brown and covered with a variable extent of gray or silvery setae. These are difficult to identify, even with a pinned specimen and good microscope. There are likely to be additional undescribed species mixed in with these forms.

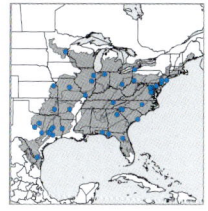

5.4.13.1—*Ephuta scrupea*

EFF-OO-TUH SCREW-PAY-UH

ETYMOLOGY Apparently from the Latin *scrupus* "sharp stone," in reference to the coarse body sculpture. **FIELD IDENTIFICATION** Very difficult. This is the only *Ephuta* male with the sides of T7 pale yellow-brown, but that feature is difficult to see in the field and not present in all specimens. **LAB IDENTIFICATION** The subantennal basin is equal in length to the clypeal basin; the mandible is slender; the tegula is coarsely rugose; and the hypopygium and often the sides of T7 are pale whitish-yellow. **FEMALE** (3.7.6.1). Generally lighter in color than the male. **NOTES** This is one of the most widespread and common species of *Ephuta*. It is the type species for this large genus.

5.4.13.2—*Ephuta argenticeps*
EFF-OO-TUH ARR-JENT-ISS-EPPS

ETYMOLOGY From the Latin *argentum* "silver" and *-ceps* "headed," in reference to the silver head color. **FIELD IDENTIFICATION** Difficult. This is apparently the only *Ephuta* in Pacific regions of California and the only all-black *Ephuta* with the setae of T3–6 black; separating *Ephuta* from other genera is difficult in the field, though.

LAB IDENTIFICATION The subantennal basin is equal in length to the clypeal basin; the mandible is slender; the ocelli are relatively large; the humeral carina is weakly developed; the tegula is sparsely sculptured; and the hypopygium is usually dull brown. **FEMALE** (3.4.17.1). Generally lighter in color than the male, with a silver head. **NOTES** This is the westernmost representative of the genus *Ephuta*. Males usually have more extensive black body setae than other species in this cluster.

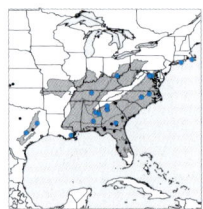

5.4.13.3—*Ephuta battlei*
EFF-OO-TUH BATTLE-EYE

ETYMOLOGY Bradley (1916) dedicated this species to "Paul Battle of Bainbridge, Georgia, my faithful companion on many a collecting trip." **FIELD IDENTIFICATION** Impossible. **LAB IDENTIFICATION** The subantennal basin is much shorter than the clypeal basin; the mandible is slender; the humeral carina is strongly developed; the tegula is moderately punctate; and the hypopygium is generally yellow-brown. **FEMALE** Not described in the literature, but Schuster (1956) said the female was known but impossible to separate from *E. slossonae* (3.6.12.1). **NOTES** This relatively common eastern species is divided into four subspecies. Two of these have the *Timulla*-like MCS (5.5.11.2). The other two fit here: *E. b. transitionalis*, from the northeast, has small, widely separated punctures on T2, and the whitish posterior band of T2 is narrow; *E. b. confusa*, from the southeast, has larger, denser punctures on T2 and has a thicker band of white setae apically on T2.

Ephuta battlei confusa   Ephuta battlei transitionalis

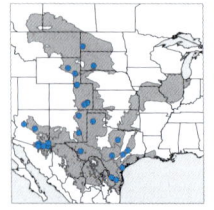

5.4.13.4—*Ephuta cephalotes*
EFF-OO-TUH SEFF-ALL-OWE-TEASE

ETYMOLOGY Derived from the Ancient Greek *kephalos* "head," in reference to the oddly conical head shape. **FIELD IDENTIFICATION** Impossible. **LAB IDENTIFICATION** This species can be immediately recognized by the unique conical head shape and tiny ocelli. The subantennal basin is equal in length to the clypeal basin; the mandible is slender; the tegula is sparsely

LEFT:
Face, note the conical vertex.

sculptured; and the hypopygium is usually yellow-brown. **FEMALE** Unknown. **NOTES** We do not yet have a good guess for the female's identity. This species has the head somewhat conical with an obscure swelling in the ocellar area. In the literature, this species is recorded from the northern Great Plains, but we have studied only specimens from Arizona.

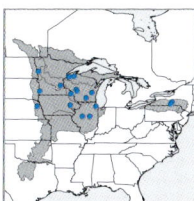

5.4.13.5—*Ephuta conchate*
EFF-OO-TUH CON-KATE-UH

ETYMOLOGY Apparently derived from the Latin *concha* "snail" or "shell," in reference to the coarse body sculpture. **FIELD IDENTIFICATION** Impossible.

LAB IDENTIFICATION The subantennal basin is as long as the clypeal basin; the mandible is slender; the humeral carina is strongly developed, making the humerus strongly dentate in dorsal view; the tegula is sparsely sculptured; and the hypopygium is dark brown. **FEMALE** (3.7.6.2). Generally lighter in color than the male. **NOTES** This is a midwestern and northeastern species with big pointy "shoulders" (the humeral carina is sharp and distinct).

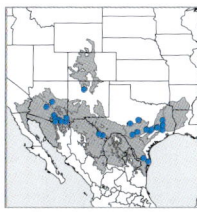

5.4.13.6—*Ephuta ecarinata*
EFF-OO-TUH EE-CAR-INN-AWE-TUH

ETYMOLOGY From the Latin *e*- "out of" or "from" and *carina* "ridge" or "keel," likely a reference to the edentate propodeal carina of this species. **FIELD IDENTIFICATION** Impossible. **LAB IDENTIFICATION** The subantennal basin is about equal in length to the clypeal basin; the mandible is slender; the lateral ocelli are smaller than the mesal ocellus; the tegula is sparsely

FAR LEFT:
Ephuta ecarinata ecarinata

LEFT:
E. e. pima

sculptured; and the hypopygium is pale whitish-yellow. **FEMALE** Unknown. **NOTES** This small-bodied southwestern species is divided into three subspecies: *E. e. ecarinata*, *E. e. neomexicana*, and *E. e. pima*. They are separated by subtle differences in the ocelli size, setal density on the propodeum, and shape of the pronotum and propodeum. We do not yet have a good guess for the female's identity. Alongside the three subspecies (which might actually be separate species), various apparently new species in the southwest loosely match *E. ecarinata*. Until a conclusive taxonomic revision is undertaken, *E. ecarinata* serves as a "dumping ground" for currently undiagnosable *Ephuta* males.

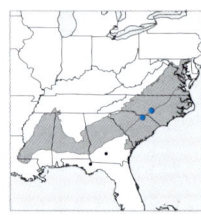

5.4.13.7—*Ephuta eurygnathus*
EFF-OO-TUH EWE-REE-NATH-USS

ETYMOLOGY From the Ancient Greek *eurys* "broad" and *gnathos* "jaw," in reference to the wide mandible. **FIELD IDENTIFICATION** Impossible. **LAB IDENTIFICATION** This species can be recognized by the wide tridentate mandible and the short pit-like subantennal basin. **FEMALE** Unknown. **NOTES** This species apparently occurs in this *Ephuta*-like MCS form and a *Timulla*-like MCS (5.5.11.3) form. This dark form seems rarer and occurs farther north; we have not seen any specimens to photograph of this darker color form. The currently unknown female will likely be similar to *E. slossonae* (3.6.12.1), based on the similar male morphology.

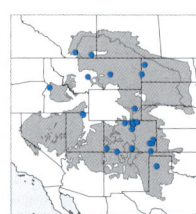

5.4.13.8—*Ephuta grisea*
EFF-OO-TUH GRISS-AY-UH

ETYMOLOGY From the Latin *griseus* "gray," in reference to the pale silvery setal color. **FIELD IDENTIFICATION** Impossible. **LAB IDENTIFICATION** The subantennal basin is equal in length to the clypeal basin; the mandible is slender; the ocelli are small; the humeral carina is weakly developed; the tegula is sparsely sculptured; and the hypopygium is blackish. **FEMALE** Unknown. **NOTES** This northwestern species is divided into two subspecies, *E. g. grisea* and *E. g. fuscosericea*, which are separated by subtle differences in the ocelli size and clypeal basin. The female might be *E. coloradella* (3.4.17.3), based on its distribution.

5.4.13.9—*Ephuta margueritae*
EFF-OO-TUH MARGUERITA-EE

ETYMOLOGY Schuster (1951) named this species after his wife, Olga Marguerite Schuster, in appreciation for her constant assistance during his study of this genus. **FIELD IDENTIFICATION** Impossible. **LAB IDENTIFICATION** The subantennal basin is equal in length to the narrow clypeal basin; the mandible is slender; the ocelli are small with the lateral ocelli smaller than

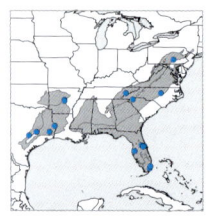

the mesal ocellus; the tegula is sparsely sculptured; and the hypopygium is pale whitish-yellow. **FEMALE** (3.4.18.2, 3.7.6.5). Generally lighter in color than the male. **NOTES** This eastern species is relatively large-bodied for *Ephuta* and includes two subspecies: *E. m. margueritae* has the propodeum with dense silver setae and the wings darker brown, while *E. m. xanthocephala* has the propodeum with sparse setae and the wings basically clear.

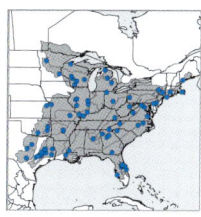

5.4.13.10—*Ephuta pauxilla*
EFF-OO-TUH POX-ILL-UH

ETYMOLOGY The name may be derived from the Latin *paucus* "few" or "little," in reference to the small body size. **FIELD IDENTIFICATION** Impossible. **LAB IDENTIFICATION** The subantennal basin is much shorter than the clypeal basin; the mandible is slender; the humeral carina is weakly developed; the tegula is coarsely sculptured; and the hypopygium is generally pale whitish-yellow.

FEMALE Unknown. **NOTES** This is a common and widespread species, with similar distribution to *E. scrupea*. Previous authors suggested that *E. puteola* (3.6.12.3, 3.7.6.7) might be the female. Relatively recently, *E. tentativa* was associated with *E. pauxilla*, but this was based on a misidentification, so *E. puteola* is once again the most likely candidate to match with *E. pauxilla*. Some specimens were reared from the spider wasp host *Dipogon sayi* by Evans & Yoshimoto (1962).

5.4.13.11—*Ephuta psephenophila*
EFF-OO-TUH SEFF-ENN-OWE-FILL-UH

ETYMOLOGY Apparently derived from the Ancient Greek *psephenos* "dark" or "obscure," in reference to the body color. **FIELD IDENTIFICATION** Impossible. **LAB IDENTIFICA-**

TION The subantennal basin is clearly longer than the narrow clypeal basin; the mandible is slender; the lateral ocelli are larger than the mesal ocellus; the tegula is sparsely sculptured; and the hypopygium is pale whitish-yellow. **FEMALE** Unknown. **NOTES** This is a rare southeastern species with large ocelli. Previous authors suggested that the female might be *E. tentativa*, but that female was recently recognized as a synonym of *E. scrupea*.

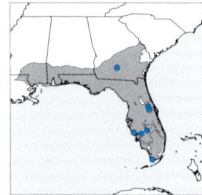

5.4.13.12—*Ephuta sabaliana*
EFF-OO-TUH SAW-BALL-EE-ANN-UH

ETYMOLOGY Named for the Sabalian Zone, a sandy coastal habitat in the southeastern USA. **FIELD IDENTIFICATION** Impossible. **LAB IDENTIFICATION** The subantennal basin is much shorter than the clypeal basin; the mandible is slender;

the humeral carina is moderately developed; the tegula is moderately punctate; and the hypopygium is black. **FEMALE** (3.6.12.2). Generally lighter in color than the male. **NOTES** This rare southeastern species includes two subspecies, *E. s. sabaliana* and *E. s. fattigi*, which are separated by subtle differences in the sculpture of the tegula, T1, and T2.

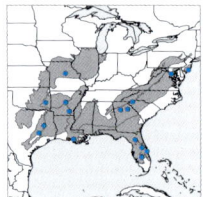

5.4.13.13—*Ephuta spinifera*
EFF-OO-TUH SPINE-IFF-ERR-UH

ETYMOLOGY Derived from the Latin *spina* "spine," in reference to the toothed hind coxa. **FIELD IDENTIFICATION** Impossible. **LAB IDENTIFICATION** This species can be immediately recognized by the small

spine on the hind coxa. The subantennal basin is much shorter than the clypeal basin; the mandible is slender; the humeral carina is moderately developed; the tegula is coarsely punctate; and the hypopygium is pale yellow-brown. **FEMALE** (3.7.6.9). Generally lighter in color than the male. **NOTES** This is a moderately common species, with a tooth on the hind coxa.

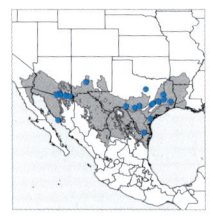

5.4.13.14—*Ephuta tegulicia*
EFF-OO-TUH TEGG-EWE-LISS-EE-UH

ETYMOLOGY Named for the unique tegula shape. **FIELD IDENTIFICATION** Impossible. **LAB IDENTIFICATION** The subantennal basin is equal in length to the clypeal basin; the mandible is slender; the tegula is mostly smooth with a sharp longitudinal carina;

and the hypopygium is blackish. **FEMALE** Unknown. **NOTES** This species can be immediately recognized by the unique tegula shape. We do not yet have a good guess for the female's identity.

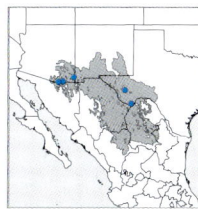

5.4.13.15—*Ephuta krombeini*

EFF-OO-TUH KROMBEIN-EYE

ETYMOLOGY This species was named for the entomologist Karl V. Krombein, who authored numerous papers about velvet ants from the 1940s to 1990s. **FIELD IDENTIFICATION** Very difficult. The head is wider than in other *Ephuta*, but separating *Ephuta* from other genera is difficult in the field. **LAB IDENTIFICATION** Immediately recognizable by the unique mandible, which is contorted with a large basal lobe dorsally, and the subantennal pit, with its sides merged and raised to form a small tooth. **FEMALE** Unknown. **NOTES** This was placed in the separate

subgenus, *Xenochile*, because of the bizarre mandible and clypeus structure. Based on the overlapping distribution, the relatively wide head, and overall strange morphological traits, *E. tumacacori* (3.4.19.1) is a good female candidate to eventually be matched with this species.

▼ 5.4.14—*Myrmosa unicolor Ephuta*-like males cluster

These species occur in the central and eastern USA and have the body cuticle entirely black and the head and mesosoma with extensive silvery setae. They could be confused for other similarly colored genera, like *Myrmosula* (5.4.16.1) and *Pseudomethoca* (5.4.6, 5.4.9.1).

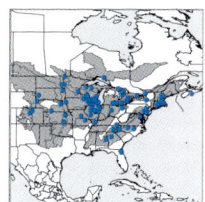

5.4.14.1—*Myrmosa unicolor*

MEER-MOE-SUH EWE-NICK-UH-LURR

ETYMOLOGY The name means "one color," but males and females have different coloration, and the males have the cuticle black and the setae mostly silver. **FIELD IDENTIFICATION** Sometimes easy. This is the only *Myrmosa* species in the eastern USA, but the genus can be difficult to recognize in the field. **LAB IDENTIFICATION** The ocelli are small; the clypeus has a strong

longitudinal carina; S1 usually has a posterolateral pit; and S1 and S2 are each armed with an anteromesal tooth. **FEMALE** (3.7.7.1). Lighter in color than male. **NOTES** This is apparently the most common *Myrmosa* species in the central and eastern USA. This species and *M. bradleyi* (5.4.15.1) belong to the nominotypical subgenus *Myrmosa* (*Myrmosa*), recognized by having a tooth on S1 and S2.

—Genus *Dasymutilla*—Eye circular, T1 shape petiolate, sculpture and setae dense

5.5.1 Species from eastern USA .... *D. archboldi, D. gibbosa*

5.5.2 T2–6 setae and cuticle uniformly orange to red ... *D. sicheliana, D. foxi, D. sophrona, D. bonita*

5.5.3 Mesonotum with silvery setae, like vertex and pronotum .... *D. monticola*

5.5.4 Mesonotum with black setae, T3–7 cuticle black,
contrasting with orange or yellow (at least
in part) T2 cuticle .... *D. fasciventroides, D. birkmani, D. iztapa, D. pulchra*

5.5.5 Mesonotum with black setae; T2–7 cuticle entirely
orange-brown, T2 sometimes with yellow patch .... *D. apicalata, D. dammersi, D. curialis,*
D. digressa, D. saetigera

—Genus *Lomachaeta*—Eye circular, axilla dentate, sculpture and setae sparse

5.5.6 Metasoma partly (sides of T2 at least) to entirely
reddish-orange, small-bodied species .... *L. hicksi, L. beadugrimi, L. crocopinna,*
L. powelli, L. ptilohyalus

—Genus *Pseudomethoca*—Eye circular, T1 shape sessile

5.5.7 T2 black, apical tergites with orange setae; central and eastern USA .... *Ps. wickhami*

5.5.8 Tergal cuticle mostly orange-brown .... *Ps. donaeanae, Ps. ajattara, Ps. russeola, Ps. flavida*

—Genus *Photomorphus*—Eye circular, mesosternal area armed with ridges

5.5.9 Body mostly black with silvery setae, except apical tergites orange .... *Ph. auriventris*

—Genus *Sphaeropthalma*—Eye circular, tergal fringes plumose

5.5.10 Body mostly black with silver or golden-brown setae, except
fringes of T2–6 golden-orange .... *S. marpesia*

—Genus *Ephuta*—Eye ovate with notch, T1 shape cylindrical

5.5.11 Metasoma partly (sides of T2 at least) to entirely reddish-orange;
usually small-bodied species .... *E. copano, E. battlei,*
E. eurygnathus, E. rufisquamis

—Genus *Timulla*—Eye ovate with notch, T1 shape cylindrical

5.5.12 Tergal setae partly black; central and eastern USA .... *T. ocellaria, T. vagans, T. hollensis*

5.5.13 Tergal setae entirely golden-orange, tarsi largely reddened;
Sonoran Desert in Arizona and California .... *T. neobule, T. tyro*

5.5.14 Tergal setae entirely golden-orange, tarsi black;
widespread .... *T. oajaca, T. suspensa, T. nitela,*
T. leona, T. navasota navasota

—Genus *Myrmosa*—Eye ovate with setae; tergal junctions constricted

5.5.15 Metasoma partly (T2 at least) to entirely reddish-orange .... *M. nocturna*

▼ 5.5.1—*Dasymutilla archboldi Timulla*-like males cluster

These are the only *Timulla*-like *Dasymutilla* males in the eastern USA. Males of the *D. birkmani* cluster (5.6.2) in the Eastern FCS seem to fit in this cluster because they often have sparse, inconspicuous gray setae intermixed with the predominantly black mesosomal coloration.

5.5.1.1—*Dasymutilla archboldi*
DAH-ZEE-MEW-TILL-UH ARCHBOLD-EYE

ETYMOLOGY This species was named for the Archbold Biological Station in Florida. **FIELD IDENTIFICATION** Easy. The color pattern, especially the dense silvery setae on the head and mesosoma, and restricted distribution, are diagnostic. **LAB IDENTIFICATION** The entire mesosomal dorsum has dense silvery setae; the tibial spurs are dark brown; S2 has a seta-filled pit mesally. **FEMALE** (3.6.4.1). Paler and balder than the male. **NOTES** The *Sphaeropthalma*-like MCS variant of *D. ursus* (5.7.4.3) was originally matched with *D. archboldi*. That male was larger and more widespread than the females, leading to some confusion. DNA sequence comparisons were used to match this male to *D. archboldi* and the previous male with *D. ursus*. This more recently associated male is small-bodied and has the same restricted distribution as the female.

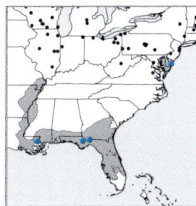

5.5.1.2—*Dasymutilla gibbosa*
DAH-ZEE-MEW-TILL-UH GIBB-OWE-SUH

ETYMOLOGY From the Latin *gibbus* "humped," in reference to the petiolate T1 shape. **FIELD IDENTIFICATION** Usually easy. This species has more extensive silvery setae than most *Dasymutilla* in the eastern USA and does not overlap with *D. archboldi*. **LAB IDENTIFICATION** The mesonotum has mostly black setae; the tibial spurs are white; and S2 lacks a seta-filled pit. **FEMALE** (3.6.2.1). Similar to male. **NOTES** Most males of *D. gibbosa*, especially the northern populations, have the cuticle entirely black, and they fit in the *Ephuta*-like MCS (5.4.1.1).

▼ 5.5.2—*Dasymutilla sicheliana Timulla*-like males cluster

These species occur mainly in Arizona and New Mexico, and this form of *D. foxi* also occurs in Texas and the Great Plains. They have the metasomal setae and cuticle uniformly orange or red, except that the first and last segments are sometimes black. They could be confused for some Texan MCS forms (5.3.4).

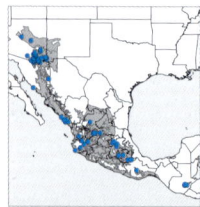

5.5.2.1—*Dasymutilla sicheliana*
DAH-ZEE-MEW-TILL-UH SICHEL-EE-ANN-UH

ETYMOLOGY This species was named for the French entomologist Frédéric J. Sichel. **FIELD IDENTIFICATION** Often easy. The metasomal setae are usually bright red, but T7 has the setae and cuticle distinctly black. The body shape is more slender than that of *D. foxi*. **LAB IDENTIFICATION** The axilla is truncate posteriorly; the mid and hind femora are rounded apically; the hind

tibia is cylindrical; and S2 has a seta-filled pit mesally. **FEMALE** (3.4.6.1). Similar to male. **NOTES** The sex association was discovered after males and females were collected together in pitfall traps (Manley & Radke 2002). Many individuals have the mesosomal setae almost entirely black and are treated in the Texan MCS (5.3.4.2). Some individuals have also been seen with the mesosomal setae partly reddish. Based on DNA comparisons, the unassociated male of *D. asteria* was discovered, but we could not find any meaningful diagnostic features to separate it from *D. sicheliana* males, even using the genitalic morphology, so that male has not been described yet.

TOP LEFT: Typical *Dasymutilla sicheliana*

LEFT: Rarer reddish *D. sicheliana*

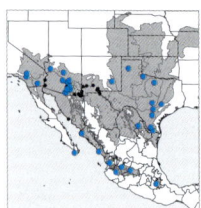

5.5.2.2—*Dasymutilla foxi*
DAH-ZEE-MEW-TILL-UH FOX-EYE

ETYMOLOGY This species was named for the entomologist, William J. Fox, who described many velvet ant species in the 1890s. This form used to be named *D. dugesii*. **FIELD IDENTIFICATION** Often easy. The metasomal setae are usually bright red and T7 has the setae partly blackened. The body shape is thicker than that of *D. sicheliana*. **LAB IDENTIFICATION** Can be immediately recognized by the hind tibia, which is apically dilated and flattened, with a concave inner surface. The axilla is dentate posteriorly; the mid and hind femora are rounded apically; and S2 has a small seta-filled pit mesally. **FEMALE** (3.1.3.3, 3.2.7.1, 3.4.6.3). Often similar to male, but sometimes with the dorsal setae entirely reddish or whitish. **NOTES** This is the dominant color pattern for Texan and Great Plains populations of this species, but it is sporadically found throughout the western states. Males from the same habitats as Desert FCS females of *D. foxi* (3.1.3.3) maintain their reddish metasomal color, like the form pictured in figure 5.5.2.2a. Additional males are known with the Western MCS (5.2.3.1).

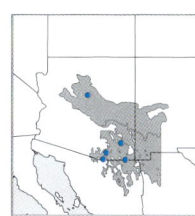

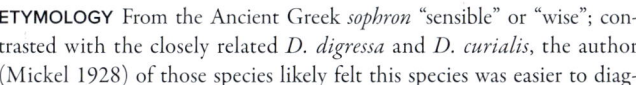

5.5.2.3—*Dasymutilla sophrona*
DAH-ZEE-MEW-TILL-UH SOE-FRONN-UH

ETYMOLOGY From the Ancient Greek *sophron* "sensible" or "wise"; contrasted with the closely related *D. digressa* and *D. curialis*, the author (Mickel 1928) of those species likely felt this species was easier to diag-

nose. **FIELD IDENTIFICATION** Often difficult. The metasomal setae are pale golden, and T7 has the cuticle orange, like that of T2–6; S2 lacks a seta-filled pit. **LAB IDENTIFICATION** The axilla is dentate posteriorly; the mid and hind femora are truncate apically; the hind tibia is cylindrical; and S2 lacks a seta-filled pit. **FEMALE** Unknown. **NOTES** Based on its truncate femoral apices and distribution in Arizona, this could be the male of *D. ferruginea* (3.4.3.2) or *D. dilucida* (3.4.3.1).

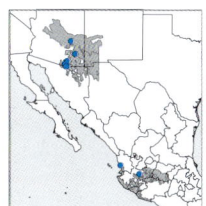

5.5.2.4—*Dasymutilla bonita*
DAH-ZEE-MEW-TILL-UH BOW-NEE-TUH

ETYMOLOGY *Bonita* is Spanish for pretty, but this species was more likely named for a town in Arizona where the type was collected. **FIELD IDENTIFICATION** Often difficult. The metasomal setae are pale golden, and T7 has the cuticle orange, like that of T2–6; S2 has a

small white, seta-filled pit mesally. **LAB IDENTIFICATION** The axilla is truncate posteriorly; the mid and hind femora are rounded apically; the hind tibia is cylindrical; and S2 has a seta-filled pit mesally. **FEMALE** (3.4.5.3). Similar to male. **NOTES** This species can be common in mainland Pacific Mexico but is rare in Arizona. None of the authors has collected this species.

5.5.3.1—*Dasymutilla monticola*
DAH-ZEE-MEW-TILL-UH MON-TICK-OWE-LUH

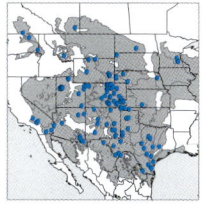

ETYMOLOGY A Latin term that means "mountain dweller." **FIELD IDENTIFICATION** Often difficult. The color pattern is unique but difficult to verify in the field. **LAB IDENTIFICATION** The mesonotum is covered with gray or silvery setae, like the vertex and pronotum; the axilla is truncate posteriorly; and S2 has a seta-filled pit mesally. **FEMALE** (3.4.4.2). Somewhat similar to male. **NOTES** This species varies in color across its range.

Dark northern
Dasymutilla monticola

Lighter southern *D. monticola*

Southwestern desert forms have more extensive reddish cuticle, especially on the mesosomal sides, and brighter silvery setal markings. Northern populations, however, are mostly black with duller gray setae.

(Photo by Jillian Cowles)

▼ 5.5.4—*Dasymutilla fasciventroides Timulla*-like males cluster

These species occur mainly in Arizona, but *D. birkmani* lives in New Mexico, Texas, and the Great Plains. They have the metasomal cuticle mostly black, with yellow or orange cuticular coloration restricted to T2 (and sometimes S2).

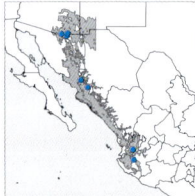

5.5.4.1—*Dasymutilla fasciventroides*
DAH-ZEE-MEW-TILL-UH FASS-IVV-ENN-TROY-DEEZ

ETYMOLOGY Named for the similarity to *Dasymutilla fasciventris*. **FIELD IDENTIFICATION** Sometimes difficult. This species has a slender body shape and broader distribution in Arizona than the similarly colored *D. iztapa*. **LAB IDENTIFICATION** The axilla is truncate posteriorly; S2 has a seta-filled pit anteromesally. **FEMALE** Unknown. **NOTES** This species was widely confused for *D. fasciventris* (5.7.2.2) in the literature, because of similarities in the anteromesal pit on S2. Only recently, it was described as a separate species. Based on clypeal and genitalic morphology, it seems closely related to the *D. quadriguttata* species-group. Females in this species-group have the vertex armed with a posterolateral tubercle, but no females from Arizona have yet been discovered with tubercles on the head.

5.5.4.2—*Dasymutilla birkmani*
DAH-ZEE-MEW-TILL-UH BIRKMAN-EYE

ETYMOLOGY This species is named after G. Birkman, who collected many important mutillid specimens in Texas in the late 1800s. This male color form used to be named *D. arcana*. **FIELD**

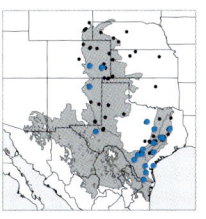

IDENTIFICATION Usually easy. The only species with this coloration in Texas and the Great Plains. **LAB IDENTIFICATION** The axilla is truncate posteriorly; S2 lacks a seta-filled pit. **FEMALE** (3.4.4.3, 3.4.5.1). Usually lighter in color than male. **NOTES** This species is smaller and lives farther east than the others in this cluster. There is a darker form of this species in the Eastern MCS (3.6.2.1).

5.5.4.3—*Dasymutilla iztapa*
DAH-ZEE-MEW-TILL-UH IZZ-TAW-PUH

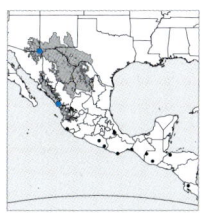

ETYMOLOGY Probably named after Iztapalapa, a borough of Mexico City. **FIELD IDENTIFICATION** Sometimes difficult. This species has a thicker body shape and more restricted distribution than the similarly colored *D. fasciventroides*. **LAB IDENTIFICATION** The axilla is dentate posteriorly; S2 has a seta-filled pit posteromesally. **FEMALE** Unknown. **NOTES** This species is widespread and variable in Mesoamerica but remarkably rare in the USA, being known from a single male collected in southeastern Arizona. Based on coloration and distribution, the female might be *D. toluca* (3.4.6.5), but there is no strong evidence for the association.

5.5.4.4—*Dasymutilla pulchra*
DAH-ZEE-MEW-TILL-UH PULL-KRUH

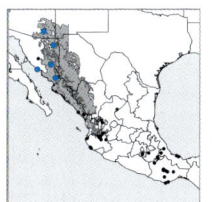

ETYMOLOGY From the Latin *pulcher* "beautiful." **FIELD IDENTIFICATION** Easy. The T2 cuticular patch is pale yellow. **LAB IDENTIFICATION** The axilla is truncate posteriorly; S2 has a seta-filled pit mesally. **FEMALE** (3.4.7.1). Similar to male. **NOTES** This species is widespread and variable in Mesoamerica but rare in the USA (or maybe even absent), being known from a single male. The locality label for this specimen says "Alamos, Arizona." It is not clear whether this is referring to the ghost town of Tres Alamos in Cochise County, or perhaps it is actually the city of Alamos in Sonora, Mexico.

▼ 5.5.5—*Dasymutilla apicalata Timulla*-like males cluster

These species occur mainly in the southwestern USA, extending into south Texas. They have the metasomal cuticle mostly orange and marked with silver and black setae.

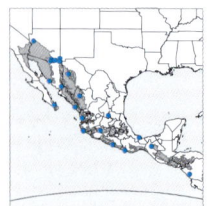

5.5.5.1—*Dasymutilla apicalata*
DAH-ZEE-MEW-TILL-UH APE-ICK-UH-LAW-TUH

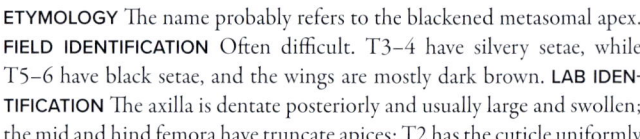

ETYMOLOGY The name probably refers to the blackened metasomal apex. **FIELD IDENTIFICATION** Often difficult. T3–4 have silvery setae, while T5–6 have black setae, and the wings are mostly dark brown. **LAB IDENTIFICATION** The axilla is dentate posteriorly and usually large and swollen; the mid and hind femora have truncate apices; T2 has the cuticle uniformly reddish-orange; S2 lacks a seta-filled pit; the tergal setae are mostly silvery with a blackish band on T5–6. **FEMALE** Unknown. **NOTES** In the literature, this species is recognized from Nicaragua north to Texas and Arizona and throughout various Mexican states. The type of this species is from Mexico (without a specific locality provided). Based on the broad distribution and some subtle morphological variation in the axilla and other features, the North American specimens of *D. apicalata* might actually belong to a separate species. Matching this species, or at least some populations of this species, with females will be a vital step to sorting out this messy taxon. Some individuals have the mesosoma partly reddened laterally and could be confused with *D. heliophila* (5.7.2.1) from the *Sphaeropthalma*-like MCS.

5.5.5.2—*Dasymutilla dammersi*
DAH-ZEE-MEW-TILL-UH DAMMERS-EYE

ETYMOLOGY This species was named in honor of Commander Charles M. Dammers, who collected many important velvet ants from southern California in the 1930s. **FIELD IDENTIFICATION** Usually easy. The color pattern is unique among species in southern California, and the wings are mostly clear, darkened only along the outer margin beyond the veins. **LAB IDENTIFICATION** The mesosoma is entirely black; the axilla is dentate posteriorly; the mid and hind femora have truncate apices; the tergal setae are mostly silvery with a blackish band on T5–6. **FEMALE** (3.4.2.1). Somewhat similar to male. **NOTES** This male was only recognized in 2020 (see female account in 3.4.2.1). This is a rarely encountered species.

5.5.5.3—*Dasymutilla curialis*
DAH-ZEE-MEW-TILL-UH CUE-REE-ALL-ISS

ETYMOLOGY Likely derived from the Latin *curiosus* "inquisitive," "odd," or "strange." **FIELD IDENTIFICATION** Often difficult. The color pattern is unique but difficult to verify in the field. **LAB IDENTIFICATION** The axilla is dentate posteriorly; the mid and hind femora have truncate

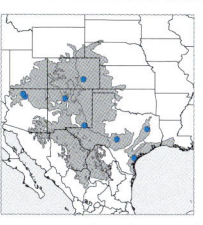

apices; T2 has the cuticle uniformly reddish-orange; S2 lacks a seta-filled pit; the setae from T3–7 are entirely black. **FEMALE** Unknown. **NOTES** This species is found in northern New Mexico. The only related female we have seen from this area is *D. snoworum* (3.4.4.1), which already has a male (5.3.4.1), so maybe this is just a color variant of that species. It is also similar to *D. scaevola* from the Eastern MCS (5.6.1.1), except for distribution and the silvery mesosomal setae.

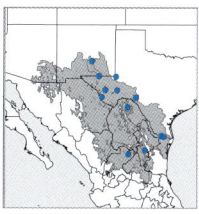

5.5.5.4—*Dasymutilla digressa*
DAH-ZEE-MEW-TILL-UH DYE-GRESS-UH

ETYMOLOGY From the Latin *digressus* "departure" or "deviation." **FIELD IDENTIFICATION** Often difficult. The color pattern is unique but difficult to verify in the field. **LAB IDENTIFICATION** The axilla is dentate posteriorly; the mid and hind femora have truncate apices; T2 has the cuticle uniformly reddish-orange; S2 lacks a seta-filled pit; T3–7 have the setae entirely silvery. **FEMALE** Unknown. **NOTES** This species is found from New Mexico to southern Texas. The only related female we have seen from this area is *D. snoworum* (3.4.4.1), which already has a male (5.3.4.1), so maybe this is just a color variant of that species. It could be confused for some species in the *D. sicheliana* cluster (5.5.2), but T2 has a somewhat distinct patch of black setae, and the apical tergites have the setae more silver-tinted than orange or golden.

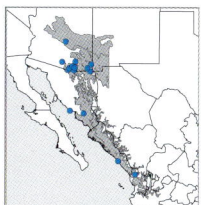

5.5.5.5—*Dasymutilla saetigera*
DAH-ZEE-MEW-TILL-UH SEE-TIJJ-AIR-UH

ETYMOLOGY From the Latin *saeta* "bristle of rough hair." The male used to be named *D. polia*. **FIELD IDENTIFICATION** Often difficult. The color pattern is unique but difficult to verify in the field. **LAB IDENTIFICATION** The axilla is truncate posteriorly and usually large and swollen; the mid and hind femora have rounded apices; T2 has a yellowish cuticular patch; and S2 has a seta-filled pit mesally; the tergal setae have somewhat irregularly mixed black and silvery setae. **FEMALE** (3.4.6.4). Similar to male. **NOTES** The sexes of *D. saetigera* were associated using DNA sequence comparisons. Even though the male and female are treated as different color syndromes, their color patterns are remarkably similar to one another.

5.5.6—*Lomachaeta hicksi Timulla*-like males cluster

These are *Lomachaeta* species with the metasoma partly orange or red. Other small-bodied genera, like *Ephuta* (5.5.11) and *Myrmosula* (5.3.7.1), often have similar coloration.

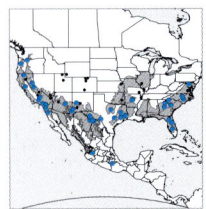

5.5.6.1—*Lomachaeta hicksi*
LOW-MUH-KEE-TUH HICKS-EYE

ETYMOLOGY This species was named after C. H. Hicks, who collected the type specimens in 1928. **FIELD IDENTIFICATION** Usually impossible, but this is apparently the only *Lomachaeta* species in the northern, central, and eastern USA. **LAB IDENTIFICATION** The mandible has a ventral tooth basally; the forewing venation is normal; the T2 fringe has long black bristles; the paramere is acuminate with mostly short setae. **FEMALE** (3.7.1.1). Usually with lighter body color than male. **NOTES** This is the most widespread and variable *Lomachaeta* species, with some populations having the *Ephuta*-like MCS (5.4.4.3). The sex association was discovered because males and females emerged from the same parasitized hosts. This species has been reared from nests of the crabronid genera *Solierella* and *Pisonopsis*. The most common color pattern has the metasoma mostly black, but T2 has the sides partly reddened. Many populations have mostly black setae on the head and mesosoma and could be confused for the Eastern MCS.

FAR LEFT: Typical *Lomachaeta hicksi.*

LEFT: Rare orange-legged *Lomachaeta hicksii* from southern California.

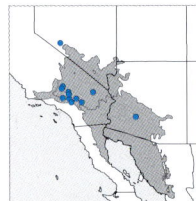

5.5.6.2—*Lomachaeta beadugrimi*
LOW-MUH-KEE-TUH BEE-DOO-GRIMM-EYE

ETYMOLOGY From the Old English *beadugrim* "war-furious," in reference to the blade-like parameres. **FIELD IDENTIFICATION** Impossible. **LAB IDENTIFICATION** The mandible is unarmed ventrally; the forewing venation is normal; the T2 fringe has simple setae only; the paramere is dorsoventrally flattened with a rounded apex. **FEMALE** (3.7.1.3). Lighter brown coloration than male. **NOTES** The sex association was recognized based on a large series of both sexes that were collected together on spurge flower mats (*Chamaesyce* sp.) by Gordon C. Snelling.

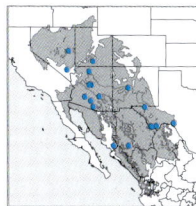

5.5.6.3—*Lomachaeta crocopinna*
LOW-MUH-KEE-TUH CROCK-OWE-PINN-UH

ETYMOLOGY From the Ancient Greek *krokos* "orange" and *pinna* "feather," in reference to the orange metasoma and brachyplumose setae. **FIELD IDENTIFICATION** Impossible. **LAB IDENTIFICATION** The mandible is unarmed ventrally; the forewing venation is normal; the T2 fringe has simple setae only; the paramere is acuminate apically, with a ventral row of long setae. Unlike *L. ptilohyalus*, the metasomal cuticle is entirely orange-brown. **FEMALE** Unknown. **NOTES** This species is widespread in the southwestern USA.

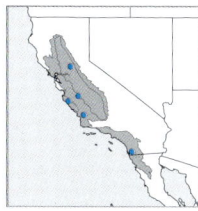

5.5.6.4—*Lomachaeta powelli*
LOW-MUH-KEE-TUH POWELL-EYE

ETYMOLOGY This species was named after J. A. Powell, who collected the type specimens in 1962. **FIELD IDENTIFICATION** Often impossible. The distribution in California's Pacific areas coupled with reduced wing venation could be diagnostic, but *L. hicksi* also occurs in this region, and the wing venation is difficult to recognize in the field or in photos. **LAB IDENTIFICA-**

TION The mandible is unarmed ventrally; the forewing venation is reduced, with the veins restricted to the basal half; the T2 fringe has simple setae only; the paramere is acuminate apically, with short setae only. **FEMALE** (3.7.1.6). Lighter brown coloration than male. **NOTES** The sex association was based on both sexes collected together as they attacked *Diodontus* wasps near a fire station in the small town of Pozo, California. Unlike the other *Lomachaeta* species with known host records, this species attacks ground-nesting, rather than twig-nesting, hosts.

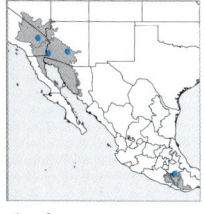

5.5.6.5—*Lomachaeta ptilohyalus*
LOW-MUH-KEE-TUH TIE-LOW-HYE-ALL-USS

ETYMOLOGY From the Ancient Greek *ptilon* "feather" and *hyalos* "glass," in reference to the brachyplumose setae and sparse punctation. **FIELD IDENTIFICATION** Impossible. **LAB IDENTIFICATION** The mandible is unarmed ventrally;

the forewing venation is normal; the T2 fringe has simple setae only; the paramere is acuminate apically with a ventral row of long setae. Unlike *L. crocopinna*, the apical tergites have the cuticle black. **FEMALE** Unknown. **NOTES** This species has a disjunct distribution. It is mostly seen in North America's hot deserts in Arizona and California but is also found in Neotropical Oaxaca, Mexico.

▼ 5.5.7.1—*Pseudomethoca wickhami*

SOO-DOE-METH-OWE-KUH WICKHAM-EYE

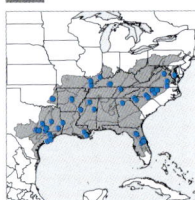

ETYMOLOGY This species was named after H. F. Wickham, who collected the type specimen in the late 1800s. The male was initially described under the name *Ps. vanduzei*. **FIELD IDENTIFICATION** Often easy. The color pattern is unique in the eastern USA. They could be confused with *Sphaerop-thalma marpesia* (5.5.10.1) or *Photomorphus auriventris* (5.5.9.1) in the central USA, but they are always diurnal and have a thicker and smoother overall look. Unlike the *Ps. donaeanae* cluster (5.5.8), the T2 cuticle is black. **LAB IDENTIFICATION** The color pattern and mandible shape (vertical anteriorly) are unique in the genus *Pseudomethoca*. The eye is circular; the mesosternal area is unarmed; and the tergal fringes have simple setae only.

FEMALE (3.5.3.1, 3.6.9.2). Lighter body color than male; the *Timulla*-like female is almost a direct inversion of the male's color pattern, with the anterior portion of the body orange and the posterior portion blackish. **NOTES** *Pseudomethoca wickhami* has many unique features in comparison with other North American species, especially in the male mandible shape, undefined pygidium in females, and postgenal armature in both sexes. Many related species, however, occur in Central and South America. This group of species might eventually be recognized as a distinct genus.

▼ 5.5.8—*Pseudomethoca donaeanae Timulla*-like males cluster

These species occur in the southwestern USA from the Sonoran Desert in California to south Texas. They have the metasomal cuticle (at least T2) orange or reddish, and the head and mesosoma with extensive silver or gray setae. They should be compared with Eastern MCS *Pseudomethoca* males (5.6.5) and some species of *Dasymutilla* with the *Timulla*-like MCS (5.5.5).

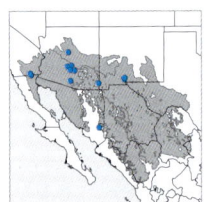

5.5.8.1—*Pseudomethoca donaeanae*

SOO-DOE-METH-OWE-KUH DOE-NYUH-ANN-EE

ETYMOLOGY This species was named for Dona Ana County, New Mexico. **FIELD IDENTIFICATION** Often difficult. This is the most common and widespread species in this cluster. It does not overlap in distribution with *Ps. russeola* or *Ps. flavida* and is more widespread than *Ps. ajattara*. **LAB IDENTIFI-**

CATION The clypeus is basically flat with two small widely separated sharp teeth; the tegula is basically flat and mostly smooth; the femur lacks a brush of long setae ventrally; the tergal fringes have separated, parallel, thick white bristles; and the paramere has simple setae only. **FEMALE** (3.4.14.1). Often similar to male. **NOTES** Manley (1999) discovered this male and identified it as *Ps. russeola* using Mickel's (1935a) key. It was later discovered, however, that these males differed from true

Texan specimens by many structural differences. The true males of *Ps. donaeanae* had not been previously known to science and were described by Williams & Pitts (2008).

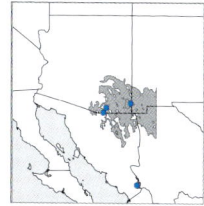

5.5.8.2—*Pseudomethoca ajatarra*
SOO-DOE-METH-OWE-KUH AJJ-UH-TAR-UH

ETYMOLOGY In Finnish mythology, Ajattara was an evil forest spirit. **FIELD IDENTIFICATION** Difficult. This rare species has slightly shaggier and denser setae than *Ps. donaeanae*. **LAB IDENTIFICATION** The clypeus is anteriorly expanded, with two closely spaced blunt teeth; the tegula is basically flat and entirely punctate; the femur lacks a brush of long setae ventrally; the tergal fringes have dense, parallel, thick white bristles; and the paramere has strange elbowed setae along the inner margin. **FEMALE** Unknown. **NOTES** This strange species was discovered during the same project wherein the sex association of *Ps. donaeanae* was resolved (Williams & Pitts 2008). It was originally described from only one specimen, and a second specimen was only recently found in 2018; both specimens are from mid-range elevations in southeastern Arizona. Various similar specimens from Mexico have been observed, but it is not yet clear whether they are new species or variants of *Ps. ajatarra*. We do not yet have a good guess about the identity of the female.

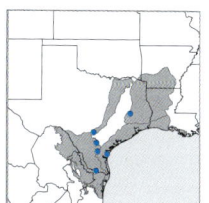

5.5.8.3—*Pseudomethoca russeola*
SOO-DOE-METH-OWE-KUH RUSS-EE-OWE-LUH

ETYMOLOGY Derived from the Latin *russus* "red," in reference to the metasomal color. **FIELD IDENTIFICATION** Sometimes difficult. This species is smaller and has smoother body sculpture than *Ps. flavida*. **LAB IDENTIFICATION** The clypeus is weakly expanded, with two blunt teeth; the tegula is strongly convex and mostly smooth; the femur lacks a brush of long setae ventrally; the tergal fringes have simple setae only; and the paramere has simple setae only. **FEMALE** Unknown. **NOTES** Since this species was "divorced" from *Ps. donaeanae* (see 5.5.8.1), there has not been a female associated with *Ps. russeola*. Since it lives in southern Texas and is structurally similar to *Ps. simillima* (5.6.5.2), it might be eventually matched with *Ps. meritoria* (3.6.10.2).

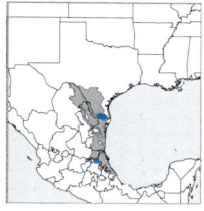

5.5.8.4—*Pseudomethoca flavida*
SOO-DOE-METH-OWE-KUH FLAVV-IDD-UH

ETYMOLOGY Derived from the Latin *flavus* "yellow," in reference to the metasomal color. **FIELD IDENTIFICATION** Often easy. This species is larger and has coarser body sculpture than *Ps. russeola*. Specimens with yellow tergal setae are easier to recognize. **LAB IDENTIFICATION** The clypeus is moderately expanded, with two sharp teeth; the tegula is coarsely rugose and strongly

FAR LEFT:
Pseudomethoca flavida from San Luis Potosi, Mexico.

LEFT:
Pseudomethoca flavida from Texas.

raised with a flat outer rim; the femur has a brush of long setae ventrally; the tergal fringes have simple setae only; and the paramere has simple setae only. **FEMALE** Unknown. **NOTES** This species is structurally similar to *Ps. oceola* (5.6.5.1, 5.7.7.2) and might just be a color variant of that species. There is some variation in the tergal setal color for this species, with some individuals having those setae yellowish and others having them interspersed black and silver.

5.5.9.1—*Photomorphus auriventris*

FOE-TOE-MORE-PHUSS ORR-IVV-ENN-TRISS

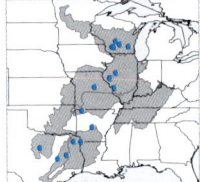

ETYMOLOGY From the Latin *aurum* "gold" and *venter* "bottom," in reference to the orange-colored apical tergites. **FIELD IDENTIFICATION** Often easy. The color pattern is unique for the genus but is similar to some species in other genera. This species is smaller than *Sphaeropthalma marpesia* (5.5.10.1) and more slender than *Ps. wickhami* (5.5.7.1).

LAB IDENTIFICATION The mandible is tridentate; the mesosternal area has a low complex of transverse ridges; and S2 has a long felt line. **FEMALE** (3.7.4.3). Lighter in color than male. **NOTES** This species was originally known only from Texas. It was more recently found as far north as Wisconsin, where the female was recognized by a fellow mutillid researcher, Craig Brabant.

5.5.10.1—*Sphaeropthalma marpesia*

S-FAIR-OPP-THAL-MUH MAR-PEE-ZHEE-UH

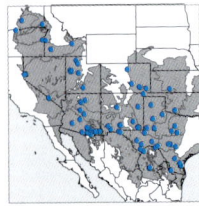

ETYMOLOGY In Ancient Greek and Roman mythology, Marpesia was the queen of the Amazons. The male used to be named *S. imperialiformis* based on its similarity to the Texan species *S. imperialis* (6.1.8b). **FIELD IDENTIFICATION** Often easy. The color pattern is unique among nocturnal species but is generally similar to *Ph. auriventris* (5.5.9.1), *Ps. wickhami* (5.5.7.1), and some other nocturnal species (6.1.8). **LAB IDENTIFICATION** The mandible is wide and vertical; the mesosternal area is unarmed; the T2 cuticle is blackish-brown; S2 has a short felt line; and the tergal fringes have plumose setae. **FEMALE** (3.4.15.1). With denser yellow or orange setae than the male. **NOTES** This widespread species is one of the easiest nocturnal velvet ants to identify because of the coloration. Even though it is frequently seen at light traps, it has also been collected in the early morning and at dusk.

Sphaeropthalma marpesia

▼ 5.5.11—*Ephuta copano Timulla*-like males cluster

These species occur in Texas and the southeastern USA. They are the only *Ephuta* species in this color syndrome. Some specimens of *E. battlei* have the mesosomal setae mostly black and could be initially confused for the Eastern MCS.

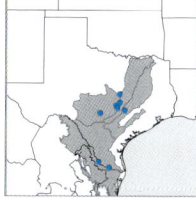

5.5.11.1—*Ephuta copano*
EFF-OO-TUH COE-PAN-OWE

ETYMOLOGY This species was named for a ghost town in Texas that was used as a port by smugglers and pirates in the 18th century. **FIELD IDENTIFICATION** Often difficult. The completely orange metasoma (including the setae) is unique for *Ephuta*, but the genus is difficult to recognize in the field. **LAB IDENTIFICATION** The subantennal basin is much shorter than the clypeal basin; the mandible is slender; the tergites are entirely orange with golden setae; and the hypopygium is generally dark brown. **FEMALE** (3.7.6.3).

Lighter body color than male. **NOTES** Because of its distinctive coloration, this species was thought to occur far south into tropical Mexico and Costa Rica, but those populations have structural differences and are separate, probably undescribed, species. The true *E. copano* occurs in southern Texas.

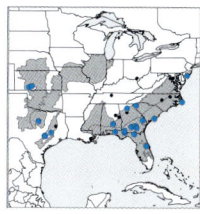

5.5.11.2—*Ephuta battlei*
EFF-OO-TUH BATTLE-EYE

ETYMOLOGY Bradley (1916) dedicated this species to "Paul Battle of Bainbridge, Georgia, my faithful companion on many a collecting trip." **FIELD IDENTIFICATION** Difficult. This species generally has a smaller body size and narrower head than other species in this cluster populations. **LAB IDENTIFICATION** The subantennal basin is much shorter than the clypeal basin; the mandible is slender; the apical tergites are mostly black with whitish setae; and the hypopygium is generally pale yellow-brown. **FEMALE** Not described in the literature, but Schuster (1956) said the female was known but impossible to separate from *E. slossonae* (3.6.12.1). **NOTES** This relatively common eastern species is divided into four subspecies. Two of these subspecies have the *Ephuta*-like MCS (5.4.13.3). The other two fit here: *E. b. microcellaria* has small ocelli and T2 with small reddish patches laterally; *E. b. battlei* has larger ocelli and T2 usually mostly reddish.

Ephuta battlei battlei

Ephuta battlei microcellata

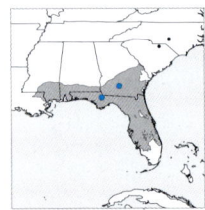

5.5.11.3—*Ephuta eurygnathus*
EFF-OO-TUH EWE-REE-NATH-USS

ETYMOLOGY From the Ancient Greek *eurys* "broad" and *gnathos* "jaw," in reference to the wide mandible. **FIELD IDENTIFICATION** Difficult. This species has a larger body size and wider head than similarly colored *E. battlei* populations. **LAB IDENTIFICATION** The subantennal basin is much shorter than the clypeal basin; the mandible is wide and tridentate; the apical tergites are mostly black with whitish setae; and the hypopygium is generally pale yellow-brown. **FEMALE** Unknown. **NOTES** This species occurs in the *Ephuta*-like MCS form (5.4.13.7) and this *Timulla*-like MCS form. This form is known mainly from the Florida panhandle area. The currently unknown female will likely be similar to *E. slossonae* (3.6.12.1), based on the similar male morphology.

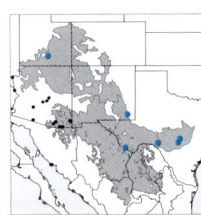

5.5.11.4—*Ephuta rufisquamis*
EFF-OO-TUH ROOF-ISS-KWAH-MISS

ETYMOLOGY From the Latin *rufus* "red" and *squama* "scale," in reference to the reddish tegula color. **FIELD IDENTIFICATION** Difficult. This species occurs farther west than others in this cluster and sometimes has orange legs, but in southern Texas, it cannot be reliably separated from *E. battlei* or *E. copano* in the field. **LAB IDENTIFICATION** The subantennal basin is as

Ephuta rufisquamis from Texas.

Ephuta rufisquamis from Utah.

long as the clypeal basin; the mandible is slender; the apical tergites are often partly darkened with whitish setae; and the hypopygium is generally yellowish brown. **FEMALE** Unknown. **NOTES** This is an atypical eastern form of *E. rufisquamis*; most other populations fit in the *Ephuta*-like MCS (5.4.12.1). One individual from Utah has the orange legs like typical *E. rufiquamis* from Arizona and California, while most specimens from Texas have the legs mostly black. The broad distribution coupled with color variation suggest that *E. rufisquamis* as defined here could eventually be revealed as a complex of multiple distinct species.

 ### 5.5.12—*Timulla ocellaria Timulla*-like males cluster

These species occur in the central and eastern USA. They are the only *Timulla*-like MCS *Timulla* species with the metasomal setae largely blackish. They can easily be confused with Eastern MCS *Timulla* species (5.6.7) based on the extent of black or gray setae on the pronotum.

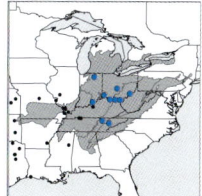

5.5.12.1—*Timulla ocellaria*
TIMM-EWE-LUH OWE-CELL-ARE-EE-UH

ETYMOLOGY Named after the enlarged ocelli. **FIELD IDENTIFICATION** Often easy. This is apparently the only nocturnal *Timulla* in the eastern USA.

LAB IDENTIFICATION The ocelli are large; the mandible has a ventral tooth basally; the scape lacks a setal brush; the mesosternal area is unarmed; and T7 has a Y-shaped carina. **FEMALE** Unknown. **NOTES** This form is less common and occurs farther east than the *Sphaeropthalma* like MCS form (5.7.16.7). The large eyes and ocelli suggest nocturnal behavior, and sometimes this species has been collected with light traps.

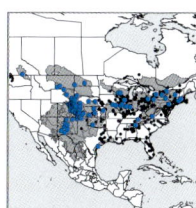

5.5.12.2- *Timulla vagans*
TIMM-EWE-LUH VAY-GUHNS

ETYMOLOGY From the Latin *vagans* "rambling." This color form used to be named *T. huntleyensis* after the city of Huntley, Montana. **FIELD IDENTIFICATION** Often difficult. This species is diurnal and has a distinct white brush on the scape; this color form is found mainly in the Great Plains. **LAB IDENTIFICATION** The ocelli are moderately large; the mandible

Timulla vagans from Kansas with black tergal fringes.

Rare *Timulla vagans* from Montana with orange tergal *fringes.*

has a ventral tooth basally; the scape has a dense white setal brush; the mesosternal area is armed with a large smooth tubercle; and T7 has a Y-shaped carina. **FEMALE** (3.5.9.1, 3.6.17.2). Different from male, with paler body color. **NOTES** This widespread species also expresses the Eastern (5.6.7.1) and *Sphaeropthalma*-like (5.7.16.10) Color Syndromes. This color form seems to be most abundant in the Great Plains and often has the metasomal setae largely golden, making it easy to confuse with the *T. oajaca* cluster (5.5.14). *Timulla vagans* has a more distinct setal brush on the scape than any species in that cluster.

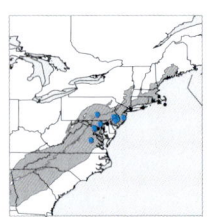

5.5.12.3—*Timulla hollensis*
TIMM-EWE-LUH HALL-ENN-SISS

ETYMOLOGY Likely named after the type locality of Woods Hole, Massachusetts. **FIELD IDENTIFICATION** Often difficult. This species is apparently diurnal, lacks a brush on the scape, and is mainly found in the northeastern USA. **LAB IDENTIFICATION** The ocelli are large; the mandible lacks a ventral tooth basally; the scape lacks a setal brush; the mesosternal area is unarmed; and T7 has a Y-shaped carina. **FEMALE** Unknown. **NOTES** *Timulla hollensis* completely lacks a ventral mandibular tooth. Some populations fit the *Sphaeropthalma*-like MCS (5.7.16.5).

 ## 5.5.13—*Timulla neobule Timulla*-like males cluster

These species occur mainly in the Sonoran Desert in California and Arizona. They have the tergal setae entirely pale golden-orange, and the tarsi (at least their setae) reddish and paler than the tibiae. The tibial color varies a bit and is sometimes darker, so these species should be compared against the *Timulla oajaca* cluster (5.5.14), as well.

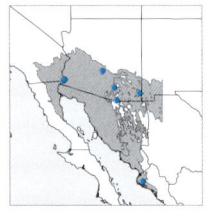

5.5.13.1—*Timulla neobule*
TIMM-EWE-LUH NAY-OWE-BOO-LAY

ETYMOLOGY In the Ancient Greek poetry of Archilochus, Neobule was a girl who was engaged to the author, before her father called off the wedding. The bitter poems of Archilochus were so poignant that she and all her family were said to have hanged themselves. **FIELD IDENTIFICATION** Impossible. **LAB IDENTIFICATION** The ocelli are small; the mandible has a ventral tooth basally; the scape lacks a setal brush; the mesosternal area has a large smooth tubercle; the mid coxa is armed with a sharp tooth; and T7 has Y-shaped carina. **FEMALE** Unknown. **NOTES** This species is similar to *T. oajaca*, *T. suspensa*, and especially *T. nitela*, but seems to occur in hotter and drier habitats than those species. The reddish tarsi are diagnostic but somewhat variable, so the characters in the key in the appendix (9.3.5.5) should be consulted. Based on their overlapping distributions, *T. nicholi* (3.5.9.4) might be the female of this species.

5.5.13.2—*Timulla tyro*
TIMM-EWE-LUH TIE-ROW

ETYMOLOGY From the Latin *tiro* "recruit" or "novice." **FIELD IDENTIFICA-TION** Impossible. **LAB IDENTIFICATION** The ocelli are small; the mandible has a ventral tooth basally; the scape lacks a setal brush; the mesosternal area is unarmed; and T7 has a pre-apical inverted U-shaped carina, and the posterior margin is emarginate mesally. **FEMALE** (3.7.17.5). Lighter in color than male. **NOTES** This species and *T. leona* (5.5.14.4) are the only species in the USA with T7 apically emarginate, but that condition occurs in many Neotropical *Timulla* species. Unlike *T. leona*, *T. tyro* has denser white setae on the tegula and slightly different T7 armature. Distribution is also useful for separating these species, since *T. leona* is a central and eastern species, while *T. tyro* lives in California and Arizona.

▼ **5.5.14—*Timulla oajaca Timulla*-like males cluster**

These species occur mainly in the central and western USA, with all but one of them occurring in Texas. They have the tergal setae entirely pale golden-orange and the legs entirely blackish. Because the color differences are often subtle and variable, these species should be compared against the *T. neobule*, *T. ocellaria*, and Texan MCS *Timulla* species (5.3.6).

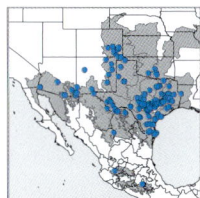

5.5.14.1—*Timulla oajaca*
TIMM-EWE-LUH WAH-HAW-KUH

ETYMOLOGY Named for the Mexican state of Oaxaca. It is confusing that this species does not actually occur in that state. **FIELD IDENTIFICATION** Impossible. **LAB IDENTIFICATION** The ocelli are small; the mandible has a ventral tooth basally; the scape lacks a setal brush; the mesosternal area has a large, smooth tubercle; the mid coxa is armed with a small blunt tubercle; and T7 has Y-shaped carina with the lateral arms of this Y much shorter than the stem. **FEMALE** (3.5.9.2, 3.6.17.1). Lighter in color than male. **NOTES** This species is widespread in the southern USA and many Mexican states, but, surprisingly, it has never been found in Oaxaca.

Timulla oajaca from Arizona.

Timulla oajaca from Texas.

5.5.14.2—*Timulla suspensa*
TIMM-EWE-LUH SUSS-PEN-SUH

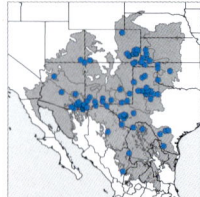

ETYMOLOGY Apparently from the Latin *suspensus* "suspended," "anxious," or "doubtful." It is not clear why this name was chosen. **FIELD IDENTIFICATION** Impossible. **LAB IDENTIFICATION** The ocelli are small; the mandible has a ventral tooth basally; the scape is apically dilated and has a sparse setal brush; the mesosternal area has a large smooth tubercle; the mid coxa is armed with a sharp tooth; and T7 has Y-shaped carina. **FEMALE** (3.5.8.1, 3.6.17.3). Lighter in color than the male. **NOTES** Males of this species are difficult to identify, even with a micro-

scope, and it is often hard to tell whether the scape is sufficiently dilated and the white setae are dense enough to be considered a "fringe." Males are variable in color and are also treated in the Texan MCS (5.3.6.3).

5.5.14.3—*Timulla nitela*
TIMM-EWE-LUH NEE-TELL-UH

ETYMOLOGY From the Latin *nitella* "splendor," perhaps in reference to the golden metasomal setae. **FIELD IDENTIFICATION** Impossible. **LAB IDENTIFICATION** The ocelli are small; the mandible has a ventral tooth basally; the scape lacks a setal brush; the mesosternal area has a large smooth tubercle; the mid coxa is armed with a sharp tooth; and T7 has Y-shaped carina. **FEMALE** Unknown. **NOTES** This species is most similar to *T. neobule*, but in the USA, it is known only from Cochise County, Arizona. It has entirely blackish legs and some other subtle structural differences (see key in the appendix for other characters: 9.3.5.5).

5.5.14.4—*Timulla leona*
TIMM-EWE-LUH LEE-OWE-NUH

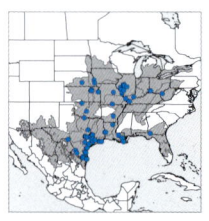

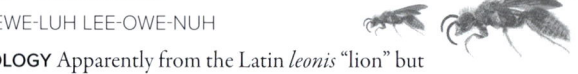

ETYMOLOGY Apparently from the Latin *leonis* "lion" but may be named for the Mexican state of Nuevo Leon, since the type came from Mexico and the author (Blake 1871) named many species after Mexican place names. **FIELD IDENTIFICATION** Impossible. **LAB IDENTIFICATION** The ocelli are small; the mandible has a ventral tooth basally; the scape lacks a setal brush; the mesosternal area is unarmed; and T7 has a pre-apical transverse arcuate carina and the posterior margin is emarginate mesally. **FEMALE** (3.5.9.3, 3.6.17.4). Lighter in color than male.

NOTES This species and *T. tyro* (5.5.13.2) are the only species in the USA with T7 emarginate apically, but that condition is commonly seen in Neotropical *Timulla* species. Unlike *T. tyro*, *T. leona* has sparser white setae on the tegula and slightly different T7 armature. Distribution is also useful for separating these species, since *T. leona* is a central and eastern species, while *T. tyro* lives in California and Arizona.

5.5.14.5—*Timulla navasota navasota*
TIMM-EWE-LUH NAW-VUH-SOE-TUH

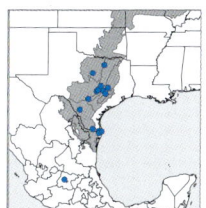

ETYMOLOGY Named after the city of Navasota, Texas. **FIELD IDENTIFICATION** Sometimes easy. This is apparently the only nocturnal *Timulla* species in this cluster. **LAB IDENTIFICATION** The ocelli are large; the scape lacks a setal brush; the mandible is unarmed ventrally; and T7 has an erect longitudinal ridge. **FEMALE** (3.6.13.1). Lighter in color than male. **NOTES** *Timulla navasota* includes two subspecies. Both occur in Texas: *T. navasota navasota* occurs in central and southern Texas, and *T. navasota coahuila* (5.3.6.2) is restricted to the Chihuahuan Desert in western Texas.

5.5.15.1—*Myrmosa nocturna rufigastra*
MEER-MOE-SUH NOCK-TURN-UH ROOF-EE-GAST-RUH

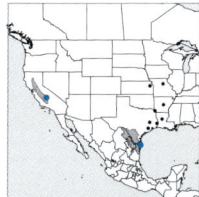

ETYMOLOGY From the Latin *nocturnus* "night active," in reference to the nocturnal behavior in some individuals. This subspecies, *M. nocturna rufigastra*, is named for the partly reddish metasoma. **FIELD IDENTIFICATION** Sometimes easy. This is apparently the only nocturnal *Myrmosa* species and the only *Myrmosa* with the metasoma reddened. **LAB IDENTIFICATION** The ocelli are large; the clypeus has a distinct longitudinal carina mesally; S1 has a slender, posteriorly acute anteromesal ridge; and S2 lacks an anteromesal tooth. **FEMALE** Unknown. **NOTES** Based on the large ocelli, this species is thought to be nocturnal, and some of the paratypes were collected in light traps. This species and *M. texana* are placed in the subgenus *Myrmosa* (*Myrmosina*), recognized by the lack of a basomesal tooth on S2 but having one on S1. The nominotypical subspecies, *M. n. nocturna* (5.4.14.2) fits in the *Ephuta*-like MCS. The original description of *M. nocturna rufigastra* from southern Texas by Krombein (1940) seems to describe the metasoma as

2 mm

entirely orange, but we have not seen any specimens that fit that description. The specimen pictured in figure 5.5.15.1a was identified as *M. n. rufigastra* by Krombein, but it has the metasoma largely black and was collected in California. Perhaps this subspecies is more widespread and variable than currently recognized in the literature, or perhaps this specimen is mislabeled or misidentified.

5.6 EASTERN MALES COLOR SYNDROME

Males of the Eastern Males Color Syndrome (MCS) are recognized by having the head, mesosoma, and apical tergites predominantly black and T2 largely orange or reddish. This color syndrome is most prevalent in the central and eastern USA. There are 20 species in four genera with the Eastern MCS.

These species could most easily be confused with the Texan, *Timulla*-like, or *Sphaeropthalma*-like MCS. Unlike the Texan MCS (5.3), Eastern males have mostly black setae on T3–6. Unlike the *Timulla*-like MCS (5.5), the head and mesosoma lack distinct patches of silvery setae, and at most they have some sparse silvery setae interspersed with mostly black setae. Additionally, Eastern MCS species always have distinct blackish setae on T3–5. Some specimens of *Lomachaeta* (5.5.6) and *Ephuta* (5.5.11) in the *Timulla*-like MCS have the head and mesosoma mostly black, but, for simplicity, no members of those genera are treated in the Eastern MCS. Some Eastern MCS species have the mesosoma obscurely reddened, but unlike the *Sphaeropthalma*-like MCS (5.7), the cuticle is mostly black with obscurely darker red portions. Most species with truly intermediate forms (like 5.6.5, 5.6.6.1) are treated under both color syndromes.

Females are recognized for 17 of the 20 (85%) Eastern MCS males. The vast majority of these associated females fit the Eastern Females Color Syndrome (FCS) (94%), and only a few are associated with species in the *Timulla*-like, Western, and Madrean FCS. Even though these males are linked with the Eastern FCS in sex association and distribution, their coloration is very different. The entire Eastern MCS and FCS are an aggregated example of dual sex-limited mimicry, with females as Müllerian mimics of one another, and males as apparent Batesian mimics of other wasp families (see 1.9.2).

Like the Texan MCS (5.3), this color syndrome is not extraordinarily common but does occur in various species around the world, like *Dasylabris scutila* from Greece and *Cephalomutilla haematodes* from Argentina. A few other examples can be seen in the later chapters of this book (7.5.2.4, 8.3.8.1).

ABOVE: *Dasylabris scutila* from Greece.

RIGHT: *Cephalomutilla haematodes* from Argentina.

—Genus *Dasymutilla*—Eye circular, T1 shape petiolate, sculpture and setae dense

5.6.1 Tergal cuticle completely reddish but covered with black setae .... *D. scaevola*

5.6.2 T3–7 cuticle black, apical tergites with sparse silvery setae .... *D. birkmani, D. canella,*
D. macilenta, D. arenneronea

5.6.3 T3–6 cuticle and setae black, S2 without setal pit or ridge .... *D. ursus, D. nigripes, D. macra*

5.6.4 T3–6 cuticle and setae black; S2 with setal pit
or ridge mesally .... *D. quadriguttata, D. asopus,*
D. bioculata, D. gentilis, D. meracula

—Genus *Pseudomethoca*—Eye circular, T1 shape sessile

5.6.5 Head and mesosoma dark reddish to black .... *Ps. oceola, Ps. simillima,*
Ps. paludata, Ps. sanbornii

—Genus *Sphaeropthalma*—Eye circular, tergal fringes plumose

5.6.6 Head and mesosoma dark reddish to black .... *S. pensylvanica scaeva*

—Genus *Timulla*—Eye ovate with notch, T1 shape cylindrical

5.6.7 Metasomal cuticle entirely reddish with moderately dense black setae .. *T. vagans, T. dubitata*

 ### 5.6.1.1—*Dasymutilla scaevola*

DAH-ZEE-MEW-TILL-UH SKEE-VOE-LUH

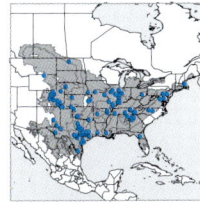 **ETYMOLOGY** This name was derived from the Latin *scaevus*, "left" or "clumsy." **FIELD IDENTIFICATION** Sometimes easy. The uniformly reddish metasomal cuticle is diagnostic. **LAB IDENTIFICATION** The axilla is dentate posteriorly; the mid and hind femora have truncate apices; S2 lacks a seta-filled pit; T3–6 have the setae black, and T7 usually has the setae  gray. **FEMALE** (3.6.6.1). Different from male, with paler body color. **NOTES** This species is widespread and abundant in the central and eastern USA. Since the metasomal color is mostly orange and the body has sparse scattered silver setae, this species could be arguably placed in the Texan or *Timulla*-like MCS, but the overall body color is generally darker than those color syndromes, and *D. scaevola* overlaps in distribution with the other Eastern MCS males.

 ### 5.6.2—*Dasymutilla birkmani* Eastern males cluster

These species occur in the central (*D. birkmani*) and eastern (three others) USA. They have silvery setae on the apical tergites. Many of them have sparse gray mesosomal setae, so they could be confused with species in the *Timulla*-like MCS (5.5.1, 5.5.4). The silvery apical tergal setae are sparse and inconspicuous, so these should also be compared with the other Eastern MCS clusters (5.6.1, 5.6.3, 5.6.4).

5.6.2.1—*Dasymutilla birkmani*

DAH-ZEE-MEW-TILL-UH BIRKMAN-EYE

ETYMOLOGY This species is named after G. Birkman, who collected many important mutillid specimens in Texas in the late 1800s. This color form used to be named *D. reclusa*. **FIELD**

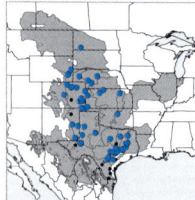

IDENTIFICATION Often difficult. This is the only species in this cluster in Texas and the central USA, but the gray tergal setae are very difficult to see in the field. **LAB IDENTIFICATION** The axilla is truncate posteriorly; the mid and hind femora have rounded apices; S2 lacks a seta-filled pit; T3–5 have the setae mostly black and T6–7 have the setae gray. The tegula is more coarsely punctate and starkly convex than the other species in this cluster. **FEMALE** (3.4.4.3, 3.4.5.1). Different from male, with patches of silvery or golden setae. **NOTES** This darker color variant of *D. birkmani* seems to be more abundant than the paler form in the *Timulla*-like MCS (5.5.4.2).

Dasymutilla birkmani from Nebraska.

Dasymutilla birkmani from Texas.

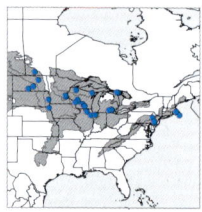

5.6.2.2—*Dasymutilla canella*
DAH-ZEE-MEW-TILL-UH CAN-ELL-UH

ETYMOLOGY Apparently from the Latin *canela* "cinnamon," in reference to the orange-brown T2 color. **FIELD IDENTIFICATION** Often difficult. This is the only species in this cluster in the northeastern USA, but the gray tergal setae are very difficult to see in the field. **LAB IDENTIFICATION** The axilla is truncate posteriorly; the mid and hind femora have rounded api-

ces; S2 has a large ovate seta-filled pit; T3–5 have the setae mostly black and T6–7 have the setae gray. The propodeal sculpture is more coarsely areolate than that of other species in this cluster. **FEMALE** (3.6.3.1). Different from male, with paler body color. **NOTES** This species occurs mainly in the northeastern USA and is usually smaller in size than other species of *Dasymutilla* in the region. The type specimen was labeled as being from Texas, but since then no specimens from Texas have been examined. It seems likely that the type was mislabeled.

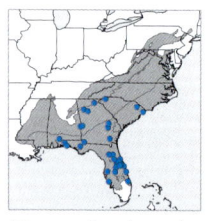

5.6.2.3—*Dasymutilla macilenta*
DAH-ZEE-MEW-TILL-UH MASS-ILL-ENN-TUH

ETYMOLOGY From the Latin *macilentus* "mean," "thin," or "meager." This name was based on the male sex and likely referred to the sparse body setae. **FIELD IDENTIFICATION** Difficult. This and *D. arenerronea* are the only species in this cluster in the southeastern USA, but the gray tergal setae are difficult to see in the field, and these species are separated by the presence (*D. macilenta*) or absence (*D. arenerronea*) of a seta-filled pit on S2. **LAB IDENTIFICATION** axilla is truncate posteriorly; the mid and hind femora have rounded apices; S2 has an indistinct

Dasymutilla macilenta

seta-filled pit; T3–7 have the setae mainly interspersed gray and black. **FEMALE** (3.6.3.2). Different from male, with paler body color. **NOTES** This species was described in the same paper as *D. canella*, and like that species, the type specimen was labeled as being from Texas, but since then, no specimens from Texas have been examined.

- -

I was lucky to catch one male and four female specimens on campus while visiting the Bug Closet collection at the University of Central Florida in Orlando. The sex association was confirmed using DNA sequences from these specimens. I will always be grateful to Sandor "Shawn" Kelly and the late Stuart Fullerton, who made this discovery possible.—KEVIN

- -

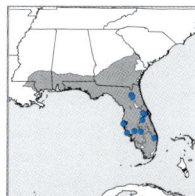

5.6.2.4—*Dasymutilla arenerronea*

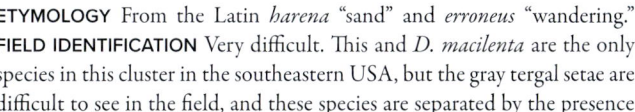

DAH-ZEE-MEW-TILL-UH UH-RENN-ERR-OWE-NEE-UH

ETYMOLOGY From the Latin *harena* "sand" and *erroneus* "wandering." **FIELD IDENTIFICATION** Very difficult. This and *D. macilenta* are the only species in this cluster in the southeastern USA, but the gray tergal setae are difficult to see in the field, and these species are separated by the presence (*D. macilenta*) or absence (*D. arenerronea*) of a seta-filled pit on S2. **LAB**

IDENTIFICATION The axilla is truncate posteriorly; the mid and hind femora have rounded apices; S2 lacks a seta-filled pit; T3–7 have the setae mainly interspersed gray and black. **FEMALE** (3.6.3.3). Different from male, with paler body color. **NOTES** When the sexes of this species were first associated by Krombein (1954), the female was still erroneously being treated as a synonym of *D. chattahoochei*, which is now recognized as a synonym of *D. macilenta*. This complicated history is explained further under the female account for this species (3.6.3.3).

▼ 5.6.3—*Dasymutilla ursus* Eastern males cluster

These species are widespread in the central and eastern USA, but *D. ursus* also expands into some northwestern areas. They have the setae of T3–7 entirely black (usually) and lack a seta-filled pit on S2. In the field, it is often impossible to separate them from the *D. quadriguttata* cluster (5.6.4). Some of these species have the cuticle partly reddened, especially in the southeastern USA, and are also included in the *Sphaeropthalma*-like MCS (5.7.4).

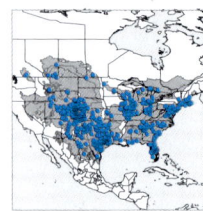

5.6.3.1—*Dasymutilla ursus*
DAH-ZEE-MEW-TILL-UH URR-SUSS

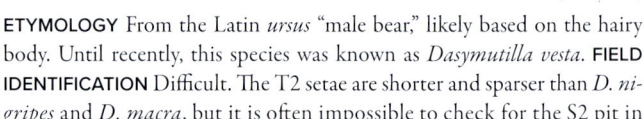

ETYMOLOGY From the Latin *ursus* "male bear," likely based on the hairy body. Until recently, this species was known as *Dasymutilla vesta*. **FIELD IDENTIFICATION** Difficult. The T2 setae are shorter and sparser than *D. nigripes* and *D. macra*, but it is often impossible to check for the S2 pit in the field. **LAB IDENTIFICATION** This species can be recognized by having the tegula more coarsely punctured than that of other Eastern MCS species and the mesonotum swollen posterolaterally (Fig. 4.1.5e); the tegula is truncate posteriorly; and S2 lacks a seta-filled pit. **FEMALE** (3.6.5.1). Different from male, with paler body color. **NOTES** This is a widespread and common species, but males are very difficult to recognize in the field.

Dasymutilla ursus from South Carolina.   *Dasymutilla ursus* from Texas.

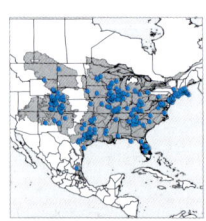

5.6.3.2—*Dasymutilla nigripes*
DAH-ZEE-MEW-TILL-UH NIH-GRIP-EEZ

ETYMOLOGY From the Latin *nigra* "black" and *pes* "foot," in reference to the black legs of some females. **FIELD IDENTIFICATION** Sometimes easy. The T2 setae are long, dense, and erect; S2 lacks a seta-filled pit and has the cuticle reddish, like T2. **LAB IDENTIFICATION** The tegula is mostly smooth; the axilla is smooth and posteriorly rounded; and S2 lacks a seta-filled pit.

Photo by Efram Goldberg

ABOVE: *Dasymutilla nigripes* from Florida.

LEFT: *Dasymutilla nigripes* from South Carolina.

FEMALE (3.6.5.4). Different from male, with paler body color. **NOTES** Females of this species are more widespread than **MALES** this male is known only from the eastern USA, while females are prevalent in the central USA as well. Most of the specimens currently treated as *D. macra* are likely really to belong to *D. nigripes*.

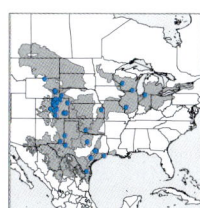

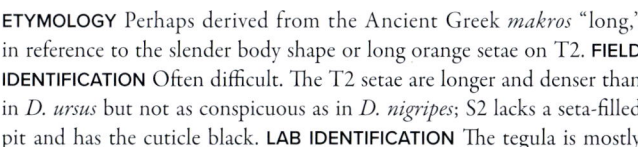

5.6.3.3—*Dasymutilla macra*
DAH-ZEE-MEW-TILL-UH MACK-RUH

ETYMOLOGY Perhaps derived from the Ancient Greek *makros* "long," in reference to the slender body shape or long orange setae on T2. **FIELD IDENTIFICATION** Often difficult. The T2 setae are longer and denser than in *D. ursus* but not as conspicuous as in *D. nigripes*; S2 lacks a seta-filled pit and has the cuticle black. **LAB IDENTIFICATION** The tegula is mostly smooth; the axilla is smooth and posteriorly rounded; and S2 lacks a seta-filled pit. **FEMALE** Un-

known. **NOTES** The female is unknown, but this species is basically identical to males of *D. nigripes*, except in S2 coloration. *Dasymutilla texanella* (3.2.8.5, 3.3.5.1), known from females only, is also structurally similar to *D. nigripes* and overlaps in distribution with *D. macra*. The current concept for *D. macra* is likely composed of males that will eventually be matched with both *D. nigripes* and *D. texanella* females, but more detailed examination of types and some DNA analyses will likely be needed to straighten out the status of this species.

 ## 5.6.4—*Dasymutilla quadriguttata* Eastern males cluster

These species are widespread in the central and eastern USA. They have the setae of T3–7 entirely black (usually) and have a seta-filled pit or ridge on S2. In the field, it is often impossible to separate them from the *D. ursus* cluster (5.6.3). Some of these species have the cuticle partly reddened, especially in the southeastern USA, and are also included in the *Sphaeropthalma*-like MCS (5.7.4).

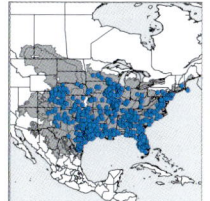

5.6.4.1—*Dasymutilla quadriguttata*
DAH-ZEE-MEW-TILL-UH QUAD-RIGG-OO-TAW-TUH

ETYMOLOGY From the Latin *quadrum* "four" and *gutta* "spot," in reference to the female coloration. **FIELD IDENTIFICATION** Usually difficult. This is one of the most common and widespread Eastern MCS species, but the S2 pit is difficult to see in the field, and the differences between this species and some others in this cluster are impossible to see without a microscope. **LAB IDENTIFICATION** The vertex is evenly rounded posteriorly; the clypeus is clearly bidentate; the tegula is mostly smooth; S2 has a mesal seta-filled pit; and T7 usually lacks a setal fringe posteriorly. **FE-MALE** (3.5.1.1, 3.6.2.4). Different from male, with paler body color. **NOTES** Depending on the region, this species is even more abundant than *D. ursus*

(5.6.3.1). They are variable in size and coloration, with some populations fitting the *Sphaeropthalma*-like MCS (5.7.4.2).

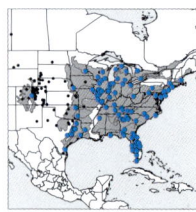

5.6.4.2—*Dasymutilla asopus*
DAH-ZEE-MEW-TILL-UH AY-SOHP-USS

ETYMOLOGY In Greek mythology, Asopus is the name of a god of rivers. This color form used to be named *D. asopus bexar*, after Bexar County in Texas. **FIELD IDENTIFICATION** Often difficult. This species has a thicker body shape than the others in this cluster, and S2 has a longitudinal setal-ridge mesally. **LAB IDENTIFICATION** The tegula is mostly smooth; the

axilla is dentate posteriorly; and S2 has a longitudinal setal-ridge mesally. **FEMALE** (3.6.5.2). Different from male, with paler body color. **NOTES** Males are variable in color, also having the Western (5.2.2.1) and *Sphaeropthalma*-like (5.7.4.1) Color Syndromes. Even though this species was originally named after Bexar County, Texas, this is the most common variant of *D. asopus* in the eastern USA. Because of the thick body shape, it can be easily confused for the genus *Pseudomethoca* (5.6.5).

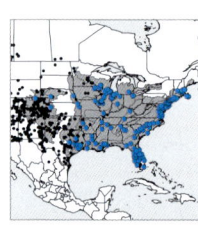

5.6.4.3—*Dasymutilla bioculata*
DAH-ZEE-MEW-TILL-UH BYE-OCK-EW-LAW-TUH

ETYMOLOGY From the Latin *bi* "two" and *oculus* "eye," in reference to the two orange spots on T2 in many males. This color form used to be named *D. lepeletierii*, after the French entomologist Amédée Louis Michel Le Peletier, who described many *Hymenoptera* species in the early 1800s. **FIELD IDENTIFICATION** Very difficult. This species is usually a bit larger and

thicker-bodied than *D. quadriguttata*. This color form is generally only found in the humid eastern USA. **LAB IDENTIFICATION** The clypeus is weakly bidentate; the tegula is mostly smooth; the axilla is truncate posteriorly; S2 has a seta-filled pit mesally; and T7 has an erect setal fringe posteriorly. **FEMALE** (3.2.8.1, 3.6.5.3). Different from male, with paler body color. **NOTES** This is the most variable and widespread *Dasymutilla* in North America, with males matching the Texan (5.3.2.3) and Western (5.2.2.2, 5.2.5.1) MCS. This form is generally matched with females with the Eastern FCS (3.6.5.3).

5.6.4.4—*Dasymutilla gentilis*
DAH-ZEE-MEW-TILL-UH JENN-TILL-ISS

ETYMOLOGY From the Latin *gentilis* "of the same clan," likely in reference to its similarity to *D. quadriguttata* and other species. **FIELD IDENTIFICATION** Impossible to separate from *D. quadriguttata* in the field. **LAB IDENTIFICATION** This species is basically identical to *D. quadriguttata*, except the tegula is coarsely punctate throughout. **FEMALE** Unknown. **NOTES** This

rare species occurs in Texas and the Great Plains. It might be the male of *D. campanula* (3.2.8.4), *D. curticeps* (3.5.1.3), *D. parski* (3.5.1.3), or yet another variant of *D. quadriguttata*.

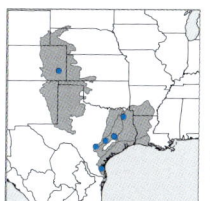

5.6.4.5—*Dasymutilla meracula*
DAH-ZEE-MEW-TILL-UH MURR-ACK-EWE-LUH

ETYMOLOGY Apparently derived from the Latin *meracus* "pure" or "unmixed." **FIELD IDENTIFICATION** Impossible to separate from *D. quadriguttata* in the field. **LAB IDENTIFICATION** This species is basically identical to *D. quadriguttata*, except the vertex is swollen posteromesally. **FEMALE** Unknown. **NOTES** This rare species occurs in Texas and the Great Plains. It might be the male of *D. campanula* (3.2.8.4), *D. curticeps* (3.5.1.3), *D. parski* (3.5.1.3), or yet another variant of *D. quadriguttata*.

 ### 5.6.5—*Pseudomethoca oceola* Eastern males cluster

These species occur in the central and eastern USA and are the only known *Pseudomethoca* with the Eastern MCS. In the field, they can be confused for similarly colored *Dasymutilla* clusters (5.6.3, 5.6.4). Many of these species take on the *Sphaeropthalma*-like MCS (5.7.7) when they are found in southern Florida.

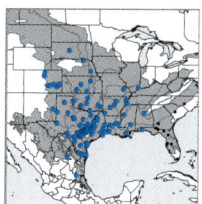

5.6.5.1—*Pseudomethoca oceola*
SOO-DOE-METH-OWE-KUH OWE-SEE-OWE-LUH

ETYMOLOGY This species was named after Osceola, an influential leader of the Seminole people in Florida. **FIELD IDENTIFICATION** Often difficult. This species is larger and has coarser body sculpture than the others in this cluster. **LAB IDENTIFICATION** The tegula is coarsely rugose and strongly

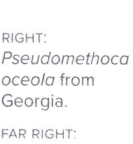

RIGHT: *Pseudomethoca oceola* from Georgia.

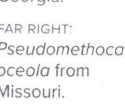

FAR RIGHT: *Pseudomethoca oceola* from Missouri.

raised with a flat outer rim; the femur has a brush of long setae ventrally; and the apical tergites usually have the setae entirely black. **FEMALE** (3.6.8.1). Different from male, with paler body color. **NOTES** Males of this species are more variable in color than the females, with some populations fitting the *Sphaeropthalma*-like MCS (5.7.7.2). *Pseudomethoca flavida* (5.5.8.4) from the *Timulla*-like MCS seems likely to be yet another color variant of this widespread species.

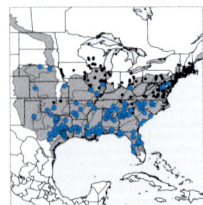

5.6.5.2—*Pseudomethoca simillima*
SOO-DOE-METH-OWE-KUH SEE-MILL-EE-MUH

ETYMOLOGY From the Latin *simillimus* "very similar." The author thought it was similar to eastern populations of *D. bioculata*. **FIELD IDENTIFICATION**

Impossible. **LAB IDENTIFICATION** The tegula is raised and mostly smooth; the femur lacks a brush of long setae ventrally; and the apical tergites usually have the setae entirely black (at most with some silvery setae on T7). **FEMALE** (3.6.10.1). Different from male, with paler body color. **NOTES** This is a widespread and common species. Many northern populations fit in the *Ephuta*-like MCS (5.4.9.1).

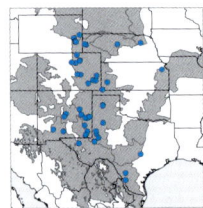

5.6.5.3—*Pseudomethoca paludata*
SOO-DOE-METH-OWE-KUH PAL-LOO-DAH-TUH

ETYMOLOGY This species was apparently named for the paludamentum, a cloak worn by military commanders in Rome. **FIELD IDENTIFICATION**

Difficult. The diagnostic silvery apical metasomal setae are inconspicuous in the field. **LAB IDENTIFICATION** The tegula is raised and entirely punctate; the femur lacks a brush of long setae ventrally; and the T6–7 are covered with silvery setae. **FEMALE** (3.2.11.2, 3.6.10.5). Different from male, with paler body color. **NOTES** This species superficially resembles *Dasymutilla birkmani* (5.6.2.1) except for the sessile T1 shape. It is the westernmost-distributed species in this cluster.

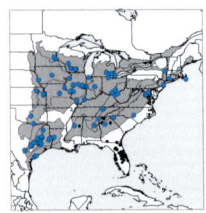

5.6.5.4—*Pseudomethoca sanbornii*
SOO-DOE-METH-OWE-KUH SANBORN-EE-EYE

ETYMOLOGY This species was named after Charles Sanborn, who collected many useful velvet ant specimens from Texas. **FIELD IDENTIFICATION** Impossible. **LAB IDENTIFICATION** The tegula is basically flat and variably punctate; the femur lacks a brush of long setae ventrally; and the apical tergites usually have the setae entirely black. **FEMALE** (3.6.10.3). Different from male, with paler body color. **NOTES** This is a widespread and common central and eastern species that frequently fits the *Sphaeropthalma*-like MCS (5.7.7.3).

Pseudomethoca sanbornii

▼ 5.6.6.1—*Sphaeropthalma pensylvanica scaeva*

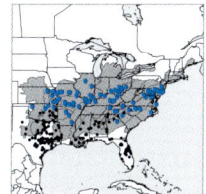

S-FAIR-OPP-THAL-MUH PENN-SILL-VANN-ICK-UH SKEE-VUH

ETYMOLOGY This species was named after the state of Pennsylvania. When the species was described by Le Peletier in 1845, he misspelled the name of the state, and now we are stuck with it. This subspecies is named after the Latin word *scaevus* "left" or "west." **FIELD IDENTIFICATION** Often easy. Has a thicker head and denser tergal setae than other Eastern MCS species. **LAB IDENTIFICATION** There are dense black plumose setae on the fringes of T2–6. **FEMALE** (3.6.11.1). Different from male, with paler body color. **NOTES** Most specimens of *S. pensylvanica* have the *Sphaeropthalma*-like MCS (5.7.11.3). This Eastern MCS form is more commonly seen in northern areas.

▼ 5.6.7—*Timulla vagans* Eastern males cluster

These species occur mainly in the central and eastern USA. They are the only Eastern MCS species with ovate emarginate eyes. There are often a few scattered gray setae on the head, and pronotum and could be confused with congeners from the *Timulla*-like MCS (5.5.12, 5.5.14).

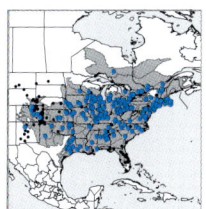

5.6.7.1—*Timulla vagans*
TIMM-EWE-LUH VAY-GUHNS

ETYMOLOGY From the Latin *vagans* "rambling." **FIELD IDENTIFICATION** Sometimes easy. This species has a distinct white brush on the scape and occurs farther north than any other Eastern *Timulla*. **LAB IDENTIFICATION** The scape has a dense white setal brush; the mandible has a distinct ventral tooth basally; S5 lacks a lateral tubercle, unlike T6–7; and T7 has a Y-shaped carina. **FEMALE** (3.5.9.1, 3.6.17.2). Different from male, with paler body color. **NOTES** This color form is more common and widespread than the more reddish Floridian form treated in the *Sphaeropthalma*-like MCS (5.7.16.10). Other populations have more

extensive gray setae on the head and mesosoma and fit the *Timulla*-like MCS (5.5.12.2). This is apparently the most common and widespread species of *Timulla*, and the only *Timulla* known from New England and eastern Canada.

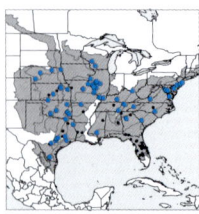

5.6.7.2—*Timulla dubitata*

TIMM-EWE-LUH DOO-BIT-AHT-UH

ETYMOLOGY From the Latin *dubitatus* "doubted." This color form used to be recognized as the subspecies *T. dubitata fugitiva*. **FIELD IDENTIFICATION** Often difficult. Unlike *T. vagans*, this species lacks a white brush on the scape. **LAB IDENTIFICATION** The scape lacks a dense white setal brush; the mandible has the ventral tooth weak or absent; S5 has a lateral

tubercle, as do S6–7; and T7 has a Y-shaped carina. **FEMALE** (3.5.10.2, 3.6.16.1). Different from male, with paler body color. **NOTES** Unlike *T. vagans*, this Eastern MCS form of *T. dubitata* is about as common as the *Sphaeropthalma*-like MCS form (5.7.16.4).

5.7 *SPHAEROPTHALMA*-LIKE MALES COLOR SYNDROME

Males with this color syndrome are recognized by having the head and/or mesosoma with the cuticle partly or entirely orange, reddish, or brown, rather than uniformly black. This color syndrome is found throughout North America. Among the diurnal forms, there are 46 species in eight genera with the *Sphaeropthalma*-like Males Color Syndrome (MCS). This is also the dominant color syndrome seen in nocturnal velvet ants (chapter 6).

This MCS is usually easy to recognize. In southern Florida, they could be confused with the Eastern MCS depending on how extensively the mesosomal cuticle is darkened; most of these species are treated in both sections. A few Western MCS species (5.2.8.1, 5.2.10.1) have the cuticle orange-brown beneath the dense yellow or orange mesonotal setae.

Females are recognized for 34 of the 46 (74%) diurnal *Sphaeropthalma*-like MCS males. In the USA, the females belong to various color syndromes: Eastern (53%), Cryptic (21%), Madrean (15%), *Timulla*-like (12%), and there are even a few species with the Western and Desert FCS. There is not a direct female analog to this color syndrome. Although only seven of the 46 species treated here belong to the genus *Sphaeropthalma*, *S. pensylvanica* (5.7.11.3) is one of the commonest species in this ring, and the vast majority of the nocturnal species (chapter 6), including about 90% of the 110 species in the genus *Sphaeropthalma*, fit this color syndrome.

This seems to be the third most common male velvet ant color syndrome around the world, behind the *Timulla*-like and *Ephuta*-like MCS. Various examples can be seen in the later chapters of this book (7.4.4.3, 7.5.1.1, 7.5.3.2, 8.2.1.1, 8.2.3, 8.2.4.1, 8.2.8.1, 8.3.10.1, 8.3.13.1, 8.3.14.1, 8.3.16.1). Like the Desert, Western, and Texan MCS, these species generally resemble their associated females.

—Genus *Dasymutilla*—Eye circular, T1 shape petiolate, sculpture and setae dense

5.7.1 Head and mesosoma entirely orange, apical tergites with distinct silvery setae; central USA .... *D. creon*

5.7.2 Head and mesosoma partly black and partly orange, tergal setae with black and silver or golden bands; Arizona and California .... *D. heliophila*, *D. fasciventris*

5.7.3 Apical tergites with setae mostly orange; central USA .... *D. hector*

5.7.4 Apical tergites all black; central and southeastern USA .... *D. asopus, D. quadriguttata, D. ursus, D. nigripes*

—Genus *Lomachaeta*—Eye circular, axilla dentate, sculpture and setae sparse

5.7.5 Body cuticle mostly light orange .... *L. calamondin*

—Genus *Pseudomethoca*—Eye circular, T1 shape sessile

5.7.6 Tergites with sparse silvery setae; Arizona .... *Ps. peremptrix*

5.7.7 Tergites with dense black setae; southeastern USA .... *Ps. oceola, Ps. oculata, Ps. sanbornii*

—Genus *Morsyma*—Eye circular, T2 fringe plumose, metasoma black

5.7.8 Metasoma black with mostly black setae, except distinct white T2 fringe .... *M. ashmeadi*

Genus *Photomorphus*—Eye circular, mesosternal area armed with ridges

5.7.9 Metasoma uniformly brown, legs usually yellow or brown .... *Ph. archboldi, Ph. spinci, Ph. paulus*

5.7.10 Metasoma partly blackened, legs usually blackened .... *Ph. alogus, Ph. banksi, Ph. impar*

—Genus *Sphaeropthalma*—Eye circular, tergal fringes plumose

5.7.11 Tergites with short, dense, bushy, plumose setae; central and eastern USA .... *S. auripilis, S. boweri, S. pensylvanica*

5.7.12 Tergites with longer, sparser, plumose setae; western USA .... *S. contracta, S. luiseno, S. unicolor, S. edwardsii*

—Genus *Ephuta*—Eye ovate with notch, T1 shape cylindrical

5.7.13 Head and mesosoma cuticle reddish, often partly black .... *E. floridana, E. slossonae, E. stenognatha*

—Genus *Timulla*—Eye ovate with notch, T1 shape cylindrical

5.7.14 Wings banded .... *T. barbata, T. ornatipennis*

5.7.15 Mesosoma entirely reddened, except ventrally .... *T. floridensis, T. subhyalina, T. tolerata*

5.7.16 Mesosoma (at least propodeum) largely black .... *T. ferrugata, T. barbigera, T. compressicornis, T. dubitata, T. hollensis, T. kansana, T. ocellaria, T. rufosignata, T. sayi, T. vagans*

▼ 5.7.1.1—*Dasymutilla creon*

DAH-ZEE-MEW-TILL-UH KREE-ONN

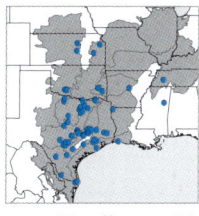

ETYMOLOGY This species could have been named after one of many important figures from Greek mythology or history named Creon. **FIELD IDENTIFICATION** Usually easy. The color pattern is unique. It could be confused for *Sphaeropthalma pensylvanica* except it has sparser and partly silvery metasomal setae. **LAB IDENTIFICATION** The hind trochanter is armed with a tooth; S2 is simply punctate; and T7

is covered with short setae. **FEMALE** (3.6.6.2). Very similar to male. **NOTES** Based on the pygidial setae and similar coloration, previous authors had guessed about the sex association. This was confirmed when James Pitts collected a mating pair in Arkansas in 2002.

▼ 5.7.2—*Dasymutilla heliophila Sphaeropthalma*-like males cluster

These species occur in California and Arizona. They have the head and mesosoma largely blackened. Superficially, they resemble species from the *Timulla*-like MCS (5.5.4, 5.5.5).

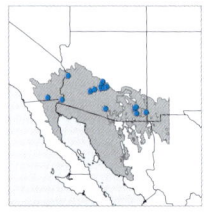

5.7.2.1—*Dasymutilla heliophila*
DAH-ZEE-MEW-TILL-UH HEE-LEE-AWE-FILL-UH

ETYMOLOGY From the Ancient Greek *helio-* "relating to the sun" and *-philos* "beloved," in reference to its prevalence in hot desert regions. **FIELD IDENTIFICATION** Usually easy. This species has T2 uniformly orange-brown and generally occurs in hot desert habitats. **LAB IDEN-**

TIFICATION The axilla is dentate posteriorly; the mid and hind femora have truncate apices; and S2 lacks a seta-filled pit. **FEMALE** (3.4.1.1). Lighter body color than male. **NOTES** The sex association was based on DNA sequence comparisons of specimens collected together in Willcox, Arizona. Some individuals of *D. apicalata* (5.5.5.1) could be confused for this species because they have the mesosoma partly reddened on the propodeal sides.

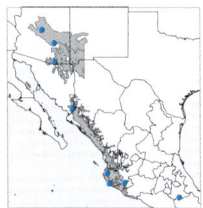

5.7.2.2—*Dasymutilla fasciventris*
DAH-ZEE-MEW-TILL-UH FASS-IVV-ENN-TRISS

ETYMOLOGY From the Latin *fascis* "bundle" and *venter* "belly," referencing the large white pit on S2. **FIELD IDENTIFICATION** Usually easy. This species has a yellow cuticular patch on T2 and is sporadically found in higher elevations in Arizona. **LAB IDEN-**

TIFICATION The axilla forms an elongate arm-like process; the mid and hind femora have rounded apices; and S2 has an elongate oval seta-filled pit anteromesally. **FEMALE** (3.4.8.3). Very similar to male. **NOTES** This species is widespread in Mexico and sporadically has been found in various moderately high-elevation areas in Arizona. *Dasymutilla fasciventroides* (5.5.4.1) was previously confused for this species because of similarities in the S2 pit shape.

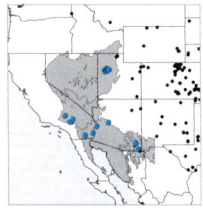

5.7.2.3—*Dasymutilla monticola*
DAH-ZEE-MEW-TILL-UH MON-TICK-OWE-LUH

ETYMOLOGY A Latin term that means "mountain dweller." **FIELD IDENTIFICATION** Usually easy. The dorsal mesosomal setae are entirely silver. **LAB IDENTIFICATION** The mesonotum is covered with gray or silvery setae, like the vertex and pronotum; the axilla is truncate posteriorly; and S2 has a small seta-filled pit mesally. **FEMALE** (3.4.4.2). Somewhat similar to male.

NOTES This species varies in color across its range. Southwestern desert forms, like this one, have more extensive reddish cuticle, especially on the mesosomal sides, and brighter silvery setal markings. Northern populations, however, are mostly black with duller gray setae and are treated in the *Timulla*-like MCS (5.5.3.1).

 ### 5.7.3.1—*Dasymutilla hector*

DAH-ZEE-MEW-TILL-UH HECK-TORE

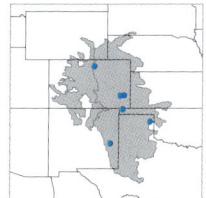

ETYMOLOGY In Greek mythology, Hector was a Trojan prince and his army's greatest warrior. **FIELD IDENTIFICATION** Usually easy. This is the only *Sphaeropthalma*-like MCS *Dasymutilla* with the apical tergal setae entirely yellow or orange. **LAB IDENTIFICATION** The axilla is dentate posteriorly and S2 has a longitudinal setal-ridge mesally. **FEMALE** Unknown. **NOTES** This species is basically identical to *D. neomexicana* (5.2.5.4) from the Western MCS, except for its reddened mesosomal cuticle. Based on structural similarity to *D. asopus* in both sexes, the female of *D. hector* might be *D. montivagoides* (3.2.7.4). Like *D. asopus* and related species, males commonly have the wings shortened.

 ### 5.7.4—*Dasymutilla asopus Sphaeropthalma*-like males cluster

These species occur mainly in the southeastern USA, especially in southern Florida. They are a good match for the Eastern MCS and are all treated there as well, but these populations have the mesosomal cuticle largely reddened.

5.7.4.1—*Dasymutilla asopus*
DAH-ZEE-MEW-TILL-UH AY-SOHP-USS

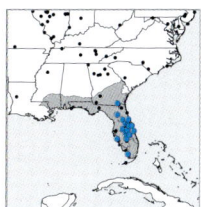

ETYMOLOGY In Greek mythology, Asopus is the name for a god of rivers. This color form used to be named *D. asopus cassandra*, after a Trojan priestess of Apollo from Greek mythology. **FIELD IDENTIFICATION** Sometimes difficult. This species has a thicker body shape than the others in this cluster, and S2 has a longitudinal setal-ridge mesally. **LAB IDENTIFICATION** The tegula is mostly smooth; the axilla is

Dasymutilla asopus from Florida.

dentate posteriorly; and S2 has a longitudinal setal-ridge mesally. **FEMALE** (3.6.5.2). Somewhat similar to male, with silver setal markings. **NOTES** Males are variable in color, also having the Western (5.2.2.1) and Eastern (5.6.4.2) Color Syndromes. This *Sphaeropthalma*-like form occurs mainly in the southeast, but some Western individuals have the mesosoma more extensively reddened, like this male from Colorado. Because of the thick body shape, it can be easily confused for the genus *Pseudomethoca* (5.7.7).

Dasymutilla asopus from Colorado.

5.7.4.2—*Dasymutilla quadriguttata*
DAH-ZEE-MEW-TILL-UH QUAD-RIGG-OO-TAW-TUH

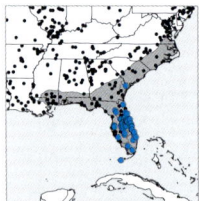

ETYMOLOGY From the Latin *quadrum* "four" and *gutta* "spot," in reference to the female coloration. **FIELD IDENTIFICATION** Often difficult. The S2 pit is diagnostic but difficult to see in the field. **LAB IDENTIFICATION** The tegula is mostly smooth; the axilla is truncate posteriorly; and S2 has a mesal seta-filled pit. **FEMALE** (3.5.1.1, 3.6.2.4). Somewhat similar to male, with silver setal markings. **NOTES** Depending on the region, this species is even more abundant than *D. ursus* (5.7.4.2). They are variable in size and coloration, with some populations fitting the Eastern MCS (5.6.4.1).

Photo by Efram Goldberg

5.7.4.3—*Dasymutilla ursus*
DAH-ZEE-MEW-TILL-UH URR-SUSS

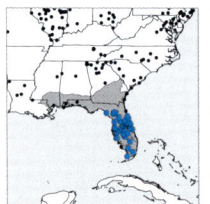

ETYMOLOGY From the Latin *ursus* "male bear," likely based on the hairy body. Until recently, this species was known as *Dasymutilla vesta*. **FIELD IDENTIFICATION** Often difficult. The T2 setae are sparse, and S2 lacks a seta-filled pit. **LAB IDENTIFICATION** The tegula is coarsely punctate; the axilla is truncate posteriorly; and S2 lacks a seta-filled pit. **FEMALE** (3.6.5.1). Somewhat similar to male, with silver setal markings. **NOTES** This male color form was previously associated with *D. archboldi* based on similarities between females of *D. ursus* and *D. archboldi* and its presence in Florida. During a molecular phylogenetic analysis of this and related

species, this male was recognized as a variant of *D. ursus*, and the female of *D. archboldi* was associated with a previously unknown male from the *Timulla*-like MCS (5.5.1.1).

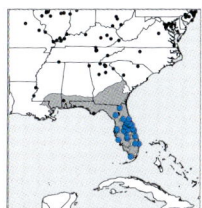

5.7.4.4—*Dasymutilla nigripes*
DAH-ZEE-MEW-TILL-UH NIH-GRIP-EEZ

ETYMOLOGY From the Latin *nigra* "black" and *pes* "foot," in reference to the black legs of some females. **FIELD IDENTIFICATION** Sometimes easy. The T2 setae are long, dense, and erect; and S2 lacks a seta-filled pit. **LAB IDENTIFICATION** The tegula is mostly smooth; the axilla is smooth and posteriorly rounded; and S2 lacks a seta-filled pit. **FEMALE** (3.6.5.4). Somewhat similar to male, with silver setal markings. **NOTES** This color form occurs mainly in southern Florida and is less common than the Eastern MCS form (5.6.3.2).

 ### 5.7.5.1—*Lomachaeta calamondin*
LOW-MUH-KEE-TUH CAHL-UH-MONN-DINN

ETYMOLOGY Named after the calamondin, a small orange-like fruit frequently grown in southern California, referring to the orangish color of this small southern Californian species. **FIELD IDENTIFICATION** Usually easy. This is the only entirely orange diurnal velvet ant in southern California but could be confused for similarly colored nocturnal species (6.1). **LAB IDENTIFICATION** The mandible is unarmed ventrally; the T2 fringe has simple setae only; the paramere is relatively straight with short setae only. **FEMALE** Unknown. **NOTES** At first glance this looks like a nocturnal species, but it has the *Lomachaeta* genus characters (p. 169) and smaller ocelli. Apparently, this rare species has been collected only once; the unique holotype specimen was collected back in 1969.

 ### 5.7.6.1—*Pseudomethoca peremptrix*
SOO-DOE-METH-OWE-KUH PURR-EMP-TRICKS

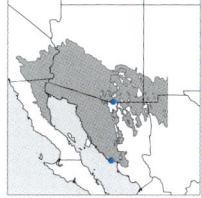

ETYMOLOGY From the Latin *peremptrix* "predator," an allusion to its similarity to *Pseudomethoca perditrix* (which also means "predator"). **FIELD IDENTIFICATION** Usually easy. This is the only entirely orange diurnal velvet ant in southern Arizona, but it could be confused for similarly colored nocturnal species (6.1). **LAB IDENTIFICATION**

The body color is unique among western species of *Pseudomethoca*. **FEMALE** (3.4.12.6). Similar to male. **NOTES** This species was discovered only recently, and the sexes were associated based on specimens collected in the same pitfall traps. In the USA, it is known only from the Coronado National Forest east of Nogales, Arizona, but some additional specimens have been seen in Sonora, Mexico.

 ### 5.7.7—*Pseudomethoca oculata Sphaeropthalma*-like males cluster

These species occur mainly in the southeastern USA, especially in Florida. Unlike *Ps. peremptrix* above, the tergal setae are dense and entirely black, and the body sculpture is generally coarse and dense. In the field, these could be confused with members of similarly colored *Dasymutilla* clusters (5.7.4).

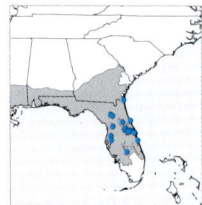

5.7.7.1—*Pseudomethoca oculata*
SOO-DOE-METH-OWE-KUH OCK-EWE-LAW-TUH

ETYMOLOGY From the Latin *oculatus* "having eyes," in reference to the spots on T2 in females. **FIELD IDENTIFICATION** Impossible. **LAB IDENTIFICATION** The tegula is raised and densely punctate; and the hind femur lacks a brush

of long setae ventrally. **FEMALE** (3.6.9.1). Similar to male but with obvious yellow-orange spots on T2 and extensive silvery setae on the metasoma. **NOTES** This species is similar to *Ps. simillima* (5.6.3.2) except that the tegula is more coarsely sculptured and the body cuticle is more extensively reddened. Additionally, *Ps. oculata* is usually larger than *Ps. simillima*.

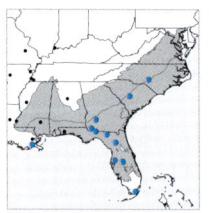

5.7.7.2—*Pseudomethoca oceola*
SOO-DOE-METH-OWE-KUH OWE-SEE-OWE-LUH

ETYMOLOGY This species was named after Osceola, an influential leader of the Seminole people in Florida. **FIELD IDENTIFICATION** Impossible. **LAB IDENTIFICATION** The tegula is coarsely rugose and strongly raised with a flat outer rim; and the hind femur has a brush of long setae ventrally. **FEMALE** (3.6.8.1). Similar to male but with an extensive silver setal pattern on metasoma. **NOTES** Males of this species are apparently more variable in color than the females, with some populations fitting the Eastern MCS (5.6.5.1). *Pseudomethoca flavida* (5.5.8.4) from the *Timulla*-like MCS seems likely to be yet another color variant of this widespread species.

Photo by Efram Goldberg

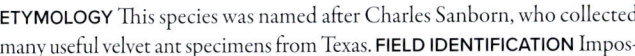

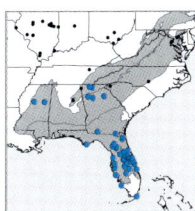

5.7.7.3—*Pseudomethoca sanbornii*

SOO-DOE-METH-OWE-KUH SANBORN-EE-EYE

ETYMOLOGY This species was named after Charles Sanborn, who collected many useful velvet ant specimens from Texas. **FIELD IDENTIFICATION** Impossible. **LAB IDENTIFICATION**

The tegula is basically flat; and the hind femur lacks a brush of long setae ventrally. **FEMALE** (3.6.10.3). Similar to male but with an extensive silver setal pattern on metasoma. **NOTES** Many of these individuals have swollen, angular axilla and they look similar to many *Dasymutilla* except for their sessile T1 shape. This species is also treated in the Eastern MCS (5.6.5.4).

▼ 5.7.8.1—*Morsyma ashmeadi*

MORE-SEE-MUH ASHMEAD-EYE

ETYMOLOGY This species was named after the Canadian entomologist William H. Ashmead, who provided the first functional treatment for velvet ant genera at the end of the 19th century. **FIELD IDENTIFICATION** Easy. This rare species can be separated from all other diurnal mutillids in California by the color pattern. **LAB IDENTIFICATION** The eyes and ocelli are relatively small; T2 has a distinct fringe of plumose white setae; and the remainder

of the tergal setae are mostly black. **FEMALE** (3.5.4.1). Similar to male. **NOTES** Males of this species can be fully winged or totally wingless. The fully winged populations seem most prevalent in the southern Coastal Range, while the wingless forms were mostly collected around the San Francisco Bay. A large series of females and winged males were studied attacking *Diodontus* (Crabronidae) nests near the town of Pozo, California by J. Powell in the 1960s.

▼ 5.7.9—*Photomorphus archboldi Sphaeropthalma*-like males cluster

These species occur mostly in the southeastern USA. Unlike the *Ph. alogus* cluster (5.7.10), the metasoma is usually uniformly brown, and the legs are generally yellow or varying shades of brown. They are similar in appearance to nocturnal species (6.1) from the central and western USA.

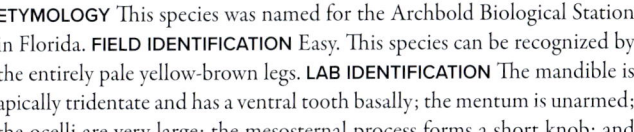

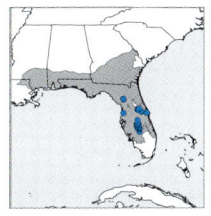

5.7.9.1—*Photomorphus archboldi*

FOE-TOE-MORE-PHUSS ARCHBOLD-EYE

ETYMOLOGY This species was named for the Archbold Biological Station in Florida. **FIELD IDENTIFICATION** Easy. This species can be recognized by the entirely pale yellow-brown legs. **LAB IDENTIFICATION** The mandible is apically tridentate and has a ventral tooth basally; the mentum is unarmed; the ocelli are very large; the mesosternal process forms a short knob; and

the tergal fringes are obscurely plumose. **FEMALE** (3.7.4.1). Similar to male. **NOTES** Based on the large ocelli, this is apparently a nocturnal species, but all the known specimens were collected using passive methods, such as malaise or pitfall traps. The unique yellow leg color made the sex association pretty obvious.

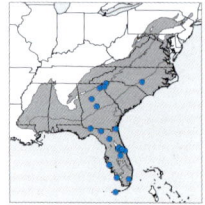

5.7.9.2—*Photomorphus spinci*
FOE-TOE-MORE-PHUSS SPINCE-EYE

ETYMOLOGY It is not clear why this species was named *spinci*; the *-I* suffix indicates it was named after a man, but there is no collector mentioned for the type material. **FIELD IDENTIFICATION** Impossible to separate from *Ph. paulus* in the field. **LAB IDENTIFICATION** The mandible is apically tridentate and has a ventral tooth basally; the mentum is unarmed; the ocelli are moderately large; the mesosternal process forms a short knob; the tergal fringes are obscurely plumose. **FEMALE** (3.7.4.2). Similar to male. **NOTES** Based on the moderately large ocelli, this is apparently a nocturnal species, but all the known specimens were collected using passive methods, such as malaise or pitfall traps. This sex association was confirmed based on the similarity to both sexes of *Ph. archboldi*.

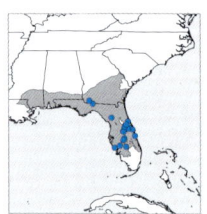

5.7.9.3—*Photomorphus paulus*
FOE-TOE-MORE-PHUSS PAW-LUSS

ETYMOLOGY It is not clear why this species was named *paulus* (originally *paula*). There is no collector mentioned for the type material. **FIELD IDENTIFICATION** Impossible to separate from *Ph. spinci* in the field. **LAB IDENTIFICATION** The mandible is apically bidentate and has a ventral tooth basally; the mentum is armed with a tongue-like process; the ocelli are moderately large; the mesosternal process forms a wide transverse ridge; the tergal fringes have simple setae only. **FEMALE** (3.7.5.4). Similar to male. **NOTES** Males and females of this species were collected by Krombein (1954) at night. We presume that the other species in this cluster are nocturnal as well. This species has larger eyes than the four other *Photomorphus* (*Photomorphus*) species and is apparently the only nocturnal species in this subgenus.

▼ 5.7.10—*Photomorphus alogus Sphaeropthalma*-like males cluster

These species occur in the central and eastern USA. Unlike the *Ph. archboldi* cluster above, the metasoma and usually the legs are largely or entirely blackened. They can be similar in appearance to some nocturnal species (6.1) from the central and western USA.

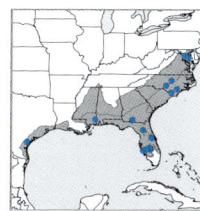

5.7.10.1—*Photomorphus alogus*
FOE-TOE-MORE-PHUSS AHL-OWE-GUSS

ETYMOLOGY Perhaps derived from the Ancient Greek *alogos* "speechless" or "irrational." **FIELD IDENTIFICATION** Impossible. **LAB IDENTIFICATION** The mandible is apically bidentate and has a weak ventral tooth basally; the mentum is armed with a transverse ridge; the ocelli are very small; the mesosternal process forms a large longitudinal row of transverse carinae; the tergal fringes have simple setae only. **FEMALE** (3.7.5.2). Similar to male. **NOTES** This southeastern species varies in color and was recently found as far west as Texas, along the Gulf Coast.

Photomorphus alogus from Florida.

Photomorphus alogus from Texas.

5.7.10.2—*Photomorphus banksi*
FOE-TOE-MORE-PHUSS BANKS-EYE

ETYMOLOGY This species was named after the entomologist Nathan Banks, who collected the type in the early 1900s. **FIELD IDENTIFICATION** Impossible. **LAB IDENTIFICATION** The mandible is apically bidentate and has a ventral tooth basally; the mentum is armed with a rounded transverse lobe; the ocelli are very small; the mesosternal area has two separate processes, the posterolateral process is erect and spine-like; and the tergal fringes have simple setae only. **FEMALE** (3.7.5.3). Similar to male. **NOTES** This species varies in color, sometimes having the body cuticle entirely black (5.4.10.2).

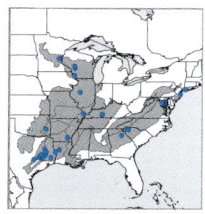

5.7.10.3—*Photomorphus impar*
FOE-TOE-MORE-PHUSS IMM-PARR

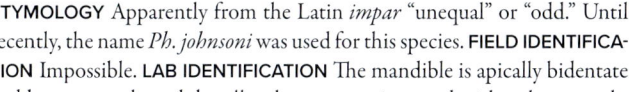

ETYMOLOGY Apparently from the Latin *impar* "unequal" or "odd." Until recently, the name *Ph. johnsoni* was used for this species. **FIELD IDENTIFICATION** Impossible. **LAB IDENTIFICATION** The mandible is apically bidentate and has a ventral tooth basally; the mentum is armed with a sharp tooth; the ocelli are small; the mesosternal process forms a transverse ridge; and

Photomorphus impar from Georgia.

Photomorphus impar from Oklahoma.

the tergal fringes have simple setae only. **FEMALE** (3.7.5.1). Similar to male. **NOTES** This species varies in color and could be confused with species in the previous cluster, especially *Ph. spinci* (5.7.9.2). This species overlaps with all four of the other *Photomorphus* (*Photomorphus*), broadly ranging throughout the central and eastern USA.

 ### 5.7.11—*Sphaeropthalma auripilis Sphaeropthalma*-like males cluster

These species occur in the central and eastern USA. They have a wide head and dense plumose setal fringes. Unlike other *Sphaeropthalma*, they almost always have the legs darkened. Most are diurnal, but they are periodically collected with light traps. Superficially, they are similar to *Dasymutilla creon* (5.7.1.1). These species are parasitoids of mud-nesting wasps and are common in and around human habitations.

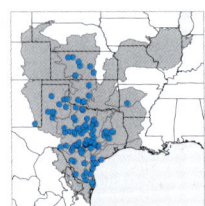

5.7.11.1—*Sphaeropthalma auripilis*
S-FAIR-OPP-THAL-MUH ORR-IPP-EYE-LISS

ETYMOLOGY From the Latin *aurum* "gold" and *pilus* "hair," in reference to the golden setae on the male's metasoma. **FIELD IDENTIFICATION** Easy. The apical tergites are covered with golden setae. **LAB IDENTIFICATION** The tergal setal color mentioned above is the only feature needed to separate the species in this cluster. **FEMALE** (3.6.11.2). Similar to male, with more blackish setae on the apical tergites. **NOTES** These species attack mud- or twig-nesting hosts and are often found in houses. They seem to have facultative nocturnal behavior, sometimes being attracted to lamps and porch lights, although they are also commonly encountered in broad daylight.

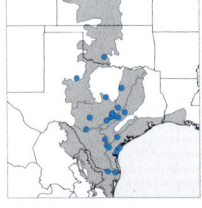

5.7.11.2—*Sphaeropthalma boweri*
S-FAIR-OPP-THAL-MUH BOWER-EYE

ETYMOLOGY Schuster (1944) named this species after his friends Noah and Henrietta Bower, who helped him during his research. **FIELD IDENTIFICATION** Often easy. The apical tergites have dark brown setae, contrasting with the grayish T2 fringe. **LAB IDENTIFICATION** The tergal setal color mentioned above is the only feature needed to separate

these species. **FEMALE** Unknown. **NOTES** This species is rarer than *S. auripilis* or *S. pensylvanica*, and its coloration is somewhat intermediate between those species. It occurs mainly in southern Texas.

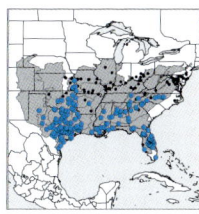

5.7.11.3—*Sphaeropthalma pensylvanica*
S-FAIR-OPP-THAL-MUH PENN-SILL-VANN-ICK-UH

ETYMOLOGY This species was named after the state of Pennsylvania. When the species was described by Le Peletier in 1845, he misspelled the name of the state, and now we are stuck with it. **FIELD IDENTIFICATION** Easy. The apical tergites have the setae entirely black. **LAB IDENTIFICATION** The tergal setal color mentioned above is the only feature needed to separate these species. **FEMALE** (3.6.11.1). Similar to male, with whitish setal pattern on the apical tergites. **NOTES** This is the most common and widespread species in the cluster. Males are divided into two subspecies: *S. p. pensylvanica*, pictured here, and *S. p. scaeva* (5.6.6.1), which is treated in the Eastern MCS. This diurnal wasp is the type species for the predominantly nocturnal genus *Sphaeropthalma*.

▼ 5.7.12—*Sphaeropthalma contracta Sphaeropthalma*-like males cluster

These species occur mainly in the Pacific and intermountain portions of the western USA. They usually have the dorsal setae golden-orange and the wings relatively dark brown. Most of them are nocturnal, but they are periodically also seen during daylight, except for *S. contracta*, which seems to be diurnal only. Numerous nocturnal genera (6.1) include species that could be confused for this cluster, so identifications of these species should be treated as tenuous.

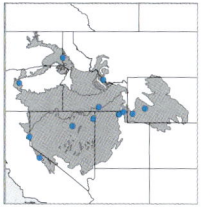

5.7.12.1—*Sphaeropthalma contracta*
S-FAIR-OPP-THAL-MUH CONN-TRACK-TUH

ETYMOLOGY From the Latin *contractus* "collected" or "accomplished." **FIELD IDENTIFICATION** Often difficult. This species is apparently diurnal only and occurs mainly in the northwestern USA. **LAB IDENTIFICATION** The mandible is broad with the anterior surface vertical; the coxae are unarmed; and S2 lacks a felt line. The eyes and ocelli are smaller than in *S. edwardsii*; T1 has the fringe composed of sparse brachyplumose setae only; and the genitalic cuspis is spatulate apically. **FEMALE** (3.6.11.3). Similar to male but with white setae on the apical tergites. **NOTES** Based on similar structure of the genitalia and mandible, this species (along with its three relatives in the *S. uro* species-group) is closely related to *S. pensylvanica* (5.7.11.3). This is the only diurnal species in the *S. uro* species-group and is the rarest in that group. The female has not been formally described, but it overlaps in distribution with the males, matches the species-group characters, and is also diurnal.

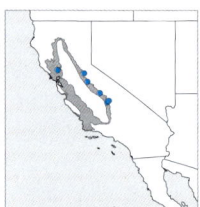

5.7.12.2—*Sphaeropthalma luiseno*
S-FAIR-OPP-THAL-MUH LOO-ISS-ENN-YO

ETYMOLOGY The species was likely named for a place name, but Schuster (1958) did not provide locality information for this species when he described the species or even mention where the type was deposited. **FIELD IDENTIFICATION** Difficult. It might be possible to recognize the large setal tufts on the coxae. **LAB IDENTIFICATION** The mandible is moderately

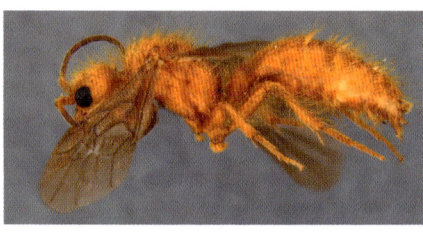

wide and apically contorted; the mid and hind coxae are armed with large setal brushes ventrally; and S2 has a felt line. **FEMALE** Unknown. **NOTES** This is a bizarre and especially rare species known from mid-range elevations near the Sequoia and Yosemite National Parks. We do not have a good guess about the female's identity and have never collected this species before.

5.7.12.3—*Sphaeropthalma unicolor*
S-FAIR-OPP-THAL-MUH EWE-NICK-UH-LURR

ETYMOLOGY Bizarrely, this name means "one color," but in both sexes, this is a remarkably variable species. Shades of brown, black, silver, and orange are frequently found together on the same individual. **FIELD IDENTIFICATION** Difficult. The body cuticle is often variably darkened laterally or ventrally, especially on the legs, and the wing is usually lighter basally with darker outer clouding. **LAB IDENTIFICATION** The mandible is slender, elongate, and contorted apically; the coxae are unarmed; and S2 has a felt line. **FEMALE** (3.2.12.1). Sometimes similar to male, usually with darker cuticle and brighter dorsal setae. **NOTES** This is one of the most commonly encountered nocturnal species in the Pacific states but is also frequently found during daylight hours, especially at high elevations.

LEFT: *Sphaeropthalma unicolor*

BELOW: *Sphaeropthalma unicolor* darker form

5.7.12.4—*Sphaeropthalma edwardsii*
S-FAIR-OPP-THAL-MUH EDWARDS-EE-EYE

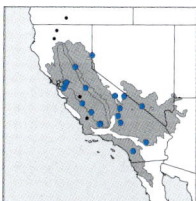

ETYMOLOGY This species was named after its collector, Henry Edwards, an English stage actor, writer, and entomologist in the late 19th century. **FIELD IDENTIFICATION** Difficult. The body size is generally large, and the dorsal setae are denser and often more red-tinged than other species in this cluster. **LAB IDENTIFICATION** The mandible is broad with the anterior surface ver-

tical; the coxae are unarmed; and S2 lacks a felt line. The eyes and ocelli are larger than in *S. contracta*; T1 has the fringe composed of plumose setae; and the genitalic cuspis is uniformly slender. **FEMALE** (3.1.7.1, 3.2.12.2). Similar to male but often with darker cuticle and denser setae. **NOTES** Many individuals have the body cuticle blackened and are treated in the Western MCS (5.2.9.1).

▼ 5.7.13—*Ephuta slossonae Sphaeropthalma*-like males cluster

These species mainly occur in Florida. They are the only *Ephuta* species in this color syndrome.

5.7.13.1—*Ephuta slossonae*
EFF-OO-TUH SLOSSON-EE

ETYMOLOGY This species was named after the American author and entomologist Annie T. Slosson, who collected the type specimen in the 1890s. **FIELD IDENTIFICATION** Difficult. This species has a larger body size and wider head than other species in the cluster. **LAB IDENTIFICATION** This species can be recognized by the wide tridentate mandible and the short pit-like subantennal basin. **FEMALE** (3.6.12.1). Similar to male. **NOTES** This species is more commonly encountered than the closely related *E. eurygnatha*, even though that species is apparently more widespread (5.4.13.7, 5.5.11.3).

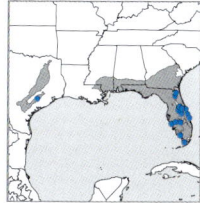

Photo by Dan Roueche

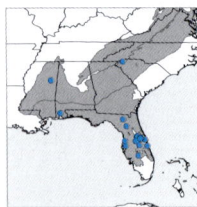

5.7.13.2—*Ephuta floridana*
EFF-OO-TUH FLOW-RIDD-ANN-UH

ETYMOLOGY Named for the state of Florida. **FIELD IDENTIFICATION** Impossible. **LAB IDENTIFICATION** The subantennal basin is as long as the clypeal basin, and the mandible is slender. **FEMALE** (3.4.18.1. 3.7.6.4). Somewhat similar to male except for the golden head. **NOTES** Structurally, males of this species are similar to *E. scrupea* (5.4.13.1), but the females are very different (3.7.6.1), especially in the head setae and postgenal carina.

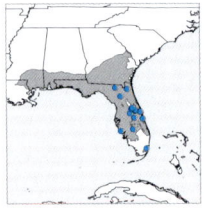

5.7.13.3—*Ephuta stenognatha*
EFF-OO-TUH STEE-NO-NAY-THUH

ETYMOLOGY From the Ancient Greek *stenos* "narrow" and *gnathos* "jaw," in reference to the slender mandible. **FIELD IDENTIFICATION** Impossible. **LAB IDENTIFICATION** The subantennal basin is much shorter than the clypeal basin, and the mandible is slender. **FEMALE** Unknown. **NOTES** This species includes two subspecies that are separated by coloration: *E. s. stenognatha* has the head and mesosoma entirely reddish, and *E. s. psenophora* has the head and part of the mesosoma blackish.

Ephuta stenognatha stenognatha

Ephuta stenognatha psenophora

 ### 5.7.14—*Timulla barbata Sphaeropthalma*-like males cluster

These species occur in the central and eastern USA. They have the mesosoma entirely orange-brown, and the wings are conspicuously banded. They could most easily be confused with the *T. floridensis* cluster (5.7.15).

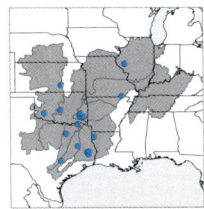

5.7.14.1—*Timulla barbata*
TIMM-EWE-LUH BAR-BAH-TUH

ETYMOLOGY From the Latin *barbus* "beard," in reference to the setal brush on the male scape. **FIELD IDENTIFICATION** Easy. The white setal brush on the scape is larger and more distinct than on any other *Timulla* species. *Timulla ornatipennis*, the only other species with banded wings, is not known to occur west of the Mississippi River. **LAB IDENTIFICATION** The mandible

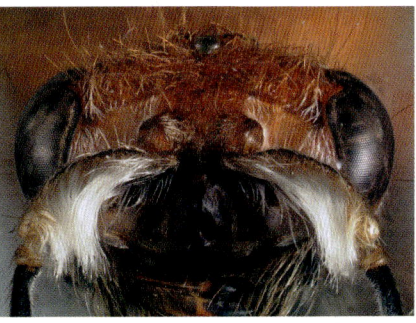

Antennal scape with long dense setal brush.

has a ventral tooth basally; the ocelli are small; the scape has a large, dense, white setal brush; the mesosternal area is unarmed; and T7 has a small curved transverse carina. **FEMALE** Unknown. **NOTES** *Timulla barbata* has banded wings and a large dense brush of white setae on the scape. Based on its structural similarity to females of the related *T. ornatipennis* and its overlapping range with *T. barbata*, *T. wileyae* (3.5.10.1) may eventually be recognized as the female of this species.

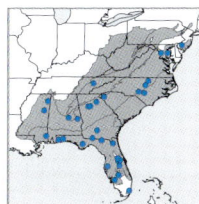

5.7.14.2—*Timulla ornatipennis*
TIMM-EWE-LUH ORR-NOT-IPP-ENN-ISS

ETYMOLOGY From the Latin *ornatus* "decorated" and -*pennis* "winged," in reference to the banded wings of the male. **FIELD IDENTIFICATION** Easy. *Timulla barbata*, the only other species with banded wings, is not known to occur east of the Mississippi River and has a more obvious white setal brush on the scape. **LAB IDENTIFICATION** The mandible is unarmed ventrally; the ocelli are small; the scape has a moderate white setal brush; the mesosternal area is unarmed; and T7 has a small curved transverse carina. **FEMALE** (3.6.16.4). Similar to male. **NOTES** *Timulla ornatipennis* has banded wings like *T. barbata*, but the scape has a sparser setal brush.

Photo by Dan Roueche

specimens have sparse, blackish dorsal setae like the pictured specimen, but some populations have denser orange setae and are treated in the Western MCS (5.2.10.1).

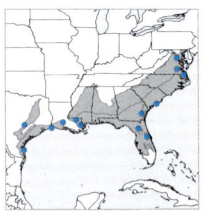

5.7.16.3—*Timulla compressicornis*

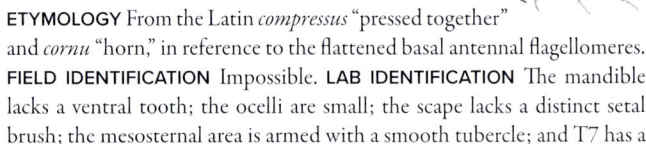

TIMM-EWE-LUH COMM-PRESS-ICK-ORR-NISS

ETYMOLOGY From the Latin *compressus* "pressed together" and *cornu* "horn," in reference to the flattened basal antennal flagellomeres. **FIELD IDENTIFICATION** Impossible. **LAB IDENTIFICATION** The mandible lacks a ventral tooth; the ocelli are small; the scape lacks a distinct setal brush; the mesosternal area is armed with a smooth tubercle; and T7 has a

Y-shaped carina. Unlike the related *T. barbigera*, the basal antennal flagellomeres (F1–2) are flattened and wider than the following flagellomeres. **FEMALE** Unknown. **NOTES** Based on overlapping distribution and large body size, this might be the male of *T. euterpe* (3.6.15.1).

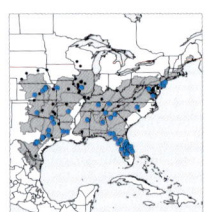

5.7.16.4—*Timulla dubitata*

TIMM-EWE-LUH DOO-BIT-AHT-UH

ETYMOLOGY From the Latin *dubitatus* "doubted." **FIELD IDENTIFICATION** Impossible. **LAB IDENTIFICATION** This is the only *Timulla* species in the USA with a lateral tooth on S5, in addition to the teeth on S6 and S7 that other species usually have. The mandible has a weak ventral tooth basally; the ocelli are small; the scape lacks a setal brush; the mesosternal area is armed with a smooth tubercle; and T7 has a Y-shaped carina. **FEMALE** (3.5.10.2, 3.6.16.1). Somewhat similar to male, with silver setal markings. **NOTES** Many northern populations fit the Eastern MCS (5.6.7.2), but this form seems to be more common.

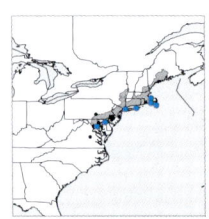

5.7.16.5—*Timulla hollensis*

TIMM-EWE-LUH HALL-ENN-SISS

ETYMOLOGY Likely named after the type locality of Woods Hole, Massachusetts. **FIELD IDENTIFICATION** Impossible. **LAB IDENTIFICATION** The mandible is unarmed ventrally; the ocelli are moderately large; the scape is simple; the tegula is blackish; the mesosternal area is unarmed; and the pygidium has a Y-shaped carina. **FEMALE** Unknown. **NOTES** This species is variable in color, with northern populations fitting the *Timulla*-like MCS (5.5.12.3).

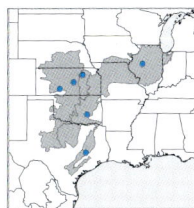

5.7.16.6—*Timulla kansana*
TIMM-EWE-LUH CAN-ZANN-UH

ETYMOLOGY This species is named for the state of Kansas. **FIELD IDENTIFICATION** Impossible. **LAB IDENTIFICATION** The mandible has a ventral tooth basally; the ocelli are moderately large; the scape lacks a setal brush; the mesosternal area is unarmed; and T7 has a Y-shaped carina. **FEMALE** Unknown. **NOTES** This is a relatively rare species in the Great Plains and is structurally similar to *T. ocellaria*, except for the small ocelli.

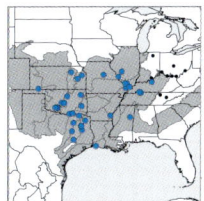

5.7.16.7—*Timulla ocellaria*
TIMM-EWE-LUH OWE-CELL-ARE-EE-UH

ETYMOLOGY Named for the enlarged ocelli. **FIELD IDENTIFICATION** Sometimes easy. This is the most commonly encountered nocturnal *Timulla* species in the central USA. **LAB IDENTIFICATION** The mandible has a ventral tooth basally; the ocelli are large; the scape lacks a setal brush; the mesosternal area is unarmed; and T7 has a Y-shaped carina. **FEMALE** Unknown. **NOTES** This species is most frequently collected at lights and is one of the most common nocturnal species of *Timulla*. Some eastern populations are treated in the *Timulla*-like MCS (5.5.12.1).

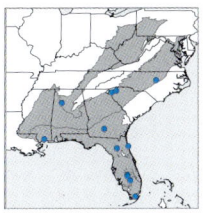

5.7.16.8—*Timulla rufosignata*
TIMM-EWE-LUH ROO-FOE-SIGG-NOT-UH

ETYMOLOGY From the Latin *rufus* "red" and *signum* "mark," in reference to the reddish mesosomal patch. **FIELD IDENTIFICATION** Impossible. **LAB IDENTIFICATION** The mandible is unarmed ventrally; the ocelli are small; the scape is simple; the mesosternal area is unarmed; and the pygidium has a Y-shaped carina. **FEMALE** Unknown. **NOTES** This Floridian species fits into a difficult complex of species known only from males in the central and eastern USA, along with *T. hollensis, T. tolerata*, and *T. sayi*. They are currently separated by often subtle differences in color, ocelli size, and body sculpture. Eventually, some of them might be recognized as synonymous color variants.

5.7.16.9—*Timulla sayi*
TIMM-EWE-LUH SAY-EYE

ETYMOLOGY This species was named after the American entomologist Thomas Say, who collected and described many velvet ant and other wasp species in the early 1800s. **FIELD IDENTIFICATION** Impossible. **LAB IDENTIFICATION** The mandible is unarmed ventrally; the ocelli are very large; the scape is simple; the tegula is reddish-brown; the mesosternal area is unarmed; and the pygidium has a Y-shaped carina. **FEMALE** Unknown. **NOTES** This rare species fits into a difficult complex of species known only from males in the central and eastern USA, along with *T. hollensis*, *T. tolerata*, and *T. sayi*. They are currently separated by often subtle differences in color, ocelli size, and body sculpture. Eventually, some of them might be recognized as synonymous color variants.

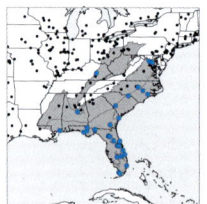

5.7.16.10—*Timulla vagans*
TIMM-EWE-LUH VAY-GUHNS

ETYMOLOGY From the Latin *vagans* "rambling." This color form used to be named *T. vagans rufinota* after the reddish mesosoma. **FIELD IDENTIFICATION** Difficult. The scape has a more distinct white setal brush than others in this cluster, but this can be difficult to see in the field. **LAB IDENTIFICATION** The mandible has a ventral tooth basally; the ocelli are moderately large; the scape has a brush of dense white setae; the mesosternal area is armed with a smooth tubercle; and T7 has a Y-shaped carina. **FEMALE** (3.5.9.1, 3.6.17.2). Somewhat similar to male, with silver setal markings. **NOTES** This reddish, southeastern form is rarer than the Eastern MCS form (5.6.7.1). Some other populations have more extensive gray setae on the head and mesosoma and fit the *Timulla*-like MCS (5.5.12.2).

Photo by Efram Goldberg

THE GRAVEYARD SHIFT— NOCTURNAL VELVET ANTS

6.1 INTRODUCTION AND GENERA

About half of the velvet ant species in the USA are nocturnal wasps in the tribe Sphaeropthalmini. In North America, the tribe includes over 200 species in 11 genera. Three of these genera, *Morsyma* (3.5.4.1, 5.7.8.1), *Protophotopsis* (3.4.9.1, 5.4.11.1), and *Stethophotopsis* (3.4.16.1), are active during daylight and include one species each; they are treated in the previous chapters only. Although some species from the genera *Dasymutilla* (3.1.3.1, 3.2.6.1, 5.1.3.1, 5.2.7.2), *Myrmosa* (5.4.14.2), and *Timulla* (3.5.5.1, 3.6.13.1, 5.3.6.2, 5.5.14.5, 5.7.16.7) can be active at night, they are not closely related to the Sphaeropthalmini and are treated in the previous chapters only. Eight genera of nocturnal velvet ants are treated in this chapter: *Sphaeropthalma* (110 species, 90 in USA), *Odontophotopsis* (65 species, 55 in USA), *Photomorphus* (45 species, 40 in USA), *Acanthophotopsis* (6 species, 5 in USA), *Acrophotopsis* (4 species, 2 in USA), *Dilophotopsis* (3 species, all in USA), *Laminatilla* (3 species, 1 in USA), and *Schusterphotopsis* (1 species, only in California, USA).

Nocturnal velvet ants are most abundant and diverse in arid habitats. The North American nocturnal fauna, however, seems to be especially rich. Fewer than 100 nocturnal forms are described from the entire Old World (mostly in the genera *Pseudophotopsis*, *Dentilla*, and *Tricholabiodes*; see 8.2.3), and about 50 are named from South America (in *Sphaeropthalma* and various Neotropical genera of Sphaeropthalmini and Pseudomethocini). With its 200 species, western North America seems to be more diverse in nocturnal forms than the rest of the world combined. This could change a bit in the future, though, as preliminary studies on the South American fauna suggest there are dozens of new species in Argentina, Chile, and Peru.

The regional success of nocturnal velvet ants in western North America also correlates with two other diverse and abundant nocturnal stinging wasp lineages. These wasp families are found in similar habitats to nocturnal velvet ants and are frequently collected using the same methods. The most species-rich genus of Chyphotidae, *Chyphotes*, includes about 50 species, and the tiphiid subfamily Brachycistidinae includes about 90 species in 10 genera. In both of these lineages, females are wingless and males are fully winged, just like mutillids. All three wasp groups are similar in size and color. They are so similar that even professional collections managers have difficulty telling them apart, so do not feel bad if you have trouble. Table 6 includes diagnostic characters, but without a lot of practice, you will likely need a specimen and microscope to be sure you have got the right family.

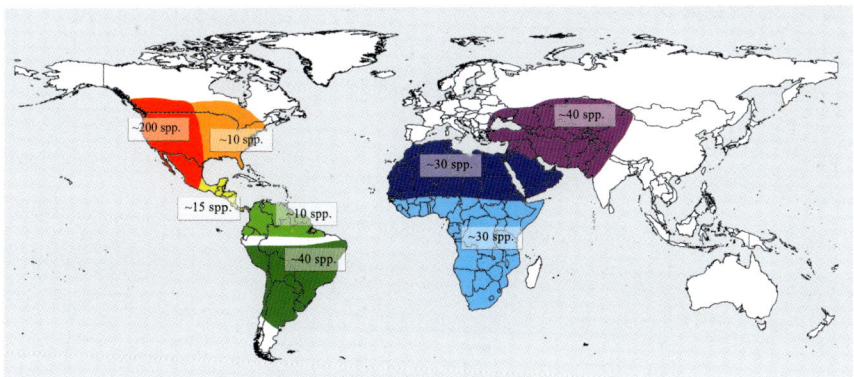

Geographic distribution of velvet ants with nocturnal behavior.

	Females	Males
Mutillidae	Mesosomal plates fused	Body "hairy" and usually brown; antenna generally shorter and curled; T1 and S1 similar in length; T2 with felt line; hypopygium unarmed
Chyphotes (Fig. 6.1.0a,b)	Mesosoma divided into two plates	Body "hairy" and usually brown; antenna generally curled; T1 only half as long as cylindrical S1; T2 with felt line; hypopygium with up-curved spine
Brachycistidinae (Fig. 6.1.0c,d)	Mesosoma divided into three plates	Body usually shiny and dark reddish; antenna long and usually straight; T1 and S1 similar in length; T2 without felt line; hypopygium with up-curved spine

6.1.0a

6.1.0b

6.1.0c

6.1.0d

Unlike diurnal velvet ants, bright warning colors are rarely useful for night-active species, so they often appear in drab red-brown attire. To some, this makes them boring—we do not hold this opinion. These insects may disappoint some in color variation, but they excel in structural variation. The remainder of this section presents a key to the genera and an account of the diversity and distribution for each genus.

In the next sections, we highlight some examples of the structural diversity of these wasps. In some cases, such as the clypeal spine of *O. unicornis* (see 6.3.4), a single unique structural feature is adequate to confirm a species diagnosis. In other cases, the general morphology of a trait is shared by over 100 species.

Although it is currently impossible to provide functional keys to all nocturnal velvet ant species in the USA, we hope the genus overviews below and the illustrations in the next sections will be useful to readers in three ways.

1. Most nocturnal velvet ants in the USA can be recognized at the genus level (with a dead specimen and microscope), and a few can be accurately diagnosed to species.
2. If readers attempt to identify nocturnal velvet ants using other sources, particularly primary taxonomic literature, these illustrations will make use of those keys easier.
3. These illustrated features can guide the sorting of nocturnal velvet ants into morphospecies, allowing readers to better estimate species richness in each trap sample or locality.

Key to genera based on females

1. Dorsum of mesosoma with specialized appressed setae that
are plumose at their base and simple at their apex (common) .... *Odontophotopsis*

2. Dorsum of mesosoma with erect bushy plumose setae mesally

    I. Pygidium convex without a plate, T1 shape slender petiolate (rare) .... *Laminatilla*

    II. With either a defined pygidial plate or broader
T1 shape (rare) .... *Sphaeropthalma*, **in part**

3. Dorsum of mesosoma with simple or brachyplumose setae mesally

    I. With the AcroDilo-mandible

        a. T2 with scattered tubercles at base, gena without carina (uncommon) .... *Acrophotopsis*

        b. T2 simply punctate, gena without carina (uncommon) .... *Dilophotopsis*

        c. T2 simply punctate, gena with carina (rare) .... *Schusterphotopsis*

    II. Mandible variable, usually without small dorsal tooth,
if dorsal tooth present, then ventral tooth weak or absent

        a. Pygidium without plate, T2 with coarse honeycomb sculpture
throughout or with basal patches of darker brown setae (rare) .... *Acanthophotopsis*

        b. Pygidium with plate, mesosoma rectangular, T1 shape sessile,
T2 with simple punctures and uniform setal color (uncommon) .... *Photomorphus*

        c. Pygidium with or without plate, mesosoma shape,
T1 shape, T2 sculpture and setae variable (common) .... *Sphaeropthalma*, **in part**

Key to genera based on males

1. Mesosternum lacking teeth or other processes

    I. Hypopygium of the DiloAcro form (uncommon) .... *Acrophotopsis*

    II. Hypopygium normal

        a. Tergal setae dense, plumose, white (uncommon) .... *Odontophotopsis*, **in part**

        b. Tergal setae usually sparser and not so obviously
plumose, variable in color (common) .... *Sphaeropthalma*

2. Mesosternum variably armed with teeth, spines, or other processes

    I. Mesosternal processes pointing posteriorly and placed near mid coxae,
always tooth or spine-like

        a. Mesosternal processes small, triangular teeth (rare) .... *Schusterphotopsis*

        b. Mesosternal processes large, conical pegs (rare) .... *Acanthophotopsis*

    II. Mesosternal processes variably shaped, usually pointing ventrally or
anteriorly and placed mesally

        a. Hypopygium of the DiloAcro form (uncommon) .... *Dilophotopsis*

        b. Hypopygium normal

        i. Mesosternum with laterally flattened thin, sharp lamellae (rare) .... *Laminatilla*

        ii. Mesosternum with 2 to 6 simple teeth; tergites with dense,
plumose setae; body punctures usually sparse (common) .... *Odontophotopsis*, **in part**

        iii. Mesosternum with "something funky": usually rows of multiple
small teeth or ridges, or a transverse process (often with a ventral
crease); tergal setae sparse and usually simple; body punctures
usually denser (uncommon) .... *Photomorphus*

6.1.1—*Acanthophotopsis*

ETYMOLOGY From the Latin word *acanth-* "with thorns" and the name *Photopsis*, an older genus name used for nocturnal velvet ants. The genus is named for its thorn-like mesosternal processes. **IDENTIFICATION** In males, the mesosternal processes of this genus are different from all the other genera; they are large conical pegs that point posteriorly. Unlike other velvet ants, the males usually have only one spine on the mesotibia. In females, the pygidium is undefined and T2 usually has dark setal patches basally. **SPECIES RICHNESS** The genus includes six species: two in the USA only, three in USA and Mexico, one in Mexico only. **NOTES** This genus is widespread in the western Nearctic region, but they are generally rare in collections and trap samples. Only three species have the sexes associated, and none of these is officially published yet. Nothing is known about their behavior or hosts. **KEYS AND CHARACTERS** The males were reviewed by Tanner et al. in 2009. Important diagnostic features include head shape, mandible shape, and coloration.

Acanthophotopsis falciformis male

Acanthophotopsis dorophora female

6.1.2—*Acrophotopsis*

ETYMOLOGY From the Neo-Latin word *acro-* "pointed," "first," or "high" and the name *Photopsis*, an older genus name used for nocturnal velvet ants. The name might refer to the wide mandibles or may be the author's indication that the genus was a highly specialized lineage (Schuster 1958). **IDENTIFICATION** Males have the AcroDilo-mandible (Fig. 6.3.4c) and AcroDilo-hypopygium (Fig. 6.3.1a), and they lack mesosternal tubercles. Females have a petiolate T1 shape, the female AcroDilo-mandible (Fig. 6.2.2a), an unarmed gena, and a bumpy multituberculate T2 disc. **SPECIES RICHNESS** There are four species: two in the Mojave and Sonoran Deserts and two in Mexico.

ABOVE: *Acrophotopsis dirce* female

LEFT: *Acrophotopsis dirce* male

NOTES The females of two species were matched up using DNA (Pitts & Wilson 2009). **KEYS AND CHARACTERS** Differences in genitalia, distribution, and coloration are useful to separate the species. The males were keyed and illustrated by Pitts & McHugh (2002). Only two females are currently recognized, and they are separated by setal color differences.

6.1.3—*Dilophotopsis*

ETYMOLOGY From the Neo-Latin word *dilo-* "evening" and the name *Photopsis*, an older genus name used for nocturnal velvet ants. The genus was likely named for its nocturnal behavior. **IDENTIFICATION** Males have the AcroDilo-mandible (Fig. 6.3.1a) and AcroDilo-hypopygium (Fig. 6.3.1a), and they have two sharp or blunt mesosternal tubercles. Females have a petiolate T1 shape, the female AcroDilo-mandible (Fig. 6.2.2a), an unarmed gena, and a simply punctate T2 disc. **SPECIES RICHNESS** There are three species, which all live in the USA and Mexico. **NOTES** While this genus has only three species, one of them (*D. concolor*) is quite widespread, and the males are commonly encountered at blacklights and porch lights. Genetic and historical biogeographic studies of this species have helped scientists better understand how the formation of North American deserts drove evolutionary processes in many arid-adapted species. **KEYS AND CHARACTERS** Males can be separated by their mandible shape, mesosternal tubercle shape, genitalia, and coloration (Wilson & Pitts 2008). Females are mainly separated by setal coloration.

Dilophotopsis concolor female

Dilophotopsis concolor male

6.1.4—*Laminatilla*

ETYMOLOGY From the Latin word *lamina* "thin membrane" and the suffix *-tilla*, a commonly used suffix for velvet ants. The genus is named for its thin, transparent mesosternal armature. **IDENTIFICATION** Males of this genus can be recognized by the unique transparent triangular mesosoternal

Laminatilla lamellifera female

Laminatilla lamellifera male

armature. Females have the pygidium undefined and the dorsum of the body covered with erect, bushy, plumose setae. **SPECIES RICHNESS** There is one species in Arizona and two in Mexico. **NOTES** Females of this genus were only recently recognized, and they have not been formally described yet. Species of this genus were originally included in *Odontophotopsis*, but Pitts (2007) realized that they belonged in a separate genus. **KEYS AND CHARACTERS** The species can be most easily separated by distribution, but Pitts (2007) wrote a good key with various structural characters.

6.1.5—*Odontophotopsis*

ETYMOLOGY From the Neo-Latin word *odonto-* "toothed" and the name *Photopsis*, an older genus name used for nocturnal velvet ants. The genus was named for the toothed mesosternal area in males. **IDENTIFICATION** In males, this is one of three genera that is too diverse and variable to perfectly define without ten characters (and all their exceptions)—frankly, it is a mess. Most *Odontophotopsis* males have simple tooth-like mesosternal processes and the tergites with dense fringes of distinctly plumose setae. The situation is a bit easier with females, though their diagnostic feature takes

TOP LEFT: *Odontophotopsis* female

MIDDLE LEFT: *Odontophotopsis arcuata* male

BOTTOM LEFT: *Odontophotopsis arcuata* female

MIDDLE RIGHT: *Odontophotopsis brunnea* male

BOTTOM RIGHT: *Odontophotopsis brunnea* female

Photo by Jillian Cowles

some practice to recognize. They have a special type of appressed mesosomal setae that are plumose basally and simple apically. **SPECIES RICHNESS** There are about 60 species, which occur mainly in hot deserts. **NOTES** This genus includes fewer species than *Sphaeropthalma*, but they are usually more abundant. In many trap samples, over half of the total specimens will belong to one single *Odonto-photopsis* species, usually *O. clypeata, O. obscura, O. melicausa,* or *O. serca.* **KEYS AND CHARACTERS** In 2007, Pitts started a revision of the genus and wrote a useful key to the species-groups and species in some of these groups. For the most diverse and common species-groups, Schuster's (1958) key is necessary, but that key lacks about 20 species that were described in the last 60 years. Microscopic differences in the shape of the clypeus, mandible, mesosternal processes, and pygidium are their most useful features. Few females are recognized, and there is no published key to separate them.

6.1.6—*Photomorphus*

ETYMOLOGY Based on the name *Photopsis*, with the Ancient Greek suffix -*morphe* "shape" or "form." The genus was named for its similarity to the formerly recognized genus *Photopsis*. **IDENTIFICATION** Like *Odontophotopsis* and *Sphaeropthalma*, this genus is too diverse and variable to define in a simple

TOP LEFT: *Photomorphus auriferus* male

ABOVE: *Photomorphus* female

LEFT: *Photomorphus quadriangulatus* male

BOTTOM LEFT: *Photomorphus obscurus* male

BELOW: *Photomorphus obscurus* female

and functional manner. In males, the mesosternal processes are usually "complicated," most often having a row of small teeth or one transverse process with a crease, and the tergites usually have sparse fringes of simple or obscurely plumose setae. In females, the setae are generally sparse and mostly simple, the mesosoma is usually rectangular, and the T1 shape is sessile. **SPECIES RICHNESS** There are about 45 published species from the USA and Mexico, but this genus includes many currently unnamed new species, and the true diversity is likely more than 60 species. **NOTES** Many species in the eastern USA are active during daylight. Of the three big genera, these are the least diverse and least common. In many samples, they are outnumbered by less diverse genera, like *Acrophotopsis*. **KEYS AND CHARACTERS** Schuster (1958) has the only available key for males. The diurnal eastern females are covered by Brabant et al. (2010) and Krombein (1954). Males are generally separated using microscopic differences in the mandible, clypeus, mesosternal processes, and pygidium. Females are separated mainly by color and pygidial sculpture.

6.1.7—*Schusterphotopsis*

ETYMOLOGY Named after Rudolf Schuster, who published an extensive treatment of nocturnal velvet ants, and the name *Photopsis*, an older genus name used for nocturnal velvet ants. **IDENTIFICATION** Males have the AcroDilo-mandible (Fig. 6.3.4c) and AcroDilo-hypopygium (Fig. 6.3.1a), and they have two posterior mesosternal tubercles directly in front of the mesocoxae. Females have a petiolate T1 shape, the female AcroDilo-mandible (Fig. 6.2.2a), a genal carina, and a simply punctate T2 disc. **SPECIES RICHNESS** One species known from a single locality in southern California. **NOTES** This species was described by Pitts in 2003 based on a single male specimen collected in 1969. No additional specimens in museums had been examined until recently, when Kevin and his dad found the species near Big Bear Lake, California. In three trips from 2016 to 2017, they collected about 150 males and two females. **KEYS AND CHARACTERS** There is only one species in the genus.

Schusterphotopsis barghesti male   *Schusterphotopsis barghesti* female

6.1.8—*Sphaeropthalma*

ETYMOLOGY From the Latin *sphaero* "ball" or "sphere" and Ancient Greek *ophthalmos* "eye." The correct spelling for "eyed" includes an extra letter *h*, but the typo from the original spelling stays with us. **IDENTIFICATION** This genus cannot be consistently defined. For males, most nocturnal species that lack mesosternal processes belong to *Sphaeropthalma*. Females are totally undefinable, so all the leftovers get squeezed into *Sphaeropthalma*, for now. **SPECIES RICHNESS** There are about 100 species, widespread throughout the Americas, including the USA, Mexico, Central America, arid regions of South America, and even one species in the Galápagos Islands. **NOTES** This is the most diverse nocturnal genus, but many of the species, especially in more humid areas, are diurnal (3.2.12, 3.6.11, 5.2.9.1, 5.7.11, 5.7.12). Many of the most colorful nocturnal species belong to *Sphaeropthalma* (like *S. marpesia*: 5.5.10.1). **KEYS AND CHARACTERS** Differences in the clypeus, mandible, wing venation, felt lines, and internal male genitalia are used to recognize most species. Some species-groups have recent functional revisions, but Schuster's key (1958) is needed for most species. Fewer than 20% of the females are recognized, so there is no useful published key for them.

ABOVE: *Sphaeropthalma arota* male and female

BELOW LEFT AND RIGHT: *Sphaeropthalma difficilis* male and female

ABOVE LEFT: *Sphaeropthalma tapio* male

ABOVE: *Sphaeropthalma mendica* male

LEFT: *Sphaeropthalma imperialis* male

ABOVE: *Sphaeropthalma ceres* female
RIGHT: *Sphaeropthalma* female

Photo by Jillian Cowles

6.2 STRUCTURAL CHARACTERS—FEMALES

Nocturnal velvet ant females are often difficult to find and collect because they cannot fly into light traps. Other than pitfall traps, the best way to find nocturnal females is by using a flashlight and a jar, an arduous process that usually yields fewer than five specimens on a given night. Altogether, nocturnal males outnumber females in museums by about 50 to one. Consequently, of the 200 Sphaeropthalmini species named in North America, fewer than 20 were initially described based on females. With recent advances largely using DNA sequences to match males with females, an additional 40 species now have published sex associations. This leaves about 140 species (70%) of nocturnal Sphaeropthalmini in the USA known from males only, and only about 60 females ever described or discussed in the published literature.

In the last few years, I have been able to borrow or collect about 13,000 nocturnal female specimens from across the USA. These have been sorted into about 160 morphospecies—100 of them are currently unknown in the literature. Based on preliminary DNA, and distributional and morphological data, about half of these can be reliably matched with males. I'm still in the early stages of this revision, and a monograph for nocturnal velvet ants in North America should come out soon but "soon" in taxonomic revision terms means "hopefully, some time in the next 20 years."—KEVIN

The features illustrated below are those that have shown to be the most useful for separating genera and species of nocturnal velvet ant females. Many of these features require a high-powered microscope, well-preserved specimens, and a good bit of practice. In addition to the general features below, presence of various carinae or tubercles, subtle differences in sculpture, and color characters in different body regions are useful for differentiating species.

Mesosomal dorsum setal types

A high-powered microscope is often needed to recognize these setal types, but after some practice and experience, a good guess can be made with many photographed specimens. When an insect has multiple types of setae, look specifically at those near the middle of the mesonotum. *Odontophotopsis* is the most commonly encountered genus in the western USA, and learning to recognize their setal type (Fig. 6.2.1a) is the most important first step in sorting nocturnal female velvet ants.

Figure	Taxon	Setal type	Prevalence
6.2.1a	*Odontophotopsis venusta*	Flat, plumose basally, brachyplumose apically	All species of *Odontophotopsis*
6.2.1b	*Laminatilla lamellifera*	Erect, bushy plumose	Occurs in all *Laminatilla* and a few *Sphaeropthalma* species
6.2.1c	*Sphaeropthalma yumaella*	Mixed plumose and brachyplumose	Some small-bodied species of *Sphaeropthalma*
6.2.1d	*Sphaeropthalma difficilis*	Uniform suberect weakly plumose	Some small-bodied species of *Sphaeropthalma*
6.2.1e	*Acrophotopsis dirce*	Mostly simple or brachyplumose	Many species in various genera
6.2.1f	*Sphaeropthalma arota*	Simple with plumose setae laterally	Some species in *Sphaeropthalma* and other genera

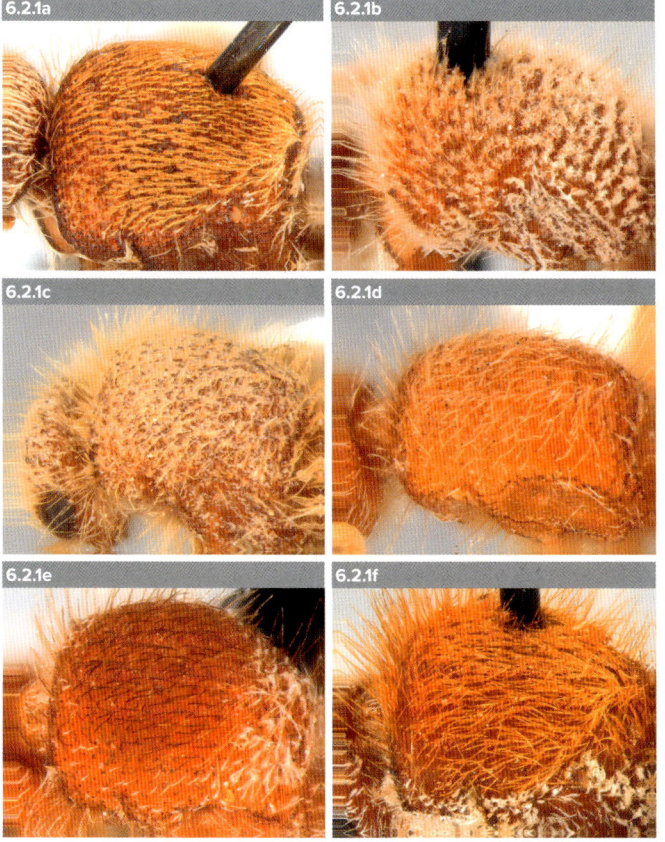

Mandible

Differences in mandible shape and structure are important for recognizing the genera and many species. The AcroDilo-mandible (Fig. 6.2.2a) is important for separating some species-poor (but often abundant) genera from the diverse genus *Sphaeropthalma*, but a high-powered microscope is absolutely necessary for finding the smaller dorsal tooth. One frustrating factor in using mandibular characters is specimen age and wear; sometimes the mandibular teeth can be completely obliterated, making identification impossible.

Figure	Taxon	Mandible type	Prevalence
6.2.2a	*Dilophotopsis stenognatha*	AcroDilo—large ventral tooth, small sharp dorsal tooth	All species of *Acrophotopsis, Dilophotopsis,* and *Schusterphotopsis*
6.2.2b	*Sphaeropthalma unicolor*	Weak or absent ventral tooth, small dorsal tooth	Some species of *Sphaeropthalma*
6.2.2c	*Odontophotopsis melicausa*	Large ventral tooth, no dorsal tooth	Many species in various genera
6.2.2d	*Odontophotopsis unicornis*	Weak ventral tooth, no dorsal tooth	Many species in various genera
6.2.2e	*Acanthophotopsis bequaerti*	No ventral tooth, no dorsal tooth	Many species in various genera
6.2.2f	*Odontophotopsis cookii*	Weak or absent ventral tooth, swollen dorsal lobe	Some species of *Odontophotopsis*

T1 shape, T2 sculpture, and T2 fringe

Just as in the diurnal velvet ants, differences in T1 shape are important for separating genera and broader groups of species. There seems to be a greater degree of variation between sessile and petiolate forms in the nocturnal species, making these somewhat subjective and difficult to interpret without lots of practice. The sculpture and types of setae on T2 are also vital for differentiating species and some genera. Most (~90%) nocturnal velvet ants have plumose tergal fringes. When a species does not seem to match any of the females from chapter 3, plumose fringes could reveal it as a nocturnal form that came out before sunset (happens frequently in colder habitats and among the more brightly colored nocturnal species).

Figure	Taxon	T1 shape	T2 sculpture	T2 fringe	Prevalence
6.2.3a	*Acrophotopsis dirce*	Narrow petiolate	Scattered basal tubercles	Plumose	All species of *Acrophotopsis*
6.2.3b	*Dilophotopsis concolor*	Narrow petiolate	Uniform punctures	Plumose	Many species in various genera
6.2.3c	*Sphaeropthalma erigone*	Short subsessile	Uniform punctures	Plumose	Many species in various genera
6.2.3d	*Photomorphus banksi*	Sessile	Uniform punctures	Simple	Sessile T1 occurs in many genera, simple T2 fringe in few genera
6.2.3e	*Sphaeropthalma difficilis*	Short subpetiolate	Uniform punctures	Plumose	Many species in various genera
6.2.3f	*Sphaeropthalma bisetosa*	Long subpetiolate	Coarse irregular punctures	Plumose	Some species of *Sphaeropthalma*

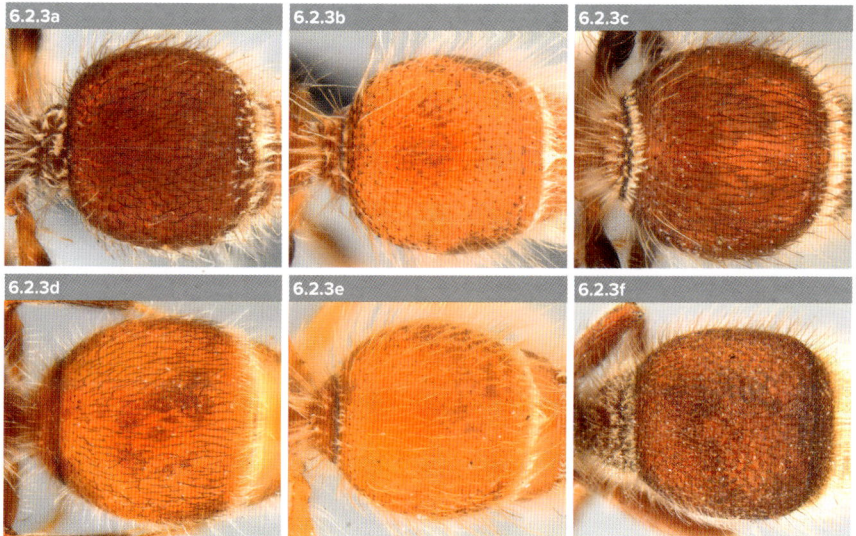

Pygidial plate

This feature seems to be an important indicator for host and habitat types. Species that lack a flat pygidial plate are usually those that attack arboreal hosts in twig or mud nests, while species with a plate usually attack ground-nesting hosts. Among the species with a defined plate (~80%), slight differences in shape and sculpture are important for separating species. There are dozens of unique pygidial forms and they cannot all be illustrated here, but the examples provided should be helpful for seeing the broader trends in how this feature varies across the group.

Figure	Taxon	Plate shape	Carinae	Sculpture	Prevalence
6.2.4a	*Sphaeropthalma bisetosa*	Undefined	Absent	Smooth	Common in some genera; associated with attacking twig- or mud-nesting hosts
6.2.4b	*Sphaeropthalma pensylvanica*	Undefined	Absent	Microreticulate	Some species of *Sphaeropthalma* that attack twig- or mud-nesting hosts
6.2.4c	*Photomorphus archboldi*	Elongate ovate	Simple	Smooth	Features occur in various genera; this combination unique to *Photomorphus*
6.2.4d	*Acrophotopsis dirce*	Short triangular	Simple	Microreticulate	Common in various genera; often associated with arid sandy habitats
6.2.4e	*Odontophotopsis unicornis*	Short triangular	Strongly raised	Rugose	Features occur in various genera; this combination unique to *Odontophotopsis*
6.2.4f	*Odontophotopsis melicausa*	Triangular	Moderately raised	Striate	Many species in various genera

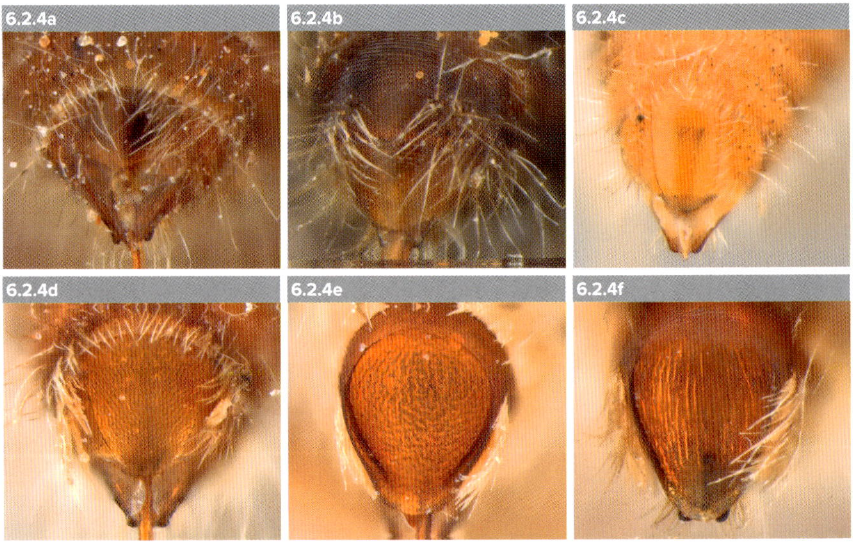

6.3 STRUCTURAL CHARACTERS—MALES

Unlike the wingless females, nocturnal males are easy to collect and are abundant in museums. Using the traps mentioned in chapter 1 (1.5.4), it is not rare to collect between 100 and 500 specimens in one night in a single trap in western deserts. Museums are full of nocturnal males, and we have seen at least 200,000 specimens. As such, most of the named species and taxonomic resources are based on males. In addition to the 180 currently named nocturnal velvet ant males in the USA, we have found at least 20 more new species, which will eventually be named and described.

The features illustrated below are those that have shown to be the most useful for separating genera and species of nocturnal velvet ant males. Many of these features require a high-powered microscope, well-preserved specimens, and a good bit of practice. In addition to the general features below, presence of various carinae or tubercles, subtle differences in sculpture, and color characters in different body regions are useful for differentiating species.

Hypopygium

Second only to the mesosternal armature below, this feature is important for recognizing genera. The AcroDilo-hypopygium (Fig. 6.3.1a) shows perfect correlation with the AcroDilo-mandible in females (Fig. 6.2.2a) at the genus level. Some other genera approximate this feature, but their lateral carinae are shorter and the posterior margin is flat or weakly concave. Most nocturnal velvet ant males have a slender hypopygium without lateral carinae.

Figure	Taxon	Hypopygium	Prevalence
6.3.1a	*Acrophotopsis dirce*	Wide, concave margin, strong lateral carinae	All species of *Acrophotopsis*, *Dilophotopsis*, and *Schusterphotopsis*
6.3.1b	*Acanthophotopsis falciformis*	Wide, convex margin, weak lateral carinae	Occurs in *Acanthophotopsis* and some *Photomorphus*
6.3.1c	*Odontophotopsis microdonta*	Narrow, convex margin, no lateral carinae	Most species in most genera

6.3.1a **6.3.1b** **6.3.1c**

Pygidium

Because males do not have to dig into the soil to find bee and wasp hosts, the pygidium structure is generally more poorly defined and shows less variation in males than females. There is often some slight correlation between the structure in conspecific males and females, but many species that have a coarsely defined pygidium in females have a nearly smooth pygidium in males.

Figure	Taxon	Pygidium	Prevalence
6.3.2a	*Odontophotopsis microdonta*	Smooth without lateral carinae	Most species in most genera
6.3.2b	*Odontophotopsis bellona*	Microreticulate without lateral carinae	Some species in *Odontophotopsis*, *Photomorphus*, and *Sphaeropthalma*
6.3.2c	*Odontophotopsis odontoloxia*	Microreticulate with lateral carina	Some species in *Odontophotopsis* and *Photomorphus*

  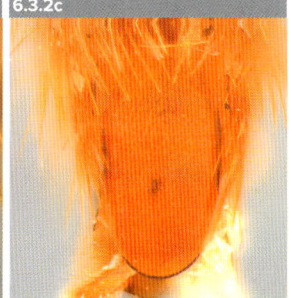

6.3.2a **6.3.2b** **6.3.2c**

Forewing

The wing color can sometimes help with recognizing a few species. The banded wing (Fig. 6.3.3d) is also seen in the non-mutillid genus *Chyphotes* (Family Chyphotidae). If an apparent nocturnal velvet ant male does not seem to fit but has a dark patch in the wing, it is worth double checking the family-level characters (6.1.0). The marginal cell character is hard to interpret at first, because the measurements being compared are not the actual total length of the marginal cell, rather the length from the tip of the stigma to the tip of the marginal cell along the costa (wing margin).

Figure	Taxon	Marginal cell	Color	Prevalence
6.3.3a	*Sphaeropthalma reducta*	Shorter than stigma	Clear	Marginal cell: many species in various genera. Color: many species in various genera
6.3.3b	*Sphaeropthalma uro*	Equal to stigma	Clear	Marginal cell: many species in various genera. Color: many species in various genera
6.3.3c	*Sphaeropthalma edwardsii*	Longer than stigma	Uniformly darkened	Marginal cell: many species in various genera. Color: some species of *Sphaeropthalma*
6.3.3d	*Sphaeropthalma parapenalis*	Longer than stigma	Clear with dark patch	Marginal cell: many species in various genera. Color: some species of *Sphaeropthalma*, *Photomorphus*, and other genera

6.3.3a

6.3.3b

6.3.3c

6.3.3d

Mandible

The male mandible shows more variation and has more useful diagnostic features than that of the female. Luckily, males have shorter life spans than females, so wear and tear on the male mandible is rarer. The view provided here also shows some useful characters of the clypeus, which also bears many useful diagnostic features.

Figure	Taxon	Mandible	Prevalence
6.3.4a	*Acanthophotopsis falciformis*	Dorsal view, large inner mandibular tooth	Unique to *Acanthophotopsis falciformis*
6.3.4b	*Acanthophotopsis bequaertii*	Dorsal view, no inner mandibular tooth	Most species in all genera
6.3.4c	*Dilophotopsis paron*	AcroDilo: Large ventral tooth, apex widened and tridentate	All species of *Acrophotopsis*, *Dilophotopsis*, and *Schusterphotopsis*
6.3.4d	*Odontophotopsis setifera*	Large ventral tooth, apex widened and quadridentate	Two *Odontophotopsis* species; dense short clypeal bristles unique to *O. setifera*
6.3.4e	*Odontophotopsis melicausa*	Large ventral tooth, apex moderate	Many species in various genera
6.3.4f	*Odontophotopsis unicornis*	Ventral tooth, long slender curved mandible	Two *Odontophotopsis* species; basal clypeal spine unique to *O. unicornis*

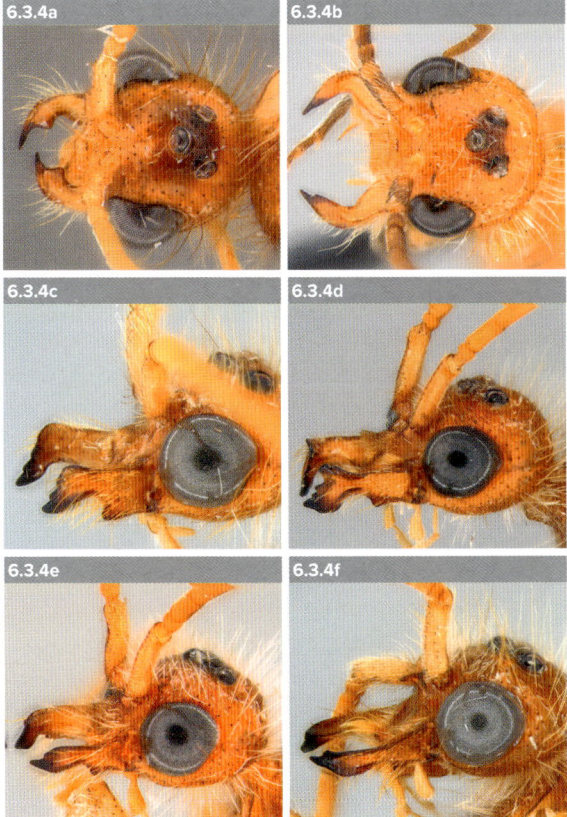

6.3.4a 6.3.4b 6.3.4c 6.3.4d 6.3.4e 6.3.4f

Figure	Taxon	Mandible	Prevalence
6.3.4g	*Odontophotopsis conifera*	Weak ventral tooth, thick vertical mandible	Some species in *Odontophotopsis* and *Sphaeropthalma*
6.3.4h	*Odontophotopsis parva*	Weak ventral tooth, moderate vertical mandible	Many species in various genera
6.3.4i	*Sphaeropthalma juxta*	Weak ventral tooth, moderate oblique mandible	Many species in various genera
6.3.4j	*Sphaeropthalma unicolor*	Weak ventral tooth, slender oblique mandible	Many species in various genera
6.3.4k	*Sphaeropthalma jacala*	Sharp ventral tooth, sharp basal dorsal tooth	Unique to *Sphaeropthalma jacala*
6.3.4l	*Sphaeropthalma ecarinata*	Large ventral tooth, apex sharp and slender	Some species in *Sphaeropthalma* and *Photomorphus*

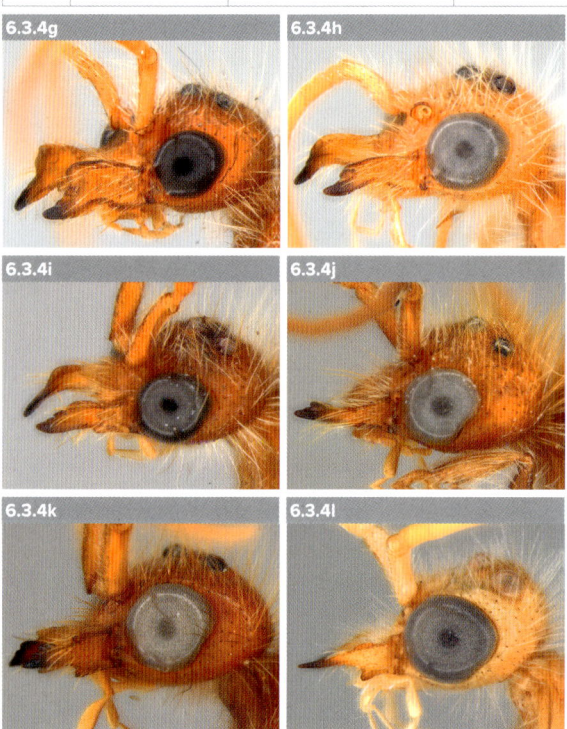

Mesosternal area

This is the most important feature for differentiating genera of nocturnal males. Some genera, like *Acanthophotopsis* (Fig. 6.3.5i) and *Laminatilla* (6.3.5h), have a unique and immediately recognizable defining feature here. Species of the abundant and diverse genera *Odontophotopsis* and *Photomorphus* are often separated by the specific shape and orientation of teeth and ridges in this area. Technically, the mesosternum is a tiny plate found between the middle legs, but the features below are found on the lower portion of the mesopleuron. We use the term "mesosternal area" because that is how they are referenced in the published literature. As used in velvet ant morphology, the term references the middle portion (*meso-*) of the underside (*sternal*) of the mesosoma.

Figure	Taxon	Mesosternal area	Prevalence
6.3.5a	*Sphaeropthalma bellerophon*	Unarmed, without teeth or ridges	Occurs in *Acrophotopsis*, *Sphaeropthalma*, and some *Odontophotopsis*
6.3.5b	*Schusterphotopsis barghesti*	Small posterior tooth near mesocoxae	Only *Schusterphotopsis barghesti*
6.3.5c	*Odontophotopsis mamata*	Longitudinal smooth swollen area	Only *Odontophotopsis mamata*
6.3.5d	*Dilophotopsis paron*	Rounded tubercles	Occurs in *Dilophotopsis*; teeth of *Odontophotopsis* sometimes similar
6.3.5e	*Odontophotopsis melicausa*	Simple teeth	Most species in *Odontophotopsis* and some in *Photomorphus*
6.3.5f	*Odontophotopsis inconspicua*	Large anteriorly pointed teeth	Some species of *Odontophotopsis*

6.3.5a

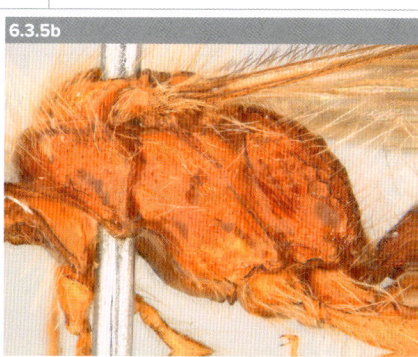
6.3.5b

6.3.5c

6.3.5d

6.3.5e

6.3.5f

Figure	Taxon	Mesosternal area	Prevalence
6.3.5g	*Odontophotopsis tenuiptera*	Curved, multidentate process	Only *Odontophotopsis tenuiptera*
6.3.5h	*Laminatilla lamellifera*	Thin sharp triangular lamelliform processes	Only *Laminatilla lamellifera*
6.3.5i	*Acanthophotopsis falciformis*	Posteriorly pointed conical pegs	Only *Acanthophotopsis*
6.3.5j	*Photomorphus obscurus*	Blunt ridge with transverse crease	Occurs in many species of *Photomorphus*
6.3.5k	*Photomorphus quadriangulatus*	Vertical transverse ridge	Occurs in some species of *Photomorphus*
6.3.5l	*Photomorphus spinciformis*	Longitudinal rows of smaller transverse ridges	Occurs in many species of *Photomorphus*

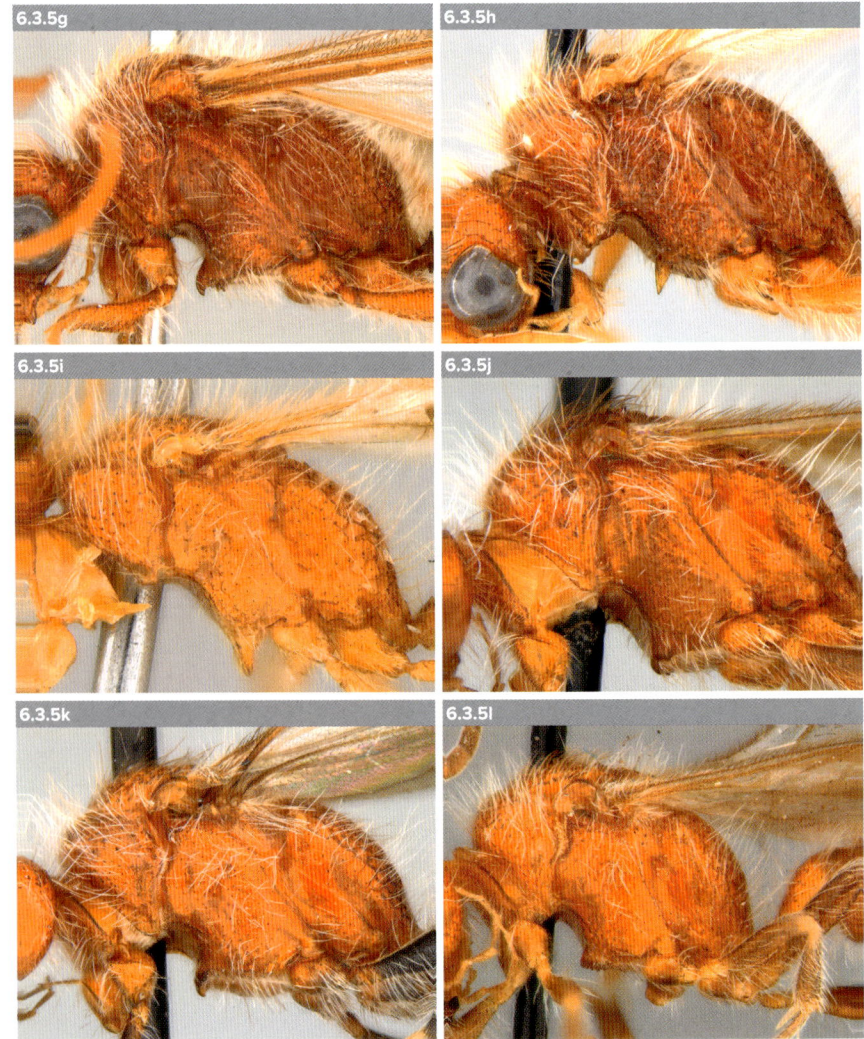

T1 shape, S2 felt line, and T2 fringe

There are some useful differences between species in T1 shape. All nocturnal velvet ants have a felt line on T2, but the presence and shape of the felt line on S2 can be helpful for recognizing species. Most (~90%) nocturnal velvet ants have plumose tergal fringes. When a species does not seem to match any of the males from chapter 5, plumose fringes could reveal it as a nocturnal form that came out before sunset (happens frequently in colder habitats and among the more brightly colored nocturnal species).

Figure	Taxon	T1 shape	S2 felt line	T2 fringe	Prevalence
6.3.6a	*Sphaeropthalma coaequalis*	Petiolate	Absent	Plumose	Features occur in many species in various genera
6.3.6b	*Sphaeropthalma hyalina*	Subsessile	Present, short	Plumose	Features occur in many species in various genera
6.3.6c	*Photomorphus alogus*	Distinctly petiolate	Present, elongate	Simple setae	Some species of *Photomorphus* and *Sphaeropthalma*
6.3.6d	*Photomorphus obscurus*	Subpetiolate	Present	Sparse plumose	Features occur in many species in various genera
6.3.6e	*Odontophotopsis unicornis*	Subpetiolate	Present	Dense plumose	Features occur in many species in various genera, especially *Odontophotopsis*
6.3.6f	*Odontophotopsis serca*	Subpetiolate	Absent	Dense plumose	Features occur in some species in various genera, especially *Odontophotopsis*

6.3.6a
6.3.6b
6.3.6c
6.3.6d
6.3.6e
6.3.6f

Genitalia

The male genitalia of nocturnal velvet ants show extreme variation in the paramere, cuspis, and penis valve. It is impossible to show the entire range of variation for all the species, but we have provided a broad range of examples here. There is an interesting morphological parallel with the two most abundant and diverse genera. *Odontophotopsis* species have relatively simple and uniform genitalic morphology (Fig. 6.3.7c) but show extreme variation in the armature of the mesosternal area (see 6.3.5). *Sphaeropthalma* species (Fig. 6.3.7e–l), however, show extreme variation in the genitalia but have the mesosternal area completely unarmed (6.3.5a). One hypothesis is that members of *Odontophotopsis* have simple genitalia because they use their mesosternal armature for various courtship rituals, and these are subject to sexual selection before mating occurs. On the other hand, *Sphaeropthalma*, with their unarmed mesosternal area, may be subject to sexual selection on genitalic morphology during mating.

Figure	Taxon	Notable features	Prevalence
6.3.7a	*Acanthophotopsis bequaertii*	Paramere flattened; cuspis out-curved	Occurs in *Acanthophotopsis*; some *Sphaeropthalma* have similar appearance
6.3.7b	*Dilophotopsis concolor*	"Elbowed" cuspis	Occurs in *Dilophotopsis*
6.3.7c	*Odontophotopsis melicausa*	Straight cylindrical paramere and cuspis	Common in various genera; dominant form of genitalia for *Odontophotopsis*
6.3.7d	*Odontophotopsis erebus*	Paramere contorted mesally; cuspis short, straight	Occurs in a few *Odontophotopsis* species
6.3.7e	*Sphaeropthalma pallida*	Paramere slender; cuspis short	Many species of *Sphaeropthalma* and *Photomorphus* are similar
6.3.7f	*Sphaeropthalma arota*	Paramere with basal setal brush; cuspis slender	Some species of *Sphaeropthalma* are similar
6.3.7g	*Sphaeropthalma coaequalis*	Paramere with apical setal brush; cuspis long, slender	Many species of *Sphaeropthalma* are similar
6.3.7h	*Sphaeropthalma orestes*	Cuspis moderate with long setae	Many species in various genera are similar
6.3.7i	*Sphaeropthalma ferruginea*	Paddle-like cuspis, setae with branched tips	Many species of *Sphaeropthalma* are similar
6.3.7j	*Sphaeropthalma abdominalis*	Paddle-like cuspis, setae with plumose tips	Many species of *Sphaeropthalma* are similar
6.3.7k	*Sphaeropthalma ceyxiodes*	Cuspis flat, contorted, with setal brushes	Many species of *Sphaeropthalma* are similar
6.3.7l	*Sphaeropthalma scudderi*	Cuspis mitten-shaped	Unique to *Sphaeropthalma scudderi*

6.3.7a

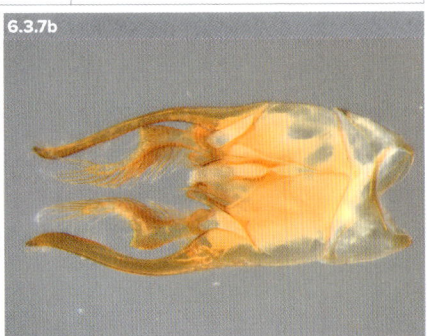

6.3.7b

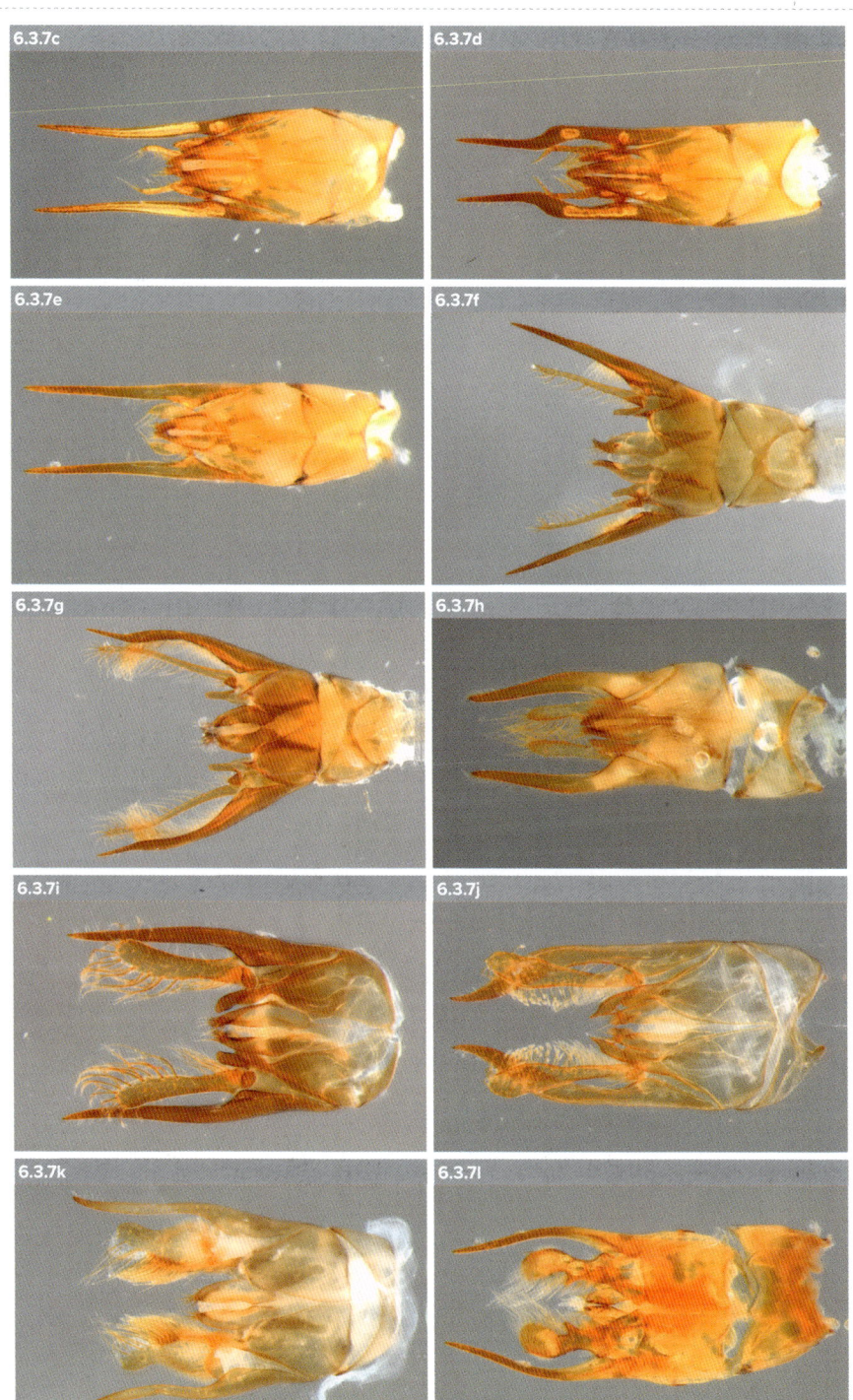

MESOAMERICA

7.1 ECOSYSTEMS AND ECOREGIONS

A brief description of ecosystems and ecoregions of the middle Americas—Mexico, the Caribbean, and Central America—is provided here to give the reader/user some background information about the environments in which velvet ants of this geographic portion of North America reside.

As one moves farther south and toward the tropical regions of the continent, a new cast of characters begins to appear. Some of the genera and species are the same as those encountered in the southern, particularly the southwestern USA, but others are different and exotic in comparison to Nearctic taxa. Although all three of the regions mentioned above, Mexico, the Caribbean, and Central America, rounded out by Canada and the USA, make up North America, these regions differ markedly from the latter two, particularly due to their geologic history, lower latitudinal position, and environmental heterogeneity.

Mexico

Mexico is a country of contrasts. Even in the southernmost mountain forests, you will find oak, pine, and fir. You will also find rainforests filled with orchids, figs, and palms. Deserts and arid scrub are abundant, but there are also mangroves, montane grasslands, chaparral, seasonally flooded wetlands, and tropical dry forests that are ghostly groves of gray and brown bare trees with flashes of bright yellows and pinks (the flowers) and emergent green palms in the dry season. If one hikes down into Copper Canyon (Barrancas del Cobre) in northwest Mexico, you will be surrounded by sycamores, alders, and other typically temperate trees, with tropical orchids and bromeliads covering their branches in the canopy—a wonderful and odd hodgepodge community!

Mexico's impressive ecosystem and community heterogeneity is due to a number of factors. One of these factors is the abundance of mountain ranges and uplands, with over half the country above 1000 m in elevation (Velásquez et al. 2000). The mountains contribute to extensive ranges in altitude, the creation of topographic features that can both capture precipitation or impede it, and can act as both thoroughfares for dispersal of highland/montane species and as allopatric barriers for lowland ones. Mexico also differs from Canada and the USA in having maritime influences that extend farther into its territory because of its narrower geographic profile—leading to reduced distances from the interior to sea. The third major factor is that the Tropic of Cancer cuts through the country, creating tropical conditions in much of the country, as well as intensifying arid influences in other portions. The mixture of topographic lowland areas surrounded by mountains and their occurrence within the tropics (e.g., Balsas Depression and the Bajío) must be a veritable wonderland for velvet ants—probably with many new species.

Grasslands are relatively rare in Mexico compared with the vast plains and prairie vegetation of the USA and Canada. Native grasslands in Mexico can be mainly found as desert and arid grasslands within the Chihuahuan Desert and Mexican Plateau regions, particularly where these ecological areas meet the lower eastern slopes of the Sierra Madre Occidental. Alpine and subalpine grasslands can be found on a number of mountains. A small portion of gulf coastal grassland also extends from southern Texas into Tamaulipas.

Arid and semi-arid deserts and scrublands dominate large portions of northern and central Mexico. Extending from southern Arizona and California and west of the Sierra Madre Occidental (including the Sky Islands), the Sonoran Desert extends to just north of Esperanza, Sonora, Mexico, and includes a portion of the northern part of the Baja California peninsula. The majority of Baja California is dominated by Baja California Desert, and xeric shrubland communities can be found along the western coastal margin of the Gulf of California. Extending from southern New Mexico and western Texas, then southward between the Sierra Madre Occidental and the Sierra Madre Oriental ranges, the Chihuahuan Desert represents the largest desert in Mexico. The southern portion of the central Mexican Plateau is dominated by matorral vegetation. Tamaulipan Mezquital, a thornscrub and brushland ecoregion, can be found in south Texas and northeastern Mexico. The Rio Grande/ Rio Bravo cuts through this community, with riparian vegetation located along its banks.

In the western deserts of Mexico, females with the Desert and Madrean Color Syndromes are most commonly encountered, but you will also find representatives of the Western and Red-headed *Timulla* here as well. The Chihuahuan Desert is interesting, with the northern portions having many more examples of the Texan, Madrean, and Red-headed *Timulla* types than can be found farther south. The Western Color Syndrome is common throughout the Chihuahuan Desert ecoregion, especially at higher elevations. The Tamaulipan Mezquital has numerous representatives of the Eastern, Texan, Red-headed *Timulla*, and Western types.

Lowland tropical moist broadleaf forests, which occur as far north as central Veracruz, are situated mainly on the eastern coastal plain, sandwiched between mountains to the west and the Gulf of Mexico to the east, and extend southward and eastward toward Central America. These biologically diverse forests receive abundant rainfall, and the sun's gifts of both warmth and light lead to luxuriant, often tall multistratal vegetation, and the novel floral and faunal niches and microhabitats that accompany it, enabling extended or multiple life cycles. Canada and the USA, excluding its tropical Atlantic and Pacific island territories, lack such habitats. Upland moist tropical broadleaf (including cloud) forests can be found in the southernmost portions of the Sierra Madre Oriental, the lower eastern slopes of the Sierra Madre de Oaxaca Mountains, and the easternmost portion of the Trans-Mexican Volcanic Belt. Whether lowland or upland, these humid tropical broadleaf forests are dominated by angiosperms and differ from the mixed conifer (typically pine)-oak forest formations more often associated with the major mountain ranges of western and southern Mexico, like the Sierra Madre Occidentals, Trans-Mexican Volcanic Belt, Sierra Madre del Sur, and northern portions of the Sierra Madre Orientals. In these jungles and tropical forests, you will find both arboreal and terrestrial velvet ants, and most females have coloration that falls within the Tropical and Black-headed Color Syndromes (introduced below: 7.4), as well as *Timulla*-like Color Syndrome.

The upland, temperate-derived, mixed conifer forest vegetation types are associated with Mexico's mountains and are in most cases characterized by a canopy layer dominated by common northern temperate tree genera (i.e., *Abies, Alnus, Pinus, Quercus*, etc.) and an understorey of herbs and shrubs that are often of more tropical affinity. While many associate oaks and pines with the deciduous forests of eastern North America, or the temperate forests of Europe and Asia, Mexico actually hosts the largest diversity of oaks in the world and is second (China is first) in pine species richness. This impressive diversity of temperate forest elements in Mexico is, in part, likely due to extensive Pleistocene glaciation in more northern portions of North America, causing the dispersal and establishment of temperate taxa southward. These forest communities host numerous mutillid color syndromes, particularly the Madrean type, but females with the Western and *Timulla*-like Color Syndromes are also common. From the Trans-Mexican Volcanic Belt southward, Tropical and Black-headed *Timulla* forms become more prevalent.

Tropical dry forest formations are characterized by a large number of deciduous tree species, displaying denuded canopies during the dry season, often limited to two strata of forest trees. In Mexico, this forest type is often recognized by the high species richness of trees in the genus *Bursera*, which are noted for their beautiful swollen trunks and shiny, flaky bark. Tropical dry/deciduous forests can be found in the lowland areas between the Gulf of California and Pacific Ocean to the west and southwest, and the mountain ranges to their east and southeast. Throughout these communities you can find female mutillids that fall in the Madrean and *Timulla*-like Color Syndromes. Tropical dry forest communities that are found within and south of the Trans-Mexican Volcanic Belt will also include numerous velvet ant species that have a Tropical or Black-headed *Timulla* veneer.

Central America

Central America contains many similar ecological communities to Mexico, albeit with their own set of unique characteristics. In broad strokes, the extreme narrowness of the isthmus allows both Atlantic and Pacific maritime influences to infiltrate into the interior even more extensively, although some areas are affected by montane rain shadow effects. This means that arid to semi-arid scrub and desert environments and grasslands are absent—forests of one type or another predominate as the original vegetation. The lower latitude of Central America also means that tropical influences and taxa increase in abundance and frequency. Central America possesses the only "true" continental tropical rainforest in North America, particularly in the Chocó-Darién Region of Panama. Females with the Tropical, Madrean, *Timulla*-like, and Black-headed Color Syndromes occur throughout Central America.

Caribbean

The Caribbean is a relatively small area in terms of landmass, consisting of nearly 237,000 km^2 of islands stretching more than 1900 km in distance from north to south, compared with the rest of North America. The vast majority (over 85%) of the Caribbean's terra firma occurs in the Greater Antillean islands of Cuba, Hispaniola, Jamaica, and Puerto Rico. The region is a wonderful geologic mixture of volcanic arcs, oceanic crust that has uplifted to the surface, carbonate (limestone) platforms, and sedimentary rocks. The vegetation is also quite diverse, from beach and mangrove vegetation to arid/semi-arid thornscrub to rainforests on Cuba and Hispaniola. The velvet ants of the Caribbean are interesting, but species richness is quite limited compared with the mainland—since the ocean is an effective dispersal barrier for terrestrial organisms, and the land area of the islands is limited.

7.2 OLD AND NEW FRIENDS—GENERA IN MESOAMERICA

Most of the velvet ant genera present in the USA are also found in Mexico and Central America. The table opposite compares the number of mutillid species in the USA and in Mesoamerica for each of the genera discussed in chapters 2 to 6. Thirteen uniquely Neotropical genera are also known from the region; they are introduced here.

Dasymutillini

In the Americas, the tribe Dasymutillini is dominated by two similarly massive genera (over 100 species each): *Dasymutilla* in North America and *Traumatomutilla* in South America. They overlap only narrowly, with two *Traumatomutilla* in Central America and four *Dasymutilla* in northern Colombia. There is one mid-sized genus, *Lomachaeta*, found throughout the Americas. After this, there are nine small genera (fewer than six species each) in South America, with only one of these, *Frigitilla*, extending north into Panama.

7.2.1—*Frigitilla*

ETYMOLOGY From the Latin *frigus* "cold," with *-tilla*, a common suffix for mutillid genera. The type species of this genus used to be named *Frigitilla frigudula*, but an older name was discovered for this species, and it is now known as *Frigitilla stimulatrix*. IDENTIFICATION Females of this genus are similar to *Dasymutilla* and *Traumatomutilla* but have the pygidium very slender and poorly defined. Males have S2 swollen and convex with a setae-free mesal pit. SPECIES RICHNESS There is one species in the Brazilian Amazon, *F. stimulatrix*, and one species in Panama, *F. panamensis*. NOTES All known specimens of *F. stimulatrix* in museums have dull yellow spots on T2. Live individuals, however, tend to have orange spots on T2. The color of T2 spots in these recently collected specimens subsequently faded to yellow after the specimen was pinned and stored (Bartholomay et al. 2015). It is not clear how many unrelated species that we currently think of as having yellow spots are

Genus	USA species	Mesoamerican species	Identification resource
Myrmosula	7	1	Wasbauer 1973
Timulla	29	74 (7 also in USA)	Mickel 1938
Ephuta	30	30 (4 also in USA)	**None available**
Dasymutilla	87	119 (47 also in USA)	Manley & Pitts 2007
Lomachaeta	15	11 (7 also in USA)	Williams, Cambra, Bartholomay et al. 2019
Invreiella	2	15 (2 also in USA)	Waldren et al. 2020
Pseudomethoca	40	77 (10 also in USA)	**None available**
Acanthophotopsis	5	3 (2 also in USA)	Tanner et al. 2009
Acrophotopsis	2	4 (2 also in USA)	Pitts & McHugh 2002
Dilophotopsis	3	3 (3 also in USA)	Wilson & Pitts 2008
Laminatilla	1	3 (1 also in USA)	Pitts 2007b
Odontophotopsis	55	30 (20 also in USA)	Schuster 1958
Photomorphus	40	15 (10 also in USA)	Schuster 1958
Protophotopsis	1	2 (1 also in USA)	Cambra & Quintero 1997
Sphaeropthalma	90	48 (28 also in USA)	Schuster 1958

different while alive, throwing some extra uncertainty into previous studies about mimicry. **KEYS AND CHARACTERS** Females can be separated by distribution and pygidial structure. Only one male is currently known, *F. stimulatrix*.

ABOVE: *Frigitilla stimulatrix* male

LEFT: *Frigitilla stimulatrix* female

7.2.2—*Traumatomutilla*

ETYMOLOGY From the Ancient Greek *trauma* "wound" and the genus name *Mutilla*, referring to the spots on T2, which looked like stab wounds to the author of the genus. **IDENTIFICATION** This genus is basically a South American version of *Dasymutilla*, and its members share most diagnostic features with that genus in both sexes. Generally their body is uniformly black with whitish setal markings, and females have obvious smooth cuticular spots on T2. **SPECIES RICHNESS** There are two species in Central America and about 100 more in South America. **NOTES** Just like in *Dasymutilla*, previously described species were separated mainly by color features that did not reflect true species-level differences, and there have been more species sunk than described in the recent treatments of both genera. **KEYS AND CHARACTERS** Species in this genus are separated by dozens of microscopic structural differences and each species-group seems to rely on different types of morphological features. The species-groups were treated by Williams et al. (2017) and keys to species in various groups are provided in the following papers: Bartholomay et al. (2018, 2020, 2021, 2022) ; Bartholomay, Williams, Lopez et al. (2019); Bartholomay, Williams, Luz et al. (2019).

Traumatomutilla dictynna female

This genus used to be a total mess, with about 150 named forms that were never incorporated into any published keys. Revising this genus was one of the goals of my PhD dissertation. I made some good progress but didn't get anywhere near publishing a revision of the giant group during my seven years of graduate school (2005–2012). Luckily, I was able to pass the torch to a talented Brazilian colleague, Pedro Bartholomay. Together, we've since been able to publish a key to the species-groups and revised 11 of the 14 species-groups, representing about 75% of the previously named forms.—KEVIN

ABOVE: *Traumatomutilla diophthalma* male

LEFT: *Traumatomutilla diophthalma* female

Pseudomethocini

In Mesoamerica, this tribe shows greater genus-level diversity than the others, and there is a good likelihood that *Pseudomethoca* will eventually be carved apart into even more genus-level groups. The genus diversity reveals a strange sort of "buddy system," whereby certain pairs of genera are linked by similarities in appearance and some unusual shared morphological features. Each of these "buddies" is separated by distinct structural features in one or both sexes. *Calomutilla* and *Pertyella* are linked by their distinct genal carina and undefined pygidium in females. *Lophomutilla* and *Lophostigma* are linked by slender body shape and undefined pygidium in females. *Hoplomutilla* and *Pappognatha* are linked by the tarsal lamella that covers the bases of the tarsal claws.

7.2.3—*Calomutilla*

ETYMOLOGY The etymology for this genus is pretty confusing. In Ancient Greek, *kalon* means a billet of wood, while *kalos* could mean "beautiful," or "rope," or "cable." As in many velvet ants, the genus name *Mutilla* is used as the suffix. **IDENTIFICATION** Like *Pertyella*, females of this genus have a sharp angular genal carina and the pygidium undefined. Unlike *Pertyella*, *Calomutilla* females have the mandible slender and apically edentate. Males of *Calomutilla* and *Pertyella* are separated from other Pseudomethocini by having the clypeus generally short and flat with small lateral teeth, the head generally broad (for a male), and the mesosomal pleura covered with short uniform setae. Differentiating males of *Calomutilla* from *Pertyella* is complicated, and useful diagnostic features are discussed by Cambra et al. (2020) and in our key to Mesoamerican genera (see 9.3.7). **SPECIES RICHNESS** There are six species in South America; one of these is also found in Panama. **NOTES** One of the Brazilian species, *C. temporalis*, was recorded as a parasite of the halictid bee *Pseudaugochlora graminea*. This genus is less diverse and less commonly collected than its apparent sister, *Pertyella*. **KEYS AND CHARACTERS** There is a good, recent key to the species in this genus by Cambra et al. (2020). The main differences

TOP LEFT AND RIGHT: *Calomutilla temporalis female*

ABOVE LEFT AND RIGHT: *Calomutilla* undetermined male

between females are color patterns and structures of the clypeus and genal carina. Only three species are known from males, and they are separated by subtle differences in the body sculpture and the genitalia.

7.2.4—*Hoplognathoca*

ETYMOLOGY From the Ancient Greek *hoplites* "armed" and *gnathos* "jaw," in reference to the large ventral mandibular tooth in most females. **IDENTIFICATION** Females are like large *Pseudomethoca*, with a small genal tooth, large postgenal tooth, and (usually) large ventral mandibular tooth; they always have a very large head and two cuticular spots on T2. Males are like large colorful *Pseudomethoca*, with a small postgenal tooth. **SPECIES RICHNESS** Five species in Central America and one additional northern South American species. **NOTES** Together with *Hoplomutilla* and a few species of *Pseudomethoca*, these are the largest-bodied Pseudomethocini in Central America. They occur mainly in tropical forests at low to moderate elevations. **KEYS AND CHARACTERS** Females are separated by coloration and various head structures. All but one species (*H. jinotega*, 7.4.2.5) is treated in a key by Suárez (1963). Only two of the males are known; they can be separated by metasomal setal color (mostly golden in *H. costarricensis* versus black and white in *H. nodifrons*).

LEFT: *Hoplognathoca guatemalteca* female

BELOW LEFT: *Hoplognathoca costarricensis* female

BELOW RIGHT: *Hoplognathoca costarricensis* male

7.2.5—*Hoplomutilla*

ETYMOLOGY From the Ancient Greek *hoplites* "armed" and the genus name *Mutilla* used as a suffix. When the genus was first described, it included members of *Hoplocrates*, which have spines under the head, but the type species and its relatives (which have the head unarmed) kept the name. **IDENTIFICATION** These are large robust velvet ants. Both sexes have the mandible mostly bare and a lamella covering the base of the tarsal claws. The females usually have two distinct spots with lots of parallel carinae on T2 and the head narrower than the mesosoma. The males usually have a large lateral swelling or tooth on the mesopleuron below the wing. **SPECIES RICHNESS** Three species in Central America and nearly 100 in South America. **NOTES** These are big and colorful species that have been associated with the large orchid bees *Eulaema meriana* and *E. nigrita*, and other unrelated large bee hosts. Many of the males, including *H. opima* from Trinidad (pictured here), have metallic bluish color reflections on the metasoma. **KEYS AND CHARACTERS** Mickel's (1939a) key is still the best for both sexes, but some names are outdated, and some more recently described species are missing from the key. Color patterns, setae and carinae on the T2 spots, and various microscopic features are useful for separating females. Microscopic differences in the mandible, antenna, and mesoscutellum, along with setal color differences, are useful for separating males.

TOP LEFT: *Hoplomutilla opima* female

ABOVE: *Hoplomutilla opima* male

LEFT: *Hoplomutilla fraterna* female

7.2.6—*Horcomutilla*

ETYMOLOGY From the Ancient Greek *horkos* "oath or solemn pledge" and the genus name *Mutilla* used as a suffix. **IDENTIFICATION** The females are similar to *Pseudomethoca*, except that they have a swollen protuberance on the frons and have T1 very wide. Males have mesosternal tubercles and a bizarrely forked genitalic paramere. **SPECIES RICHNESS** There are 13 species in South America, one of which also occurs in Panama. **NOTES** This genus is usually rare in collections. Some of the South

American species have large antler-like frons projections, while others have them reduced to shallow swollen bumps. **KEYS AND CHARACTERS** The genus was recently revised by Cambra et al. (2022). Females are separated mainly by the structure of head armature and color patterns. The males are separated mainly by their genitalia, but mandible color and shape of the mesosternal tubercles can also be useful.

TOP LEFT: *Horcomutilla krombeini* female

ABOVE: *Horcomutilla krombeini* male genitalia

LEFT: *Horcomutilla krombeini* male

Horcomutilla piala

Horcomutilla glabriceps

7.2.7—*Lophomutilla*

ETYMOLOGY From the Ancient Greek *lophos* "crest" and the genus name *Mutilla*, referencing the raised carinae on T2 in many species. **IDENTIFICATION** Females of this genus and *Lophostigma* are generally slender *Pseudomethoca* relatives with the gena unarmed and the pygidium undefined. *Lophomutilla* has a dilated, generally tridentate mandible and the T1 shape subsessile, while *Lophostigma* has a slender mandible and the T1 shape disciform. Males of *Lophomutilla* and *Lophostigma* are generally slender and have the epaulets raised to sharp teeth; in *Lophomutilla* the hypopygium is rounded, while *Lophostigma* has a notch in the hypopygium. **SPECIES RICHNESS** Three species in Central America and over 20 in South America. **NOTES** *Lophomutilla* species have been associated with various genera of sweat bee (Halictidae) hosts. The genus is named after the large crests on T2 in females, but this feature is seen in only about half of the species. **KEYS AND CHARACTERS** Keys to the South American females were written by Fritz (1990) and Fritz & Pagliano (1993). The Central American

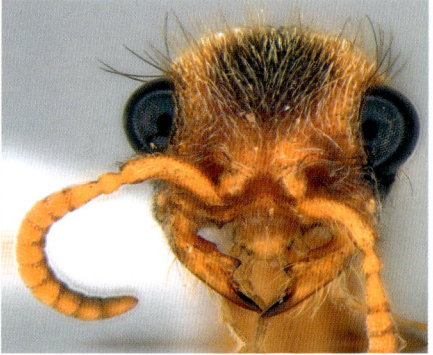

Lophomutilla female from Costa Rica; note the dilated mandible (right).

Lophomutilla calva male

Lophomutilla calva female

species were not included in these, or other, officially published keys. In fact, most of them have not been formally transferred into the genus (they were treated as members of *Lophomutilla* in an unpublished graduate thesis by Cambra [1996]). Only three males are recognized, only one is formally associated with a female, and no keys have been published to separate these wasps.

7.2.8—*Lophostigma*

ETYMOLOGY From the Ancient Greek *lophos* "crest" and *stigma* "mark or brand," referencing the raised carinae located in contrasting color patches on T2 in many species. **IDENTIFICATION** Females of this genus and *Lophomutilla* are generally slender *Pseudomethoca* relatives with the gena unarmed and the pygidium undefined. *Lophomutilla* has dilated, generally tridentate mandible and the T1 shape subsessile, while *Lophostigma* has a slender mandible and the T1 shape disciform. Males of *Lophomutilla* and *Lophostigma* are generally slender and have the epaulets raised to sharp teeth; in *Lophomutilla* the hypopygium is rounded, while *Lophostigma* has a notch in the hypopygium. **SPECIES RICHNESS** Three Central American species and about five more in South America. **NOTES** In both sexes, *Lophostigma* are similar to the preceding genus in general body shape, coloration, and distribution. *Lophostigma* are less abundant and less diverse than *Lophomutilla*. In *Lophostigma cincta*, the intraspecific color variation of females is remarkable. This species has been recorded as a parasite of the nocturnal bee genus *Megalopta*. **KEYS AND CHARACTERS** The Central American species were keyed by Cambra & Quintero (1996). They are separated mainly by sometimes subtle, and often microscopic, differences in metasoma shape and mesosomal punctation. Only one male, *L. cincta*, has been associated with a female and described.

Lophostigma cincta female

Lophostigma cincta female

TOP LEFT: *Lophostigma* male

TOP RIGHT: *Lophostigma* female; note the slender mandible

RIGHT: *Lophostigma* male; note the dentate epaulets

7.2.9—*Pappognatha*

ETYMOLOGY From the Ancient Greek *pappo* "woolly" and *gnathos* "jaw," in reference to the densely setose mandible. **IDENTIFICATION** Both sexes can be recognized by having the mandible entirely covered with short, dense setae. They are similar to *Hoplomutilla* in appearance and even have a lamella covering the tarsal claws like that genus. **SPECIES RICHNESS** Two species in Central America and 13 more in South America. **NOTES** *Pappognatha* species are parasites of orchid bees (Apidae: Euglossini). One of the Central American species, *Pa. myrmiciformis* (8.2.7.1), is a convincing co-mimic with the ant species *Camponotus sericeiventris*. The other Central American species, *Pa. panamensis*, looks like many *Hoplomutilla* and is pictured in figure 7.4.5.2. **KEYS AND CHARACTERS** The best key for females was written by Mickel (1939b). Females are separated by differences in color, and the setae and sculpture of the T2 spots. Males were keyed by Quintero & Cambra (2005); they are separated by numerous microscopic differences, especially in the clypeus and mandible, and by some color differences.

7.2.10—*Pertyella*

ETYMOLOGY This genus was named after the German entomologist Maximilian Perty. **IDENTIFICATION** Like *Calomutilla*, females of this genus have a sharp, angular genal carina and the pygidium undefined. Unlike *Calomutilla*, *Pertyella* females have the mandible broad and tridentate apically. Males of these genera are harder to tell apart, with subtle differences in the mandible shape and genitalia. **SPECIES RICHNESS** Four species in Central America, and nearly 20 more species in South America. **NOTES** This genus is more commonly encountered and more diverse than the related *Calomutilla*. Even though they are widespread in the Neotropical region and relatively common in collections, very little is known about their biology. **KEYS AND CHARACTERS** Females are separated mainly by color differences. Fritz & Pagliano (1993) has a key that includes most of the Central American species. Very few males are recognized, none is formally associated with a female, and no keys have been published.

Pappognatha patruelis:
female (ABOVE),
male (ABOVE RIGHT), and
densely setose mandible (RIGHT)

BELOW LEFT AND RIGHT: *Pertyella salutatrix* female

BOTTOM LEFT AND RIGHT: *Pertyella* undetermined male

7.2.11—*Silvorientilla*

ETYMOLOGY From the Latin *silva* "forest" and *orient* "east," with a common suffix for mutillid genera. This genus was jointly named for the tropical forests of eastern Mexico, where the type occurs, and in honor of film actor and director Clint Eastwood. **IDENTIFICATION** Females of this genus have the pygidium undefined and T1 clearly disciform, usually with a distinct dorsomesal tooth. Males are strange, with a flattened head and bizarre genitalia that have a strange swelling near the paramere apex. **SPECIES RICHNESS** Six species, all known from Central America. **NOTES** This genus is apparently restricted to Central America, with species known only from tropical Mexico to Panama. So far, no hosts have been recognized for these insects. **KEYS AND CHARACTERS** Females can usually be reliably separated by color differences, while subtle antennal and genitalic characters must be relied upon for diagnosis of males. There is a usable key in Williams, Cambra, Waldren et al. (2019).

The treatment of *Silvorientilla* is one of my favorite papers and was a good opportunity to build and revise a genus from scratch, since six of the seven species were new to science. I named the genus after Clint Eastwood and named each of the new species using themes from his movies. The etymology sections below are quoted from the original manuscript (Williams, Cambra, Waldren et al. 2019). One species, *S. philobeddoe*, is really special to me, because on a trip to Los Tuxtlas Biological Research Station in Veracruz State, Mexico, I collected the males and females together on a single unidentified bush. That week felt like a failure at first, since I collected a total of fewer than 15 mutillids at the station. About half of these velvet ants, however, belonged to this new species, and the ones I caught at Los Tuxtlas are apparently the only specimens of *S. philobeddoe* known to science.—KEVIN

Silvorientilla philobeddoe (below)

ETYMOLOGY Named for the fictional character Philo Beddoe, played by Clint Eastwood in the 1978 film *Every Which Way but Loose*.

Silvorientilla dasylymatos (right)

ETYMOLOGY From the Ancient Greek *dasy-* "hairy" and *lymatos* "dirt"; a play on words based on the 1971 film *Dirty Harry*.

Silvorientilla bonus (below left)

ETYMOLOGY From the Latin *bonus* "good" and simultaneously from the colloquial American English term *bonus*, meaning something welcomed and usually unexpected that accompanies something that is itself already good. This was the final new species added to this study after most of the manuscript was already written. It is named for the 1966 film *The Good, the Bad, and the Ugly*, wherein Clint Eastwood's character, Blondie, was "the good."

Silvorientilla sinenomine (above right)

ETYMOLOGY From the Latin *sine* "without" and *nomen* "name"; named for the character of the "Man with No Name" played by Clint Eastwood in the *Dollars* trilogy, which was directed by Sergio Leone in the 1960s.

Silvorientilla incondinatus (below left)

ETYMOLOGY From the Latin prefix *in-* "not" and *condinatus* "pardonable"; named for the 1992 film *Unforgiven*, in which Clint Eastwood served as director and leading actor.

Silvorientilla prosarasoror (above right)

ETYMOLOGY From the Latin terms *pro* "for" and *soror* "sister," and the name Sara; based on the 1970 film *Two Mules for Sister Sara*, starring Shirley MacLaine and Clint Eastwood and based in Mexico; also, an allusion to the type series (two males for sister Sara).

Sphaeropthalmini

Most species in this tribe are nocturnal forms from arid regions. Most of these species (and genera) occur in North American deserts. Other unique nocturnal species and genera are found in South American deserts, especially in Chile and Argentina. The tribe also includes another seven apparently diurnal genera across various habitats of South America. Three of these extend north into Mesoamerica: *Protophotopsis*, *Nanotopsis*, and *Xystromutilla*.

7.2.12—Nanotopsis

ETYMOLOGY From the Latin *nanus* "dwarf" and a common suffix used for nocturnal velvet ant genera. The name references the small body size of these wasps. **IDENTIFICATION** These are strange looking velvet ants with shallow broad reticular sculpture. The females have a very broad T1 shape, and males have T1 nearly as broad as females but strangely elongated and nearly petiolate in some views. **SPECIES RICHNESS** Two species in Central America and four in South America. **NOTES** These are rarely collected small-bodied species and females are especially rare. Of all the thousands of velvet ant specimens studied by the authors, only one *Nanotopsis* female has been examined. Males are slightly more commonly seen because of malaise traps. Still, we have seen fewer than 20 *Nanotopsis* specimens and never seen any from Central America; this male of *H. pananae* actually came from Colombia. **KEYS AND CHARACTERS** Cambra & Quintero (2010) reviewed the genus. Females are separated mainly by coloration and mesosoma shape. Males are separated by coloration and body sculpture. Distribution is also useful for recognizing these species.

ABOVE: *Nanotopsis isolatrix* female

RIGHT: *Nanotopsis pananae* male

When I was working in Curitiba, Brazil in 2013, I was able to borrow and photograph this specimen of *N. isolatrix* from the National Museum of Brazil in Rio de Janeiro. Tragically, a fire destroyed much of that collection in 2018. Luckily, this specimen and some others had not yet been returned to the museum and were still safely housed in Curitiba during the disaster. The specimen was saved and can be safely returned to the museum when facilities are available.—KEVIN

7.2.13—Xystromutilla

ETYMOLOGY From the Ancient Greek *xyster* "scraper" and *-mutilla*, a common suffix for velvet ant genera. The name is probably a reference to the bumpy, raspy base of T2 in some of the females. **IDENTIFICATION** In Central America, these are generally black diurnal species with whitish patterns of extremely bushy plumose setae and the pygidium undefined. Many of the South American species are generally dull brown in color and have obvious clusters of erect tubercles at the base of T2. **SPECIES RICHNESS** Two Central American species and 14 South American species. **NOTES** Females of this genus form two conspicuous groups, based on the presence or absence of scattered raspy tubercles

Xystromutilla turrialba

at the base of T2. All the known Mesoamerican species fit into the group without tubercles. Members of the genus have been recognized as parasites of mud- or crevice-nesting Hymenoptera in the families Apidae, Crabronidae, Pompilidae, and Sphecidae. **KEYS AND CHARACTERS** The females were keyed out by Casal (1969), and the males were keyed out by Cambra & Quintero (2004).

VELVET ANT SPECIES IN MESOAMERICA

Altogether, over 460 species have been recorded from Mesoamerica, though the total is likely much higher. Of these, about 140 also occur in the USA. For genera that were recently reviewed, new species greatly outnumbered already known species: five of six species were new to science in *Silvorientilla*, and 11 of 15 species were new to science in *Invreiella*. Most of the diverse genera have available keys to the species (though some are outdated), but there has never been a usable key published for Mesoamerican *Pseudomethoca* or *Ephuta* species.

For those genera, more than half of examined specimens do not seem to fit any of the currently described species and are apparently new to science. Among genera with older published revisions, like *Timulla* and *Sphaeropthalma*, numerous undescribed species are also found in collections. Even in the widely studied genus *Dasymutilla*, which was reviewed in 2007, a new species was described in 2016 (*Dasymutilla paraparadoxa*: 7.4.3.1), and, since then, two more new species have been found from Mexico, which await description.

Even if there were an additional 200 pages in this book, it would be impossible to provide a guide for all the Mesoamerican species. Instead, we provide photographs and information on a relatively small sample of species. Firstly (7.3), we present some species that match North American color syndromes and extend southward into mostly Nearctic portions of Mesoamerica. Secondly (7.4), we introduce some species clusters in Neotropical Mexico and Central America with color syndromes that do not occur in North America. Thirdly (7.5), we provide examples of three broadly defined clusters on the Caribbean islands. We hope these accounts are fun to look at and might allow a few of the more common species to be recognized by collectors. We also hope this encourages readers to jump into studying velvet ants in Mesoamerica.

7.3 NORTH AMERICAN COLOR SYNDROMES IN MESOAMERICA

CRYPTIC FEMALES COLOR SYNDROME (FCS)—Small-bodied and nocturnal species around the whole world often express the Cryptic FCS. They are generally impossible to identify from photographs, and the Mesoamerican Cryptic species are not illustrated here.

EASTERN FCS—This color syndrome does not seem to expand into Mesoamerica, except in a few cases near the Mexican border with Texas. Some ostensibly Eastern-looking forms occur in Mesoamerica, but they are intermediate with the Madrean or *Timulla*-like Color Syndromes there.

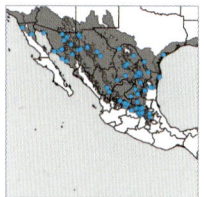

TEXAN FCS—Members of the Texan Color Syndrome occurring in the USA frequently expand south into Mexico, particularly in the Chihuahuan Desert, such as *Dasymutilla klugii*. There are no known Texan Color Syndrome species that are unique to Mesoamerica.

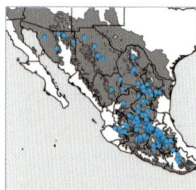

WESTERN FCS—Many Western species extend south into mountainous areas of Mexico, especially *D. vestita*. Furthermore, the most commonly encountered species in Mexico, *D. erythrina* (below left), is only rarely seen in mountainous areas of Arizona. Lastly, the genus *Invreiella* (below right) has multiple Western FCS species in mountainous areas of Mexico.

DESERT FCS—Most species from the Desert Color Syndrome, which were treated in chapter 3, are also found in Mesoamerica, like *D. gloriosa* and *D. magna*. Mexico has extensive unique arid habitats, however, and there are uniquely Mesoamerican Desert form species in Baja California Sur, Sinaloa, and Oaxaca, such as *D. scabra* (below left), *D. pallene* (below right), and *D. melanargyrea*.

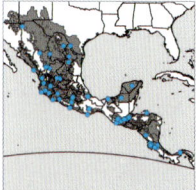

TIMULLA-LIKE FCS—There are only about 15 *Timulla*-like FCS species in the USA, mostly occurring in Texas and the Great Plains. Many more Mesoamerican species in the genera *Timulla*, *Pseudomethoca*, and *Invreiella* occur from subtropical portions of Mexico south into Central America.

Pseudomethoca chontalensis female   *Timulla belti* female

MADREAN FCS—Most of the Madrean species from chapter 3 are also found in Mexico, and numerous additional Madrean species are found throughout Mesoamerica. The Madrean Color Syndrome generally shows greater variation than the other color syndromes. They can be arranged in three broad categories.

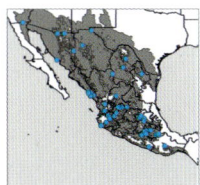

1) Black body with whitish setae and partly reddish metasoma, like *D. sicheliana* (3.4.6.1). Most species of this sort are also found in Arizona, but a few are unique to subtropical regions of Mexico, like *D. monstrosa* (below left) and *D. fimbriata* (below right).

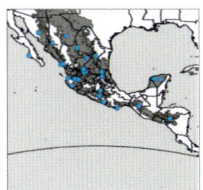

2) Reddish-brown body with golden head, like *D. dilucida* (3.4.3.1) or *Ps. contumax* (3.4.11.1). These species are abundant and widespread in mountainous areas of Mexico, with some similarly colored species in Central America. Some notable endemic Mexican examples include *D. formosa* (below left) and *T. rhanis* (below right).

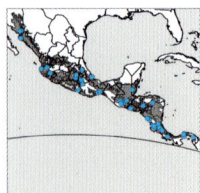

3) Leftovers with variable patterns of silver, white, gold, yellow, orange, red, brown, and/or black markings, like *D. heliophila* (3.4.1.1) and *Ps. bethae* (3.4.13.1). Females treated here are often somewhat intermediate with the Eastern and *Timulla*-like Color Syndromes. There are dozens of species that fit this loose category throughout Mexico and Central America, like *Ephuta cinaloa* (below left), *Lophostigma cincta* (7.2.8), *Pertyella salutatrix* (7.2.10), and *Pseudomethoca petricola* (below right).

7.4 NEW COLOR SYNDROMES IN NEOTROPICAL MESOAMERICA

In their study of Müllerian mimicry in North America, Wilson et al. (2015) recognized eight mimicry rings, now treated as Female Color Syndromes (FCS), excluding the cryptic forms. Six of these rings are abundant in the USA, and their species are treated in chapter 3. Two of the mimicry rings, however, occur only in the Neotropical portions of North America. These include the Black-headed and Tropical Color Syndromes.

7.4.1 *Pertyella beata* cluster

7.4.1.1—*Pertyella beata*

ETYMOLOGY From the Latin *beatus* "happy" or "fortunate."

This is an uncommon species, and I felt fortunate to study and identify the specimen photographed here.—KEVIN

HOW TO RECOGNIZE THIS CLUSTER These are the typical form for the Black-headed Color Syndrome in Central America. In Mesoamerica, there are about 30 *Timulla* and 20 *Pseudomethoca* species, and a few species from other genera with this general coloration. This generalized form is the most commonly seen color syndrome in female velvet ants around the world. Studies in East Asia recognized over 220 species loosely fitting this FCS (8.3.7); in Africa, over 160 species with this FCS (8.3.8) have been studied. Hundreds more species like this are known from other continents as well. **MALE** Relatively few males are recognized in these groups, but they generally match the *Timulla*-like or *Ephuta*-like MCS. **NOTES** This is a rare species; it was not even included in the best available key to *Pertyella* species by Fritz & Pagliano (1993).

7.4.1.2—*Pseudomethoca masneri*

ETYMOLOGY This species was named after Lubomir Masner, a superstar parasitic Hymenoptera researcher. **NOTES** This common species has two color forms: most specimens have silver setal spots on T2, but some have bright golden metasomal markings.

7.4.1.3—*Dasymutilla colorado*

ETYMOLOGY From the Spanish *colorado* "red colored," in reference to the mesosomal color and the type locality of Barro Colorado Island in Panama. **NOTES** There is some uncertainty about whether this species belongs to *Dasymutilla* or *Traumatomutilla*, but it was placed in *Dasymutilla* because that's an older name, most *Traumatomutilla* have a black mesosoma, and *Dasymutilla* is more common in Central America.

7.4.1.4—*Timulla tumidula*

ETYMOLOGY From the Latin *tumidus* "swelling" or "tumor," likely in reference to the male's raised mesoscutellum. **NOTES** This is one of about 30 similarly colored *Timulla* females in Mesoamerica; there are at least 40 more in South America. The female of this species was only recently recognized, based on mating pairs found in museum collections.

7.4.2 *Dasymutilla cressoni* cluster
7.4.2.1—*Dasymutilla cressoni*

ETYMOLOGY This species was named after the entomologist Ezra T. Cresson, who described many velvet ants in the 19th century. **HOW TO RECOGNIZE THIS CLUSTER** These species have a black body with extensive yellow to golden setal or cuticular markings. They range from tropical Mexico south to Panama. This color pattern is most commonly seen in *Dasymutilla* (about 13 species)

7.4.3.4—*Timulla gaumeri*

ETYMOLOGY This species was named after George F. Gaumer, the entomologist who collected the holotype. **NOTES** All but three *Timulla* species in the New World have the mesosoma pale reddish, and all but about 10 have silvery or white setal markings. The combination of a dark mesosoma and golden setal markings are totally unique in this genus and almost certainly came about from Müllerian mimicry.

7.4.3.5—*Pseudomethoca cleonica*

ETYMOLOGY This species was named after Cleon, an Athenian general during the Peloponnesian War. **NOTES** While most species in this cluster, including the three *Dasymutilla* species above and *Hoplognathoca guatemalteca* (7.2.4), have their orange T2 spots formed by cuticle color, this species and *T. gaumeri* have the spots formed from golden-orange setae.

7.4.4—*Dasymutilla araneoides* cluster
7.4.4.1—*Dasymutilla araneoides*

ETYMOLOGY From the Latin *araneus* "spider" and *-oides* "similar to." The name simultaneously references the general spider-like appearance and its close relationship to *D. arachnoides* (3.4.7.2). **HOW TO RECOGNIZE THIS CLUSTER** These species have a black body with orange anterior markings and white posterior markings. They are found mainly in Costa Rica, Nicaragua, and Panama. **MALE** Some males in this cluster are known; most of them are similar to the *Timulla*-like MCS (like *D. araneoides*), but at least one (*D. spilota*) has the *Sphaeropthalma*-like MCS. **NOTES** This is the most common species with this pattern. It is closely related to *D. arachnoides* (3.4.7.2), *D. asteria* (3.4.6.2), and *D. sicheliana* (3.4.6.1).

7.4.4.2—*Dasymutilla canina*

ETYMOLOGY From the Latin *caninus* "pertaining to a dog." In some Spanish-speaking countries, velvet ants are referred to as *perritos de Dios* "God's little dogs." **NOTES** This species is widespread in Mexico and Central America and expresses a lot of color variation. Some populations look more like *D. cressoni* (7.4.2.1), and others are similar to *D. monstrosa* (p. 323). This is the most typical color form in the southern part of their range.

This was the first velvet ant to sting me. It was not pleasant, but not as bad as I expected. There was sharp pain for about 30 seconds; then, after two minutes, there was no lingering trace of pain or itching.—KEVIN

Dasymutilla canina

7.4.4.3—*Dasymutilla spilota*

ETYMOLOGY From the Ancient Greek *spilota* "spotted," in reference to the whitish setal spots on T2. The male used to be named *D. ionothorax* in reference to the purple-reddish mesosomal tint in some populations. **NOTES** There is color variation in both males and females of this species, with northern populations having paler golden setae in females and a more extensively blackened mesosoma in males. The male was only recently recognized in a treatment of the species from Panama.

7.4.4.4—*Pseudomethoca sermenoi*

ETYMOLOGY This species was named after the El Salvadorean entomologist Jose M. Sermeño, who collected some of the paratypes. **NOTES** This species has strange body proportions, with a more slender head than most *Pseudomethoca*; it is one of many species that might eventually merit their own new genus placement.

7.4.5—*Hoplomutilla xanthocerata* cluster

7.4.5.1—*Hoplomutilla xanthocerata*

ETYMOLOGY From the Ancient Greek *xanthos* "yellow" and *keratos* "horn," in reference to the yellow antennae. **HOW TO RECOGNIZE THIS CLUSTER** This wasp has a black body with small whitish or pale yellow markings. They are mostly known from Costa Rica and Panama. **MALE**

Males are known for only a few of the species in this cluster, they generally match the *Ephuta*-like MCS, like the male of *H. excentrica*. **NOTES** This large species is a parasitoid of the orchid bee *Eulaema meriana*. The male is not yet recognized, but it should be pretty easy to figure out based on the large body size and the limited number of Central American *Hoplomutilla* species.

Hoplomutilla xanthocerata is the only species in Central America with yellow antennae, but two species from Colombia have yellow antennae, as well: *H. insignis* and *H. aurigena*.

7.4.5.2—Pappognatha panamensis

ETYMOLOGY This species is named for being from Panama. **NOTES** As in other *Pappognatha* species, males are less common than females, even in areas where malaise traps are widely used. To our knowledge, only one male has ever been collected (we have never seen one), but about 15 females have been studied.

7.4.5.3—Pseudomethoca tetraspilota

ETYMOLOGY From the Ancient Greek *tetra* "four" and *spilota* "spotted," in reference to the four yellow spots on T2. **NOTES** This species is superficially similar to *Hoplomutilla* but has four yellow spots on T2.

This is one of the first "exotic" velvet ants that I tried to identify as an undergraduate student at UC Davis back in 2003. I found a good key to the genus *Hoplomutilla* and went to work. Needless to say, I didn't get far. Luckily, I didn't give up and was finally able to identify the creature after Cambra & Quintero (2008) wrote a good treatment of some *Pseudomethoca* species in Central America.—KEVIN

7.4.5.4—Hoplomutilla excentrica

ETYMOLOGY From the Latin *ex-* "out of" and *centrum* "midpoint" or "spine." It is not clear whether the author, Cameron (1895), thought the species was something spectacular or if he was referencing the spines on the male's mesosoma. The female of this species used to be named *H. panamensis*, after the country of origin. **NOTES** Quintero & Cambra (2001a) recognized the sex association

ABOVE RIGHT: *Hoplomutilla excentrica* female

BELOW RIGHT: *Hoplomutilla excentrica* male

BELOW LEFT: *Hoplomutilla gabbii* female

by placing a female in a wire cage and capturing a male that flew to her. In Central America, two of the four *Hoplomutilla* females have an all-black head. They can be separated by antenna color (black in *H. excentrica* and yellow in *H. xanthocerata*). For the other two species, *H. gabbii* is identical to *H. excentrica* except for the white spots on the head, and *H. fraterna* (7.2.5) can be recognized by the larger oblique yellow spots on T2 and the divided white setal spots on the mesonotum.

7.5 CARIBBEAN CLUSTERS

Coloration of members of the Caribbean clusters is interesting. Some appear to follow the Black-headed form, others look somewhat like the Madrean type, others resemble the Tropical forms of Central and South America, and yet a few others vary drastically from all of the typical syndromes on the continents. The Caribbean is a vibrant mix of many syndrome types, and most islands do not have a large velvet ant fauna to strictly adhere to a single color syndrome. Even when they do look similar, there are embellishments like different coloration of fringing setae, for example, that distinguish them from mainlanders. One of our papers (Pan et al. 2017) looked at some Caribbean *Dasymutilla* color patterns and provided some interesting evidence that their flashy "fashion" may provide an aposematic warning system suited toward alerting iguanian lizards, particularly anoles (a common and speciose group of visually attuned sit-and-wait invertebrate predators of the islands), that mutillids are not ants and should be left alone. We strongly encourage others to do more work in the Caribbean on velvet ants, perhaps with a net in one hand and a daiquiri in the other.

7.5.1—*Timulla senex* cluster

7.5.1.1—*Timulla senex*

ETYMOLOGY From the Latin *senex* "elderly," likely in reference to the pale "hair" on the head, which looks shaggier and whiter on males. Similarly, the female used to carry the name *palliceps* (pale headed). **HOW TO RECOGNIZE THIS CLUSTER** These are the Madrean FCS species that occur in Caribbean islands. Most of them have a distinctly golden-tinted head and reddish-brown body with silvery and blackish markings. **MALE** Males are recognized for *Ps. argyrocephala*, which shows the *Ephuta*-like Color Syndrome, and *T. senex*, which somewhat resembles its female (and the other females in this complex). **NOTES** This species has strange coloration for the genus *Timulla*, with its bright golden head and black basal spot on T2 in females. The male's coloration is equally strange, and, to us, males and females of *T. senex* are more similar to each other than are matched sexes of any other *Timulla* species. Numerous other *Timulla* species are known from Caribbe-

an islands, but they all have more typical coloration for the genus: females with the Black-headed Color Syndrome (like *T. tumidula*, 7.4.1.4) and males with the typical *Timulla*-like MCS (5.5.14).

7.5.1.2—*Ephuta festata*

ETYMOLOGY From the Latin *festus* "joyful," likely in reference to the lively and colorful appearance. **NOTES** This is a really beautiful little wasp. The co-mimicry between this and *T. senex* (7.5.1.1) is remarkable: the entirely pale golden head, the same extent of darkening around the sides of the

mesosoma, the apically darkened metasoma with similar setal marks, and the small black spot on T2 with lateral silver streaks; even the mesosoma shape of *E. festata* is expanded posteriorly, unlike that of other *Ephuta* species—this is a textbook example of perfect mimicry among distantly related taxa.

7.5.1.3—*Pseudomethoca salti*

ETYMOLOGY This species was named after George Salt, who collected the type of this species and many other important mutillid specimens in Cuba, including the male of *T. senex* (7.5.1.1). **NOTES** This is yet another Cuban species that shares similar coloration with *E. festata* and *T. senex*.

7.5.1.4—*Pseudomethoca argyrocephala*

ETYMOLOGY From the Ancient Greek *argyros* "silver" and *kephale* "head," referencing the head color, although in most specimens it appears more gold than silver. **NOTES** This species has been found on multiple Caribbean islands, including St. John, Guana Island, Isla de Mona, and Puerto Rico. The males and females were collected together as they inspected nests of various ground-nesting bees and spheciform wasps in Puerto Rico.

7.5.2—*Dasymutilla militaris* cluster

7.5.2.1—*Dasymutilla militaris*

ETYMOLOGY From the Latin *militaris* "of a soldier" or "warlike"; it is not clear why the author chose this name. One of the two subspecies is named *D. militaris nigriceps*, in reference to the black head, even though both subspecies have a black head; this subspecies is unique in having the

Dasymutilla militaris militaris female   *Dasymutilla militaris nigriceps* female

Dasymutilla militaris male

Dasymutilla dominica fossil

metasoma with mostly black setae. **HOW TO RECOGNIZE THIS CLUSTER** These are mostly orange-red wasps with black heads and various black and whitish patterns on the body. **MALE** Most of the males are known; some are similar to the Eastern Color Syndrome males from chapter 5, while others have the *Ephuta*-like Color Syndrome. **NOTES** This is a beautiful wasp, which ranges from a dark, almost purple-red to a faint pinkish color. The fossil species *Dasymutilla dominica*, found in Dominican amber (~20 million years old), is nearly identical to the extant males of *D. militaris*.

7.5.2.2—*Dasymutilla melancholica*

ETYMOLOGY From the Ancient Greek *melank-holos* "melancholic" or, more literally, "dipped in black bile." The sad name of this species is somewhat confusing, since the erect brushes of setae on the propodeum are pretty flashy, but in many specimens the body color is a somewhat subdued dark reddish-brown. **NOTES** This species is basically the same as *D. militaris* but has long white setae around the propodeum. Currently, the male is unknown, but one field guide showed a museum specimen of the female alongside a large dark-colored male. This male does not look like anything we have seen in the Caribbean area; rather, it looks eerily similar to many of the *Traumatomutilla* species that are found in Argentina.

7.5.2.3—*Dasymutilla insulana*

ETYMOLOGY From the Latin *insulanus* "of islands," in reference to its presence in Cuba. **NOTES** This species and *D. bouvieri* are close relatives of *D. bioculata* (3.2.8.1, 5.6.4.3), which is one of the most widespread and variable species in North America.

7.5.2.4—*Dasymutilla bouvieri*

ETYMOLOGY This species was named after the French entomologist Eugène L. Bouvier. **NOTES** This species from Hispaniola is closely related to *D. insulana* but has darker color and longer white setae, making it superficially similar to *D. melancholica* specimens from Haiti.

	Mesonotum with tooth	Mesonotum rounded
Propodeum with sparse setae	*Dasymutilla militaris*	*Dasymutilla insulana*
Propodeum with dense white setae	*Dasymutilla melancholica*	*Dasymutilla bouvieri*

7.5.3—*Dasymutilla cargilli* cluster

7.5.3.1—*Dasymutilla cargilli*

ETYMOLOGY The species was named after one Dr. Cargill, who collected the type specimen. **HOW TO RECOGNIZE THIS CLUSTER** These are leftover Caribbean species that are generally dark-bodied, or at least have the mesosoma blackish. **MALE** Males are known for some of these: *Ephuta prima* approximates the *Sphaeropthalma*-like Color Syndrome from chapter 5, while *D. cargilli* and *Ps. olgae* are similar to the *Ephuta*-like Color Syndrome. **NOTES** This species has been treated in five different genera throughout history: *Sphaeropthalma*, *Ephuta*, *Dasymutilla*, *Pseudomethoca*, and even its own monotypic genus, *Jamaitilla*. The most recent treatment by Quintero & Cambra (2001a) returned this species to *Dasymutilla*, but *D. cargilli* seems to be missing some important diagnostic features for that genus. The genus placement seems likely to change again in the future.

7.5.3.2—*Ephuta prima*

ETYMOLOGY From the Latin *primus* "first," in reference to this being the first *Ephuta* species that was recognized from the Dominican Republic. **NOTES** Males and females of this species are very different in appearance, with males being largely reddish but females entirely black, except the yellowish head. This species is somewhat similar in appearance to a fossil species, *Ephuta clavigera*, from Dominican amber (~20 million years old).

7.5.3.3—*Pseudomethoca olgae*

ETYMOLOGY The describer of this species named it after his wife, Olga M. Schuster. **NOTES** This species was discovered on the island of St. Croix in the US Virgin Islands.

7.5.3.4—*Pseudomethoca merengue*

ETYMOLOGY This species was named after a musical genre, merengue, from the home country of this species, the Dominican Republic. **NOTES** This species does not fit obviously into any previously studied color syndrome. The color combination of a red head, black mesosoma, and red metasoma is not really seen anywhere else in the

world, which is strange given that the combination of a black head, reddish mesosoma, and black metasoma is a color syndrome shared by thousands of velvet ants around the world.

VELVET ANTS AROUND THE WORLD

8.1 GLOBAL DIVERSITY AND DISTRIBUTION

8.1.1—Biogeographic realms

Naturalists divide the world's terrestrial life forms across eight biogeographic realms. Velvet ants are absent from two of these realms, the Antarctic and Oceanian (depending on how the borders are drawn—a few velvet ants can be found in Vanuatu and on other Australasian islands). The velvet ants from temperate North America treated in this book are often considered the most famous and attractive, but less than 10% of the world's known velvet ant species occur in the USA (largely representative of the Nearctic realm). Table 8.1.1 shows an approximate number of named species and subspecies in each realm (this number fluctuates frequently, with each new published paper). How each subfamily and tribe is applied to the table, and additional information, can be found on the following pages (see 8.1.2).

The best treatment for Nearctic species is the book you are reading now (hard copies can also be used as a paperweight or to hide emergency cash—no one will suspect that you have stashed a one hundred dollar bill between pages 46 and 47), with the caveat that the nocturnal species are not fully keyed out here. James Pitts and various coauthors have written keys to species of some nocturnal genera, species-groups within larger genera, or regional faunas that are discussed in chapter 6 (6.1).

The Neotropical species were most recently cataloged by Nonveiller et al. (1990). Many genera were revised by Clarence Mickel in the 1930s and 1940s or Osvaldo Casal in the 1960s, and those keys work relatively well. Since the 1990s many authors have published useful identification resources for various other genera. There are about 60 recognized genera, which were keyed most recently by Brothers (2006).

The Australasian mutillids are perhaps the most neglected, but inroads are being made by various authors. The only identification keys available treat either small new genera (e.g., Brothers 1971,

Biogeographic realm	Total species	Subfamilies and tribes, sorted by approximate number of species (Sph.=Sphaeropthalminae, Mut.=Mutillinae)
Nearctic (NA)	455	Sph.Sphaeropthalmini (200), Sph.Dasymutillini (110), Sph.Pseudomethocini (50), Sph.Ephutini (40), Mut.Trogaspidiini (40), Myrmosinae (15)
Neotropical (NT)	1325	Sph.Pseudomethocini (525), Sph.Dasymutillini (250), Sph.Ephutini (250), Sph.Sphaeropthalmini (150), Mut.Trogaspidiini (150)
Australasian (AU)	276	Sph.Dasymutillini (260), Mut.Trogaspidiini (10), Odontomutillinae (5), Dasylabrinae (1)
Afrotropical (AT)	1525	Mut.Trogaspidiini (500), Dasylabrinae (275), Mut.Ctenotillini (180), Mut.Smicromyrmini (150), Mut.Mutillini (135), Myrmillinae (130), Odontomutillinae (85), Ticoplinae (30), Rhopalomutillinae (35), Pseudophotopsidinae (5)
Oriental (OR)	720	Mut.Trogaspidiini (300), Mut.Smicromyrmini (210) Myrmillinae (90), Odontomutillinae (40), Dasylabrinae (20), Mut.Ctenotillini (15), Myrmosinae (10), Mut.Mutillini (10), Sph.Dasymutillini (5), Sph.Sphaeropthalmini (5), Ticoplinae (5), Rhopalomutillinae (5), Pseudophotopsidinae (5)
Palaearctic (PA)	599	Mut.Smicromyrmini (245), Dasylabrinae (125), Myrmillinae (75), Mut.Mutillini (45), Myrmosinae (40), Mut.Trogaspidiini (25), Pseudophotopsidinae (30), Ticoplinae (10), Mut.Ctenotillini (2), Sph.Sphaeropthalmini (2)
TOTAL	4900	Mut.Trogaspidiini (1025), Sph.Dasymutillini (625), Mut.Smicromyrmini (605), Sph.Pseudomethocini (575), Dasylabrinae (421), Sph.Sphaeropthalmini (357), Myrmillinae (295), Sph.Ephutini (290), Mut.Ctenotillini (197), Mut.Mutillini (190), Odontomutillinae (130), Myrmosinae (65), Ticoplinae (45), Pseudophotopsidinae (40), Rhopalomutillinae (40)

1994), specific taxa in specific regions (e.g., Krombein 1971), or only the new species of their author (e.g., André 1901; Turner 1914). Recently, Brothers (2022) published an excellent treatment of the type specimens for Australian mutillids, with lots of great photographs. Twelve genera are described, but the large genus *Ephutomorpha* will eventually be split into at least 20 additional genera.

The Afrotropical realm is either the most diverse for mutillids (if you count subspecies) or second most diverse, behind the Neotropical, if you count full species only. There are about 80 described genera. The most recent and complete review of the Afrotropical fauna was by Bischoff (1920), where keys to the genera and species known at the time were published. His monograph provides at least a modicum of hope of identifying species from all genera and regions. From 1973 to 1999, Guido Nonveiller revised many of the genera, providing additional useful keys to some taxa. Furthermore, a mimicry study, Wilson et al. (2018), is available that provides habitus photos for over 260 females of African species.

The Palaearctic species were most recently cataloged by Lelej (2002), who keyed the approximately 50 recognized genera. Lelej's (1985) treatment of the mutillids of Russia and Central Asia is excellent, but it is difficult for us to use since it was written in Russian—our ignorance of the Cyrillic alphabet is appalling. Other useful keys have been written for many countries or genera in Western Europe (e.g., Lelej & Schmid-Egger 2005; Pagliano 2007; Turrisi et al. 2015), but the most diverse region, North Africa, is relatively poorly understood—some habitus photos for these taxa can be found in Wilson et al. (2018).

Arkady Lelej and various coauthors have presented useful keys to many of the species of the Oriental biogeographic realm species. Most recently, these species were cataloged by Lelej (2005), who also keyed the approxiately 60 recognized genera. Other than these modern revisions, Chen's (1957) key to the Chinese species and Mickel's (1935b) key treating the Pacific Islands are the most useful papers. A study on color syndromes in Eastern Asia (Okayasu et al. 2018) discusses the dominant color patterns and includes habitus photos for nearly 60 species, including most of the genera in the region.

8.1.2—Subfamilies

Currently, the world's velvet ant fauna is subdivided into eight subfamilies, based on the morphological phylogenetic studies published by Brothers & Lelej (2017) and the molecular phylogeny by Waldren et al. (2021 in press). These reconstructions are relatively similar at the subfamily level except for three major differences. The molecular study treats Myrmosidae as a separate family outside the Mutillidae (subfamily Myrmosinae in morphological study), treats Odontomutillinae as a separate subfamily (part of tribe Mutillini in subfamily Mutillinae in morphological study), and treats Ephutini as a tribe within subfamily Sphaeropthalminae (part of tribe Mutillini in subfamily Mutillinae in morphological study). In this book, we adopt a hybrid classification, which includes nine subfamilies by maintaining the Myrmosinae within Mutillidae and recognizing Odontomutillinae as a distinct subfamily.

Four subfamilies include 65 or fewer species, three include fewer than 500 species, and two have more than 1800 species. Furthermore, five of the subfamilies were divided into tribes and two tribes were subdivided into subtribes. In the appendix (9.3.8), a key to subfamilies and tribes is presented, with a few caveats. These keys are expected to work for only about 90% of specimens; many structural exceptions cannot be incorporated into such a key without expanding it by dozens of couplets and characters. Since such groups have a relatively manageable size, tribes are not keyed for subfamilies with fewer than 500 species, and subtribes are ignored (these criteria are also applied to Table 8.1.1).

Myrmosinae (~65 species) is generally a Holarctic group, but a few species occur in the Oriental realm. They are considered the most "primitive" subfamily, always being recovered at the base of the mutillid tree in phylogenetic studies. They are considered a totally separate family by some authors, including the newest molecular phylogenetic studies. The most conspicuous primitive trait is that females have the mesosoma divided, with the pronotum forming a separate plate (see 2.0.1). The male wing venation is also considered primitive. Additionally, none of the genera has felt lines in either sex. Brothers & Lelej (2017) recognized two tribes: Myrmosini and Kudakrumiini. These tribes can be differentiated by various microscopic features, especially in the female clypeus and male pygidium. Lelej's catalogs (2002, 2005) are good for recognizing the Old World genera. All of the New World genera and most species are discussed in previous chapters of this book (see 2.6, 4.6). Lastly, most Old World fossil velvet ants belong to the subfamily Myrmosinae, with nine species from Baltic amber (~44 million years old) recognized in the extinct genus *Protomutilla*.

Pseudophotopsidinae (~40 species) are most commonly seen in the Palaearctic realm, but a few species are known from the Afrotropical and Oriental realms. Only one genus, *Pseudophotopsis*, is recognized. They exhibit many apparently primitive traits. Males, for example, have a jugal lobe in the hindwing. The females are the only mutillids, except some Myrmosinae, to retain ocelli in the female sex. Females also have the pronotum differentiated from the mesonotum by a transverse carina, giving them a somewhat intermediate appearance between Myrmosinae and the other "higher" mutillid subfamilies. The best identification resource for *Pseudophotopsis* was provided by Lelej (1980). *Pseudophotopsis* is one of three dominant groups of nocturnal mutillids in the Old World, which are discussed further in the next section (see 8.2.3).

The Ticoplinae (~45 species) subfamily is widespread but not common in the Old World. They are considered a "primitive" subfamily, being recovered near the base of the tree in most phylogenetic studies. They are presumably diurnal, based on their coloration and eye size, but their biology is poorly understood. None of the genera has a felt line on T2, but many have a sternal felt line. Male aptery or brachyptery is common in this subfamily. Brothers & Lelej (2017) recognized two tribes: Ticoplini and Smicromyrmillini. These tribes can be differentiated by the presence (Smicromyrmillini) or absence (Ticoplini) of a felt line on the second mesosomal sternite, in addition to other differences. Afrotropical and Palaearctic genera of Ticoplinae can be identified using Mitchell & Brothers (2002); the Oriental genera can be identified using Lelej (2005). This subfamily includes some of the smallest-bodied velvet ants in the world; an example of the genus *Nanomutilla* is pictured in the next section (see 8.2.4.3).

Subfamily Rhopalomutillinae (~40 species), is restricted to the Afrotropical and Oriental ecozones. Until recently, the only recognized genus was *Rhopalomutilla*. Brothers (2015) published the first part of a revision of Rhopalomutillinae, wherein he described three new genera. This initial review has great keys and information about the subfamily. Rhopalomutillinae exhibit extreme sexual dimorphism in coloration, structure, and size. Males, which carry the females during courtship and mating, are significantly larger than females. Males have large emarginate eyes and slender antennae, while females have minute eyes and short clublike antennae. According to Brothers (2015), there are numerous undescribed species in each of the four genera. Therefore, it is unwise to use any of the older keys to attempt species identification. Examples of Rhopalomutillinae members can be found in the next section (see 8.2.4.1–2).

Odontomutillinae (~130 species) subfamily members are widespread and common in the Old World, except the Palaearctic. They are all apparently diurnal, and all described males are fully winged (we have seen one bizarre short-winged male from Africa). Until recently, they were treated within the Mutillinae. They are generally similar to mutillines in appearance, but the females have the mesosoma shape similar to Dasylabrinae and Sphaeropthalminae, and the males always have spines or teeth on the mesoscutellum. Various examples of Odontomutillinae are presented in the following sections (8.2.9, 8.3.8.5, 8.3.13.3, 8.3.16.4).

Myrmillinae (~295 species) species are widespread and relatively common in the Old World, except Australia. All myrmillines appear to be diurnal. Male aptery or brachyptery is common among the Myrmillinae, with nearly half of the genera having wingless or short-winged males. Generally, male Myrmillinae can be recognized by their ovate, weakly emarginate eyes, sessile first metasomal segment, and small tegulae. Female myrmillines generally have a large quadrate head, mesopleural lamella anterior to the mid-coxa, sessile first metasomal segment, and undefined pygidium. Myrmillinae is the apparent sister group to the mega-diverse Mutillinae, and many genera are difficult to place into one subfamily or the other. Examples of Myrmillinae can be found in the next sections (see 8.2.2, 8.3.7.2, and 8.3.15.4).

The Dasylabrinae (~420 species) subfamily is widespread and common in the Old World, except Australia. Most are diurnal, except for *Tricholabiodes*, which is the most species-rich nocturnal genus in the Old World (see 8.2.3e,f). Dasylabrinae members are superficially similar to those of Sphaeropthalminae, especially in female mesosoma shape. Like the Sphaeropthalminae, members of this subfamily seem to show greater color variation, especially in females, than the other subfamilies. Various examples of Dasylabrinae members are presented in the following sections (8.2.3.4–5, 8.3.8.2–3, 8.3.10.1, 8.3.11.1–2, 8.3.12.4, 8.3.13.2, 8.3.14.1–2, 8.3.15.1–3).

Sphaeropthalminae (~1845 species) is the dominant subfamily in the New World and Australia. They usually have circular eyes in both sexes, and most of the species have brachyplumose or plumose setae in various regions of the body. They are subdivided into four tribes, which are discussed in chapters 2 and 4. Currently, all the Australian Sphaeropthalminae are placed together in the tribe Dasymutillini, although many of these species lack some diagnostic features for that tribe. It is not yet clear whether they will remain in that tribe after further studies. Various examples of Dasymutillini (see 8.2.6, 8.2.10, 8.3.1, 8.3.2, 8.3.3.3, 8.3.5.1,5, 8.3.6.2–3, 8.3.9.4–6), Ephutini (see 8.3.4.1, 8.3.3.5), Pseudomethocini (see 8.2.5,7, 8.3.3.1–2,4, 8.3.4.2–4, 8.3.5.2–4, 8.3.6.1,4, 8.3.9.2–3), and Sphaeropthalmini (see 8.3.4.5, 8.3.6.5) are presented on the following pages.

Mutillinae (~2150 species) is the most diverse subfamily of Mutillidae, and they are dominant in the Old World, except Australia. There have been some minor adjustments to the tribal configuration in the recent molecular phylogeny, but they are not as cleanly organized as the tribes suggested in the morphological study. For simplicity, we discuss this subfamily using the older classification, except for the Mutillini that have been moved to other subfamilies and were treated above. They generally have ovate eyes, which are emarginate in males. Four tribes are currently recognized. Trogaspidiini is the most diverse and members occur in every biogeographic realm; females usually have two spots positioned side-by-side on T2 (examples: 2.4, 3.5.5–10, 3.6.13–17, 8.3.7.1, 8.3.8.1, 8.3.12.1–2, 8.3.14.4–5). Smicromyrmini members are abundant and diverse in the Old World, except Australia; they are generally similar to Trogaspidiini, except on T2 the females usually have one or three setal spots (examples: 8.3.7.4, 8.3.9.7–8, 8.3.10.3–5, 8.3.11.5, 8.3.13.5, 8.3.15.5–6, 8.3.16.5). Ctenotillini is a recently recognized tribe that is difficult to define, being somewhat intermediate between Myrmillinae and the other tribes of Mutillinae (examples: 8.3.7.3, 8.3.7.5, 8.3.12.5, 8.3.14.3); most of the females have a row of spines on the dorsoposterior mesosomal margin. Finally, Mutillini are widespread in the Old World, except Australia; females have a widely sessile T1 shape (examples: 8.2.1, 8.3.9.9, 8.3.10.2, 8.3.11.4, 8.3.13.4).

8.2 HIGHLIGHTED TAXA

8.2.1.1—*Mutilla europaea*

ETYMOLOGY The genus name *Mutilla* might be derived from the Latin *mutilus* "mutilated," based on the lack of wings in females. The specific epithet is based on the continent of Europe, where this species lives. **NOTES** This is the type species for the genus *Mutilla* and, consequently, the type species for all velvet ants in the family Mutillidae. It was one of the first six velvet ants described by Linnaeus in 1758, the official starting point for zoological nomenclature. These six Linnaean species are historically important, and we have made a special effort to include photographs of them all in the book (see 3.6.1.1, 5.2.1.1, 8.2.6, 8.3.5.1, 8.3.10.1, 8.3.10.2). In a surprising twist, two additional species were included in *Mutilla* by Linnaeus, but they are actually parasitic wasps in the family Ichneumonidae, now known as *Gelis acarorum* (Fig. 1.9.1c) and *Gelis formicarium*.

Mutilla europaea is one of the largest and most common species in Western Europe and is a parasite of bumblebees. In 1791, back before people understood the parasitic nature of mutillids, Johann Ludwig Christ wrote this,

> First I must mention that all the Mutillid nests I have been able to find have contained bumblebees. ... In each of these combined dissimilar societies there were males, females, and young of each species in the nest. The young of both the bumblebee and the Mutillid were found together in the cells, like children of one family, so that I was much pleased with this brotherly harmony in two so different appearing species of insects, and I would have liked to have given them the name of Damon and Pythias, if they had not already been named by Linne. (Translation found in Mickel 1928)

SIMILAR SPECIES The second most common *Mutilla* species in Europe, *M. marginata*, has a longer mesosoma and shaggier setae than *M. europaea*. A rarer European species, *M. laevigata*, has much smoother body sculpture than the others. *Mutilla mikado* is the only *Mutilla* species in East Asia; it is the largest-bodied mutillid in Japan and Korea. Other examples of *Mutilla* species can be seen in the next section (see 8.3.8.4, 8.3.11.3, 8.3.12.3, 8.3.13.1).

8.2.1.2—*Mutilla marginata* (below left)

8.2.1.3—*Mutilla laevigata* (below middle)

8.2.1.4—*Mutilla mikado* (below right)

8.2.2.1—*Myrmilla capitata*

ETYMOLOGY The genus name *Myrmilla* is likely derived from their similarity to ants, based on the Ancient Greek *myrmex* "ant." The specific epithet is likely derived from the Latin *caput* "head," in reference to the black head. **NOTES** This is one of many velvet ants with wingless males, and the commonest wingless male in Europe. About half of the species in *Myrmilla* and about half of the myrmilline genera have wingless males. Sexual dimorphism is common in species with fully winged males, but there is generally little to any difference in color between males and females in species with wingless males. This can lead to difficulty in accurately sexing these wasps; the wingless male of *Morsyma ashmeadi* from California (3.5.4.1) was even mistaken for a female by its author. With a dead specimen on hand, wingless males can be recognized the same way that most stinging Hymenoptera are sexed: females have six metasomal tergites, with seven in males; and ten antennal flagellomeres, with eleven in males. Without a microscope, it can be tougher, but there are a few hints in the body shape. Flightless males usually have longer and straighter antennae, they usually have discrete, separated plates on the mesosoma (and often have tiny wing "buds"), and, at least in Myrmillinae and Mutillinae, the mesosomal sides are more distinctly pinched-in, like the species pictured. **SIMILAR SPECIES** There are about 30 additional *Myrmilla* species in Europe and Russia. Some examples from other Myrmillinae genera in Europe are pictured here: *Bidecoloratilla leopoldina* (8.2.1.2) and *Sigilla dorsata* (8.2.1.3).

8.2.2.2—*Bidecoloratilla leopoldina*

8.2.2.3—*Sigilla dorsata*

8.2.3.1—*Pseudophotopsis komarovii*

ETYMOLOGY The genus name is based on the Ancient Greek *pseudes* "false" and *Photopsis*, a common genus name for nocturnal velvet ants. The species was named after one General Komarov, who participated in the Russian conquest of Turkmenistan in the late 19th century. **NOTES** This is the type species for the genus *Pseudophotopsis*. In arid portions of the Old World, excluding Australia, numerous unrelated velvet ant lineages are nocturnal, including *Pseudophotopsis*. One of the creepiest observations about velvet ant behavior involves an African *Pseudophotopsis* species, *Ps. continua*. In Egypt, Mellor (1927) observed a female enter the nest of a sleeping sand wasp, *Bembix olivacea* (as *Bembex mediterranea*) and bite the sleeping wasp in the neck to drink its hemolymph (blood). It is not clear whether other *Pseudophotopsis* species, or any other nocturnal mutillids, are insectine vampires too. **SIMILAR SPECIES** Other nocturnal mutillids from the Old World have similar coloration to *Pseudophotopsis,* including *Dentilla* and some other members of Smicromyrmini and *Tricholabiodes* from the subfamily Dasylabrinae. They have all converged on similar color patterns to unrelated nocturnal species in North America (see chapter 6).

8.2.3.2—*Dentilla* undetermined (below left)

8.2.3.3—*Dentilla irana* (below right)

8.2.3.4—*Tricholabiodes craspedopygius* (bottom left)

8.2.3.5—*Tricholabiodes aegyptiacus* (bottom right)

8.2.4.1—*Rhopalomutilla anguliceps*

ETYMOLOGY The genus name is based on the Ancient Greek *rhopalon* "club" and the name *Mutilla*, likely in reference to the female antennal shape. The specific epithet comes from the Latin *angulatus* "angular" and *-ceps* "headed," in reference to the male's head shape. **NOTES** Rhopalomutillinae species show some of the most extreme sexual dimorphism seen in velvet ants. The females are tiny and have especially small eyes. Females are rarely collected alone, and there is no clear evidence of their host selection. The tiny eyes and dull color suggest they spend most of their lives underground and might be parasites inside ant nests. If not for the males carrying females in flight during courtship and feeding, even less would be known of these bizarre creatures. Most of the males are black, like this *Pherotilla* from Thailand, but some African species have a reddish mesosoma, like *R. anguliceps*.

8.2.4.2—*Pherotilla* undetermined (right)

8.2.4.3—*Nanomutilla parila*

ETYMOLOGY The genus name is based on the Latin *nanus* "dwarf" and the name *Mutilla*, likely in reference to the small body size. The specific epithet may be based on the Latin *parilis* "equal" or "like," because the author, Nagy (subsequently Argaman), thought it was similar to the other species in the genus. **NOTES** Species in this genus are consistently among the smallest velvet ants in the world. Unlike the Rhopalomutillinae with their tiny females, both sexes of *Nanomutilla* are equally tiny, measuring less than 2.5 mm in body

Nanomutilla parila

length. Members of other genera in Ticoplinae are relatively small, but most of them have similar appearance to other velvet ants. A female of this species was observed attacking a leaf beetle host (Chrysomelidae: *Pachybrachis scripticollis*) by Argaman (formerly Nagy) (1988).

8.2.5.1—*Euspinolia militaris*—Panda Ant

ETYMOLOGY The genus was named after the Italian entomologist Maximilian Spinola. The specific epithet comes from the Latin *militaris* "of a soldier" or "warlike." Online, the common name "Panda Ant" has been used for this species because of the black and white coloration, and the general cuteness of this insect—although we think all velvet ants are adorable. **NOTES** At this point, no information is known about the male or the host association for this species. Based on its similar size, color, and distribution, *E. canescens* is a good candidate for the male of this species, but there are other unassociated *Euspinolia* females and males known from the same region. **SIMILAR SPECIES** The genus *Euspinolia* includes 14 species in Argentina, Chile, and Peru. Most are generally similar to *E. militaris*, but others are redder, have a nearly bald head, or have both the head and mesosoma covered in dense orange setae. This is one of the most common genera west of the Andes in South America, especially in Chile.

8.2.5.2—*Euspinolia canescens* (below right)

8.2.5.3—*Euspinolia chilensis* (below left)

8.2.5.4—*Euspinolia clypeata* (bottom left)

8.2.5.5—*Euspinolia rufula* (bottom right)

8.2.6.1—*Traumatomutilla americana*

ETYMOLOGY The genus name comes from the Ancient Greek *trauma* "wound" and the genus name *Mutilla*, referring to the spots on T2 that looked like stab wounds to the author of the genus. The specific epithet is named after the continent of South America. **NOTES** This is one of the "original six" species described by Linnaeus (1758). It took over two centuries for this name to be correctly applied to a modern species concept. For decades, the name *T. dubia* was more widely used. A recent revision by Bartholomay, Williams, Lopez, et al. (2019) recognized that *T. americana*, *T. dubia*, and three other named forms all belonged to this species. It is one of the most common species in northern South America, and the male is easy to identify because of its coloration. In females, there is a great degree of color variation in the extent of white setal stripes on the mesosoma and the color and number of spots on T2. One more fun fact: the type of *T. dubia* is a gynandromorph specimen, wherein the head, mesosoma, and T1–2 have male morphology and T3–6 have female morphology. **SIMILAR SPECIES** Many *Traumatomutilla* in northern South America have similar coloration to *T. americana*, and the two other species in the *T. americana* species-group are even more widespread and variable than this one (Bartholomay, Williams, Lopez, et al. 2019).

8.2.7.1—*Pappognatha myrmiciformis*

ETYMOLOGY The genus name comes from the Ancient Greek *pappo* "woolly" and *gnathos* "jaw," in reference to the densely setose mandible, while the specific epithet is from the Ancient Greek *myrmex* "ant" and the suffix *-formis* "shaped." **NOTES** This species is a convincing co-mimic with the ant

species *Camponotus sericeiventris*, even going as far as to gain a "pseudo-petiole" by having a ridge on T1. The male is more similar in appearance to the other *Pappognatha* species (7.2.9, 7.4.5.2), providing yet another example of dual sex-limited mimicry. The species has been reared from an aerial nest of the host orchid bee *Euglossa dodsoni*. **SIMILAR SPECIES** *Dasymutilla paradoxa* and *Ephuta championi* have similar coloration and are likely also co-mimics of *Camponotus sericeiventris* in Central America.

8.2.7.2—*Dasymutilla paradoxa* (below left)

8.2.7.3—*Ephuta championi* (below right)

8.2.8.1—*Dolichomutilla sycorax*

ETYMOLOGY The genus name comes from the Ancient Greek *dolichos* "long" and the genus name *Mutilla*, referring to the elongate body shape. The specific epithet is apparently based on a powerful witch from Shakespeare's play, *The Tempest*. **NOTES** This is apparently the most commonly seen species in southern Africa. In addition to their broad distribution and large size, their behavior

makes them more conspicuous to people. Like *Sphaeropthalma pensylvanica* (3.6.11.1), this species attacks mud-nesting wasps, like *Sceliphron*, that are commonly found in and around human habitations. Females of this species, and many others in the genus, have numerous spines on the underside of the metasoma, including three on S1, one on S2, and six on S6. Unlike many mutillids, males and females of *Dolichomutilla* species are remarkably alike in coloration. Surprisingly for such a commonly encountered wasp, males of *Dolichomutilla* are especially rare. In one parasitized nest examined by Bayliss & Brothers (2001) they found one male and nine females. Since wasp mothers can choose the sex of their offspring when they lay eggs (1.2.2), *Dolichomutilla* may purposely skew their offspring in preference for females. **SIMILAR SPECIES** There are about 15 species in the genus *Dolichomutilla*. All occur in sub-Saharan Africa.

8.2.8.2—*Dolichomutilla kibotonensis* (right)

8.2.8.3—*Dolichomutilla heterodonta* (below, left and right)

8.2.9.1—*Cockerellidia sohmi*

ETYMOLOGY The genus was named after the entomologist T. D. R. Cockerell, who described over 5000 new species of mainly bees and wasps in the 1890s and 1900s, including this one. **NOTES** This bizarre wasp is one of the rarest and most mysterious velvet ants in the world (and one of our

favorites!). To our knowledge, prior to 2019, the species was represented in museums and literature only by the unique type specimen. The type locality, "Siam near Mecatin," is pretty ambiguous, and it was not clear where in Thailand the species actually occurred. **SIMILAR SPECIES** *Karlidia peterseni* has similar long setae, and it also occurs in Thailand; it has a narrower black head.

From 2017 to 2019, I joined three expeditions to Thailand, primarily to sample agriculturally significant fruit flies. The third expedition, in March 2019, was undertaken during the dry season, and we collected many fewer *mutillids* than in the previous two years. On the last field day of our last expedition, I was shocked and elated to collect a female of this species. Even though we came home with fewer velvet ants than on the other trips, it took months for my smile to fade. Using DNA from that specimen, we matched it to the male pictured above. Happily, there were about 20 males from six different localities available in museum material. We now know much more about the distribution of this strange beast, and, unbeknownst to me at the time, I had already caught this species: one male from a malaise trap during the second expedition in 2018.—KEVIN

8.2.9.2—*Karlidia peterseni*

RIGHT: *Karlidia peterseni*

8.2.10.1—*Ephutomorpha paradisiaca*

ETYMOLOGY The genus name is based on the apparent similarity to the genus *Ephuta*. The specific epithet comes from the Latin *paradisus* "delightful spot." **NOTES** This is a large and beautiful wasp

from New Guinea. The orange antennae and lateral mesonotal teeth in females are shared by three species in New Guinea and *Ephutomorpha ruficornis* (8.3.1.5) in Australia. Strangely, some unrelated large-bodied species from other regions also have bright, contrasting orange antennae, such as *Hoplomutilla xanthocerata* (7.4.5.1). **SIMILAR SPECIES** New Guinea has a few other large and brilliant metallic-colored velvet ants, such as *E. fulgida* and *E. mirabilis*. Australia also has many metallic-colored species (8.3.2).

Ephutomorpha paradisiaca

Ephutomorpha paradisiaca

8.2.10.2—*Ephutomorpha fulgida* (below left)

8.2.10.3—*Ephutomorpha mirabilis* (below right)

8.3 EXAMPLES AROUND THE WORLD

8.3.1.1—*Bothriomutilla rugicollis*

DISTRIBUTION Species with this coloration are widespread in Australia. Unrelated but somewhat similarly colored species are sporadically found in South America and Africa. **MALE** Most of these species are known from females only. The apparent males (and the male of *B. rugicollis*) are generally similar in appearance to their females. **NOTES** These are generally large-bodied, dark-colored species in the tribe Dasymutillini (Sphaeropthalminae), which have the head and metasoma with white to golden setal markings, but the mesosoma uniformly dark. We have examined about 30 similarly colored species, and more than half are apparently new to science. *Bothriomutilla rugicollis* is one of the largest and commonest velvet ants in Australia, and the female can be recognized by having three pairs of lateral mesosomal spines (on the pronotum, mesonotum, and propodeum).

8.3.1.2—*Australotilla* undetermined (below left)

8.3.1.3—*Ephutomorpha depressa* (below right)

8.3.1.4—*Ephutomorpha formicaria* (bottom left)

8.3.1.5—*Ephutomorpha ruficornis* (bottom right)

8.3.2.1—*Aglaotilla ignita*

DISTRIBUTION Species with this coloration are widespread in Australia and New Guinea. A few other metallic-colored species are sporadically found in Chile, India, and West Africa. **MALE** Most of these species are known from females only. The apparent males (like *Aglaotilla mira* pictured here) are generally similar in appearance to their females. **NOTES** These are species with predominantly metallic blue, green, or purple cuticular color. We have examined about 50 similarly colored species, and more than half are apparently new to science.

Aglaotilla ignita female

Aglaotilla mira male

8.3.2.2—*Ephutomorpha picta*

8.3.2.3—*Ephutomorpha lauta* (right)

8.3.2.4—*Ephutomorpha aurata* (below left)

8.3.2.5—*Ephutomorpha princeps* (below right)

8.3.3.1—*Hoplomutilla spinosa*

DISTRIBUTION Species with this coloration occur mainly in the Atlantic rainforest region of Brazil. **MALE** Males are recognized for about half of these species, including *H. spinosa*. They are similar to the females, except that they usually lack the orange markings on T2. **NOTES** These are dark-bodied species with the propodeum covered with gray to pale golden setae, and T2 with large orange markings. We have examined 20 species in 10 genera with this color pattern. This complex can help us better understand the effect of phylogenetic constraints on mimicry. Across their ranges throughout South America, all *Hoplocrates* species have their dominant T2 markings composed of a mesal setal patch, *Traumatomutilla* species usually have four cuticular spots, and *Hoplomutilla* species usually have only two cuticular spots. These similarly colored species used different "tools," derived from their distantly related lineages, to approximate a shared color pattern right near the heart of the "*cidade maravilhosa*" (a.k.a. Rio de Janeiro, Brazil).

8.3.3.2—*Hoplocrates cephalotes* (below right)

8.3.3.3—*Traumatomutilla quadrinotata* (below left)

**8.3.3.4—*Lophomutilla prionophora*
(bottom right)**

**8.3.3.5—*Ephuta chrysodora*
(bottom left)**

8.3.4.1—*Ephuta grauna*

DISTRIBUTION Species with this coloration occur mainly in the Atlantic rainforest region of Brazil. **MALE** Of the species pictured below, only *Ps. spixi* is known from both sexes; its male looks like the *Hoplomutilla spinosa* male. The Atlantic rainforest region, however, has many bright golden and orange males, like *Ephuseabra morra*, that are good candidates to eventually be matched with some of the species pictured here. **NOTES** These species loosely fit with the Black-headed Color Syndrome seen in Mesoamerica (see 7.4.1) and other regions (see 8.3.7–9), but they have a brighter orange mesosomal color and extensive gold-tinted metasomal markings. There are about 60 species in 10 different genera with this color pattern.

*Ephuseabra
morra*

*Ephuta
grauna*

8.3.4.2—*Lophomutilla chrysomalla* (below right)

8.3.4.3—*Seabratilla baleia* (below left)

8.3.4.4—*Pseudomethoca spixi* (bottom right)

8.3.4.5—*Ptilomutilla* undetermined (bottom left)

8.3.5.1—*Traumatomutilla indica*

DISTRIBUTION Species with this coloration occur mainly in the Amazon rainforest region of Brazil and other countries in northern South America. **MALE** Most of these species are known from females only. The few known males vary widely in color but usually have more extensive white or silvery setae than males in the *Atillum charoneum* cluster on the next page. **NOTES** These females are dark-bodied with extensive whitish setal markings on the body and usually with pale yellow spots on T2. There are at least 60 similarly colored species in northern South America. *Traumatomutilla indica* is one of the six original species described in *Mutilla* by Linnaeus back in 1758; it was originally thought to have come from India, hence the name.

8.3.5.2—*Hoplocrates smithi* (below left)

8.3.5.3—*Hoplomutilla lanata* (below right)

8.3.5.4—*Mickelia harypyia* (bottom left)

8.3.5.5—*Suareztilla bicolorata* (bottom right)

8.3.6.1—*Atillum charoneum*

DISTRIBUTION Species with this coloration are usually found in arid or semi-arid grasslands in southern South America, especially in Argentina. **MALE** Males are recognized for fewer than half of these species. They are similar to the females, but they usually lack the orange markings on T2, except in *A. charoneum*. **NOTES** These are dark-bodied species with mostly black setae and dark reddish-orange markings on T2. We have seen about 40 similarly colored species in southern South America.

8.3.6.2—*Cephalomutilla graviceps* (right)

8.3.6.3—*Traumatomutilla alhuampa* (below)

8.3.6.4—*Darditilla usta* (right)

8.3.6.5—*Tallium sefene* (below right)

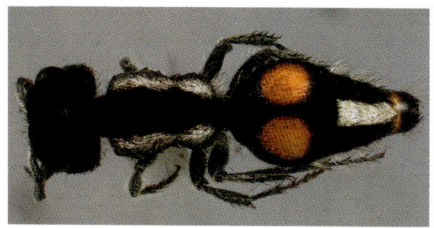

8.3.7.1—*Wallacidia oculata*

DISTRIBUTION Similarly colored species to these live on literally every continent (except Antarctica). The species pictured here were all collected in southern Thailand. **MALE** Fewer than half of these species are known from both sexes; the apparent males are actually more variable in color than their females. Some, like *W. oculata* pictured here, match the *Timulla*-like Males Color Syndrome from chapter 5 (5.5), while others seem to fit the *Ephuta*-like (5.4) or *Sphaeropthalma*-like (5.7) Color Syndromes. **NOTES** These species were labeled as the Black-Headed Color Syndrome (BHCS) in a study about color patterns in the eastern Oriental and Palaearctic regions (Okayasu et al. 2018). In that paper, 224 similarly colored species were discussed. Not only is this color syndrome the most diverse in Eastern Asia, it is also the most abundant.

During three collecting expeditions to Thailand from 2017 to 2019, I obtained 711 female velvet ants. Almost 700 of these matched the color pattern seen here. While it's initially disappointing to be unable to separate these insects in the field, the diversity that's revealed at home under the microscope is truly rewarding. I was happy to name two of these species after native Thai animals: *Bischoffitilla tokay* after the Tokay Gecko (*Gekko gecko*) and *Smicromyrme helarctos* after the Sun Bear (*Helarctos malayanus*).—KEVIN

8.3.7.2—*Bischoffitilla tokay* (below left)

8.3.7.3—*Ctenotilla guangdongensis* (below right)

8.3.7.4—*Smicromyrme helarctos* (bottom left)

8.3.7.5—*Zeugomutilla saepes* (bottom right)

8.3.8.1—*Lobotilla leucospila*

DISTRIBUTION These are examples of the Black-headed *Timulla* Color Syndrome (see 7.4. 1) from sub-Saharan Africa. **MALE** The apparent males are even more variable than the Oriental species with this color syndrome. The male of *L. leucospila* is somewhat reminiscent of the Eastern Color Syndrome in chapter 5 (see 5.6). **NOTES** These species were treated as members of the Pan-African mimicry ring in a project about Müllerian mimicry in Africa (Wilson et al. 2018). That study examined 187 similarly colored species, but the true number of species in the Pan-African mimicry ring is likely greater than 700.

8.3.8.2—*Dasylabroides willowmoorensis*

8.3.8.3—*Stenomutilla schulthessi*

8.3.8.4—*Mutilla parallela*

8.3.8.5—*Odontomutilla calida*

Dasylabroides willowmoorensis

Stenomutilla schulthessi

Mutilla parallela

Odontomutilla calida

8.3.9—Miscellaneous black-headed color syndrome

NOTES This is the commonest color form for velvet ant females around the world. In addition to the Oriental (8.3.7) and Afrotropical (8.3.8) assortments on the previous pages, examples are shown below from South America, Australia, and Europe. There are likely more than 1,500 species in the world with this general pattern.

South America

8.3.9.1—*Timulla mediata* (right)

8.3.9.2—*Darditilla gabrielae* (below)

8.3.9.3—*Lynchiatilla hoplites* (below right)

Australia

8.3.9.4—*Ephutomorpha strigosa* (right)

8.3.9.5—*Ephutomorpha morosa rufithorax* (below left)

8.3.9.6—*Ephutomorpha turneri convergens* (below right)

Europe

8.3.9.7—*Smicromyrme burgeri* (right)

8.3.9.8—*Physetopoda fusculina* (below left)

8.3.9.9—*Tropidotilla littoralis* (below right)

8.3.10.1—*Dasylabris maura*

DISTRIBUTION This color pattern is seen mostly around the Mediterranean Sea in Europe and North Africa, and extends east into Palaearctic Asia and more rarely deeper south into Africa. **MALE** Many of these species are known from both sexes, including *D. maura*. **NOTES** This species is generally similar to the Black-headed Color Syndrome (see 7.3.4.1 and 8.3.7–9) but with a distinct whitish setal patch on the head. They are basically intermediate between that pattern and the Mediterranean-Steppe mimicry

ring (see 8.3.11). The final two of the original six Linnaeus (1758) species are pictured here: *D. maura* and *Ronisia barbara*.

Dasylabris maura male

8.3.10.2—*Ronisia barbara* (below left)

8.3.10.3—*Nemka viduata* (below right)

8.3.10.4—*Physetopoda punctata* (bottom left)

8.3.10.5—*Smicromyrme partita* (bottom right)

8.3.11.1—*Dasylabris lugubris*

DISTRIBUTION These species are known from the Mediterranean region, especially Spain and North Africa, and extend east into Central Asia. There are similarly colored species in arid and semi-arid regions of South America. **MALE** Males of these species, when known, are generally similar to the females or entirely black, like *D. lugubris*. **NOTES** These females are

generally dark-bodied but are covered with extensive markings of white to coppery setae. Nine species with this color pattern were diagnosed in the Mediterranean-Steppe mimicry ring in a project about mimicry in Africa (Wilson et al. 2018). About 20 additional species with this pattern are known from Europe and Asia.

8.3.11.2—*Stenomutilla argentata* (below)

8.3.11.3—*Mutilla erschoffi* (right)

8.3.11.4—*Ronisia maculosa* (middle right)

8.3.11.5—*Smicromyrme mareotica* (bottom right)

8.3.12.1—*Trispilotilla africana*

DISTRIBUTION Species with this coloration are nearly as widespread, though less common, as the Black-headed Color Syndrome (see 7.4.1, 8.3.7–9). The species pictured here are all from more or less humid regions in sub-Saharan Africa. **MALE** Relatively few males are known; many are similar to their females, like *T. africana*. **NOTES** These species are dark-bodied with generally sparse dorsal setae, except for the white to golden metasomal setal markings. They were treated as the Pan-African mimicry ring in a project about mimicry in Africa (Wilson et al. 2018). That study examined 34 similarly colored species, but we have studied at least 200 species like this around the world.

8.3.12.2—*Seriatopsidia biseriata* (right)

8.3.12.3—*Mutilla astarte ignava* (below)

8.3.12.4—*Stenomutilla kohli* (middle right)

8.3.12.5—*Pristomutilla octacantha* (bottom right)

8.3.13.1—*Mutilla quinquemaculata*

DISTRIBUTION Species with this coloration are nearly as widespread, though less common, as the Black-headed *Timulla* Color Syndrome (see 7.4.1, 8.3.7–9). The species pictured here are all from more or less arid regions in Europe and Africa. **MALE** Males of these species are variable in color, but many are similar to their females, like *M. quinquemaculata*. **NOTES** These species are a good match for the *Timulla*-like Color Syndrome in North America (see 3.5). Many of these were treated as the Arid mimicry ring in a project about mimicry in Africa (Wilson et al. 2018). That study examined over 50 similarly colored species, but we have studied at least 350 species like this around the world.

8.3.13.2—*Dasylabroides inconspicua* (right)

8.3.13.3—*Odontomutilla ovata* (below)

8.3.13.4—*Macromyrme sinuata* (bottom left)

8.3.13.5—*Sulcotilla sulcata* (bottom right)

8.3.14.1—*Stenomutilla freyi*

DISTRIBUTION The species treated here are all from Madagascar, but similarly colored species are known from various regions around the world. **MALE** Males, when known, usually look similar to their females, like *S. freyi*. **NOTES** These Malagasy species are similar to the preceding cluster (see 8.3.13) but generally have brighter golden metasomal markings. This is the dominant color pattern for female velvet ants in Madagascar; we have examined about 40 species like this and fewer than 20 with different coloration.

8.3.14.2—*Dasylabris seyrigi*

8.3.14.3—*Pristomutilla pauliani*

8.3.14.4—*Aureotilla madecassa*

8.3.14.5—*Trogaspidia seyrigiana*

LEFT: *Dasylabris seyrigi*

ABOVE: *Pristomutilla pauliani*

BELOW LEFT: *Aureotilla madecassa*

BELOW: *Trogaspidia seyrigiana*

8.3.15.1—*Dasylabris rugosa*

DISTRIBUTION These species are known mainly from India and Sri Lanka. They are loosely similar to some members of the Madrean Color Syndrome (see 3.4) in North America and a handful of South American species. **MALE** Most of the species with this coloration are not known from both sexes. *Dasylabris kraciva* is probably the male of either *D. rugosa* or *D. argentipes*. Additionally, two of the other species pictured here (*Spilomutilla cotesii* and *Indratilla gynandromorpha*) are flightless males. **NOTES** These wasps have some combination of blackish, silvery, and bright orange color, with the mesosoma always orange. To some degree they seem like an Old World version of the Madrean Color Syndrome in North America. We have seen about 10 species with similar coloration in India, but there are likely many more.

8.3.15.2—*Dasylabris kraciva*

Dasylabris kraciva

8.3.15.3—*Dasylabris argentipes* (below left)

8.3.15.4—*Spilomutilla cotesii* (below right)

8.3.15.5—*Indratilla gynandromorpha* (bottom left)

8.3.15.6—*Smicromyrme* undetermined (bottom right)

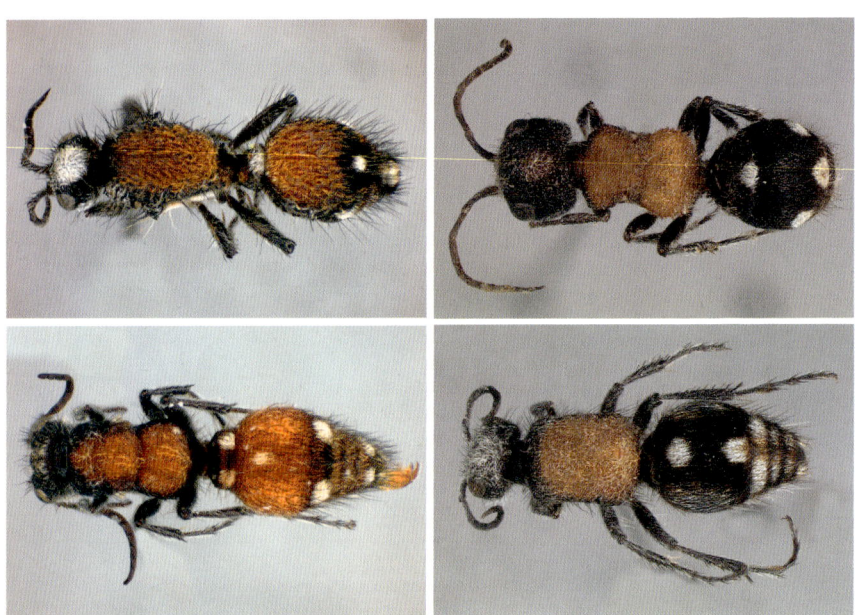

8.3.16.1—*Wallacidia opulenta*

DISTRIBUTION These species are known mainly from India and Sri Lanka. They are loosely similar to some members of the Tropical Color Syndrome (see 7.4) in Mesoamerica and a handful of South American species. **MALE** Few males are recognized for species with this coloration, including *Wallacidia opulenta*. Other males for these species might also have similar color to their females. **NOTES** These are generally dark-bodied species with reddish, orange, or golden markings on the head and legs, and bright golden markings on the metasoma. We have seen about 10 species with similar coloration in India, but there are likely many more.

8.3.16.2—*Wallacidia humbertiana* (below left)

8.3.16.3—*Odontomutilla* undetermined (below right)

8.3.16.4—*Smicromyrme* undetermined (bottom left)

8.3.16.5—*Storozhenkotilla aurofasciata* (bottom right)

APPENDICES

9.1 MORPHOLOGY

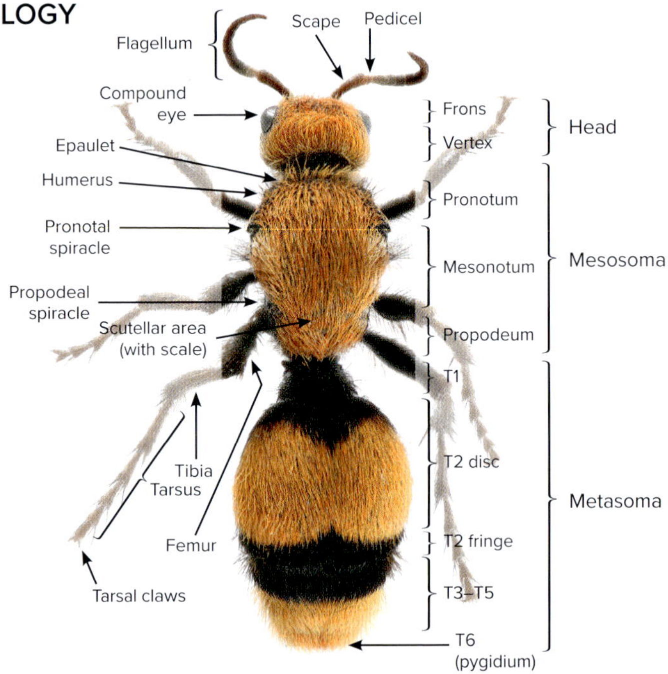

RIGHT:
Dasymutilla occidentalis female dorsal morphology

Flagellum
Scape
Pedicel
Compound eye
Frons
Vertex
Head
Epaulet
Humerus
Pronotum
Pronotal spiracle
Mesonotum
Mesosoma
Propodeal spiracle
Scutellar area (with scale)
Propodeum
T1
Tibia
Tarsus
T2 disc
Femur
Metasoma
Tarsal claws
T2 fringe
T3–T5
T6 (pygidium)

BELOW:
D. occidentalis female lateral morphology

Head
Mesosoma
Metasoma
Vertex
Pronotum
Mesonotum
Propodeum
T1
T2 disc
T2 fringe
T3–T5
Frons
T6 (pygidium)
Gena
Compound eye
Mandible
Trochanter
Humerus
Coxa
Femur
S1
Tibia
Tibial spurs
S2–S6
S2 disc
S2 fringe
Felt line
Sting
Tarsal claws
Tarsus (Tarsomeres 1–5)

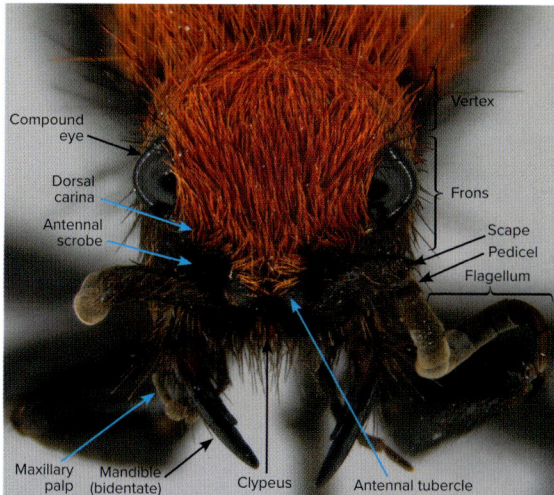

D. occidentalis female face morphology

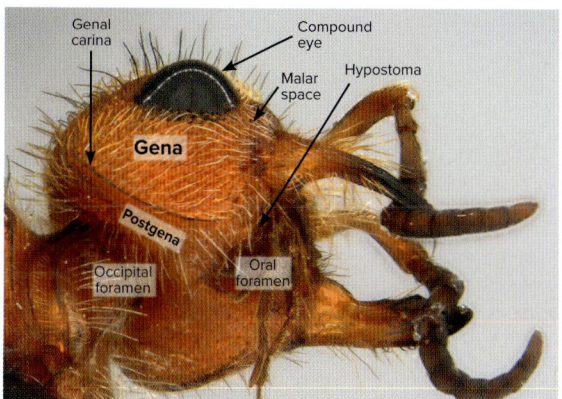

Pseudomethoca oceola female head morphology

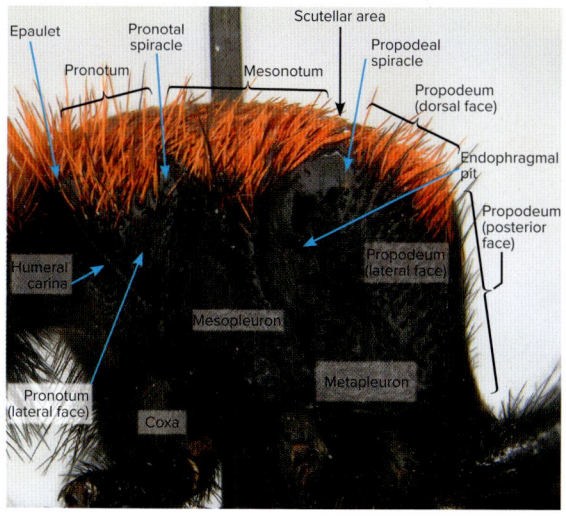

D. occidentalis female mesosoma morphology

Dasymutilla occidentalis male dorsal morphology

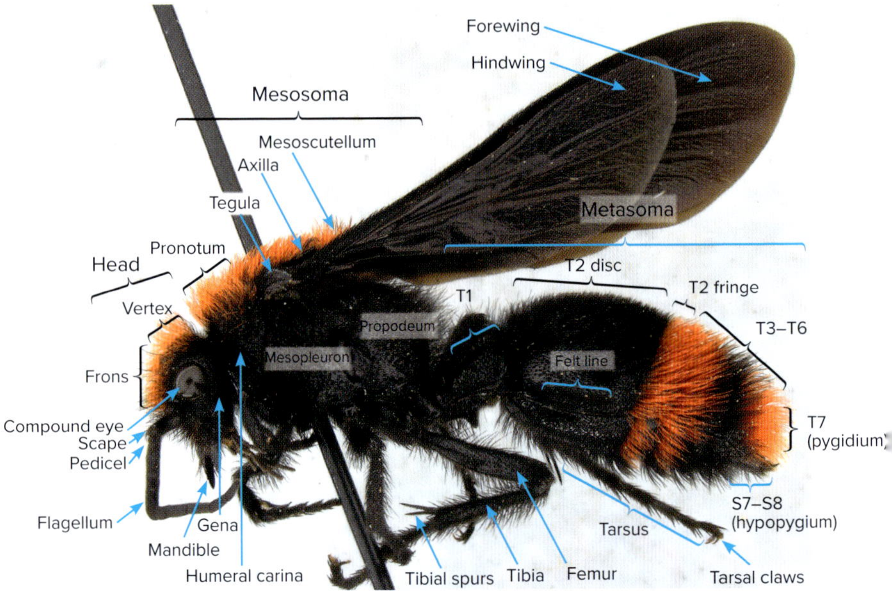

D. occidentalis male lateral morphology

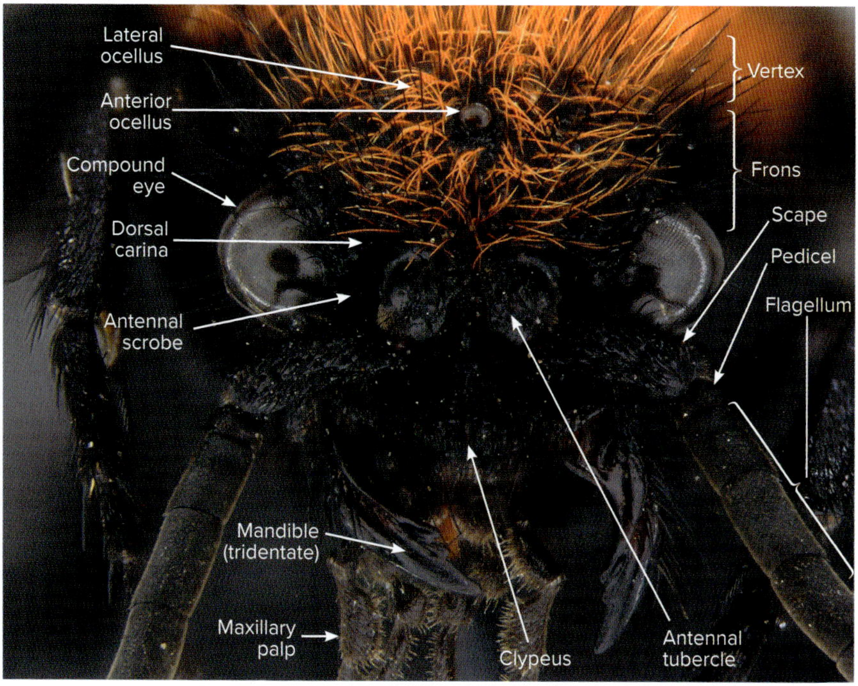

Lateral ocellus

Anterior ocellus

Compound eye

Dorsal carina

Antennal scrobe

Vertex

Frons

Scape

Pedicel

Flagellum

Mandible (tridentate)

Maxillary palp

Clypeus

Antennal tubercle

D. occidentalis male face morphology

Free length of paramere

Cuspis

Digitus

Penis valve

Basal lobe of paramere

Basal ring

D. occidentalis male genitalia

9.2 GLOSSARY

Many of the definitions in this glossary are based on their utility for velvet ant diagnostics and may have different meanings in different scientific fields or contexts.

AcroDilo-hypopygium—in nocturnal male velvet ants, three genera possess this diagnostic feature, wherein the hypopygium is wider than long, with its posterior margin concave, and with longitudinal carinae along the lateral margin (Fig. 6.3.1a).

AcroDilo-mandible—in nocturnal female velvet ants, three genera possess this diagnostic feature, wherein the mandible has a large basal tooth ventrally and a small sharp basal tooth dorsally (Fig. 6.2.2a).

Aculeata—official name for the stinging wasps, which also includes bees, ants, hornets, and, of course, velvet ants.

Acuminate—tapering down to a point at the apex of a given structure.

Antennal scrobe—the basin between the base of the antenna and the compound eye. This basin often has a dorsal carina, which is an especially important diagnostic feature in *Dasymutilla*.

Antennal tubercle—the structure expanding out over the base of each antenna.

Anterior—a technical term for the front portion of a given structure, or the entire insect.

Apex (adj. apical, pl. apices)—the tip of a given structure, or the portion of a given structure farthest from the midpoint of the body. Some scientific papers use the term "distal," instead of "apical."

Aposematism—a technical term for warning coloration, typically consisting of contrasting patterns of black, white, yellow, or red.

Appressed—referring to setae that are flattened against the body.

Apterous—wingless.

Areolate—body sculpture that is composed of large, often irregular areas that generally each have one seta. In other works, this type of sculpture is often referred to as reticulate.

Axilla (pl. axillae)—the portion of the mesonotum between the mesoscutellum, mesoscutum, and tegula. In *Dasymutilla* males, the presence and shape of projections on the axilla are important.

Base (adj. basal)—the base of a given structure, or the portion of a given structure nearest to the midpoint of the body.

Batesian mimicry—when a harmless organism gains protection by resembling a harmful organism. Many male velvet ants are considered Batesian mimics of the stinging females.

Bidentate—having two teeth, especially in reference to the mandible's apex.

Brachyplumose—these are setae that have short inconspicuous barbs along their length.

Brachypterous—having shortened wings.

Bristle—a seta that is especially thick, at least compared with the surrounding setae.

Carina—a low ridge; this term is used for various body structures. If a carina is present, a given structure is referred to as carinate.

Cluster—in this book, a species-cluster (cluster for short) is a group of species in the same genus and same color syndrome that share some combination of diagnostic features. This is an ecological group name and does not necessarily include closely related species.

Clypeal basin—in *Ephuta* males, the clypeal basin is a concave area on the lower portion of the clypeus.

Clypeus—a plate on the lower portion of the face, between the antennal tubercles and mandibles.

Color syndrome—in this book, a color syndrome is a broadly defined color pattern shared by numerous species in multiple different genera.

Compound eye—the large multifaceted eye on each side of the head. Usually we just say "eye," but the ocelli are technically eyes as well, so "compound eye" is more specific and useful in certain circumstances.

Conspecific—belonging to the same species.

Constriction—the state of having the sides distinctly narrowed or pinched-in near a structure or between two different structures.

Costa—the anterior margin of a wing.

Coxa—the most basal segment of each leg.

Crepuscular—active at dusk; basically, between diurnal and nocturnal.

Cryptic species—a complex of species that are known to be different, usually based on DNA evidence, but that cannot be reliably separated using external body features.

Cuspis—usually the second largest appendage of the male genitalia, and usually second farthest to the outer edge, next to the paramere.

Cuticle (adj. cuticular)—our preferred term to refer to the exoskeleton color. Other papers often use the term "integument."

Dentate—having a tooth (regarding the mandible) or tooth-like expansion or process (for other body regions).

Description—when describing a new species, this is the section where an author describes various features of the creature.

Diagnosis—the science of identifying things. When describing a new species, this is the section where an author explains how to tell the new species apart from all the others.

Digitus—a component of the male genitalia, usually situated between the cuspis and penis valve.

Dilated—becoming wider in a certain direction, usually toward the apex.

Dimorphism—having two different forms. For velvet ants, this usually applies to sexual dimorphism, where the males and females are very different from one another.

Disc—when referring to metasomal tergites and sternites, the disc is the larger portion of the structure, usually encompassing the anterior 80–90% of the tergite or sternite. This is especially important for T2, which is the largest metasomal segment and often has different color on the T2 disc and T2 fringe.

Disciform—disc-shaped. Usually used in reference to the shape of T1, and meaning that the structure is somewhat rounded and the anterior surface is flatter and set apart from the dorsal surface by an angle or carina.

Diurnal—active during daylight.

Dorsum (adj. dorsal)—a technical term for the top side of a given structure, or whole creature.

Dual sex-limited mimicry—a phenomenon where males and females of a given species participate in different mimicry rings. Usually, the female is a Müllerian mimic of other velvet ants, and the male is a Batesian mimic of different velvet ant species or other types of wasps.

Ecological group names—a collection of organisms that are linked by some feature of their biology (such as activity period or color pattern) but are not necessarily related in the phylogenetic sense.

Emarginate—having a portion cut out or notched, rather than being flat or simply convex.

Entire—this term applies to a structure that has the margin flat or evenly rounded, without an emargination or notch; or to a setal or cuticular band that is uniformly one color, rather than interrupted mesally with a different color.

Entomology—the study of insects.

Epaulet—this is a swelling on the pronotum, sublaterally near the juncture of the dorsal and anterior surfaces, that has a small clump of short setae and usually an attached small tooth, tubercle, or ridge.

Erect—a structure, usually a seta, that is sticking up at an angle of more than 45 degrees from the body surface.

Etymology—the study of naming things.

Face—the term "face" is often used instead of "surface," i.e., dorsal face of propodeum.

Felt line—a row of short, dense setae situated on T2 and/or S2, near the juncture of these plates.

Femur—the third major portion of the leg, between the trochanter and tibia; usually the longest portion of the leg.

Fiddle-shaped—regarding the female mesosoma shape, it is considered fiddle-shaped when the propodeum is narrower than the mesonotum, and there is a constriction near the propodeal spiracle.

Flagellomere (F1, F2, etc.)—the remainder of the antenna beyond the scape and pedicel is called the flagellum; each article of the flagellum is called a flagellomere, starting with F1 at the base.

Foramen—generally, a depression on the body near the connection point of another structure. For velvet ants, this is really only used for the oral foramen (surrounds the mouth) and occipital foramen (surrounds the underside of the head where it connects to the mesosoma). In other papers, sometimes referred to as a fossa.

Fringe—contrasting with the term "disc" above, this term refers to the posterior portion of a given tergite or sternite and the row of setae along each posterior margin. The term "fringe" can more loosely refer to any row of setae on a given structure.

Frons—the front surface of the head, between the clypeus, compound eyes, and vertex.

Gena—the "cheek" of an insect, or the area posterior to the mandible and ventral to the eye on the side of the head.

Granulate—this is a type of sculpture where there are many small, relatively uniform, bead-like bumps. Historically, this term was used instead of "microreticulate" for sculpture that included tiny uniformly dense pits or divots.

Gynandromorph—a specimen that has the body tissue partly male and partly female.

Head—come on, you do not need our help with this one, right? Okay, since you asked, this could technically be referred to as the prosoma if we wanted to be consistent with the terms "mesosoma" and "metasoma" used in Hymenoptera, but "head" works fine for us.

Holotype—the original, designated specimen that carries the name for a species.

Hot deserts—this term applies to the Mojave, Sonoran, Chihuahuan, and Baja California Deserts.

Humeral carina—a vertical carina that runs between the anterior and lateral surfaces of the pronotum and usually ends near the epaulet.

Humerus—the shoulder, or anterolateral corner of the mesosoma.

Hymenoptera—the order that includes sawflies, parasitoid wasps, ants, bees, and velvet ants.

Hypopygium—the last visible metasomal sternite (technically, it includes the last two sternites, since S7 and S8 are often somewhat fused together).

Hypostoma—the margin that runs along the oral foramen; it is almost always defined by a carina (hypostomal carina), and this carina is sometimes armed with a tooth or prominent ridge.

Interantennal prominence—an elevation of the frons between and behind the antennal tubercles.

Intervals—the areas on the body cuticle between each puncture; the sculpture, shape, or presence of setae on these intervals in specific body regions can be diagnostic. Some previous works call these interspaces.

Irregular—basically, referring to body sculpture that is not uniformly dense or uniformly pointing in the same direction.

Lamella—basically, a large and see-through carina.

Lateral—to the side.

Longitudinal—directed along the longitudinal body access—basically, running anterior to posterior, rather than laterally.

Madrean Archipelago—in the USA, this ecoregion occurs only in the southeastern corner of Arizona and the southwestern corner of New Mexico. Some papers place this region in the Sonoran or Chihuahuan Desert, but the mountain elevation and summer monsoons make this a hotbed of velvet ant diversity. Over 25% of all velvet ant species in the USA are known to occur in this region.

Malaise trap—an insect trap that looks like a tent with a bottle at the top of one end. It is called a malaise trap because the collector can set it up and leave it for weeks at a time while it catches the insects for them.

Mandible—that bitey-pinchy thing at the side of the mouth; there are two of them.

Manuscript name—an unpublished species name. Often when a scientist first discovers that an insect specimen belongs to a new species, they provide a name and place labels onto the material before the species is officially named. Sometimes, the project never gets written, and many manuscript names can be found in museums from researchers who have been dead for over 100 years.

Margin—the edge, usually outer edge, of a given structure.

Marginal cell—in the forewing, this is the cell that is found beyond the stigma along the costa.

Mesal—middle; other papers have used "medial," but that term is more suited to mathematics than insect diagnostics.

Meso-—a prefix that means "middle." In velvet ants, it is most often applied to various structures of the mesothorax (between the prothorax and metathorax). The mesofemur, for example, is another name for the femur of the middle leg.

Mesonotum—the top plates of the mesothorax. In females, this is loosely defined as the dorsal area between the pronotal and propodeal spiracles. In males, this includes the mesoscutum, mesoscutellum, axilla, and tegula.

Mesopleuron—the large plate along the side of the mesothorax.

Mesosoma—the middle body region. In most insects, this would just be the thorax, but in most Hymenoptera, it includes the thorax and the propodeum (which is technically the first segment of the abdomen).

Mesosternal area—broadly speaking, this is the middle of the underside of the mesosoma, between the fore and mid legs. This region is technically the underside of the mesopleuron but has been used for ever this way in velvet ant papers, and we do not feel like changing it now. The true mesosternum is a tiny plate tucked away between the mid and hind coxae.

Meta-—a prefix that means "last." In velvet ants, it is most often applied to various structures of the metathorax (between the mesothorax and propodeum). The metafemur, for example, is another name for the femur of the hind leg.

Metasoma—the hind body region. In most insects, this would just be called the abdomen, but in most Hymenoptera, the true first abdominal segment is fused with the thorax to form the mesosoma.

Micropunctate—a type of body sculpture that includes many small separated punctures, especially if they are smaller than the surrounding punctures on a given body structure.

Microreticulate—a type of body sculpture composed of a relatively uniformly tight, net-like arrangement. This used to be called granulate in papers in the past.

Mimicry—the phenomenon when different organisms resemble one another for some benefit.

Mimicry ring—a group of similarly colored species that co-occur geographically and share some defensive benefit by their resemblance.

Monophyletic group—a taxon that includes all of the descendants from a shared common ancestor.

Morphology—body structure.

Morphospecies—similar to a manuscript name, this is an informally recognized group of organisms that are thought to be a distinct species but are not yet officially described.

Müllerian mimicry—a phenomenon whereby multiple well-defined species converge on a shared color pattern in order to more efficiently train predators to avoid them.

Nocturnal—active during night.

Nominotypical (or nominal)—when a species is divided into subspecies, this is the one that keeps the original species name as its subspecies name.

Oblique—this refers to a body structure that is contorted at a different angle near the apex.

Ocellus (adj. ocellar, pl. ocelli)—these are three small eyes found on the vertex of the head, mostly in males.

Ovate—oval-shaped.

Ovipositor—the egg-laying structure in insects; in many Hymenoptera, it is modified to deliver venom and referred to as the sting.

Paramere—the outermost portion of the male genitalia, composed of two portions: the basal lobe, surrounding the genitalic capsule, and the free length, the longest and outermost article of the genitalia posteriorly. When the term "paramere" is used alone in the text, we are referring to the free length of the paramere.

Paraphyletic group—a taxon that includes only some of the descendants of a common ancestor, such as reptiles when birds are removed.

Pear-shaped—regarding the female mesosoma shape, it is considered pear-shaped when the propodeum is narrower than the mesonotum and there is not a distinct constriction near the propodeal spiracles. In some other papers, this shape is called pyriform or subpyriform.

Penis valve—the innermost sclerotized article of the male genitalia.

Petiolate—in reference to the T1 shape, it is considered petiolate when the posterior margin is narrower than the middle of T1, showing constriction between T1 and T2.

Phoretic copulation—a type of mating wherein the male carries the female, often in flight.

Phylogeny (phylogenetics)—the field of reconstructing historical relationships between different taxa.

Phylogeny—a visual representation of phylogenetic relationships, almost like a family tree.

Pitfall trap—a type of insect trap that consists of a cup, bowl, or trench dug into the soil that insects will accidentally fall into; one of the best traps for catching female velvet ants.

Plate—a term for the different sclerotized portions of the exoskeleton; they are generally separated by sutures.

Pleuron (adj. pleural, pl. pleura)—a general term for the lateral sides of a given structure or body region.

Plumose—a type of seta that has long barbs along its length, making it appear bushy.

Polyphyletic group—a taxon that includes phylogenetically unrelated organisms.

Posterior—a technical term for the back portion of a given structure, or the entire insect.

Postgena—the underside of the head, found between the gena, hypostoma, and occipital foramen.

Pro-—a prefix that means "first." In velvet ants, it is most often applied to various structures of the prothorax (between the head and mesothorax). The profemur, for example, is another name for the femur of the front leg.

Process—any sclerotized structure raised up above the surrounding body cuticle.

Pronotum (adj. pronotal)—the most anterior plate of the dorsum of the mesosoma.

Propodeum (adj. propodeal)—technically, the first abdominal segment, but this is fused to the thorax in most Hymenoptera, forcing us to use the terms "mesosoma" and "metasoma," instead of "thorax" and "abdomen."

Puncture (adj. punctate)—a pit on the body cuticle with a seta emerging from it. When body sculpture is described as being punctate, it has many separated, usually circular, punctures.

Pygidial plate—a flattened area on the dorsum of the pygidium that is usually bare, usually defined by lateral carinae, and usually having sculpture different from the rest of that tergite.

Pygidium (adj. pygidial)—technically, this is just a different term for the last visible metasomal tergite: T6 in females or T7 in males. Previous authors used this term interchangeably with "pygidial plate," so for males and females without a defined pygidial plate, we have tried to use the terms "T7" and "T6" instead of "pygidium."

Quadrate—more or less rectangular in shape.

Resurrected status—a species that was previously synonymized with another but later determined to be a valid recognized species.

Rounded—we all know what rounded means, but the context differs slightly depending on the body structure and genus being discussed. Sometimes, it means "not marked by a sharp, carinate angle," and sometimes it means "more smoothly rounded than angular."

Rugose—this refers to sculpture, especially on the pygidial plate, that has numerous wrinkles or irregular wavy ridges.

Scape—the first, and longest, article of the antenna, which connects the antenna to the head.

Sclerotized—made up of hardened cuticle that retains its shape after the specimen dies and dries out.

Sculpture—the structure of the basic body cuticle on a given region of the body.

Scutellar area—in males, the mesoscutellum is a small plate located between the axillae, anterior to the propodeum. In females, all these plates are fused, so this refers to the general area where the mesoscutellum would be—basically, the area between the propodeal spiracles.

Scutellar scale—a transverse carina or ridge found mesally in the scutellar area.

Sentinel species—the species for which a given species-cluster is named.

Sessile—in reference to T1 shape, this means broadly and smoothly rounded into the contours of T2, with the posterior portion wider than the middle portion.

Seta (adj. setal, pl. setae)—because hair is a structural feature that is unique to mammals, the term "seta" is used for all the different types of hair-like structures on a velvet ant, or other non-mammal fauna.

Sex association—the discovery of which male and female are conspecific.

Simple seta—a seta without barbs.

Species-cluster—in this book, a species-cluster (cluster for short) is a group of species in the same genus and same color syndrome that share some combination of diagnostic features. This is an ecological group name and does not necessarily include species that are closely related.

Species-complex—an informal taxonomic name for a group of species that have been recognized as distinct based on behavior or genetic lineages but that cannot be reliably separated using external body features.

Species-group—an informal taxonomic name for a group of related species within a genus; often used instead of the formal category of subgenus because of its flexibility.

Spiracle—a pit or cavity in the body through which the insect collects air to breathe. In velvet ants, the pronotal and propodeal spiracles often carry diagnostic differences in shape, and they serve as good indicators for comparing various body proportions.

Sternite (S1, S2, etc.)—the ventral plate of each metasomal segment.

Stigma—also called pterostigma; a specialized cell along the costa of the forewing that is located between the costal veins and marginal cell; when this cell is completely sclerotized (Fig. 4.0.2a, 4.0.2c), this is referred to as a "full stigma," and when this cell lacks sclerotization between the veins (Fig. 4.0.2b), this is referred to as an "empty stigma."

Sting—the venom-delivering apparatus of wasps.

Striate—sculpture that consists of numerous parallel carinae, most commonly used for the pygidial plate sculpture; each single carina is called a striation.

Subantennal basin—in males of *Ephuta*, this is a pit or longitudinal furrow found at the base of the clypeus between and below the antennal tubercles.

Subfamily—a formal taxonomic group used to subdivide the family into more manageable chunks; there are usually multiple tribes and genera within a subfamily.

Subsessile—referring to the T1 shape, this is intermediate between petiolate and sessile, with the posterior portion not clearly wider or narrower than the middle portion.

Subspecies—a formal taxonomic category for populations within a species that can be reliably separated from each other and live in separate regions. Historically, this term has had a broader application and been somewhat sloppily applied, so many subspecies have been synonymized in recent years.

Suture—the connecting lines between each body plate.

Synonym (adj. synonymous)—a formerly recognized species that was later discovered to be conspecific with another already named form.

Synonymy—the taxonomic act of recognizing that two or more species are not actually distinct.

Systematics—the scientific field of classifying life, generally thought to include phylogenetics and taxonomy as subdisciplines.

T1 shape—the shape of the first metasomal tergite.

Tarsus (adj. tarsal, pl. tarsi)—the apical portion of each leg, consisting of multiple tarsomeres.

Taxon (pl. taxa)—a name used for an organism or a taxonomic unit.

Tegula (pl. tegulae)—a scale-like structure covering the base of the forewing.

Tergite (T1, T2, etc.)—the dorsal plate of each metasomal segment.

Tibia (pl. tibiae)—the penultimate article of each leg, between the femur and tarsi.

Tibial spur—a relatively long and stout articulating spine at the tibial apex. Nearly all velvet ant species have two tibial spurs on each middle and hind tibia.

Transverse—oriented side to side perpendicular to the longitudinal body axis.

Tribe—a formal taxonomic group used to subdivide a subfamily into more manageable chunks; there are usually multiple genera within a tribe.

Tridentate—possessing three teeth or tooth-like structures.

Trochanter—the second article of each leg, between the coxa and femur.

Truncate—flattened at the apex or posterior margin of a structure.

Tubercle—a raised bump on the body cuticle.

Type species—the single species that a given genus is based upon; this species can never be removed from that genus when things are rearranged, except by synonymizing the whole genus.

Type specimen—a voucher specimen with special status regarding the definition of a named organism. The most important kind is the holotype; when a specimen is referred to just as the "type" it usually means the holotype.

Undefined pygidium—a condition whereby the last tergite is convex and lacks a pygidial plate defined by lateral carinae.

Venter (adj. ventral)—a technical term for the bottom side of a given structure, or whole creature.

Vertex—the top portion of the head, between the dorsal margins of the compound eyes, the frons, and the posterior head margin.

Vertical—in reference to a body structure, usually the mandible, it is considered vertical when the apical portion is not contorted and actually remains on the same plane as the basal portion.

9.3 KEYS, KEYS, KEYS

9.3.1—Keys to separate velvet ants from other *Hymenoptera*

Wingless or short-winged forms—wings absent or shorter than mesosoma length

1.	First, and sometimes second, metasomal tergite transformed into a node-like structure; antennal scape more than half as long as the flagellum	**Formicidae**—true ants
	Metasomal tergites not transformed into a node-like structure; antennal scape less than half as long as the flagellum	**2**
2 (1)	With one or more of the following features: antenna with more than 13 articles or fewer than 12 articles; antenna with a discrete apical club of especially thickened flagellomeres; or metasoma with fewer than 6 tergites	**various families of "Parasitica" or superfamily Chrysidoidea**
	Antenna with 12 or 13 articles (including scape and pedicel); flagellomeres relatively uniform in width; metasoma with 6 or 7 tergites	**3**
3 (2)	Mesosomal dorsum with all plates fused together; T2 usually with felt line	**Mutillidae females**, subfamilies Mutillinae and Sphaeropthalminae (176 of the 188 species treated in chapter 3 and nocturnal species in chapter 6)
	Mesosomal dorsum with separated plates, at least with pronotum separated from mesonotum by a suture; felt line present or absent	**4**
4 (3)	Mesosomal dorsum with only 2 plates: mesonotum, metanotum, and propodeum fused; wings or tegula never present	**5**
	Mesosomal dorsum with at least 3 plates; small wings or tegula sometimes present	**6**
5 (4)	T2 without felt line; T1 shape sessile	**Mutillidae females**, subfamily Myrmosinae (12 of the 188 species treated in chapter 3)
	T2 with felt line; T1 shape more or less petiolate	**Chyphotidae**
6 (4)	T2 with felt line; male with 13 antennal articles and 7 metasomal tergites	**Mutillidae flightless males**, subfamilies Mutillinae and Sphaeropthalminae (three of the 188 species treated in chapter 3 and a handful of nocturnal species in chapter 6)
	T2 without felt line; males or females	**various families of stinging wasps**

Fully winged forms—wings longer than mesosoma

1	T2, and sometimes S2, with felt line present	**2**
	Metasoma without felt lines	**3**

2 (1)   Antennal insertion raised to form antennal tubercle .... **Mutillidae**, subfamilies Mutillinae and Sphaeropthalminae (151 of the 189 species treated in chapter 5 and all nocturnal species in chapter 6)

Antennal insertion flat and open on frons .... **Chyphotidae**

3 (1)   First, and sometimes second, metasomal tergite transformed into a node-like structure; antennal scape more than half as long as the flagellum .... **Formicidae**—true ants

Metasomal tergites not transformed into a node-like structure; antennal scape less than half as long as the flagellum **4**

4 (3)   With one or more of the following features: antenna with either more than or less than 13 articles (including scape and pedicel); forewing with fewer than 7 closed cells; or metasoma with 6 or fewer visible tergites .... **various families of "Parasitica," superfamily Chrysidoidea, or females of other stinging wasps**

Antenna with exactly 13 articles (including scape and pedicel); forewing with 8 or more closed cells; metasoma with 7 tergites .... **5, male stinging wasps**

5 (4)   Pronotum forming a narrow transverse stripe in dorsal view, with lateral lobe below tegula .... **Apoidea**

Pronotum broader with more rounded posterior margin, without lobe below tegula **6**

6 (5)   T1 shape short, slender, cylindrical; hypopygium flat with simple posterior margin, 2 down-curved, spine-like genitalic parameres often visible; body usually with distinct short dense black and silvery setae .... **Mutillidae**, genus *Ephuta* (22 of the 189 species treated in chapter 5)

T1 shape variable, not short cylindrical; hypopygium variable, if 2 spines visible, they are usually not down-curved; body usually with longer, sparser setae **7**

7 (6)   Apex of metasoma with 3 straight spines; forewing margin corrugated .... **Scoliidae**

Apex of metasoma without 3 spines, often with 2 sharp genitalic parameres visible; forewing margin simple **8**

8 (7)   Hypopygium forming a raised up-curved mesal spine .... **Tiphiidae and Thynnidae, in part**

Hypopygium not forming up-curved spine **9**

9 (8   Antennal insertion flat and open on frons .... **males of various stinging wasp families**

Antennal insertion raised to form antennal tubercle **10**

10 (9)   S1 truncated posteriorly with distinct constriction between S1 and S2; tergites often with raised, heavily sculptured areas, not interleaved with sternites in lateral view .... **Mutillidae**, subfamily Myrmosinae (seven of the 189 species treated in chapter 5)

S1 basically flat, overlapping with base of S2; tergites smoothly convex, broadly interleaved with sternites in lateral view .... **Sapygidae**

9.3.2—Subfamilies, tribes, and genera of Nearctic females (and flightless males)

1   Pronotum divided from mesonotum as an articulating plate, females with 6 tergites and 10 flagellomeres .... **2 (Myrmosinae)**

Pronotum not divided from mesonotum, pronotum differentiated only in flightless males (7 tergites and 11 flagellomeres) **4**

2 (1)   Clypeus with basomesal tooth or longitudinal ridge; ocelli usually present; T2 without pale yellow cuticular spots .... *Myrmosa* (p. 46)

            Clypeus without basomesal tooth; ocelli absent; T2 usually with pale yellow cuticular spots **3**

3 (2)   Head shape round; mandible tridentate without ventral lamella basally ... *Leiomyrmosa* (p. 49)

            Head shape more or less quadrate; mandible bidentate with ventral lamella basally .... *Myrmosula* (p. 46)

4 (1)   T1 shape slender cylindrical; T2 without felt line .... *Ephuta* (p. 45)

            T1 shape variable but not cylindrical; T2 with felt line **5**

5 (4)   Mesosoma rectangular, pronotum and/or propodeum as wide or wider than mesonotum; eye large, longitudinally ovate .... *Timulla* (p. 45)

            Mesosoma ovate or pear-shaped, mesonotum wider than pronotum and/or propodeum; if mesosoma apparently rectangular, then eye small, transversely ovate **6**

6 (5)   Head much wider than mesosoma, and gena armed with ventral tooth **7**

            Head shape variable, gena not armed with ventral tooth **9**

7 (6)   Gena and postgena each armed with large ventral tooth; in females, propodeum as wide as pronotum; in males, short wing "buds" present .... *Myrmilloides* (p. 39)

            Postgena usually unarmed; propodeum narrower than pronotum **8**

8 (7)   Medium to large-bodied species with genal tooth thick and angular .... *Invreiella* (p. 39)

            Small-bodied species with genal tooth small and sharp .... ***Pseudomethoca*, in part** (p. 41)

9 (6)   T1 shape sessile; head shape usually quadrate; mesosomal sides constricted at propodeal spiracles .... ***Pseudomethoca*, in part** (p. 41)

            T1 shape variable; head shape usually rounded; mesosomal sides not so constricted **10**

10 (9)   T2 fringe composed of plumose setae **11**

            T2 fringe composed of simple or obscurely brachyplumose setae **16**

11 (10)   Metasoma blackish, or at least with distinct black patches basally on T2, contrasting with pale reddish-orange or brown mesosoma **12**

            Metasoma background reddish-orange or brown, concolorous with mesosoma **14**

12 (11)   Body uniformly orange-brown with sparse setae, except with two black cuticular spots basally .... *Stethophotopsis* (p. 45)

            Color not as above, with metasoma entirely dark, with setae denser, or T2 with continuous black band basally **13**

13 (12)   Setae of T2 disc and T3–5 mostly black, contrasting with distinct plumose white T2 fringe (coastal areas of California) .... *Morsyma* (p. 41)

            Metasomal setae not as above, setae of T2–5 usually concolorous white, yellow, or orange **14**

14 (11, 13)   Mesosoma elongate subrectangular; T1 shape sessile .... *Photomorphus*, **in part** (p. 42)

            Mesosoma generally pear-shaped, often as wide as long; T1 shape variable **15**

15 (14)   Diurnal species with darkened legs or long brightly colored dorsal setae; mesonotum with simple or brachyplumose setae .... *Sphaeropthalma* (p. 42)

            Nocturnal species with dull brown color; mesonotum often with flattened setae that are plumose basally .... **various nocturnal genera** (see chapter 6, p. 285)

16 (10)   T6 convex without pygidial plate **17**

T6 with pygidial plate defined by lateral carinae **18**

17 (16)   T1 shape broad sessile; head covered with golden setae;
T3 with two pale yellow cuticular spots .... *Protophotopsis* (p. 42)

T1 shape narrow disciform or subsessile; color pattern not
as above, dull brown with sparse setae .... *Lomachaeta* (p. 39)

18 (16)   T1 shape petiolate; mesosoma pear-shaped .... *Dasymutilla* (p. 36)

T1 shape sessile; mesosoma elongate subrectangular .... **Photomorphus, in part** (p. 42)

9.3.3—Females Color Syndromes—including flightless males

1   Head and mesosoma generally with dark cuticle and covered with somewhat
uniformly long dense setae dorsally **2**

Head and mesosoma variable in color, generally with sparser setae or
multicolored setal patches **4**

2 (1)   Dorsal setae of head and mesosoma black,
contrasting with yellow, orange, or reddish dorsal
metasomal setae or cuticle .... **Texan FCS (3.3:** 15 species in two genera)

Dorsal setae of head and mesosoma concolorous
with those of metasoma (at least T2), generally white,
yellow, orange, or reddish **3**

3 (2)   Dorsal setae of head, mesosoma, and T2
predominantly white, gray, or pale yellow;
species from generally arid habitats .... **Desert FCS (3.1:** 17 species in two genera)

Dorsal setae of head, mesosoma, and T2
predominantly yellow, orange, or red;
species from throughout western
North America and the Great Plains .... **Western FCS (3.2:** 30 species in four genera)

4 (1)   Head, mesosoma, and metasoma (at least T2) predominantly
reddish-orange to brown; head and mesosoma generally with
uniformly sparse short whitish, reddish, brown, or pale yellow setae **5**

Body with distinct color contrast between head, mesosoma, and/or metasoma;
usually with extensive silver or golden setae on the head and/or mesosoma,
with head color different from mesosoma, with metasoma mostly black, or
with mesosoma having contrasting patches of different setal colors **6**

5 (4)   Body color usually reddish-orange, legs usually blackened;
metasoma with more or less distinct contrasting pattern
of yellow to orange cuticle, black setae, and silvery to
white setae; generally medium to large-bodied
diurnal species from eastern USA .... **Eastern FCS (3.6:** 40 species in six genera)

Body color usually dull brown, legs usually concolorous
with mesosoma; metasoma without distinct contrasting
pattern, at most with sparse silvery setal spots
or fringes; generally small-bodied or nocturnal
species from all regions .... **Cryptic FCS (3.7:** 35 species in seven genera)

6 (4)   Mesosoma uniformly reddish-orange; metasoma
black with distinct white to pale yellow setal or
cuticular spots or bands; head usually concolorous
with mesosoma, sometimes partly or entirely
blackened .... **Timulla-like FCS (3.5:** 12 species in four genera)

Body color variable, usually with extensive
silver or golden setal patches on the head
or mesosoma, or metasoma (at least T2)
predominantly orange-tinted .... **Madrean FCS (3.4:** 55 species in eight genera)

9.3.3.1—Desert Females Color Syndrome

1	Fringe of T2 composed of simple setae	**2**
	Fringe of T2 composed of plumose setae	**20**
2 (1)	T1 shape sessile; mesosomal sides constricted near propodeal spiracle (western: Central Valley in California) .... *Pseudomethoca anthracina* (3.1.6.1)	
	T1 shape petiolate; mesosomal sides not so constricted .... **3** (*Dasymutilla*)	
3 (2)	Entire body, including legs, covered with dense white setae	**4**
	Body with black setae, at least on legs, venter, or tergites	**6**
4 (3)	Mesosoma without scutellar scale; mandible with large pre-apical dorsal tooth (western: Sonoran Desert and Madrean Archipelago) .... *D. thetis* (3.1.1.3)	
	Mesosoma with scutellar scale; mandible with inner tooth small, inconspicuous	**5**
5 (4)	Scutellar scale broad and flat with distinct anterior and lateral carinae (usually visible through the setae); pygidium uniformly rugose to rugostriate; cuticle usually reddish (western: widespread in hot deserts and adjacent areas) .... *D. gloriosa* (3.1.1.1)	
	Scutellar scale narrow and erect without distinct anterior carinae (weak carinae sometimes visible when setae removed); pygidium with raised distinct parallel striae basally; cuticle usually blackish (western: widespread in hot deserts and adjacent areas) .... *D. pseudopappus* (3.1.1.2)	
6 (3)	Setae of head and mesosoma glittering silver; T2 with yellowish cuticular spots (western: Mojave, Sonoran, and Great Basin Deserts) .... *D. monticola* (3.4.4.2)	
	Dorsal setae generally white, gray, or pale yellow; T2 without cuticular spots	**7**
7 (6)	Pronotum and/or mesonotum with large patch of black setae .... **see couplet 35 in Madrean FCS key** (p. 387)	
	Pronotum and mesonotum with uniformly white, gray, or pale yellow setae	**8**
8 (7)	Mesosoma without scutellar scale	**9**
	Mesosoma with scutellar scale	**13**
9 (8)	Mesosoma clearly longer than wide; T2–3 with setae white, T4–6 with setae black (western: Algodones dunes area in California) .... *D. atricauda* (3.1.3.2)	
	Mesosoma as wide as long; T2–5 setae white, gray, or pale yellow, at most with mesal black patches on T2 and T3–4	**10**
10 (9)	Head clearly wider than mesosoma (western: Central Valley and southern Sierra Nevada regions in California) .... *D. aureola* (3.1.4.4)	
	Head not clearly wider than mesosoma	**11**
11 (10)	S2 with scabrous sculpture laterally (Sonoran Desert in California and Arizona) .... *D. eminentia* (3.1.2.3)	
	S2 with simply punctate sculpture	**12**
12 (11)	T2–5 with uniformly whitish-yellow setae; legs with setae entirely black (western: transitional areas between southern California mountains and deserts) .... *D. albiceris* (3.1.5.1)	
	T2–4 with mesal patch of black setae; legs with extensive silvery setae (western: Mojave and Sonoran Deserts in California and Arizona) .... *D. foxi* (3.1.3.3)	

13 (8) T2 with white setae; T3–6 with setae entirely black (western: Algodones dunes area in California) .... *D. nocturna* (3.1.3.1)

T2–5 with setae entirely white, gray, or pale yellow **14**

14 (13) Fringes of S2–5 with setae black mesally (Central Valley and Sierra Nevada regions in California) **15**

Fringes of S2–5 concolorous with tergal setae, white or pale yellow (generally found in hot deserts) **17**

15 (14) Vertex armed with posterolateral tubercle (western: southern Sierra Nevada in California) .... *D. californica* (3.1.4.2)

Vertex unarmed posteriorly **16**

16 (15) Antennal scrobe without dorsal carina; head as wide or slightly wider than mesosoma (western: Pacific) .... *D. coccineohirta* (3.1.4.3)

Antennal scrobe with dorsal carina; head clearly narrower than mesosoma (western: Pacific) .... *D. sackenii* (3.1.4.1)

17 (14) Apices of mid and hind femora truncate (western: Sonoran Desert in Arizona) .... *D. connectens* (3.1.2.2)

Apices of mid and hind femora rounded **18**

18 (17) Gena armed with carina; propodeum with setae black (western: San Diego area and Sonoran Desert in California and Arizona) .... *D. magna* (3.1.2.1)

Gena ecarinate; propodeal dorsum with setae white to pale yellow **19**

19 (18) Dorsal body setae pale yellow; mesosoma clearly longer than wide (western: Sonoran Desert in California and Arizona) .... *D. satanas* (3.1.2.4)

Dorsal body setae stark white; mesosoma about as wide as long (western: Algodones dunes area in California) .... *D. imperialis* (3.1.5.2)

20 (1) Mesonotum covered with specialized flattened setae that are plumose basally and simple or brachyplumose apically .... *Odontophotopsis* (see chapter 6, p. 288)

Mesonotum with simple or uniformly brachyplumose setae **21**

21 (20) Body cuticle mostly blackened; T1 shape petiolate; dorsal body setae dense whitish (western: eastern slopes of Sierra Nevada in California) ..... *S. edwardsii* (3.1.7.1)

Body cuticle lighter brown; or T1 shape sessile; or dorsal body setae sparser .... **various nocturnal genera** (see chapter 6, p. 285)

9.3.3.2—Western Females Color Syndrome

1 Fringe of T2 composed of plumose setae **2**

Fringe of T2 composed of simple setae **6**

2 (1) T3–5 with setae brown or whitish, generally contrasting with setae of head, mesonotum, and T2 (at least posterior half) .... **various nocturnal genera** (see chapter 6, p. 285)

T3–5 setae yellow, orange, or red, concolorous with setae of head, mesonotum, or T2 (at least posterior half) **3**

3 (2) Mesonotum covered with specialized flattened setae that are plumose basally and simple or brachyplumose apically .... *Odontophotopsis* (see 6.1.5, p. 288)

Mesonotum with simple or uniformly brachyplumose setae .... **4** (*Sphaeropthalma*)

4 (3) Basal half of T2 with transverse band of black setae (central, western: widespread) .... *S. marpesia* (3.4.15.1)

Setae on basal half of T2 concolorous with apical half: yellow, orange, or red **5**

5 (4)   T1 shape sessile; mandible with dorsal carina basally (western: Pacific) .... *S. unicolor* (3.2.12.1)

T1 shape petiolate; mandible without such a dorsal carina
(western: Pacific) .... *S. edwardsii* (3.2.12.2)

6 (1)   T1 shape petiolate; mesosomal sides not constricted
at propodeal spiracle .... **7** (*Dasymutilla*)

T1 shape sessile; mesosomal sides pinched near propodeal spiracle **32**

7 (6)   Mesosoma with divided plates and small black wings;
male with 7 tergites and 11 antennal flagellomeres ...... **see couplet 9 in Western MCS** (p. 401)

Dorsal setae of head, mesosoma, and T2 generally yellow, orange, or red **8**

8 (7)   Setae of head and mesosoma glittering silver
or golden; T2 disc setae largely blackish and
silvery or golden .... **see couplet 20 in Madrean FCS** (p. 386)

Dorsal setae of head, mesosoma, and T2 generally yellow, orange, or red **9**

9 (8)   Mesosoma without scutellar scale **10**

Mesosoma with scutellar scale **19**

10 (9) Mesosoma clearly longer than wide **11**

Mesosoma as wide as long, or nearly so **13**

11 (10 Setae of mesonotum and T2 sparse; T2–6 with silvery setae laterally
(central: western Great Plains; western: Chihuahuan Desert) .... *D. texanella* (3.2.8.5)

Setae of mesonotum and T2 dense; T2–6 with black
and yellow or orange setae only **12**

12 (11) T2–3 with yellow or orange setae, T4–6 with black setae
(western: Mojave and Sonoran Deserts) .... *D. atricauda* (3.2.6.2)

T2–5 setae uniformly yellow or orange (central, western:
Great Plains and Texas to eastern Arizona) .... *D. stevensi* (3.2.3.3)

13 (10) S2 with scabrous sculpture laterally **14**

S2 with simply punctate sculpture **15**

14 (13) Setae of T2–5 uniformly yellow or orange, without
black patches (western: Sonoran Desert) .... *D. eminentia* (3.2.2.2)

Tergal setae largely orange, fringe of T2 and T3 mesally
with patch of black setae (western: Cochise County, Arizona) .... *D. furina* (3.2.7.2)

15 (13) Head clearly wider than mesosoma (western: Pacific and Northwest) .... *D. aureola* (3.2.1.1)

Head narrower than mesosoma, or nearly so **16**

16 (15) Mandible tridentate; T2 with distinct patch of yellow-orange
cuticle (central: Texas and western Great Plains; western:
Chihuahuan Desert) .... *D. montivagoides* (3.2.7.4)

Mandible bidentate; T2 without distinct yellow-orange cuticular patch **17**

17 (16) Antennal scrobe with dorsal carina weak or absent; usually with
contrasting patch of black and silvery setae on fringe of T2 and T3
mesally (western: Chihuahuan Desert and Madrean Archipelago) .... *D. foxi* (3.2.7.1)

Antennal scrobe with dorsal carina distinct; T2–5 generally with uniformly
yellow, orange, or red setae, at most with small black spot on T2 fringe **18**

18 (17) Gena with coarse sculpture; dorsal setae bright scarlet
red (western: Arizona mountains) .... *D. erythrina* (3.2.2.3)

Gena with weaker sculpture; dorsal setae yellow, orange,
or red (central, western: widespread, mostly absent from
hot deserts and Pacific areas) .... *D. vestita* (3.2.2.1, 3.2.7.3)

19 (9) Vertex armed with posterolateral tubercle **20**

Vertex unarmed posteriorly **21**

20 (19) Metasoma mostly black with large distinct yellow cuticular patch on T2 disc, tergal setae largely black, pale yellow over T2 patch, and silver apically (central: Great Plains) .... *D. campanula* (3.2.8.4)

T2–6 with black cuticle, usually covered entirely with yellow, orange, or red setae, rarely with black setal patch on T2 fringe and T3 mesally (western: Pacific and northwestern areas) .... *D. californica* (3.2.4.1, 3.2.8.6)

21 (19) T3–5 with more or less extensive patches of black or silvery setae **22**

T3–5 with uniformly yellow, orange, or red setae **26**

22 (21) T2 fringe and T3 with setae black; T4–6 with yellow or orange setae (eastern, central, western: widespread in eastern and central with scattered populations in northwestern USA) .... *D. occidentalis* (3.2.8.2, 3.6.1.1)

Not matching the above coloration, usually with T3–6 setae entirely black or with some degree of silvery setae on apical tergites **23**

23 (22) Dorsal body setae long and shaggy; setae of T3–6 and S3–6 entirely black; pygidium rugose or rugostriate **24**

Dorsal body setae generally shorter and smoother; tergal or sternal setae partly silvery; pygidium striate **25**

24 (23) Pygidium with irregular rugae; eye size moderate (western: Great Basin Desert) .... *D. scitula* (3.2.6.3)

Pygidial rugae often arranged longitudinally to apparent striae; eye size large (western: Mojave and Sonoran Deserts) .... *D. arenivaga* (3.2.6.1)

25 (23) Mesosoma with thick transverse wavy carina anterior to scutellar scale (central, western: widespread, mostly absent from hot deserts and Pacific areas) .... *D. bioculata* (3.2.8.1)

Mesosoma lacking carina anterior to scutellar scale (central, western: Chihuahuan Desert southwestern Great Plains) .... *D. radkei* (3.2.8.3)

26 (21) Pygidium generally striate, at least with rugae arranged into apparent striae (Great Plains and Texas) **27**

Pygidium irregularly rugose (western USA, mostly in California) **29**

27 (26) Head clearly wider than mesosoma (central: widespread; western: Chihuahuan Desert) .... *D. leda* (3.2.3.2)

Head narrower than mesosoma **28**

28 (27) Antennal scrobe with distinct dorsal carina; propodeum covered with black setae (central: widespread; western: Chihuahuan Desert) .... *D. calorata* (3.2.3.1)

Antennal scrobe with dorsal carina inconspicuous or absent; propodeum covered with yellow or orange setae (central: southern Texas) .... *D. bioculata* (3.2.3.4)

29 (26) Antennal scrobe completely lacking dorsal carina; head as wide or slightly wider than mesosoma (western: Pacific and northwestern areas) .... *D. coccineohirta* (3.2.4.2)

Antennal scrobe with dorsal carina, sometimes weakly defined; head narrower than mesosoma **30**

30 (29) Fringes of S2–5 with setae entirely orange (western: Mojave and Sonoran Deserts) .... *D. satanas* (3.2.5.1)

Fringes of S2–5 largely interrupted with black setae mesally **31**

31 (30) Antennal scrobe dorsal carina distinct; body cuticle usually distinctly black; pygidium clearly coarsely rugose (western: Pacific) .... *D. sackenii* (3.2.4.4)

Antennal scrobe dorsal carina weaker; body cuticle dark reddish, especially beneath reddish setae; pygidium largely microreticulate between sparser rugae (western: Pacific and northwestern areas) .... *D. flammifera* (3.2.4.3)

32 (6) Gena armed with large blunt tooth (western: Madrean Archipelago) .... *Invreiella manleyi* (3.2.9.1)

Gena unarmed .... **33 (*Pseudomethoca*)**

33 (32) Dorsal setae of head and mesosoma sparse, mostly flattened, not obscuring punctation **34**

Dorsal setae of head and mesosoma dense, largely erect, obscuring punctation **35**

34 (33) T2 fringe usually mostly black; humeral carina weakly defined (central, western: widespread but absent from Pacific area, Mojave and Sonoran Deserts) .... *Ps. propinqua* (3.2.11.1)

T2 fringe mostly silver; humeral carina sharp, distinct (central: widespread) .... *Ps. paludata* (3.2.11.2)

35 (33) Head distinctly wider than mesosoma; setae of T3–5 generally uniformly yellow, orange, or red (western: Pacific) .... *Ps. anthracina* (3.2.10.1)

Head only scarcely wider than mesosoma; setae of T3–5 usually partly black or silver **36**

36 (35) Dorsal setae generally bright red; setae of T3–5 mostly black mesally, with narrow silvery margins laterally (western: Madrean Archipelago) .... *Ps. flammigera* (3.2.10.3)

Dorsal setae generally pale orange, setae of T3–5 more loosely interspersed black and silvery (central: widespread; western: Madrean Archipelago and Chihuahuan Desert) .... *Ps. aureovestita* (3.2.10.2)

9.3.3.3—Texan Females Color Syndrome

1 Short-winged Males mesosoma with divided separated plates, small wing "buds" present **2**

Wingless females mesosoma with plates entirely fused, no trace of wing "buds" **3**

2 (1) Head much wider than mesosoma; T2 with sparse blackish setae and two subcircular silvery setal patches (central: southern Texas) .... *Myrmilloides grandiceps* (3.3.6.1)

Head narrower than mesosoma; T2 with dense orange setae (central: Texas) .... *D. waco* (5.3.2.1)

3 (1) T1 shape sessile; mesosomal sides constricted near propodeal spiracle .... **4 (*Pseudomethoca*)**

T1 shape petiolate; mesosomal sides not so constricted .... **5 (*Dasymutilla*)**

4 (3) Dorsal setae of head and mesosoma sparse; gena with weakly defined carina (central: Texas; western: Chihuahuan Desert) .... *Ps. brazoria* (3.3.7.1)

Dorsal setae of head and mesosoma dense; gena with strong distinct carina (central: southern Texas; western: Chihuahuan Desert) .... *Ps. pigmentata* (3.3.7.2)

5 (3) Mesosoma without scutellar scale **6**

Mesosoma with scutellar scale **9**

6 (5) Mesosoma clearly longer than wide **7**

Mesosoma as wide as long **8**

7 (6) Dorsal mesosomal setae sparse, often with scattered orange or silvery setae; T3–5 with setae largely black or silvery (central: southern Texas) .... *D. texanella* (3.3.5.1)

Dorsal mesosomal setae dense, entirely black; T2–5 setae uniformly yellow to orange (central: Texas and southern Great Plains) .... *D. nupera* (3.3.4.1)

8 (6) Mandible tridentate; T2 with distinct orange cuticular patch beneath orange setae (central: Texas and southern Great Plains) .... *D. waco* (3.3.2.1)

Mandible bidentate; T2 without distinct orange cuticular patch beneath orange setae (central, western: Mojave Desert east to southern Texas) .... *D. vestita* (3.3.2.2)

9 (5) Vertex armed with posterolateral tubercles **10**

Vertex unarmed posteriorly **11**

10 (9) Body with sparse setae, metasoma with setae mostly black and T2 with mesal yellow patch (sometimes reduced to absent) (central: south Texas) .... *D. uniguttata* (3.3.5.2)

Body with dense setae, T2–6 with setae yellow to orange, T2 with inconspicuous orange cuticular patch covered with orange setae (central: southern Texas) .... *D. wileyae* (3.3.4.2)

11 (9) Apices of mid and hind femora truncate (western: Madrean Archipelago and Arizona mountains) .... *D. nogalensis* (3.3.4.5)

Apices of mid and hind femora rounded **12**

12 (11) Head clearly wider than mesosoma (central: Texas and southern Great Plains) .... *D. gorgon* (3.3.3.1)

Head narrower, at most equally wide as mesosoma **13**

13 (12) T2 with orange cuticular patch beneath orange setae; head clearly narrower than mesosoma (central: southern Texas) .... *D. bioculata* (3.3.4.3)

T2 without discrete orange cuticular patch, sometimes with most of tergite dark reddish; head scarcely narrower than mesosoma, if at all **14**

14 (13) Antennal scrobe lacking dorsal carina (central: Texas and southern Great Plains) .... *D. klugiodes* (3.3.4.4)

Antennal scrobe with distinct dorsal carina **15**

15 (14) Metasoma with distinct black setal patch from apex of T2 to T3 (central: transition areas between dry and humid areas in central and eastern Texas and Oklahoma) .... *D. clotho* (3.3.1.3)

T2–5 with uniformly yellow to red setae **16**

16 (15) Fringes of S2–5 mostly red; genal carina distinct and sharp, gena coarsely sculptured; T2 cuticle often dark reddish beneath reddish setae; dorsal setae more extensively erect and shaggy (western: Mojave and Sonoran Deserts, Madrean Archipelago and Arizona mountains) .... *D. magnifica* (3.3.1.1)

Fringes of S2–5 mostly black mesally; genal carina weaker, gena more finely sculptured; T2 cuticle generally distinctly black; dorsal setae usually more appressed and smoother (central, western: Madrean Archipelago and mountains in Arizona east to Texas and Great Plains) .... *D. klugii* (3.3.1.2)

9.3.3.4—Madrean Females Color Syndrome

1 Pronotum divided from mesonotum as an articulating plate **2**

Pronotum fused with mesonotum **3**

2 (1) Female; metasoma with 6 tergites; antenna with
10 flagellomeres .... ***Myrmosula*, see couplet 8
in Cryptic FCS key (p. 395)

Male; metasoma with 7 tergites; antenna with 11 flagellomeres **12**

3 (1, 2) T1 shape short subcylindrical; T2 lacking felt line .... **4 (*Ephuta*)**

T1 shape variable; T2 with felt line **12**

4 (3) T6 convex, with pygidial plate scarcely defined or completely covered by dense setae **5**

T6 with distinct, although slender, pygidial plate not obscured by setae **7**

5 (4) Pygidial area punctate mesally, completely covered by dense
curly setae; body color darker reddish-brown (western: Madrean
Archipelago) .... *E. tumacacori* (3.4.19.1)

Pygidial area smooth mesally, usually visible through slightly sparse
and straighter setae; body color usually lighter reddish-orange **6**

6 (5) Frons and vertex with uniformly dense bright golden setae; setal
spots of T2 large and distinct (central: southern Texas) .... *E. sudatrix* (3.4.19.4)

Head setae usually paler golden and sparser, frons or vertex
usually with erect setae darker brown; setal spots of T2 less
distinct (eastern: widespread) .... *E. margueritae* (3.4.18.2)

7 (4) Postgenal carina distinct and entire, connecting hypostoma and occipital foramen **8**

Postgenal carina weakly defined, often mostly obliterated between hypostoma
and occipital foramen **10**

8 (7) T2 disc with setal spots weak or absent; hypopygium with distinct
transverse carina basally (western: Madrean Archipelago) .... *E. baboquivari* (3.4.19.2)

T2 disc with distinct whitish setal spots; hypopygium without transverse carina basally **9**

9 (8) Head setae white; hypopygium truncate posteriorly (central: Texas) .... *E. albiceps* (3.4.17.2)

Head setae golden-yellow; hypopygium quadridentate posteriorly
(central: Texas) .... *E. auricapitis* (3.4.19.3)

10 (7) Head setae golden-yellow; T2 disc lacking setal spots (eastern: Florida–Gulf) ..
E. floridana (3.4.18.1)

Head setae silvery-white; T2 disc usually having two silver or golden setal spots **11**

11 (10) Hypopygium with two separated teeth or tubercles basally;
silvery setae on frons sparser than those on vertex (western:
Rocky Mountains) .... *E. coloradella* (3.4.17.3)

Hypopygium with transverse ridge basally; frons and vertex each with dense
silvery setae (western: Pacific) .... *E. argenticeps* (3.4.17.1)

12 (2, 3) T6 convex, lacking distinct pygidial plate **13**

T6 with distinct pygidial plate defined by lateral carina **15**

13 (12) Body uniformly pale orange, except T2 with black basal
patches; head and mesosoma with uniformly sparse
orange setae (western: Madrean Archipelago) .... *Stethophotopsis maculata* (3.4.16.1)

Body color variable, usually darker reddish-brown, T2 without black basal
patches; head setae golden, contrasting with brownish mesosomal setae **14**

14 (13) Head narrower, rounded; T2 uniformly orange-brown;
T3 orange-brown with two pale yellow cuticular patches
basally (central, western: widespread, mostly absent
from hot deserts and Pacific areas) .... *Protophotopsis venenaria* (3.4.9.1)

Head large, quadrate; T2 with four pale pale yellow
cuticular spots; T3 uniformly orange-brown
(western: Madrean Archipelago) .... *Pseudomethoca quadrinotata* (3.4.11.4)

15 (12) Mesosoma quadrate, pronotum and propodeum wider than
mesonotum or mesosomal sides basically flat; eye large, vertically ovate .... **16 (*Timulla*)**

Mesonotum generally pear-shaped, mesonotum generally
wider than pronotum or propodeum **17**

16 (15) Metasoma blackish, clearly darker than mesosoma .... ***Timulla*, in part, see couplet 14**
in *Timulla*-like FCS key (p. 390)

Metasoma pale orange or brown, basically concolorous
with mesosoma .... ***Timulla*, in part, see couplet 6**
in Eastern FCS key (p. 391)

17 (15) T2 fringe composed of plumose setae **18**

T2 fringe composed of simple setae **19**

18 (17) T2 with broad transverse basal band of black setae (central,
western: widespread) .... *S. marpesia* (2.4.17.1)

T2 without black basal band .... **various nocturnal genera** (see chapter 6, p. 285)

19 (17) T1 shape petiolate; mesosomal sides not constricted at
propodeal spiracle .... **20 (*Dasymutilla*)**

T1 shape sessile; mesosomal sides constricted at propodeal spiracle **39**

20 (19) Mesosoma without scutellar scale **21**

Mesosoma with scutellar scale **29**

21 (20) Apices of mid and hind femora truncate; pygidium uniformly microreticulate **22**

Apices of mid and hind femora rounded; pygidium largely rugose or striate **26**

22 (21) Genal carina distinct, extending posteriorly onto vertex, head angulate
posterolaterally **23**

Genal carina weaker, not reaching vertex, head rounded posterolaterally **24**

23 (22) Fringe of T2 with mesal silver setal patch (western: Madrean
Archipelago) .... *D. dilucida* (3.4.3.1)

Fringe of T2 entirely black mesally (western: Madrean Archipelago) .... *D. ferruginea* (3.4.3.2)

24 (22) Antennal scrobe lacking dorsal carina; head and mesosoma
covered entirely with dense golden setae (central, western:
Madrean Archipelago east to Texas and Great Plains) .... *D. snoworum* (3.4.4.1)

Antennal scrobe having dorsal carina; head and/or mesosoma
with sparser silvery or pale orange setae **25**

25 (24) Vertex covered with erect white setae, contrasting with sparser
short gray mesonotal setae (western: desert-adjacent mountains
and canyons in California and Nevada) .... *D. dammersi* (3.4.2.1)

Vertex and mesosoma with similar pale orange setae, except
propodeum with distinct mesal patch of white setae (western:
Sonoran Desert) .... *D. heliophila* (3.4.1.1)

26 (21) T2 disc uniformly areolate with sparse erect setae only (western:
Madrean Archipelago) .... *D. dionysia* (3.4.8.1)

T2 disc variably punctured with at least some patches of dense appressed setae **27**

27 (26) Metasoma entirely black, T2 disc covered with dense reddish setae
(central, western: Madrean Archipelago east to Texas and Great Plains) .... *D. foxi* (3.4.6.3)

Metasoma black or reddish with sparsely setose yellow or orange
cuticular markings on T2 disc (rare species in Arizona) **28**

28 (27) Mesosoma clearly longer than wide; mesosoma with distinct
Y-shaped pale golden setal patch, sides of propodeum with
distinct patches of black setae (western: Arizona mountains) .... *D. fasciventris* (3.4.8.3)

Mesosoma as wide as long or nearly so; mesosoma with posterior half, including propodeum, covered with pale golden setae (western: Madrean Archipelago) .... *D. toluca* (3.4.6.5)

29 (20) Antennal scrobe with no trace of dorsal carina **30**

Antennal scrobe with dorsal carina, sometimes poorly defined **34**

30 (29) Propodeum nearly bare, sculpture consisting of many scattered tubercles (eastern: sandhills in central Florida) .... *D. archboldi* (3.6.4.1)

Propodeum generally covered with dense silvery or golden setae (central and western USA) **31**

31 (30) F1 about twice as long as wide, nearly as wide as F2 + 3 combined; T2 disc with coarse sparsely setose sculpture except for mesal black spot surrounded by silver setae **32**

F1 not much longer than wide, much shorter than F2 + 3 combined; T2 disc with somewhat uniformly dense sculpture and setae, without black mesal spot **33**

32 (31) Pronotum with distinct patch of black setae, mesonotum and propodeum with silvery setae (western: Madrean Archipelago) .... *D. bonita* (3.4.5.3)

Mesosomal dorsum covered entirely with silvery setae, setae sparse mesally (central: Texas) .... *D. eurynome* (3.4.5.2)

33 (31) Mesosomal dorsum covered entirely with dense silvery or golden setae (central, western: widespread but absent from Pacific areas) ...... *D. monticola* (3.4.4.2)

Pronotum and mesonotum (at least anteriorly) with sparse brown or blackish setae, contrasting with dense silvery or golden propodeal setae (central: widespread; western: Chihuahuan Desert) .... *D. birkmani* (3.4.5.1)

34 (29) Head and mesosoma with orange-brown cuticle; T2 fringe and T3 mostly black with mesal patch of white setae (western: Madrean Archipelago) .... *D. cirrhomeris* (3.4.8.2)

Head and mesosoma with black or dark reddish-brown cuticle; T2 fringe and T3 with different pattern, usually with setae mostly whitish or pale yellow **35**

35 (34) Vertex covered with blackish setae; T2 disc with coarse sparsely setose sculpture except for anteromesal black spot surrounded by silver setae (western: Madrean Archipelago) .... *D. saetigera* (3.4.6.4)

Head setae entirely white or pale yellow; T2 disc with somewhat uniformly dense sculpture and setae, without black and silver spot mesally **36**

36 (35) Mesonotum armed with lateral tubercle; T3–6 with dense whitish setae (western: Madrean Archipelago) .... *D. pulchra* (3.4.7.1)

Mesonotum without lateral tubercle; T3–6 (at least T5–6) with black setae mesally **37**

37 (36) T2 with basal patch of bright red setae; propodeum and T3 with black setae mesally (western: Madrean Archipelago) .... *D. asteria* (3.4.6.2)

T2 with black and white setae only; propodeal dorsum and T3 entirely whitish or pale yellow **38**

38 (37) Cuticle and legs entirely black (western: Madrean Archipelago) .... *D. arachnoides* (3.4.7.2)

Cuticle (especially T2) and legs largely reddish-brown (western: Madrean Archipelago) .... *D. sicheliana* (3.4.6.1)

39 (19) Gena with large blunt ventral tooth; body longer than 7 mm (western: Madrean Archipelago) .... *Invreiella cephalargia* (3.4.10.1)

Gena usually unarmed; if genal tooth present, then tooth acute and body length less than 6 mm .... **40 (*Pseudomethoca*)**

40 (39) Gena, postgena, or hypostoma armed with ventral tooth
(body size less than 6 mm) **41**

Head unarmed ventrally, if hypostomal carina raised to weak tooth,
then body longer than 8 mm **45**

41 (40) Head clothed with sparse largely brownish setae .... ***Ps. frigida* cluster, see
couplet 35 in Cryptic FCS key (p. 397)**

Head clothed with (usually) dense golden setae, sometimes more silver-tinted **42**

42 (41) Gena unarmed, hypostoma with small tooth (western: Madrean
Archipelago) .... *Ps. peremptrix* (3.4.12.6)

Gena armed with ventral tooth, postgena sometimes armed **43**

43 (42) Mandible bidentate, inner tooth small or obliterated; hypostoma with
raised tooth-like ridge, making gena appear bidentate (western:
Chihuahuan and Sonoran Deserts, Madrean Archipelago) .... *Ps. toumeyi* (3.4.12.1)

Mandible generally tridentate, inner tooth large and distinct;
postgena unarmed, gena clearly with one tooth only **44**

44 (43) Mesosoma as wide as long, dorsal propodeal surface indistinct
and smoothly rounding into posterior surface (central, western:
widespread, mostly absent from hot deserts and Pacific areas) .... *Ps. bequaerti* (3.4.12.2)

Mesosoma longer than wide, dorsal propodeal surface distinct
and somewhat distinctly perpendicular to posterior surface
(central, western: Rocky Mountains and Great Plains in Colorado) .... *Ps. klotsi* (3.4.12.3)

45 (40) Scutellar area with raised truncate ridge (western: Sonoran
and Chihuahuan Deserts) .... *Ps. donaeanae* (3.4.14.1)

Scutellar area unarmed, simply punctate or areolate **46**

46 (45) T2 posterior margin with two subrectangular pale yellow cuticular spots;
head mostly reddish-brown with sides black, dorsal setae sparse **47**

T2 cuticle lacking pale yellow spots; head generally uniformly
pale reddish-brown, often covered with dense golden setae **48**

47 (46) Antennal scrobe carina thick, distinct; hypostomal carina thickened
near mandible base, slender near midpoint of head venter; vertical
mesopleural carina largely obliterated in ventral portion (western:
Madrean Archipelago) .... *Ps. bethae* (3.4.13.1)

Antennal scrobe carina faint, inconspicuous; hypostomal carina
evenly slender throughout; vertical mesopleural carina complete
from mesosomal dorsum to mid-coxa (central: southern Texas) .... *Ps. mulaiki* (3.4.13.2)

48 (46) Frons and vertex with sparse blackish setae, not obscuring head cuticle **49**

Frons and vertex with entirely golden setae, usually dense
and obscuring head cuticle **50**

49 (48) Fringes of T1 and T2 black; body length greater than 8 mm
(western: Madrean Archipelago) .... *Ps. sonorae* (3.4.14.2)

Fringe of T1 and/or T2 whitish; body length less than 6 mm .... ***Ps. frigida* cluster, see
couplet 35 in Cryptic FCS key (p. 397)**

50 (48) Fringe of T2 entirely composed of pale golden setae; pygidial
plate sculpture uniformly microreticulate (western: Mojave
Desert, Colorado Plateau, Madrean Archipelago) .... *Ps. perditrix* (3.4.12.5)

Fringe of T2 with black setae, at least mesally; pygidial plate
sculpture generally rugose **51**

51 (50) T2 with two large subcircular lateral spots of silvery setae; body
length generally less than 5 mm (western: Chihuahuan Desert,
Colorado Plateau, Madrean Archipelago) .... *Ps. scaevollela* (3.4.12.4)

T2 pattern variable, not as above; body length generally greater than 8 mm **52**

52 (51) Disc of T2 broadly covered with reddish-orange setae
(western: Sonoran Desert and Madrean Archipelago) .... *Ps. praeclara* (3.4.11.2)

Disc of T2 with contrasting pattern of black and silvery setae **53**

53 (52) T2 disc with hourglass shape of black setae, laterally with
large silvery triangular setal patches; T2 fringe with obvious
silvery setae laterally (western: Mojave Desert) .... *Ps. connectens* (3.4.11.3)

T2 disc with ⊥⊥-shaped pattern of silvery setae; T2 fringe
almost entirely black (central, western: widespread, mostly
absent from hot deserts and Pacific areas) .... *Ps. contumax* (3.4.11.1)

9.3.3.5—*Timulla*-like Females Color Syndrome

1   Pronotum divided from mesonotum as an articulating plate **2**

     Pronotum fused with mesonotum **3**

2 (1)   Female; metasoma with 6 tergites; antenna with 10
flagellomeres .... **Myrmosinae, see couplet 3
in Cryptic FCS key** (p. 395)

     Male; metasoma with 7 tergites; antenna with 11 flagellomeres **3**

3 (1, 2) Fringe of T2 composed of distinctly plumose whitish setae **4**

     Fringe of T2 composed of simple or weakly brachyplumose setae **5**

4 (3)   Pronotum differentiated from mesonotum as articulating
plate (males) or by distinct arcuate transverse carina
(western: Pacific) .... *Morsyma ashmeadi* (3.5.4.1)

     Pronotum not so differentiated from remainder of mesosoma .... **various nocturnal genera**
(see chapter 6, p. 285)

5 (3)   Head much wider than mesosoma, gena and postgena
armed with ventral teeth (central: widespread) .... *Myrmilloides grandiceps* (3.5.2.1)

     Head at most scarcely wider than mesosoma, usually
unarmed ventrally, at most with small postgenal tooth only **6**

6 (5)   T6 mesally smooth, convex, without defined pygidial plate **7**

     T6 with distinctly sculptured pygidial plate defined by lateral carina **8**

7 (6)   Mesosomal sides constricted at propodeal spiracle; postgena
armed with small tooth; T2 marked with four pale golden setal
spots: one basomesal, one apicomesal, and two lateral
(central: widespread) .... *Pseudomethoca wickhami* (3.5.3.1)

     Mesosomal sides not so constricted at propodeal spiracle;
postgena unarmed; T2 lacking discrete setal spots .... **Lomachaeta, see
couplet 42 in Cryptic FCS key** (p. 398)

8 (6)   T1 shape slender subcylindrical .... **Ephuta, see couplet 15 in
Cryptic FCS key** (p. 396)

     T1 shape somewhat broader, subpetiolate or subsessile **9**

9 (8)   Mesosomal sides constricted at propodeal spiracle;
T2 black with whitish lateral setae and pale yellow
posterolateral cuticular spots and mesal patch
of reddish-brown cuticle .... **Pseudomethoca bethae species-cluster,
see couplet 47 in Madrean FCS key** (p. 388)

     Mesosomal sides not so constricted; T2 cuticle generally
entirely black marked with pale yellow cuticular spots of
white to pale golden setal patches **10**

10 (9) Mesosoma pear-shaped; eye circular and somewhat bulging .... **11** (*Dasymutilla*)

Mesosoma rectangular; eye longitudinally ovate, basically flat .... **14** (*Timulla*)

11 (10) Head and mesosoma with dense golden setae;
vertex lacking posterolateral tubercle .... **D. snoworum species-cluster, see couplet 20 in Madrean FCS key** (p. 386)

Head and mesosoma with sparse pale orange setae;
vertex armed with posterolateral tubercle **12**

12 (11) Vertex posterolaterally angular, tubercles sharp comma-shaped,
more distantly spaced than epaulets (central: southern Texas) .... *D. curticeps* (3.5.1.3)

Vertex posterolaterally rounded, tubercles smooth sublinear,
more closely spaced than epaulets **13**

13 (12) Propodeum with dense brush of thick pale orange setae
(central: southern Texas) .... *D. parksi* (3.5.1.2)

Propodeum without setal brush, setae of similar length and
density as on mesonotum (central: widespread) .... *D. quadriguttata* (3.5.1.1)

14 (10) Propodeum clearly narrower than pronotum and mesonotum,
scutellar scale absent (western: Madrean Archipelago and
Chihuahuan Desert) .... *T. navasota coahuila* (3.5.5.1)

Propodeum as wider or wider than pronotum and mesonotum,
scutellar scale present or absent **15**

15 (14) Propodeum clearly wider than pronotum (eastern, central:
widespread) .... *T. ferrugata* (3.5.6.1)

Propodeum generally equally wide as pronotum **16**

16 (15) Pygidial sculpture with distinct parallel striae continuous to
posterior margin (central: widespread) .... *T. leona* (3.5.9.3)

Pygidial sculpture variable, usually rugose or granulate, especially apically **17**

17 (16) T2 disc without whitish setal stripes or bands basally
(western: Utah southeast to western Texas) .... *T. grotei* (3.5.7.1)

T2 disc with whitish setal stripes or bands basally **18**

18 (17) Setal stripes of T2 continuous from base to T2 fringe (central,
western: Great Plains, Chihuahuan Desert, Madrean Archipelago) .... *T. suspensa* (3.5.8.1)

Setal stripes of T2 restricted to basal portion of tergite **19**

19 (18) Scutellar scale narrow, posteriorly rounded, without furrow or
scattered tubercles anterior to scale; humerus generally rounded **20**

Scutellar scale broad, posteriorly straight or slightly wavy,
usually with transverse furrow or row of scattered tubercles
anterior to scale; humerus generally angular **21**

20 (19) Mesonotum not constricted (central: widespread) .... *T. wileyae* (3.5.10.1)

Mesonotum constricted (central: widespread) .... *T. dubitata* (3.5.10.2)

21 (19) Mesonotum not constricted (western: Sonoran Desert and
Madrean Archipelago) .... *T. nicholi* (3.5.9.4)

Mesonotum constricted **22**

22 (21) Punctures of T2 coarse, intervals generally bare and shining (central:
widespread; western: Chihuahuan Desert and Madrean Archipelago) .... *T. oajaca* (3.5.9.2)

Punctures of T2 generally weaker, intervals generally micropunctate
and covered with short setae (eastern: widespread; central: Great P
lains; western: rare in some northwestern areas) .... *T. vagans* (3.5.9.1)

9.3.3.6—Eastern Females Color Syndrome

1	Head wider than mesosoma, gena and/or postgena armed with ventral tooth **2**
	Head usually not, or only scarcely, wider than mesosoma, unarmed ventrally **4**
2 (1)	Head much wider than mesosoma, gena and postgena both armed with ventral tooth (central: widespread) .... *Myrmilloides grandiceps* (3.6.7.1)
	Head slightly wider than mesosoma, usually with only one genal or postgenal tooth **3**
3 (2)	T6 convex, without defined pygidial plate; postgena armed with small tooth; T2 marked with two distinct lateral spots of orange cuticle and setae (eastern: widespread) .... *Pseudomethoca wickhami* (3.6.9.2)
	T6 with defined pygidial plate; usually with tooth on gena; T2 with setal spots silvery or inconspicuous .... **Pseudomethoca, in part, see couplet 35 in Cryptic FCS key (p. 397)**
4 (1)	Pronotum divided from mesonotum as an articulating plate .... **Myrmosinae, see couplet 3 in Cryptic FCS key (p. 395)**
	Pronotum not so differentiated from mesonotum **5**
5 (4)	Mesosoma rectangular, pronotum and/or propodeum as wide or wider than mesonotum; eye large, longitudinally ovate .... **6 (*Timulla*)**
	Mesosoma ovate or pear-shaped, mesonotum wider than pronotum or propodeum; if mesosoma apparently rectangular, then eye small, transversely ovate **20**
6 (5)	Fringe of T1 composed of black setae **7**
	Fringe of T1 composed mostly or entirely of whitish setae **9**
7 (6)	T3–6 black or orange, covered with black and white setae (central and eastern: widespread; western: some northwestern areas) .... *T. dubitatiformis* (3.6.15.3)
	T3–6 pale orange, covered with pale orange setae **8**
8 (7)	Mesonotum constricted; T2 with anterolateral black setal patch (central: southern Texas; eastern: widespread) .... *T. euterpe* (3.6.15.1)
	Mesonotum not constricted; T2 without anterolateral black setal patch (central: southern Texas; eastern: Florida–Gulf) .... *T. euphrosyne* (3.6.15.2)
9 (6)	Propodeum clearly narrower than pronotum and mesonotum, scutellar scale absent (central: Texas) .... *T. navasota navasota* (3.6.13.1)
	Propodeum as wide as or wider than pronotum and mesonotum, scutellar scale present or absent **10**
10 (9)	Scutellar area unarmed, without scale or transverse carina **11**
	Scutellar area with distinct scale or transverse wavy carina **12**
11 (10)	Mesonotum not constricted; legs entirely black (central: southern Texas) .... *T. contigua* (3.6.13.3)
	Mesonotum constricted; legs largely reddish-orange (central, eastern: widespread) .... *T. barbigera* (3.6.16.2)
12 (10)	Propodeum clearly wider than pronotum **13**
	Propodeum generally equally wide as pronotum **14**
13 (12)	Pygidium longitudinally striate; head pale orange-brown (central, eastern: widespread) .... *T. ferrugata* (3.6.14.2)
	Pygidium faintly microreticulate; head often darker reddish-orange (eastern: southern areas) .... *T. floridensis* (3.6.14.1)
14 (12)	Pygidial sculpture with distinct parallel striae continuous to posterior margin **15**
	Pygidial sculpture variable, usually rugose or granulate, especially apically **16**

15 (14) Legs pale orange-brown throughout; lateral propodeal margin weakly serrate (western: Sonoran Desert) .... *T. tyro* (3.6.17.5)

Legs variably darkened brown, at least on femoral apices; lateral propodeal margin with row of distinct teeth (central: Great Plains) .... *T. leona* (3.6.17.4)

16 (14) Setal stripes of T2 continuous from base to T2 fringe (western: Chihuahuan Desert) .... *T. suspensa* (3.6.17.2)

Setal stripes of T2 restricted to basal portion of tergite **17**

17 (16) Scutellar scale narrow, apically rounded, without furrow or scattered tubercles anterior to scale **18**

Scutellar scale broad, apically straight or slightly wavy, usually with transverse furrow or row of scattered tubercles anterior to scale **19**

18 (17) Mesosoma with sides basically parallel, mesonotum scarcely narrower than pronotum and propodeum; T2 with setal spots inconspicuous (eastern: widespread) .... *T. ornatipennis* (3.6.16.4)

Mesonotum distinctly pinched-in, much narrower than pronotum and propodeum; T2 with setal spots usually distinct (eastern: widespread) .... *T. dubitata* (3.6.16.1)

19 (17) Punctures of T2 coarse, sculpture clearly visible beneath setae (central: southern Texas; western: Chihuahuan Desert and Madrean Archipelago) .... *T. oajaca* (3.6.17.3)

Punctures of T2 generally weaker, largely obscured by setae (eastern: widespread) .... *T. vagans* (3.6.17.1)

20 (5) T1 shape short subcylindrical; T2 lacking felt line .... **21 (*Ephuta*)**

T1 shape variable; T2 with felt line **22**

21 (20) Frons and vertex with short, dense silver or golden setae between larger punctures .... **see couplet 4 in Madren FCS key** (p. 385)

Frons and vertex bare between punctures .... **see couplet 15 in Cryptic FCS key** (p. 396)

22 (20) T1 shape subpetiolate; T2 fringe composed of simple setae; pygidium with defined plate .... **23 (*Dasymutilla*)**

T1 shape usually sessile; if more slender, then T2 fringe composed of plumose setae or pygidium convex without defined plate **38**

23 (22) Body cuticle entirely black, covered with dense red or orange setae; metasomal markings composed of black and reddish setae only (eastern: widespread) .... *D. occidentalis* (3.6.1.1)

Body cuticle largely orange or brownish, especially dorsally; metasomal marking composed at least in part of silvery setae **24**

24 (23) Mesosomal dorsum, at least propodeum, with distinct spot or patch of silvery, golden, or white setae .... **see couplet 20 in Madrean FCS key** (p. 386)

Mesosomal dorsum with more or less uniformly pale orange setae, lacking distinct spot or patch **25**

25 (24) Mesosoma without scutellar scale **26**

Mesosoma with scutellar scale **28**

26 (25) Apices of mid and hind femora truncate; T1 with stark discrete white setal patch (central, eastern: widespread) .... *D. scaevola* (3.6.6.1)

Apices of mid and hind femora rounded; T1 with generally sparse black and whitish setae **27**

27 (26) Mesosoma as wide as long; mandible tridentate
(central, eastern: widespread) .... *D. asopus* (3.6.5.2)

Mesosoma much longer than wide; mandible bidentate
(central, eastern: widespread) .... *D. nigripes* (3.6.5.4)

28 (25) Vertex rounded posterolaterally, without tubercle or angle **29**

Vertex armed posterolaterally with tubercle or angle **32**

29 (28) Pygidial plate covered with short setae; head clearly wider
than mesosoma (central: southeastern Great Plains and Texas) .... *D. creon* (3.6.6.2)

Pygidial plate striate or rugose, without setae on plate; head
usually narrower than mesosoma **30**

30 (29) Mesosoma with thick transverse wavy carina anterior to
scutellar scale (central, eastern: widespread) .... *D. bioculata* (3.6.5.3)

Mesosoma lacking carina anterior to scutellar scale **31**

31 (30) Head covered with silvery setae; propodeal sculpture having
many scattered tubercles (eastern: sandhills in central Florida) .... *D. archboldi* (3.6.4.1)

Head setae mostly orange, like mesosomal setae; propodeal
sculpture areolate or punctate without tubercles (central, eastern,
western: widespread except Pacific and western hot deserts) .... *D. ursus* (3.6.5.1)

32 (28) Vertex simply angular, lacking discrete smooth posterolateral
tubercle; propodeum with many scattered tubercles and largely
obliterated intervals or with broad shallow areolations **33**

Propodeum generally equally wide as pronotum **35**

33 (32) Propodeum with broad and shallow complete areolations
(eastern: northeastern areas) .... *D. canella* (3.6.3.1)

Propodeum with numerous scattered tubercles (southeastern USA) **34**

34 (33) Lateral margin of head directly behind eye curving inwardly; mesal
black spot on fringe of T2 narrow; lateral face of propodeum
densely clothed with silver setae (eastern: Florida–Gulf) .... *D. arenerronea* (3.6.3.3)

Lateral margin of head directly behind eye virtually straight;
mesal black setae on fringe of T2 broad, covering at least
0.25 × fringe width; lateral face of propodeum sparsely
clothed with silver setae (eastern: southern areas) .... *D. macilenta* (3.6.3.2)

35 (32) Tibial spurs whitish; antennal scrobe with dorsal carina
(eastern: widespread but more common in northeast) .... *D. gibbosa* (3.6.2.1)

Tibial spurs black or reddish, concolorous with tibia; antennal
scrobe lacking dorsal carina **36**

36 (35) Tubercles on vertex linear, generally weak, head usually
rounded posterolaterally (central, eastern: widespread) .... *D. quadriguttata* (3.6.2.4)

Tubercles on vertex somewhat comma-shaped, large,
head clearly angular posterolaterally (rare) **37**

37 (36) Propodeum with brush of thick dark reddish-brown setae,
much denser than mesonotal setae (eastern: Florida–Gulf) .... *D. rubicunda* (3.6.2.3)

Propodeum with sparse short orange setae, similar to
mesonotum (central: southeastern Great Plains) .... *D. angulata* (3.6.2.2)

38 (22) Fringe of T2 composed of plumose setae **39**

Fringe of T2 composed of simple of brachyplumose setae **43**

39 (38) T6 with pygidial plate defined by lateral carinae **40**

T6 convex without defined pygidial plate **41**

40 (39) Mesosoma elongate subrectangular (eastern USA) .... ***Photomorphus*, in part, see couplet 25 in Cryptic FCS key** (p. 397)

Mesosoma usually shorter, pear-shaped .... **various nocturnal genera** (see chapter 6, p. 285)

41 (39) T1 shape short sessile, with long sparse weakly plumose setae (western: northwestern areas) .... *S. contracta* (3.6.11.3)

T1 shape broadly subpetiolate, with patch or fringe of short dense plumose setae **42**

42 (41) T1 with discrete small mesal patch of white plumose setae (central, eastern: widespread) .... *S. pensylvanica* (3.6.11.1)

T1 with wide band of white plumose setae (central: widespread) .... *S. auripilis* (3.6.11.2)

43 (38) Mesosoma elongate subrectangular; head generally small, rounded posteriorly; gena without carina (eastern USA) .... ***Photomorphus*, in part, see couplet 31 in cryptic FCS key** (p. 397)

Mesosoma usually shorter, pear-shaped, constricted at propodeal spiracle; head often large, quadrate; gena armed with carina **44**

44 (43) Humeral carina weakly defined, slightly interrupted by adjacent punctures or becoming obsolete ventral to epaulet; margin between posterior and lateral propodeal surfaces simply punctate **45**

Humeral carina sharp, distinct, complete to epaulet; margin between posterior and lateral propodeal surfaces with irregular row of small teeth or tubercles **47**

45 (44) T2 disc with two distinct spots of paler orange cuticle (eastern: Florida–Gulf) .... *Ps. oculata* (3.6.9.1)

T2 disc relatively unicolorous, without distinct pale orange spots **46**

46 (45) Metapleuron covered with short silvery setae; dorsal propodeal surface indistinct and smoothly rounding into posterior surface; sides of body, and often legs, pale orange-brown, concolorous with dorsum of body (central, eastern: widespread) .... *Ps. sanbornii* (3.6.10.3)

Metapleuron mostly bare; dorsal propodeal surface distinct and somewhat distinctly perpendicular to posterior surface; legs and sides of body often partly blackened (central, western: widespread but absent from Pacific and western hot deserts) .... *Ps. propinqua* (3.6.10.4)

47 (44) Humeral carina forming distinct obtuse angle (central, eastern: widespread) .... *Ps. oceola* (3.6.8.1)

Humeral carina smoothly rounded **48**

48 (47) Pygidium longitudinally striate (central, eastern: widespread) .... *Ps. simillima* (3.6.10.1)

Pygidium irregularly rugose **49**

49 (48) Apex of T2 with thin band of black setae and extensive silvery setae pre-apically; legs generally blackish-red (central: widespread) .... *Ps. paludata* (3.6.10.4)

Apex of T2 with distinct thick band of black setae; legs pale orange-brown (central: Texas) .... *Ps. meritoria* (3.6.10.2)

9.3.3.7—Cryptic Females Color Syndrome

1 Pronotum divided from mesonotum as an articulating plate **2**

Pronotum fused with mesonotum **14**

2 (1) Male; metasoma with 7 tergites; antenna with 11 flagellomeres; fringe of T2 usually with plumose setae .... **various nocturnal genera** (see chapter 6, p. 285)

Female; metasoma with 6 tergites; antenna with 10 flagellomeres; fringe of T2 with simple setae **3**

3 (2) Clypeus with basomesal raised carina; foretarsus without rake of long flattened setae; ocelli usually present; T2 without pale yellow cuticular spots .... **4 (*Myrmosa*)**

Clypeus without basomesal carina; foretarsus with rake of long flattened setae; ocelli absent; T2 usually with pale yellow cuticular spots **7**

4 (3) Head narrow, vertex with lateral margins convergent directly posterior to eye (central, eastern: widespread) .... *M. unicolor* (3.7.7.1)

Head wider, vertex with lateral margins parallel directly posterior to eye **5**

5 (4) Antennal scrobe with dorsal carina; propodeum with posterior row of thick erect teeth (central: Kansas) .... *M. peculiaris* (3.7.7.2)

Antennal scrobe without dorsal carina; propodeum with smaller narrower tubercles **6**

6 (5) Humeral angle and posterolateral corner of propodeum each with large sharp tooth (eastern: New York, Virginia, and maybe Georgia) .... *M. blakei* (3.7.7.3)

Humeral angle rounded; posterolateral corner of propodeum with small weak tooth (western: Pacific) .... *M. bradleyi* (3.7.7.2)

7 (3) Head narrow, scarcely wider than mesonotum, rounded posteriorly; mandible tridentate without ventral lamella basally (western: Sonoran Desert near Colorado River) .... *Leiomyrmosa spilota* (3.7.8.1)

Head usually clearly wider than mesonotum, subquadrate posteriorly; mandible bidentate with ventral lamella basally .... **8 (*Myrmosula*)**

8 (7) Antennal tubercles separated, frons between them not raised into interantennal prominence **9**

Antennal tubercles fused into raised interantennal prominence **10**

9 (8) Antennal tubercle with small dorsal tooth, distance between antennal tubercles narrower than tubercle width (central: widespread; western: Alberta [Canada]) .... *M. peregrinatrix* (3.7.9.2)

Antennal tubercle unarmed, distance between antennal tubercles subequal to tubercle width (central, eastern: widespread) .... *M. parvula* (3.7.9.1)

10 (8) Postgenal area unarmed, or at least not visible in lateral view; interantennal prominence forming two large posteriorly divergent ridges (western: Pacific) .... *M. rutilans* (3.4.20.1, 3.7.9.5)

Postgena armed with tooth that is clearly visible below ventral mandibular lamella in lateral view; interantennal prominence variable, not as above **11**

11 (10) Ventral mandibular lamella low convex; interantennal prominence slender, erect **12**

Ventral mandibular lamella concave or raised basally; interantennal prominence broader, flatter **13**

12 (11) Interantennal prominence a slender bifurcated wedge, in lateral view with anterior margin nearly flat (western: Sonoran Desert) .... *M. boharti* (3.4.20.2)

Interantennal prominence stouter with small sharp parallel carinae, in lateral view with anterior margin convex (western: Sonoran Desert) .... *M. nasuta* (3.4.20.3, 3.7.9.6)

13 (11) Interantennal prominence with short widely separated ridges (western: Pacific) .... *M. pacifica* (3.7.9.4)

Interantennal prominence with long closely spaced ridges (western: Pacific) .... *M. exaggerata* (3.7.9.3)

14 (1) T1 shape short subcylindrical; T2 without felt line .... **15 (*Ephuta*)**

T1 shape variable, usually petiolate, disciform, or sessile; T2 with felt line **23**

15 (14) Frons and vertex with intervals micropunctate with short, dense setae; T2 disc often marked with two small spots of white setae (not to be confused with the whitish posterior setal bands) **16**

Frons and vertex with intervals smooth, bare; T2 disc never having setal spots, whitish setal patches restricted to fringe area or posterior bands **17**

16 (15) Vertex setae dark brown; T2 disc with small white setal patches; T7 with defined pygidial plate (central, eastern: northern areas) .... *E. conchate* (3.7.6.2)

Vertex setae white, silver, or golden; other characters variable (generally from southeastern, southern, or western USA) .... **Ephuta, in part, see couplet 4 in Madrean FCS key** (p. 385)

17 (15) Postgenal carina entire, connecting hypostoma and occipital foramen **18**

Postgenal carina mostly obliterated between hypostoma and occipital foramen **19**

18 (17) Pygidium slender with parallel lateral carinae, largely micropunctate (central: southern Texas; western: Chihuahuan Desert and Madrean Archipelago) .... *E. minuta* (3.7.6.6)

Pygidium broader with lateral carinae convex, mostly smooth (central, eastern: widespread) .... *E. scrupea* (3.7.6.1)

19 (17) T3–4 with dense bands of entirely black setae; antenna (including scape and pedicel) and legs (except coxae and trochanters) blackish (eastern: Florida–Gulf) .... *E. sabaliana* (3.6.12.2)

Tergal setae variable in color; when apparently darkened, they are generally sparse and brown with interspersed silver or golden-brown setae, at least on T4; antenna (at least scape) and legs (at least femora basally) lighter brown or reddish-orange **20**

20 (19) Eye large and bulging, distance between eyes on frons less than 1.2 × greatest eye diameter, vertex relatively short with the posterolateral margin somewhat flattened (central: southern Texas; eastern: widespread) .... *E. slossonae* (3.6.12.1, 3.7.6.8)

Eye generally smaller, distance between eyes on frons usually greater than 1.3 × greatest eye diameter, vertex somewhat longer with the posterolateral margin more convex **21**

21 (20) T2 with many punctures longitudinally oblique, some transverse intervals obliterated (central, eastern: widespread) .... *E. puteola* (3.6.12.3, 3.7.6.7)

T2 with punctures uniformly circular, without any obliterated transverse intervals **22**

22 (21) Cuticle of T3–5 darker brown than orange-brown T2 and often T6; head punctures separated, even on frons; T3–4 with setae dark brown (eastern: widespread) .... *E. spinifera* (3.7.6.9)

Cuticle of T2–6 concolorous pale orange-brown; head punctures (at least on vertex) confluent; T3–4 with pale golden setae (central: southern Texas) .... *E. copano* (3.7.6.3)

23 (14) Fringe of T2 composed of plumose setae **24**

Fringe of T2 composed of simple or obscurely brachyplumose setae **27**

24 (23) Mesosoma elongate subrectangular; T1 shape sessile; species from eastern USA .... **25 (*Photomorphus*, in part)**

Mesosoma or T1 shape not as above, species from central or western USA .... **various nocturnal genera** (see chapter 6, p. 285)

25 (24) Femora darker than mesosoma, blackish-brown; pygidium with distinct striae basally (central: widespread; eastern: northern areas) .... *Ph. auriventris* (3.7.4.3)

Femora yellow-brown or orange-brown, concolorous with mesosoma; pygidium smooth or obscurely microparticulate **26**

26 (25) Entire body pale yellow-brown; body setae uniformly silvery-white (eastern: Florida–Gulf) .... *Ph. archboldi* (3.7.4.1)

Body darker orange-brown; mesonotum and T2 disc with some appressed brown setae (eastern: mostly southern areas) .... *Ph. spinci* (3.7.4.2)

27 (23) T6 with pygidial plate defined by lateral carinae **28**

T6 convex without defined pygidial plate **41**

28 (27) Mesosoma rectangular; eye longitudinally ovate, basically flat .... ***Timulla*, see couplet 6 in Eastern FCS Key** (p. 391)

Mesosoma pear-shaped; eye generally circular **29**

29 (28) T1 shape petiolate .... ***Dasymutilla*, see couplet 23 in Eastern FCS Key** (p. 392)

T1 shape sessile **30**

30 (29) Head rounded, not wider than mesosoma; mesosoma longer than wide, subrectangular .... **31 (*Photomorphus*, in part)**

Head quadrate, wider than mesosoma; if head narrower and rounded, then mesosoma as wide as long .... **35 (*Pseudomethoca*)**

31 (30) Pygidial sculpture uniformly microreticulate (central: widespread) .... *Ph. myrmicoides* (3.7.5.5)

Pygidial sculpture striate **32**

32 (31) Mandible with distinct ventral tooth basally **33**

Mandible with ventral tooth absent or minute, inconspicuous **34**

33 (32) Body pale orange-brown, legs pale yellow-brown; setae of T3–5 entirely pale whitish (eastern: Florida–Gulf) .... *Ph. paulus* (3.7.5.4)

Body orange-brown, legs often darker brown; setae of T4–5 darker brown than silvery setae of T3 (central, eastern: widespread) .... *Ph. impar* (3.7.5.1)

34 (32) Setae of T3–5 entirely pale whitish (central, eastern: Florida west to Texas) .... *Ph. alogus* (3.7.5.2)

Setae of T4–5 darker brown than silvery setae of T3 (eastern: widespread) .... *Ph. banksi* (3.7.5.3)

35 (30) Gena armed with ventral tooth **36**

Gena unarmed (postgena sometimes with small tooth) **39**

36 (35) Antennal tubercle with raised lamella (western: Pacific) .... *Ps. athamas* (3.7.3.5)

Antennal tubercle unarmed, lacking raised lamella **37**

37 (36) Gena and postgena each having a sharp ventral tooth (central: southern Texas) .... *Ps. nephele* (3.7.3.3)

Postgena unarmed **38**

38 (37) T1 fringe with setae blackish; mesonotum lateral margin with small sharp tooth anterior to propodeal spiracle (central, eastern: widespread) .... *Ps. frigida* (3.7.2.1)

T1 fringe with setae whitish; mesonotum lateral margin with weak blunt tubercle anterior to propodeal spiracle (eastern: Florida–Gulf) .... *Ps. torrida* (3.7.2.2)

39 (35) Postgena armed with ventral tooth (central: southern Texas) .... *Ps. dentigula* (3.7.2.4)

Postgena unarmed **40**

40 (39) Head rounded posteriorly, narrower than mesosoma; mesosoma wider than long (central: widespread) .... *Ps. gila* (3.7.3.1)

Head quadrate posteriorly, wider than mesosoma; mesosoma longer than wide (central: widespread; western: Chihuahuan Desert and Colorado Plateau) .... *Ps. nudula* (3.7.2.6)

41 (27) T1 shape sessile, T1 nearly as wide as T2 (central: widespread; western: Chihuahuan Desert and Madrean Archipelago) .... *Protophotopsis venenaria* (3.4.9.1)

T1 shape disciform or subsessile, much narrower than T2 .... **42 (*Lomachaeta*)**

42 (41) Mesosoma and T2 disc with matching thickened posteriorly directed subparallel appressed silvery to golden setae **43**

Mesosoma and T2 disc without such thickened parallel bristles **44**

43 (42) Mesosoma elongate; head wider than mesosoma (central, western: Arizona east to Texas) .... *L. vacamuerta* (3.7.1.7)

Mesosoma more compact; head as wide as mesosoma (western: Mojave Desert, Colorado Plateau) .... *L. argenta* (3.7.1.2)

44 (42) Mandible with basoventral subhyaline lamella, mandible narrowed beyond lamella; pronotal and propodeal spiracles swollen and apparently tuberculate; T1 shape subdisciform **45**

Mandible without such a basal lamella; pronotal and propodeal spiracles usually lower and flatter; T1 shape subsessile **46**

45 (44) Body (except legs) unicolorous pale orange-brown, except T6 blackened; T3–6 with extensive sparse blackish setae (western: widespread) .... *L. cirrhomeris* (3.7.1.4)

Body color variable, with shades of dark and/or light brown; T3–6 with interspersed silvery and blackish setae (central, eastern, western: widespread) .... *L. hicksi* (3.7.1.1)

46 (44) Mesosoma elongate; mesosomal areolations deep and distinct, all intervals distinct; frons with mostly complete areolations (western: Madrean Archipelago) .... *L. osita* (3.7.1.5)

Mesosoma more compact; mesosomal areolations shallower, many intervals obliterated; frons generally with obscure, often incomplete punctures **47**

47 (46) T2 sculpture mostly smooth between punctures; dorsal setae of mesonotum and T2 disc blackish, many setae nearly as long as scape (western: Pacific) .... *L. powelli* (3.7.1.6)

T2 sculpture largely microreticulate between punctures; dorsal setae of mesonotum and T2 generally whitish, shorter than scape **48**

48 (47) Body, including legs, orange-brown, at most with T6 blackened (western: Mojave and Sonoran Deserts) .... *L. beadugrimi* (3.7.1.3)

Femora and T2–5 variably darkened brown, clearly darker than orange-brown mesosoma (western: Arizona) .... *L. warneri* (3.7.1.8)

9.3.4—Subfamilies, tribes, and genera of Nearctic winged males

1 Eye ovate with setae; forewing with veins reaching outer margin .... **2 (Myrmosinae)**

Eye ovate or circular, without setae; forewing with veins not reaching outer margin **3**

2 (1) S1 without anteromesal tooth or ridge; T2–6 junctures not constricted ...... *Myrmosula* (p. 178)

S1 with distinct tooth or ridge anteromesally; T2–6 junctures constricted laterally .... *Myrmosa* (p. 178)

3 (1) Eye vertically ovate with anterior emarginate notch **4**

Eye generally circular, without anterior emarginate notch **5**

4 (3)   T1 shape slender cylindrical; T2 without felt line .... *Ephuta* (p. 175)

　　　　　T1 shape sessile; T2 with felt line .... *Timulla* (p. 175)

5 (3)   Axilla armed posteriorly with tooth or lobe; T1 shape petiolate, disciform,
　　　　　or narrow subsessile; mesosternal area unarmed; S2 without felt line **6**

　　　　　Axilla unarmed; T1 shape variable, often sessile; mesosternal area
　　　　　often armed with teeth or ridges; S2 often with felt line **7**

6 (5)   T1 shape petiolate; body sculpture mostly with dense coarse
　　　　　punctures; body often with bright setal color patterns .... *Dasymutilla* (p. 167)

　　　　　T1 shape disciform or subsessile; body sculpture sparse and
　　　　　shallow; body dully colored with sparse setae .... *Lomachaeta* (p. 169)

7 (5)   Mesosternal area armed with teeth or series of ridges **8**

　　　　　Mesosternal area unarmed **9**

8 (7)   Species from central and eastern USA; T2 fringe with simple
　　　　　or sparse weakly plumose setae; S2 with felt line elongate .... *Photomorphus* (p. 173)

　　　　　Species from western and central USA; T2 fringe
　　　　　often with dense distinctly plumose setae; S2
　　　　　with felt line usually short or absent .... **various nocturnal genera** (see chapter 6, p. 285)

9 (7)   T2 fringe composed of plumose setae **10**

　　　　　T2 fringe composed of simple or obscurely brachyplumose setae **12**

10 (9)   Head and mesosoma reddish-orange, metasoma blackish,
　　　　　setae of T2 disc and T3–5 mostly black, contrasting with
　　　　　distinct plumose white T2 fringe (coastal areas of California) .... *Morsyma* (p. 173)

　　　　　Coloration not as above, setae of T2–5 usually concolorous
　　　　　black, white, yellow, or orange **11**

11 (10) Diurnal species with darkened legs or long brightly colored
　　　　　dorsal setae .... *Sphaeropthalma* (p. 174)

　　　　　Nocturnal species with generally dull
　　　　　brown coloration .... **various nocturnal genera** (see chapter 6, p. 285)

12 (9)   Wings short, not extending beyond mesosomal apex;
　　　　　head wide, gena and postgena armed with ventral teeth .... *Myrmilloides* (p. 169)

　　　　　Fully winged species; head unarmed ventrally **13**

13 (12) T1 shape thick petiolate; S2 with felt line .... *Protophotopsis* (p. 173)

　　　　　T1 shape sessile; S2 without felt line .... *Pseudomethoca* (p. 169)

9.3.5—Males Color Syndromes—fully winged males

1   Head and mesosoma generally with blackish cuticle and covered almost
　　　entirely with black setae **2**

　　　Head and mesosoma with cuticle or setae largely red, yellow, brown, silvery, or whitish **4**

2 (1)   Metasomal setae and cuticle
　　　　　entirely black, at most with some
　　　　　scattered silvery setae .... ***Ephuta*-like MCS, in part** (**5.4:** six species in three genera)

　　　　　Metasoma with extensive yellow to orange-red setae or cuticle **3**

3 (2)   Reddish-orange metasomal color mostly
　　　　　restricted to T1–2 (and sometimes
　　　　　S1–2), T3–6 setae mostly black .... **Eastern MCS** (**5.6:** 20 species in three genera)

　　　　　Reddish-orange metasomal color more extensive,
　　　　　T3–6 (at least) with setae and/or cuticle
　　　　　predominantly reddish .... **Texan MCS** (**5.3:** 19 species in four genera)

4 (1)   Mesonotum with dense white, yellow, orange, or reddish dorsal setae; metasoma generally with dense white, yellow, orange, or reddish setae (at least on T2 or T3–6); body cuticle mostly blackish **5**

Mesonotum usually with black or sparse brownish dorsal setae; if mesonotal setae silvery, then metasoma predominantly reddish-orange; if mesonotal setae orange, then mesosomal cuticle largely reddish-orange or brown **6**

5 (4)   Pale dorsal setae of mesosoma and metasoma predominantly white to pale yellow; species from generally arid habitats .... **Desert MCS (5.1:** 11 species in *Dasymutilla*)

Pale dorsal setae of mesosoma and T2 predominantly yellow to red; species from throughout western North America and the Great Plains .... **Western MCS (5.2:** 29 species in four genera)

6 (4)   Entire body cuticle black or dark brown, at most with legs or tegula pale orange; body setae generally sparse silver and/or black .... ***Ephuta*-like MCS, in part (5.4:** 45 species in eight genera)

Mesosoma and/or metasoma (at least T2 mesal patch) with cuticle largely yellow, reddish-orange, or brown or metasoma with dense yellow to reddish setae **7**

7 (6)   Head and mesosoma with cuticle predominantly black; metasoma with extensive reddish-orange to brown setal or cuticular coloration .... ***Timulla*-like MCS (5.5:** 43 species in nine genera)

Mesosoma (and often head) with cuticle extensively reddish-orange to brown; metasomal color variable .... ***Sphaeropthalma*-like MCS (5.7:** 42 species in nine genera)

9.3.5.1—Desert Males Color Syndrome

1   Hind trochanter dentate; S2 scabrous laterally (western: Sonoran Desert) .... *D. eminentia* (5.1.1.3)

Hind trochanter unarmed; S2 simply punctate laterally **2**

2 (1)   S2 with mesal seta-filled pit or patch of dense microsetae **3**

S2 unmodified, lacking seta-filled pit or patch of microsetae **7**

3 (2)   S2 mesally flattened with central patch of dense microsetae (western: Sonoran Desert in Arizona) .... *D. connectens* (5.1.1.2)

S2 usually convex, with distinct seta-filled pit **4**

4 (3)   Metasomal cuticle (at least T2 and S2) pale orange-brown with sparse silvery and blackish setae .... **see couplet 42 in Timulla-like MCS** (p. 411)

Metasomal cuticle black, or dark reddish-brown with denser black and whitish setae **5**

5 (4)   Head quadrate, as wide as mesosoma; axilla dentate (western: Central Valley and southern Sierra Nevada regions in California) .... *D. aureola* (5.1.1.5)

Head rounded, narrower than mesosoma; axilla armed with truncate lobe **6**

6 (5)   T7 with distinct posterior fringe of erect setae; tergal setae usually distinctly yellow-tinted (western: San Diego area and Sonoran Desert in California and Arizona) .... *D. magna* (5.1.1.1)

T7 without fringe of erect setae; tergal setae generally grayish-white (western: Pacific) .... *D. sackenii* (5.1.2.1)

7 (2)   Eyes and ocelli large, ocellar diameter as large or larger than distance between ocelli (western: Algodones dunes area in California) .... *D. nocturna* (5.1.3.1)

Eyes and ocelli smaller, ocellar diameter much less than distance between ocelli **8**

8 (7) T2 setae almost entirely black, at most with some scattered whitish setae on apical fringe; hypopygium nearly as wide as long (western: transitional areas between southern California mountains and deserts) .... *D. albiceris* (5.1.2.2)

T2 setae with more extensive whitish setae, posterior margin of T2 entirely covered with white setae; hypopygium clearly longer than wide **9**

9 (8) Axilla truncate, with posterior fringe of short setae (Pacific and mountain areas of California) **10**

Axilla dentate, without setal fringe (Sonoran Desert and Madrean Archipelago) **11**

10 (9) S2 evenly convex; F1 distinctly shorter than F2; vertex laterally compressed posteriorly, expanded mesally; pronotum anteromesally with smooth emarginate furrow (western: southern Sierra Nevada in California) .... *D. californica* (5.1.2.3)

S2 mesally flattened; F1 nearly as long as F2; vertex evenly rounded posteriorly; pronotum without anteromesal furrow (western: southern Sierra Nevada in California) .... *D. coccineohirta* (5.1.2.4)

11 (9) T7 without posterior fringe; pygidial surface with some distinct longitudinal ridges (western: Sonoran Desert and Madrean Archipelago) .... *D. thetis* (5.1.3.3)

T7 usually with posterior fringe of erect setae; pygidial surface usually smooth **12**

12 (11) Antennal scrobe lacking dorsal carina; dorsal body setae uniformly stark white (western: Algodones dunes area in California) .... *D. atricauda* (5.1.3.2)

Antennal scrobe with low dorsal carina; dorsal body setae pale yellowish-gray (western: Madrean Archipelago) .... *D. dionysia* (5.1.1.4)

9.3.5.2—Western Males Color Syndrome

1 Eye vertically ovate with emarginate notch (central: Great Plains) .... *Timulla barbigera* (5.2.10.1)

Eye circular without emarginate notch **2**

2 (1) T2 fringe composed of plumose setae (western: Pacific) .... *S. edwardsii* (5.2.9.1)

T2 fringe composed of simple or obscurely brachyplumose setae **3**

3 (2) T1 shape petiolate; axilla armed with posteriorly directed tooth or lobe .... **4 (Dasymutilla)**

T1 shape sessile; axilla not armed with distinct tooth or lobe .... **30 (Pseudomethoca)**

4 (3) Hind trochanter dentate; S2 scabrous laterally **5**

Hind trochanter unarmed; S2 simply punctate laterally **6**

5 (4) Body cuticle entirely black (western: Sonoran Desert and Madrean Archipelago) .... *D. eminentia* (5.2.3.2)

Body cuticle partly reddish-brown, especially on T2 (western: Cochise County, Arizona) .... *D. furina* (5.2.3.3)

6 (4) Hind tibia posteriorly dilated, often flattened **7**

Hind tibia simply cylindrical **8**

7 (6) Hind tibia flattened, inner face convex (western: Chihuahuan Desert and Madrean Archipelago) .... *D. foxi* (5.2.3.1)

Hind tibia subcylindrical with apex flared (western: Arizona mountains) .... *D. erythrina* (5.2.5.3)

8 (6) Hypopygium wider than long, armed with posterolateral teeth (central, western: widespread but mostly absent from hot deserts and Pacific areas) .... *D. vestita* (5.2.6.1)

Hypopygium longer than wide, unarmed posteriorly **9**

9 (8) Setae of T3–6 entirely black **10**

Setae of T3–6 partly or entirely yellow, orange, or red **12**

10 (9) S2 with longitudinal row of stiff setae; axilla dentate posteriorly
(central: Great Plains) .... *D. asopus* (5.2.2.1)
S2 unmodified or with mesal seta-filled pit **11**

11 (10) Dorsum of head and mesosoma with long erect setae; S2 seta-filled
pit large, circular (western: Great Basin) .... *D. scitula* (5.2.2.3)
Dorsum of head and mesosoma with shorter flattened setae; S2 seta-filled
pit small and ovate, scarcely defined, or totally obliterated (central:
widespread; western: Chihuahuan Desert, northwestern areas) .... *D. bioculata* (5.2.2.2)

12 (9) S2 with mesal seta-filled pit, patch of dense microsetae,
or longitudinal row of stiff setae **13**
S2 unmodified, lacking seta-filled pit, patch of microsetae, or row of stiff setae **20**

13 (12) S2 with longitudinal row of stiff setae (central: Great Plains,
western: Chihuahuan Desert) .... *D. neomexicana* (5.2.5.4)
S2 with seta-filled pit or patch of dense microsetae **14**

14 (13) S2 mesally flattened with central patch of dense microsetae
(western: Madrean Archipelago) .... *D. nogalensis* (5.2.5.2)
S2 with seta-filled pit, usually convex **15**

15 (14) Head quadrate, as wide as mesosoma; axilla dentate (western:
Pacific and northwestern areas) .... *D. aureola* (5.1.2.5)
Head rounded, narrower than mesosoma; axilla armed with truncate lobe **16**

16 (15) T7 with distinct posterior fringe of erect setae **17**
T7 without fringe of erect setae **19**

17 (16) T2 with orange cuticular patch beneath orange setae (western:
Intermountain West and Madrean Archipelago) .... *D. bioculata* (5.2.5.1)
T2 with cuticle uniformly black **18**

18 (17) Apical tergites with transverse band of black setae (central, eastern:
widespread; western: scattered records from northwestern areas) ...... *D. occidentalis* (5.2.1.1)
T2–6 with setae uniformly yellow or orange (central: widespread;
western: Chihuahuan Desert) .... *D. calorata* (5.2.6.2)

19 (16) S2 pit circular (western: Pacific) .... *D. sackenii* (5.2.4.6)
S2 pit longitudinally ovate (western: Mojave and Sonoran Deserts) .... *D. gloriosa* (5.2.7.1)

20 (12) Eyes and ocelli large, ocellar diameter as large or larger than
distance between ocelli (western: Mojave and Sonoran Deserts) .... *D. arenivaga* (5.2.7.2)
Eyes and ocelli smaller, ocellar diameter much less than distance between ocelli **21**

21 (20) T7 with distinct posterior fringe of erect setae (widespread) **22**
T7 without fringe of erect setae (only known from California and northwestern USA) **28**

22 (21) T2 with orange cuticular patch beneath orange setae
(western: Intermountain West and Madrean Archipelago) .... *D. bioculata* (5.2.5.1)
T2 with cuticle uniformly black **23**

23 (22) Tegula covered with fine punctures and yellow setae throughout
(central: widespread; western: Chihuahuan Desert) .... *D. stevensi* (5.2.6.4)
Tegula mostly smooth and bare **24**

24 (23) T2 with at least posterior half covered with yellow, orange, or red setae;
antennal scrobe with dorsal carina, sometimes indistinct **25**
T2 setae usually mostly black except for narrow apical band of yellow
or orange setae; antennal scrobe without dorsal carina **27**

25 (24) Antennal scrobe dorsal carina low, nearly reaching inner eye margin; smaller-bodied species, usually shorter than 10 mm; dorsal setae usually pale yellow or orange (western: Madrean Archipelago) .... *D. dionysia* (5.2.7.4)

Antennal scrobe dorsal carina an erect truncate lobe widely separated from eye margin; larger-bodied species, usually longer than 12 mm; dorsal setae usually orange or bright red **26**

26 (25) Fringes of S2–6 mostly black, with a few scattered reddish setae laterally; F1 nearly as long as F2; cuspis length about 0.65 × free paramere length (western: Pacific and northwestern areas) .... *D. flammifera* (5.2.4.5)

Fringes of S2–6 largely reddish laterally; F1 much shorter than F2; cuspis length less than half free paramere length (western: Mojave and Sonoran Deserts) .... *D. pseudopappus* (5.2.7.5)

27 (24) Carina of S1 distinct, with large posterior tooth (central: widespread; western: Chihuahuan Desert) .... *D. myrice* (5.2.6.3)

Carina of S1 lower, without distinct posterior tooth (western: Mojave and Sonoran Deserts) .... *D. atricauda* (5.2.7.3)

28 (21) Hypopygium apically truncate, punctate throughout (western: Pacific and northwestern areas) .... *D. testaeceiventris* (5.2.4.4)

Hypopygium apically rounded, posterior third smooth **29**

29 (28) S2 evenly convex; F1 distinctly shorter than F2; vertex laterally compressed posteriorly, expanded mesally; pronotum anteromesally with smooth emarginate furrow (western: Pacific and northwestern areas) .... *D. californica* (5.2.4.1)

S2 mesally flattened; F1 nearly as long as F2; vertex evenly rounded posteriorly; pronotum without anteromesal furrow (western: Pacific and northwestern areas) .... *D. coccineohirta* (5.2.4.2)

30 (3) Mesosomal dorsum with cuticle pale orange-brown (central, western: widespread but absent from Pacific area, Mojave and Sonoran Deserts) .... *Ps. propinqua* (5.2.8.1)

Mesosomal dorsum with cuticle black (central: widespread; western: Madrean Archipelago and Chihuahuan Desert) .... *Ps. aureovestita* (5.2.8.2)

9.3.5.3—Texan Males Color Syndrome

1 Eye vertically ovate with emarginate notch .... **Timulla, see couplet 6 in Timulla-like MCS** (p. 408)

Eye circular or ovate, without emarginate notch **2**

2 (1) Eye vertically ovate and covered with setae; forewing veins continuous to outer wing margin (western: Nevada) .... *Myrmosula rufiventris* (5.3.7.1)

Eye circular and bare; forewing veins not reaching outer wing margin **3**

3 (2) T1 shape sessile; axilla not armed with distinct tooth or lobe .... **4 (Pseudomethoca)**

T1 shape petiolate; axilla armed with posteriorly directed tooth or lobe .... **5 (Dasymutilla)**

4 (3) Head quadrate, as wide as mesosoma; tibial spurs whitish (central: southern Texas; western: Chihuahuan Desert) .... *Ps. pigmentata* (5.3.5.1)

Head rounded, narrower than mesosoma; tibial spurs blackish (central: Texas; western: Chihuahuan Desert) .... *Ps. brazoria* (5.3.5.2)

5 (3) S2 with mesal seta-filled pit, patch of dense microsetae, or longitudinal row of stiff setae **6**

S2 unmodified, lacking seta-filled pit, patch of microsetae, or row of stiff setae **15**

6 (5) S2 with longitudinal row of stiff setae; axilla dentate posteriorly (central: Texas) .... *D. waco* (5.3.2.1)

S2 with seta-filled pit or patch of dense microsetae; axilla usually armed with truncate lobe **7**

7 (6) S2 mesally flattened with central patch of dense microsetae (western: Madrean Archipelago and Arizona mountains) .... *D. nogalensis* (5.3.3.2)

S2 usually convex, with distinct seta-filled pit **8**

8 (7) Cuticle of T2–6 entirely reddish-orange (western: Madrean Archipelago) .... *D. sicheliana* (5.3.4.2)

Cuticle of T2–6 mostly or entirely black, at most with obscure orange cuticular patch on T2 .... **9**

9 (8) Hypopygium wider than long, armed with posterolateral teeth (central, western: Mojave Desert east to southern Texas) .... *D. vestita* (5.3.1.3)

Hypopygium longer than wide, unarmed posteriorly **10**

10 (9) T2 with orange cuticular patch beneath orange setae **11**

T2 with cuticle uniformly black **12**

11 (10) Clypeus weakly bidentate anteriorly, teeth connected by transverse lamella; T7 with distinct setal fringe apically (central: southern Texas) .... *D. bioculata* (5.3.2.3)

Clypeus deeply emarginate and bidentate anteriorly, teeth not connected by lamella; T7 without setal fringe apically (central: southern Texas) .... *D. serenitas* (5.3.2.2)

12 (10) T7 with distinct posterior fringe of erect setae **13**

T7 without fringe of erect setae **14**

13 (12) Fringes of S2–5 mostly red (western: Mojave and Sonoran Deserts, Madrean Archipelago and Arizona mountains) .... *D. magnifica* (5.3.3.1)

Fringes of S2–5 mostly black mesally (central, western: Madrean Archipelago and mountains in Arizona east to Texas and Great Plains) .... *D. klugii* (5.3.1.1)

14 (12) Vertex laterally compressed posteriorly, expanded mesally; pronotum anteromesally with smooth emarginate furrow (western: Chihuahuan Desert and areas adjacent to western hot deserts) .... *D. gloriosa* (5.3.3.4)

Vertex evenly rounded posteriorly; pronotum without anteromesal furrow (western: Mojave and Sonoran Deserts) .... *D. satanas* (5.3.3.3)

15 (5) Apices of mid and hind femora truncate; cuticle of T3–7 reddish, like T2 (central: southwestern Great Plain; western: Chihuahuan Desert, southern Rocky Mountains) .... *D. snoworum* (5.3.4.1)

Apices of mid and hind femora rounded; cuticle of T3–7 generally black, contrasting with reddish-orange T2 .... **16**

16 (15) Hypopygium wider than long, armed with posterolateral teeth (central, western: Chihuahuan Desert east to southern Texas) .... *D. vestita* (5.3.1.3)

Hypopygium longer than wide, unarmed posteriorly **17**

17 (16) T2 with orange cuticular patch beneath orange setae (central: southern Texas) .... *D. bioculata* (5.3.2.3)

T2 with cuticle uniformly black **18**

18 (17) T7 with distinct posterior fringe of erect setae; antennal scrobe with distinct dorsal carina (western: Chihuahuan Desert and areas adjacent to western hot deserts) .... *D. pseudopappus* (5.3.3.5)

T7 without posterior fringe of erect setae; antennal scrobe without distinct dorsal carina (central: Texas and southern Great Plains) .... *D. gorgon* (5.3.1.2)

9.3.5.4—*Ephuta*-like Males Color Syndrome

1.   Eye vertically ovate with emarginate notch; T1 shape cylindrical .... **2 (*Ephuta*)**
Eye circular or ovate, without emarginate notch **17**

2 (1)   Subantennal carinae merged to form a tooth-like projection; mandible
broad, recurved downward, with subbasal rounded lamella dorsally
(western: Madrean Archipelago and Chihuahuan Desert) .... *E. krombeini* (5.4.13.15)
Subantennal carinae parallel or divergent, forming pit or trough; mandible
basically straight, without subbasal lamella, slender or broad, bi- or tridentate **3**

3 (2)   Mandible broadly dilated, tridentate; subantennal basin pit-like,
much shorter than clypeal basin (eastern: North Carolina) .... *E. eurygnathus* (5.4.13.7)
Mandible slender, bidentate; subantennal basin variable **4**

4 (3)   Subantennal basin pit-like, much shorter than clypeal basin; gena with coarse
raised carina; frons with short longitudinal carinae proceeding from antennal tubercles **5**
Subantennal basin trough-like, as long as or longer than clypeal basin; genal
carina inconspicuous; frons without longitudinal carinae proceeding from
antennal tubercles **8**

5 (4)   Hind coxa with posteriorly directed tooth (central, eastern:
widespread) .... *E. spinifera* (5.4.13.13)
Hind coxa unarmed **6**

6 (5)   Hypopygium black **and** ocelli small, least distance between eye margin
and lateral ocellus more than 3 × diameter of lateral ocellus; tegula folded
with distinct dorsal and lateral faces; subantennal and clypeal basins
not separated by raised transverse carina (eastern: Florida–Gulf) .... *E. sabaliana* (5.4.13.12)
Hypopygium usually yellow or brown; **if** hypopygium dark, **then** ocelli larger;
tegula and clypeus variable **7**

7 (6)   Transverse clypeal ridge weak or absent; humeral carina weakly developed;
tegula coarsely sculptured except for narrow smooth margin; body uniformly
black; hypopygium pale yellow-white (central, eastern: widespread) .... *E. pauxilla* (5.4.13.10)
Transverse clypeal ridge usually sharp, distinct; humeral carina strongly
developed; tegula less coarsely sculptured; T2 often partly reddened;
hypopygium yellow to brown (eastern: widespread) .... *E. battlei* (5.4.13.3)

8 (4)   Tegula densely and coarsely punctate; lateral margins of T7 usually
whitish-yellow (central, eastern: widespread) .... *E. scrupea* (5.4.13.1)
Tegula more sparsely punctate; T7 entirely black **9**

9 (8)   Legs and scape pale orange-brown (western: widespread) .... *E. rufisquamis* (5.4.12.1)
Legs and scape blackish **10**

10 (9)   Vertex conical in ocellar area; hypopygium yellowish to brown;
ocelli minute (central: widespread; western: Madrean Archipelago
and Chihuahuan Desert) .... *E. cephalotes* (5.4.13.4)
Vertex simply rounded; other characters variable **11**

11 (10)   Setae of T3–6 almost entirely blackish; ocelli moderately large,
least distance between eye margin and lateral ocellus less
than 2 × diameter of lateral ocellus; hypopygium yellow-brown
(western: Pacific) .... *E. argenticeps* (5.4.13.2)
Setae of T3–6 almost entirely whitish; ocelli and hypopygium variable; not
present in Pacific region **12**

12 (11)   Hypopygium stark whitish or hyaline, at most with dark apex **13**
Hypopygium predominantly brown to black, at most yellowish-brown basal area **15**

13 (12) Posterior ocelli large, clearly larger than anterior ocellus; subantennal basin clearly longer than clypeal basin (eastern: Florida–Gulf) .... *E. psephenophila* (5.4.13.11)

Posterior ocelli usually smaller, as small or smaller than anterior ocellus in central and eastern USA (some western populations of *E. ecarinata* have large ocelli); subantennal basin usually subequal in length to clypeal basin **14**

14 (13) Subantennal basin with lateral margins weakly expanded, only slightly narrower than clypeal basin, lacking discrete apicolateral teeth; ocelli small; legs and tegula black (central, eastern: widespread) .... *E. margueritae* (5.4.13.9)

Subantennal basin with lateral margins generally straight and converging dorsally, much narrower than clypeal basin, usually with discrete apicolateral teeth; ocelli variable in size; legs and tegula often dark brown or partly yellow-brown (central: widespread; western: Madrean Archipelago, southern Rocky Mountains, Chihuahuan and Sonoran Deserts) .... *E. ecarinata* (5.4.13.6)

15 (12) Tegula strongly carinate dorsally (central: southern Texas; western: Madrean Archipelago) .... *E. tegulicia* (5.4.13.14)

Tegula simply convex **16**

16 (15) Lateral pronotal carina thick and distinct, humeral angle in dorsal view sharply dentate (central, eastern: northern areas) .... *E. conchate* (5.4.13.5)

Lateral pronotal carina weakly defined, humeral angle in dorsal view weakly angular with indistinct apparent tooth (western: Madrean Archipelago, Rocky Mountains, northwestern areas) .... *E. grisea* (5.4.13.8)

17 (1) Eye vertically ovate, usually setose; forewing veins continuous to outer wing margin **18**

Eye circular and bare; forewing veins not reaching outer wing margin **22**

18 (17) S1 without anteromesal tooth; T2–6 junctures not constricted (central, eastern: widespread, mostly from northern states) .... *Myrmosula parvula* (5.4.16.1)

S1 with distinct tooth or ridge anteromesally; T2–6 junctures constricted laterally .... **19 (*Myrmosa*)**

19 (18) S2 with distinct tooth anteromesally **20**

S2 without tooth anteromesally **21**

20 (19) Body setae predominantly silvery (central, eastern: widespread) .... *Myrmosa unicolor* (5.4.14.1)

Body setae predominantly black (western: Pacific) .... *Myrmosa bradleyi* (5.4.15.1)

21 (19) Eyes and ocelli especially large; mesal carina of S1 slender blade-like, posterior face sharp (central: widespread) .... *Myrmosa nocturna* (5.4.14.2)

Eyes and ocelli smaller; mesal carina of S1 thickened, posterior face flat (central: widespread) .... *Myrmosa texana* (5.4.14.3)

22 (17) T2 and S2 each with lateral felt line **23**

T2 with felt line, S2 without felt line **26**

23 (22) Mesosternal area armed with large transverse ridge .... **24 (*Photomorphus*)**

Mesosternal area unarmed **25**

24 (23) Mesosternal area with two separate complexes of transverse carinae, posterolateral complex erect and slender (eastern: widespread) .... *Ph. banksi* (5.4.10.2)

Mesosternal area with single anterior complex of transverse carinae (central: widespread) .... *Ph. quintilis* (5.4.10.1)

25 (23) T2 fringe composed of simple or brachyplumose setae;
T1 shape thick petiolate; sculpture of head and mesosoma
generally densely punctate (central: widespread; western:
Chihuahuan Desert and Madrean Archipelago) .... *Protophotopsis venenaria* (5.4.11.1)

T2 fringe composed of plumose setae;
T1 shape variable; sculpture of head and
mesosoma sparse and shallow .... **various nocturnal genera** (see chapter 6, p. 285)

26 (22) T1 shape sessile, generally broad; axilla unarmed .... **27 (*Pseudomethoca*)**

T1 shape petiolate, disciform, or narrowly subsessile; axilla armed with
posterior tooth or truncate lobe **39**

27 (26) Head quadrate posteriorly with sharp angular posterolateral corners **28**

Head generally rounded posteriorly, without sharp posterolateral corners **29**

28 (27) T7 and S7 reddish-brown; punctures of T2 sparse, intervals about
2 × puncture diameter (eastern: Florida–Gulf) .... *Ps. torrida* (5.4.6.2)

T7 (and often S7) blackish; punctures of T2 denser, intervals
subequal to puncture diameter (central, eastern: widespread) .... *Ps. frigida* (5.4.6.1)

29 (27) Body setae entirely black **30**

Body with silvery setae present, at least on vertex, pronotum, or metasomal sternites **31**

30 (29) Lateral surface of propodeum with coarsely punctate
sculpture (western: Pacific) .... *Ps. anthracina* (5.4.5.1)

Lateral surface of propodeum with transversely striate
sculpture (western: Madrean Archipelago) .... *Ps. nigricula* (5.4.5.2)

31 (29) Fringes of T2–4 with parallel thickened bristles **32**

Fringes of T2–4 composed of mesally convergent simple setae **34**

32 (31) Tegula semi-hemispherical with distinct posterior face;
forewing uniformly dark brown (central: southern Texas) .... *Ps. carbonaria* (5.4.8.3)

Tegula shallowly convex; forewing usually mostly clear **33**

33 (32) Antennal tubercle mostly smooth; parameres parallel apically (central, western:
widespread, mostly absent from hot deserts and Pacific areas) .... *Ps. contumax* (5.4.8.1)

Antennal tubercle with dense small punctures; parameres divergent
apically (western: Sonoran Desert and Madrean Archipelago) .... *Ps. praeclara* (5.4.8.2)

34 (31) Medium-sized species, usually longer than 7 mm **35**

Small-sized species, usually shorter than 6 mm **36**

35 (34) Setae of T3–7 with setae mostly black (western: Madrean Archipelago) ... *Ps. sonorae* (5.4.9.2)

Setae of T3–7 with setae mostly silvery (central, eastern: widespread,
mostly in northern areas) .... *Ps. simillima* (5.4.9.1)

36 (34) Head narrowly rounded; clypeus anteriorly with transverse
pre-apical furrow; mandible with inner pre-apical tooth small,
parallel with outer tooth (central: southern Texas) .... *Ps. gila*, in part (5.4.7.3)

Head broadly subquadrate; clypeus anteriorly without transverse furrow;
mandible with inner tooth large, nearly perpendicular from outer tooth **37**

37 (36) Hypopygium with dense lateral tuft of long white setae; paramere
paddle-shaped clearly dorsoventrally flattened, apically divergent
(Sonoran Desert in Arizona and California) .... *Ps. toumeyi* (5.4.7.2)

Hypopygium generally without dense lateral tuft; paramere more
or less acuminate to apex **38**

38 (37) Marginal cell as long as or longer than stigma (central: widespread) .. *Ps. gila*, in part (5.4.7.3)

Marginal cell much shorter than elongate stigma (western: Pacific) .... *Ps. athamas* (5.4.7.1)

39 (26) T1 shape petiolate; body punctures generally dense .... **40 (*Dasymutilla*)**

T1 shape disciform or subsessile; body sculpture generally sparse
and shallow .... **43 (*Lomachaeta*)**

40 (39) Tibial spurs white; body setae mostly silvery (eastern: northeastern areas) *D. gibbosa* (5.4.1.1)

Tibial spurs black; body setae entirely black **41**

41 (40) S2 lacking seta-filled pit mesally (western: northern Pacific Coast) .... *D. californica* (5.4.2.3)

S2 having mesal seta-filled pit **42**

42 (41) S2 pit circular (western: Algodones dunes in California) .... *D. imperialis* (5.4.2.2)

S2 pit longitudinally ovate (central: southern Texas) .... *D. nigra* (5.4.2.1)

43 (39) Mandible armed with ventral tooth basally **44**

Mandible unarmed ventrally **45**

44 (43) Legs and tegula pale orange-brown (western: widespread) .... *L. cirrhomeris* (5.4.3.1)

Legs and tegula blackish (central, eastern, western: widespread) .... *L. hicksi* (5.4.4.3)

45 (43) Paramere broadly flattened throughout, rounded apically
(western: Sonoran Desert) .... *L. snellingella* (5.4.4.5)

Paramere acuminate apically, at most laterally compressed **46**

46 (45) Paramere lacking long setae, all setae shorter than paramere
width (western: Madrean Archipelago) .... *L. polemomechana* (5.4.4.4)

Paramere with at least some ventral or apical setae longer than mid-paramere width **47**

47 (46) Entire paramere with continuous ventral setal row (western: Mojave
and Sonoran Deserts) .... *L. ilex* (5.4.4.2)

Long ventral setae of paramere absent from basal half **48**

48 (47) Paramere narrow throughout, long cylindrical, apically convergent,
and evenly curving ventrally throughout length (western: Madrean
Archipelago) .... *L. litosisyra* (5.4.4.1)

Paramere evenly tapering toward apex, somewhat laterally
compressed, subparallel, and virtually straight to apex or
scarcely downcurving (central, western: Arizona east to Texas) .... *L. vacamuerta* (5.4.4.6)

9.3.5.5—*Timulla*-like Males Color Syndrome

1 Eye vertically ovate with emarginate notch **2**

Eye circular or ovate, without emarginate notch **19**

2 (1) T1 shape cylindrical; T2 without felt line .... **3 (*Ephuta*)**

T1 shape sessile; T2 with felt line .... **6 (*Timulla*)**

3 (2) Mandible broad, dilated apically (eastern: Florida, South Carolina) .... *E. eurygnathus* (5.5.11.3)

Mandible slender, acuminate apically **4**

4 (3) Subantennal basin as long as or longer than clypeal basin (western:
Colorado Plateau and Chihuahuan Desert; central: southern Texas) ... *E. rufisquamis* (5.5.11.4)

Subantennal basin pit-like, much shorter than clypeal basin **5**

5 (4) T2–7 entirely orange (central: southern Texas) .... *E. copano* (5.5.11.1)

T3–7 mostly black, T2 partly or entirely reddish-orange (central,
eastern: widespread but mostly southern) .... *E. battlei* (5.5.11.2)

6 (2) Posterior margin of T7 emarginate mesally **7**

Posterior margin of T7 convex **8**

7 (6) Tegula with dense whitish setae; T7 with pre-apical inverted U-shaped
carina (western: Sonoran Desert) .... *T. tyro* (5.5.13.2)

	Tegula with sparse whitish setae; T7 with pre-apical transverse arcuate carina (central, eastern: widespread) .... *T. leona* (5.5.14.4)
8 (6)	T7 with pre-apical erect vertical keel **9**
	T7 with pre-apical Y-shaped carina **10**
9 (8)	Tibial spurs dark brown; vertex and pronotum with mostly black setae (western: Madrean Archipelago and Chihuahuan Desert) .... *T. navasota coahuila* (5.3.6.2)
	Tibial spurs whitish; vertex and pronotum with mostly whitish setae (central: southern Texas) .... *T. navasota navasota* (5.5.14.5)
10 (8)	S5–7 each with sharp lateral tubercle (central, eastern: widespread) .... *T. dubitata* (5.6.7.2)
	S5 unarmed, S6 sometimes with small tubercle, S7 with distinct tubercle **11**
11 (10)	Mandible unarmed ventrally (eastern: northeastern areas) .... *T. hollensis* (5.5.12.3)
	Mandible armed with ventral tooth basally **12**
12 (11)	Scape apically dilated, with brush of silvery setae in apical half **13**
	Scape not dilated apically, with short uniform setae throughout **15**
13 (12)	Ocelli large, subequal in size, diameter of lateral ocellus as long as or longer than distance between ocelli; setae of T2–7 usually black (eastern: widespread; central: Great Plains; western: rare in some northwestern areas) .... *T. vagans* (5.5.12.2, 5.6.7.1)
	Ocelli smaller, lateral ocelli much smaller than mesal ocellus, diameter of lateral ocellus shorter than distance between ocelli; setae of T2–7 always golden-orange **14**
14 (13)	Scapal brush distinct with long dense setae (western: Utah southeast to western Texas) .... *T. grotei* (5.3.6.1)
	Scapal brush weakly defined (central, western: Great Plains, Chihuahuan Desert, Madrean Archipelago) .... *T. suspensa* (5.3.6.3, 5.5.14.2)
15 (12)	Ocelli very large, diameter of lateral ocellus much longer than distance between ocelli; mesosternal area lacking smooth tubercle or ridge (eastern: mostly northern) .... *T. ocellaria* (5.5.12.1)
	Ocelli small to moderate, diameter of lateral ocellus usually shorter than distance between ocelli; mesosternal area with large blunt smooth tubercle or ridge **16**
16 (15)	Mesocoxa armed with small tubercle; Y-shaped carina of T7 with lateral arms much shorter than apical stem (central: widespread; western: Chihuahuan Desert and Madrean Archipelago) .... *T. oajaca* (5.5.14.1)
	Mesocoxa armed with sharp tooth; Y-shaped carina of T7 with lateral arms nearly as long as apical stem **17**
17 (16)	Scape apically dilated; T1 cuticle usually black (central, western: Great Plains, Chihuahuan Desert, Madrean Archipelago) .... *T. suspensa* (5.3.6.3, 5.5.14.2)
	Scape equally wide throughout length; T1 cuticle orange **18**
18 (17)	Mid and hind tarsi reddish-orange; lateral propodeal sculpture similar to posterior sculpture, areolate; smooth posterior convex portion of clypeus quadrate, much higher than greatest mandibular width (western: Madrean Archipelago) .... *T. nitela* (5.5.14.3)
	Mid and hind tarsi blackish; lateral propodeal sculpture weaker than posterior sculpture, partially punctate; smooth posterior convex portion of clypeus transverse, scarcely higher than greatest mandibular width (western: Sonoran Desert) .... *T. neobule* (5.5.13.1)
19 (1)	Eye vertically ovate, usually setose; forewing veins continuous to outer wing margin **20**
	Eye circular and bare; forewing veins not reaching outer wing margin **21**

20 (19) S1 with distinct tooth anteromesally; T2–6 junctures constricted laterally
(central: southern Texas; one dubious record from California) ..... *Myrmosa nocturna* (5.5.15.1)

S1 without anteromesal tooth; T2–6 junctures not constricted
(western: Nevada) .... *Myrmosula rufiventris* (5.3.7.1)

21 (19) Mesosternal area armed with large transverse ridge; mandible
with large rounded tooth ventrally (central: widespread;
eastern: northern areas) .... *Photomorphus auriventris* (5.5.9.1)

Mesosternal area unarmed; mandible usually without ventral tooth basally **22**

22 (21) Fringe of T2 composed of plumose setae; eyes and
ocelli especially large (central, western: widespread) ..... *Sphaeropthalma marpesia* (5.5.10.1)

Fringe of T2 composed of simple or weakly brachyplumose setae **23**

23 (22) T1 shape sessile, generally broad; axilla unarmed .... **24 (*Pseudomethoca*)**

T1 shape petiolate, disciform, or narrowly subsessile; axilla armed with
posterior tooth or truncate lobe **28**

24 (23) Metasomal cuticle black; mandible vertical throughout its length
(central, western: widespread) .... *Ps. wickhami* (5.5.7.1)

Metasomal cuticle, at least T2, orange to reddish-brown; mandible apically oblique **25**

25 (24) Femora with ventral brush of long setae; tegula coarsely rugose and
strongly raised with a flat outer rim (central: southern Texas) .... *Ps. flavida* (5.5.8.4)

Femora with simple setae only; tegula smooth or punctate **26**

26 (25) Fringes of T2–4 composed of mesally convergent simple setae
(central: Texas) .... *Ps. russeola* (5.5.8.3)

Fringes of T2–4 with parallel thickened bristles **27**

27 (26) Clypeus anteromesally expanded with two closely spaced teeth;
paramere with elbowed setae along inner margin (western: Madrean
Archipelago) .... *Ps. ajattara* (5.5.8.2)

Clypeus basically flat with two widely separated teeth; paramere without
elbowed setae (western: Sonoran and Chihuahuan Deserts) .... *Ps. donaeanae* (5.5.8.1)

28 (23) T1 shape disciform or subsessile; body sculpture generally shallow,
body setae generally sparse .... **29 (*Lomachaeta*)**

T1 shape petiolate; body sculpture generally coarse, body setae
generally dense .... **33 (*Dasymutilla*)**

29 (28) Mandible armed with ventral tooth basally (central, eastern, western:
widespread) .... *L. hicksi* (5.5.6.1)

Mandible unarmed ventrally **30**

30 (29) Paramere broadly flattened throughout, rounded apically
(western: Mojave and Sonoran Deserts) .... *L. beadugrimi* (5.5.6.2)

Paramere acuminate apically, at most laterally compressed **31**

31 (30) Wing venation reduced, cells contained in basal half of forewing;
paramere lacking row of elongate setae ventrally (western: Pacific) .... *L. powelli* (5.5.6.4)

Wing venation normal, cells extending beyond basal half of forewing;
paramere with somewhat uniform row of long setae ventrally **32**

32 (31) Pronotum and mesoscutum with widely separated punctures;
metasoma blackish with orange color restricted to T2 and T3
(western: Sonoran Desert) .... *L. ptilohyalus* (5.5.6.5)

Pronotum and mesoscutum with denser punctures; metasoma
mostly orange (western: widespread, except Pacific and
western hot deserts) .... *L. crocopinna* (5.5.6.3)

33 (28) S2 without seta-filled pit **34**

S2 with seta-filled pit mesally **41**

34 (33) Apices of mid and hind femora rounded; axilla armed with truncate lobe **35**

Apices of mid and hind femora truncate; axilla armed with tooth **36**

35 (34) Tibial spurs black (central: widespread; western: Chihuahuan Desert) .... *D. birkmani* (5.5.4.2)

Tibial spurs white (eastern: scattered records from various localities) .... *D. gibbosa* (5.5.1.2)

36 (34) Setae of T2–7 (including T2 disc) covered with golden-orange setae
(western: Madrean Archipelago) .... *D. sophrona* (5.5.2.3)

Setae of T2–7 composed of silvery and/or blackish setae... **37**

37 (36) Setae of T2–7 black, at least dorsomesally (central: widespread;
western: Chihuahuan Desert and Colorado Plateau) .... *D. curialis* (5.5.5.3)

Setae of T2–7 largely silvery, at least T3–4 with entirely silvery bands **38**

38 (37) Setae of T3–7 entirely silvery (central: southern Texas; western:
Chihuahuan Desert) .... *D. digressa* (5.5.5.4)

Apical tergites, at least T5–6, with setae blackish **39**

39 (38) Mesosomal pleura with cuticle largely orange-brown (western:
Sonoran Desert) .... *D. heliophila* (5.7.2.1)

Mesosomal pleura mostly black, at most with lateral propodeal surface dark reddish **40**

40 (39) Axilla large, dorsally smooth; forewing largely darkened brown, with
distinct clouding around veins (western: Arizona mountains and
Madrean Archipelago) .... *D. apicalata* (5.5.5.1)

Axilla smaller, more extensively punctate; forewing clear around veins,
only margin beyond veins dark brown (western: desert-adjacent
mountains and canyons in California and Nevada) .... *D. dammersi* (5.5.5.2)

41 (33) Hind tibia flattened, inner face convex (central: widespread; western:
widespread but absent from Pacific and northwestern areas) .... *D. foxi* (5.5.2.2)

Hind tibia not modified as above, cylindrical **42**

42 (41) Mesoscutum with setae silvery, like pronotum and vertex **43**

Mesoscutum with setae black, contrasting with silvery pronotum and vertex **44**

43 (42) Propodeum having lateral patches of dense appressed setae
obscuring punctation; free length of paramere weakly kinked
basally and slightly recurved inwardly, making paramere apices
only slightly divergent (eastern: Florida–Gulf) .... *D. archboldi* (5.5.1.1)

Setae of propodeum erect throughout, propodeal reticulations
unobscured; free length of paramere strongly kinked basally,
making paramere apices obviously divergent (central,
western: widespread but absent from Pacific areas) .... *D. monticola* (5.5.3.1, 5.7.2.3)

44 (42) Setae of T2–6 entirely golden-orange or reddish **45**

Setae of T2–6 with contrasting pattern of black and white, silvery, or golden setae **46**

45 (44) T7 with cuticle and setae orange, concolorous with T2–6
western: Madrean Archipelago) .... *D. bonita* (5.5.2.4)

T7 with cuticle and setae black, contrasting with reddish T2–6
(western: Madrean Archipelago) .... *D. sicheliana* (5.5.2.1)

46 (44) T2 and S2 black, except T2 with large pale yellow cuticular patch
(western: Madrean Archipelago) .... *D. pulchra* (5.5.4.4)

T2 and S2 predominantly orange-brown, sometimes with yellowish patch on T2 .... **47**

47 (46) Cuticle of T3–7 orange-brown, concolorous with T2 background
cuticle (western: Madrean Archipelago) .... *D. saetigera* (5.5.5.5)

Cuticle of T3–7 blackish, contrasting with orange-brown T2 .... **48**

48 (47) Pit on S2 situated posteromesally; axilla with sharp posterior tooth
(western: Madrean Archipelago) .... *D. iztapa* (5.5.4.3)

Pit on S2 situated anteromesally; axilla variable, with truncate posterior surface **49**

49 (48) Axilla with short truncate lobe; T2 uniformly orange-brown
(western: Madrean Archipelago) .... *D. fasciventroides* (5.5.4.1)

Axilla forming long curved "arm"; T2 reddish-brown with large
pale yellow cuticular patch (western: Arizona mountains) .... *D. fasciventris* (5.7.2.2)

9.3.5.6—Eastern Males Color Syndrome

1 Eye vertically ovate with emarginate
notch .... **Timulla, see couplet 6 in Timulla-like MCS** (p. 408)

Eye generally circular without emarginate notch **2**

2 (1) Fringe of T2 composed of plumose setae (eastern:
mostly in northern areas) .... *Sphaeropthalma pensylvanica* (5.6.6.1)

Fringe of T2 composed of simple or brachyplumose setae **3**

3 (2) T1 shape petiolate; axilla armed with posteriorly directed tooth or lobe .... **4 (Dasymutilla)**

T1 shape sessile; axilla not armed with distinct tooth or lobe .... **16 (Pseudomethoca)**

4 (3) S2 with mesal seta-filled pit or longitudinal row of stiff setae **5**

S2 with relatively uniform sculpture, lacking seta-filled pit or row of stiff setae **11**

5 (4) S2 with longitudinal row of stiff setae; axilla dentate posteriorly
(eastern: widespread) .... *D. asopus* (5.6.4.2, 5.7.4.1)

S2 with seta-filled pit; axilla usually armed with truncate lobe **6**

6 (5) Apical tergites (at least T5–6) with inconspicuous gray or silvery setae **7**

Apical tergites with setae entirely black, at most with T7 having some silvery setae **8**

7 (6) Propodeal areolations deep, well-defined; S2 usually black, darker
than orange T2; cuspis short, its length about half free paramere
length (eastern: northern areas) .... *D. canella* (5.6.2.2)

Propodeal areolations shallower, poorly defined; S2 usually reddish-orange,
like T2; cuspis longer, its length about 0.65 × free paramere length
(eastern: mostly in southern areas) .... *D. macilenta* (5.6.2.3)

8 (6) Tegula punctate and setose throughout (central: scattered records
from Kansas and Texas) .... *D. gentilis* (5.6.4.4)

Tegula largely smooth and bare mesally **9**

9 (8) Clypeus weakly bidentate anteriorly, teeth connected by transverse
lamella; T7 with distinct setal fringe apically (eastern: widespread) .... *D. bioculata* (5.6.4.3)

Clypeus deeply emarginate and bidentate anteriorly, teeth not
connected by lamella; T7 usually without setal fringe apically **10**

10 (9) Vertex laterally compressed posteriorly, expanded mesally;
pronotum anteromesally with smooth emarginate furrow
(central: scattered records from Kansas, Oklahoma, and Texas) .... *D. meracula* (5.6.4.5)

Vertex evenly rounded posteriorly; pronotum without anteromesal
furrow (common throughout central and eastern USA) .... *D. quadriguttata* (5.6.4.1, 5.7.4.2)

11 (4) Apices of mid and hind femora truncate; cuticle of T3–7 reddish,
like T2 (widespread in central and eastern USA) .... *D. scaevola* (5.6.1.1)

Apices of mid and hind femora rounded; cuticle of T3–7 generally
black, contrasting with reddish-orange T2 .... **12**

12 (11) Apical tergites (at least T5–6) with inconspicuous gray or silvery setae **13**

Apical tergites with setae entirely black, at most with T7 having some silvery setae **14**

13 (12) Tegula usually punctate throughout; cuspis long, length greater than 0.8 × free paramere length (central: widespread; western: Chihuahuan Desert) .... *D. birkmani* (5.6.2.1)

Tegula mostly smooth, at least on posterior half; cuspis moderate, length less than 0.65 × free paramere length (eastern: Florida–Gulf) .... *D. arrenneronea* (5.6.2.4)

14 (12) Tegula punctate throughout; posterolateral corner of mesonotum forming rounded lobe (central, eastern, western: widespread except Pacific and western hot deserts) .... *D. ursus* (5.6.3.1, 5.7.4.3)

Tegula mostly smooth; posterolateral corner of mesonotum angular **15**

15 (14) S2 with cuticle reddish-orange, like T2 (central, eastern: widespread) .... *D. nigripes* (5.6.3.2, 5.7.4.4)

S2 with cuticle black, darker than reddish-orange T2 patch (central: widespread) .... *D. macra* (5.6.3.3)

16 (3) Tegula coarsely rugose and strongly raised with a flat outer rim; hind femur with ventral brush of long setae (central, eastern: widespread) .... *Ps. oceola* (5.6.5.1)

Tegula largely smooth or punctate without flat outer rim; hind femur with normal setae **17**

17 (16) Tegula shallowly convex, basically flat (central, eastern: widespread) ...... *Ps. sanbornii* (5.6.5.4)

Tegula "humped" with distinct dorsal and posterior faces **18**

18 (17) Tegula dorsal surface mostly smooth; T6–7 setae mostly black (central, eastern: widespread) .... *Ps. simillima* (5.6.5.2)

Tegula dorsal surface with dense punctures; T6 and/or T7 setae largely silvery **19**

19 (18) Head and mesosoma with cuticle entirely black (central: widespread) ..... *Ps. paludata* (5.6.5.3)

Head and mesosoma with cuticle largely reddish-orange (eastern: Florida–Gulf) .... *Ps. oculata* (5.7.7.1)

9.3.5.7—*Sphaeropthalma*-like Males Color Syndrome

1 Eye vertically ovate with emarginate notch **2**

Eye generally circular, without emarginate notch **19**

2 (1) T1 shape cylindrical; T2 without felt line .... **3 (*Ephuta*)**

T1 shape sessile; T2 with felt line .... **5 (*Timulla*)**

3 (2) Mandible broadly dilated, tridentate; subantennal basin pit-like, much shorter than clypeal basin (eastern: Florida–Gulf) .... *E. slossonae* (5.7.13.1)

Mandible slender, bidentate; subantennal basin variable **4**

4 (3) Subantennal basin pit-like, much shorter than clypeal basin; gena with coarse raised carina; frons with short longitudinal carinae proceeding from antennal tubercles (eastern: Florida–Gulf) .... *E. stenognatha* (5.7.13.3)

Subantennal basin trough-like, as long as or longer than clypeal basin; genal carina inconspicuous; frons without longitudinal carinae proceeding from antennal tubercles (eastern: Florida–Gulf) .... *E. floridana* (5.7.13.2)

5 (2) Forewing conspicuously banded, dark brown with pre-apical colorless band **6**

Forewing not banded, more or less uniformly light or dark brown **7**

6 (5) Scape with dense brush of long white setae; F1 distinctly flattened, clearly wider than apical flagellomeres (central: widespread) .... *T. barbata* (5.7.14.1)

Scape without white setal brush; F1 somewhat flattened, not wider than apical flagellomeres (eastern: widespread) .... *T. ornatipennis* (5.7.14.2)

7 (5) T7 basically smooth pre-apically, without distinct Y-shaped carina (eastern, central: widespread) .... *T. ferrugata* (5.7.16.1)

T7 with pre-apical Y-shaped carina **8**

8 (7) S5–7 each with sharp lateral tubercle (eastern, central: widespread, more common in southern areas) .... *T. dubitata* (5.7.16.4)

S5 unarmed, S6 sometimes with tubercle, S7 with distinct tubercle **9**

9 (8) Mandible unarmed ventrally **10**

Mandible armed with ventral tooth basally **16**

10 (9) Mesosternal area armed with large smooth tubercle; S6 armed with lateral tubercle **11**

Mesosternal area unarmed; T6 without lateral tubercle **12**

11 (10) F1–2 flattened; scape without setal brush (central: southern Texas; eastern: mostly southern areas) .... *T. compressicornis* (5.7.16.3)

F1–2 cylindrical; scape with sparse brush of whitish setae (central, eastern: widespread) .... *T. barbigera* (5.7.16.2)

12 (10) Ocelli smaller, diameter of mesal ocellus shorter than distance between ocelli **13**

Ocelli large, diameter of mesal ocellus as long as or longer than distance between ocelli **14**

13 (12) Propodeum deeply and coarsely areolate; legs with setae whitish (eastern: Florida–Gulf) .... *T. rufosignata* (5.7.16.8)

Propodeum with shallower sculpture; legs with setae black (central: Great Plains) .... *T. tolerata* (5.7.15.3)

14 (12) Forewing very pale brown, nearly colorless; mesosoma almost entirely orange-brown (western: northwestern areas) .... *T. subhyalina* (5.7.15.2)

Forewing dark brown; mesosoma largely black, at least propodeum entirely blackish **15**

15 (14) Tegula pale orange-brown, concolorous with mesonotum; distance between eye margin and lateral ocellus subequal to lateral ocellar diameter (central: southern Texas) .... *T. sayi* (5.7.16.9)

Tegula blackish, darker than reddish-orange mesonotum; distance between eye margin and lateral ocellus clearly greater than lateral ocellar diameter (eastern: northern areas) .... *T. hollensis* (5.7.16.5)

16 (9) Scape apically dilated, with brush of silvery setae in apical half; mesosternal area armed with large smooth tubercle (eastern: Florida–Gulf) .... *T. vagans* (5.7.16.10)

Scape not dilated apically, with short uniform setae throughout; mesosternal area without distinct smooth tubercle **17**

17 (16) Ocelli very large, diameter of lateral ocellus much longer than distance between ocelli (central: widespread) .... *T. ocellaria* (5.7.16.7)

Ocelli small to moderate, diameter of lateral ocellus usually shorter than distance between ocelli **18**

18 (17) T2–7 covered entirely with pale golden setae; smooth posterior face of clypeus subrectangular, transversely concave (eastern: Florida–Gulf) .... *T. floridensis* (5.7.15.1)

T2–7 with pale whitish-yellow setae, except T2 fringe and T3 with blackish setae; smooth posterior face of clypeus subtriangular, basically flat (central: Great Plains) .... *T. kansana* (5.7.16.6)

19 (1) Axilla armed with posteriorly directed tooth or lobe; T1 shape petiolate or narrowly subsessile; S2 without felt line; T2 fringe composed of simple or obscurely brachyplumose setae, and mesosternal area unarmed **20**

Axilla not armed with distinct tooth or lobe; with one or more of the following characters: T1 shape sessile, S2 with felt line, T2 fringe composed of plumose setae, or mesosternal area armed with teeth or ridges **26**

20 (19) T1 shape disciform or subsessile; body sculpture
generally sparse and shallow (southern California) .... *Lomachaeta calamondin* (5.7.5.1)

T1 shape petiolate; body punctures generally dense .... **21 (*Dasymutilla*)**

21 (20) Hind trochanter armed with sharp tooth **22**

Hind trochanter unarmed **23**

22 (21) T7 surface covered with setae; S2 sculpture simply punctate
(central: widespread) .... *D. creon* (5.7.1.1)

T7 surface generally bare; S2 sculpture scabrous laterally
(western: Cochise County, Arizona) .... *D. furina* (5.2.3.3)

23 (21) T3–6 covered with black setae, at most with a few pale
gray setae on T5–6 (southeastern USA) .... **see couplet 4 in Eastern MCS key** (p. 412)

T3–6 with setae orange, or with distinct bands
of whitish or silver setae on pre-apical tergites **24**

24 (23) S2 with longitudinal row of stiff setae (central: Great Plains;
western: Chihuahuan Desert) .... *D. hector* (5.7.3.1)

S2 without longitudinal row of setae, simply punctate or with seta-filled pit mesally **25**

25 (24) Apices of mid and hind femora truncate;
S2 without seta-filled pit .... **see couplet 36 in *Timulla*-like MCS key** (p. 411)

Apices of mid and hind femora rounded;
S2 with seta-filled pit .... **see couplet 41 in *Timulla*-like MCS key** (p. 411)

26 (19) Mesosternal area unarmed; T1 shape sessile;
S2 without felt line; T2 fringe composed of silvery setae .... **27 (*Pseudomethoca*)**

With one or more of the following characters: mesosternal
area armed with teeth or ridges, T1 shape petiolate, S2
with felt line, or T2 fringe composed of plumose setae **28**

27 (26) Small-bodied species, shorter than 5 mm; T2–6 with interspersed
silvery and blackish sparse setae (western: Madrean Archipelago) ...... *Ps. peremptrix* (5.7.6.1)

Generally medium-bodied species, longer
than 6 mm; T2–6 generally dense black setae
only (southeastern USA) .... **see couplet 16 in Eastern MCS key** (p. 413)

28 (26) Mesosternal area armed with teeth or transverse complexes of ridges **29**

Mesosternal area unarmed **35**

29 (28) Mandible thick, vertical to bidentate apex; mentum carinate
or tuberculate; tergal fringes composed of simple setae
(central and eastern USA only) .... **30 (*Photomorphus*, in part)**

Mandible generally more slender, with apex oblique and usually
tridentate; mentum unarmed or simply carinate; tergal fringes
usually with plumose setae (widespread) **33**

30 (29) Mentum with simple erect tubercle arising in anterior half;
mesosternal complex basically transverse, wider than long
(eastern, central: widespread) .... *Ph. impar* (5.7.10.3)

Mentum with anterior-facing process arising in posterior half;
mesosternal complex variable **31**

31 (30) Mesosternal complex broadly transverse, much wider than long;
eyes and ocelli large, diameter of lateral ocellus subequal to least
distance between lateral and anterior ocelli; generally nocturnal
species (eastern: Florida–Gulf) .... *Ph. paulus* (5.7.9.3)

Mesosternal complex generally as long as or longer than wide; eyes
and ocelli smaller, diameter of lateral ocellus much less than least
distance between anterior and lateral ocelli; generally diurnal species **32**

32 (31) Mesopleuron with two separate complexes of transverse carinae, posterolateral complex erect and slender (eastern: Florida–Gulf) .... *Ph. banksi* (5.7.10.2)

Mesopleuron with single anterior complex of transverse carinae (central: southern Texas; eastern: Florida–Gulf) .... *Ph. alogus* (5.7.10.1)

33 (29) Species from western and central USA ...... **various nocturnal genera** (see chapter 6, p. 285)

Species from eastern USA .... **34 (*Photomorphus*, in part)**

34 (33) Legs brown to blackish, concolorous with or darker than brown mesosoma (eastern: widespread) .... *Ph. spinci* (5.7.9.2)

Legs pale yellow-brown, lighter than brown mesosoma (eastern: Florida–Gulf) .... *Ph. archboldi* (5.7.9.1)

35 (28) Metasoma, at least T2, distinctly blackened, contrasting with reddish-orange mesosoma **36**

Metasoma, at least T2, concolorous with mesosoma, reddish-orange or brown **37**

36 (35) T2 fringe composed of a distinctly plumose white band, contrasting with dark brown setae of T2 disc and T4–7; diurnal species with small eyes and ocelli (western: Pacific) .... *Morsyma ashmeadi* (5.7.8.1)

T2 fringe generally not so conspicuously dense, usually concolorous with sparse whitish setae of T2 disc and T4–7; usually nocturnal species with larger eyes and ocelli (western and central USA) .... **various nocturnal genera** (see chapter 6, p. 285)

37 (35) Forewing basically colorless; T3–7 setae usually pale whitish; nocturnal species with body color generally pale or dark brown (western and central USA) .... **various nocturnal genera** (see chapter 6, p. 285)

Forewing generally dark brown; T3–7 with golden, dark brown, or black setae; body color generally partly blackish and brighter orange or reddish-orange (widespread) **38**

38 (37) T2–7 with generally short erect setae, fringe of T2 with short bushy distinctly plumose setae (central and eastern USA) **39**

T2–7 with many longer flatter setae, fringe of T2 with sparser and flatter plumose setae (western USA) **41**

39 (38) T3–7 with pale golden setae (central: widespread) .... *S. auripilis* (5.7.11.1)

T3–7 with dark brown or blackish setae **40**

40 (39) Fringe of T2 pale gray, T3–7 setae dark brown (central: southern Texas) .... *S. boweri* (5.7.11.2)

Setae of T2 fringe and T3–7 uniformly black (central, eastern: widespread) .... *S. pensylvanica* (5.7.11.3)

41 (38) Mid and hind coxae with distinct ventral brush of bright golden setae (western: western Sierra Nevada in California) .... *S. luiseno* (5.7.12.2)

Mid and hind coxae without setal brushes **42**

42 (41) Mandible slender and apically oblique; forewing usually mostly pale yellow with dark brown streak (western: Pacific) .... *S. unicolor* (5.7.12.3)

Mandible thicker, vertical to apex; forewing usually uniformly dark brown **43**

43 (42) T1 fringe with sparse brachyplumose setae only; body cuticle uniformly pale orange-brown; genitalic cuspis spatulate with plumose setae (western: northwestern areas) .... *S. contracta* (5.7.12.1)

T1 fringe with plumose setae; body cuticle often blackened; genitalic cuspis slender cylindrical with simple setae only (western: Pacific and northwestern) .... *S. edwardsii* (5.7.12.4)

9.3.6—Genera of nocturnal velvet ants in the USA

These keys are expected to work for only about 90% of specimens; many structural exceptions cannot be incorporated into such a key without expanding it by dozens of couplets and characters.

Females—wings absent, antenna with 10 flagellomeres, metasoma with 6 tergites

1	Dorsum of mesosoma with plumose setae mesally	**2**
	Dorsum of mesosoma with setae simple or brachyplumose mesally	**4**
2 (1)	Dorsal setae of head and mesosoma appressed, each seta plumose basally and simple apically	*Odontophotopsis* (6.1.5)
	Plumose dorsal setae erect, each seta plumose throughout its length	**3**
3 (2)	Pygidium convex without a plate; T1 shape slenderly petiolate	*Laminatilla* (6.1.4)
	With either a pygidial plate or wider T1 shape	*Sphaeropthalma*, **in part** (6.1.8)
4 (1)	Mandible with a large ventral tooth basally and a small sharp dorsal tooth basally (of the DiloAcro form)	**5**
	Mandible not as above, usually without small dorsal tooth basally	**7**
5 (4)	T2 disc with scattered raised tubercles basally	*Acrophotopsis* (6.1.2)
	T2 disc simply punctate	**6**
6 (5)	Genal carina present	*Schusterphotopsis* (6.1.7)
	Genal carina absent	*Dilophotopsis* (6.1.3)
7 (4)	Pygidium convex without a plate	**8**
	Pygidial plate present	**9**
8 (7)	T2 disc with basal patches of darker setae, or T2 disc with uniformly coarse honeycomb sculpture; T1 shape subsessile	*Acanthophotopsis* (6.1.1)
	T2 disc color and sculpture variable, but not as above; T1 shape variable	*Sphaeropthalma*, **in part** (6.1.8)
9 (7)	Mesosomal shape elongate subrectangular; T1 shape sessile; small, dull-colored insects	*Photomorphus* (6.1.6)
	Mesosomal shape usually pear-shaped; T1 shape variable; size and coloration variable	*Sphaeropthalma*, **in part** (6.1.8)

Males—wings usually present, antenna with 11 flagellomeres, metasoma with 7 tergites

1	Mesosternal area unarmed, surface simply punctate, areolate, or smooth	**2**
	Mesosternal area armed with teeth, ridges, or longitudinal irregularly denticulate swellings; small teeth sometimes located adjacent to mesocoxae	**4**
2 (1)	Hypopygium posteriorly emarginate, anterolaterally with longitudinal carina (of the DiloAcro form)	*Acrophotopsis* (6.1.2)
	Hypopygium rounded or truncate apically, without anterolateral carinae	**3**
3 (2)	Tergal setae dense, plumose, white (uncommon and restricted to hot deserts)	*Odontophotopsis*, **in part** (6.1.5)
	Tergal fringes usually with plumose setae sparser or variable in color (common and widespread)	*Sphaeropthalma* (6.1.8)
4 (1)	Hypopygium posteriorly emarginate, anterolaterally with longitudinal carina (of the DiloAcro form)	**5**
	Hypopygium rounded or truncate apically, without anterolateral carinae	**6**
5 (4)	Mesosternal armature forming posteriorly directed teeth directly adjacent to the mesocoxae	*Schusterphotopsis* (6.1.7)
	Mesosternal armature forming ventrally directed teeth or tubercles widely separated from the mesocoxae	*Dilophotopsis* (6.1.3)

6 (4) Mesosternal processes large posteriorly directed conical
pegs; mesotibia usually with only one spur .... *Acanthophotopsis* (6.1.1)

Mesosternal processes variable, generally ventrally or anteriorly directed **7**

7 (6) Mesosternum with laterally flattened thin, sharp lamellae (rare) .... *Laminatilla* (6.1.4)

Mesosternal processes different from above, usually with
simple teeth or blunter processes **8**

8 (7) Mesosternal area with 2–6 simple teeth; tergites with
dense plumose setae; body sculpture generally sparse
and faint .... *Odontophotopsis*, **in part** (6.1.5)

Mesosternal area variably armed, with rows of multiple
small teeth or ridges or with a blunt transverse ridge; tergites
with sparser, less distinctly plumose setae; body sculpture
often coarse and dense .... *Photomorphus* (6.1.6)

9.3.7—Genera of Mesoamerican velvet ants

Females—wings absent, antenna with 10 flagellomeres, metasoma with 6 tergites

1 Pronotum divided from mesonotum as an articulating plate
(northern Mexico) .... ***Myrmosula*** (2.6, p. 46)

Pronotum not divided from mesonotum **2**

2 (1) T1 shape slender cylindrical; T2 usually without felt line (widespread) ...... ***Ephuta*** (2.4, p. 45)

T1 shape variable but not cylindrical; T2 with felt line **3**

3 (2) Mesosoma rectangular, pronotum and/or propodeum as wide or wider than
mesonotum; eye large, longitudinally ovate (widespread) .... ***Timulla*** (2.5, p. 45)

Mesosoma ovate or pear-shaped, mesonotum wider than pronotum and/or
propodeum; if mesosoma apparently rectangular, then eye small, transversely ovate **4**

4 (3) Apical tarsomere with lamella covering base of tarsal claws **5**

Apical tarsomere without such a lamella, base of tarsal claws exposed **6**

5 (4) Mandible clothed with dense microsetae (tropical Central America) ...... ***Pappognatha*** (7.2.9)

Mandible with scattered normal setae (tropical Central America) .... ***Hoplomutilla*** (7.2.5)

6 (4) Pygidial plate present **7**

Pygidium convex without a plate **14**

7 (6) T2 fringe composed of plumose setae;
generally nocturnal species
with dull-brown coloration .... **various nocturnal genera** (see chapter 6, p. 285)

T2 fringe composed of simple setae; generally diurnal species with bright coloration **8**

8 (7) T1 shape petiolate; mesosoma pear-shaped **9**

T1 shape sessile; mesosoma fiddle-shaped **10**

9 (8) T2 disc without spots or, if spots present, they are
composed of setae or are densely punctate like black
or reddish background cuticle; tibial spurs usually
black or dark reddish (widespread) .... ***Dasymutilla***, **in part** (2.1, p. 36)

T2 disc black with 2–4 yellowish cuticular spots, spots
usually with much sparser punctures than black background
cuticle; tibial spurs white (tropical Central America) .... ***Traumatomutilla*** (7.2.2)

10 (8) Gena and/or postgena armed with a ventral tooth **11**

Gena usually unarmed; pronotum without longitudinal carina; procoxa unarmed **13**

11 (10) Gena and postgena each with ventral tooth, postgenal tooth much larger than genal tooth; mandible usually with ventral lamella basally (tropical Mexico and Central America) .... *Hoplognathoca* (7.2.4)

With tooth on gena or postgena only; mandible without ventral lamella basally **12**

12 (11) Gena with large ventral tooth; pronotum usually with longitudinal carina on dorsolateral margin; procoxa armed with tubercle (widespread in Mexico) .... *Invreiella* (2.2, p. 39)

Gena usually unarmed, postgena often with small tooth; pronotum without longitudinal carina; procoxa unarmed (widespread) .... *Pseudomethoca*, **in part** (2.2, p. 41)

13 (10) Frons with swollen protuberance; T1 nearly as wide as T2 (tropical Central America) .... *Horcomutilla* (7.2.6)

Frons lacking swollen protuberance; T1 generally clearly narrower than T2 (widespread) .... *Pseudomethoca*, **in part** (2.2, p. 41)

14 (6) T2 fringe composed of plumose setae **15**

T2 fringe composed of simple or vaguely brachyplumose setae **17**

15 (14) Body uniformly orange-brown with sparse setae, except with 2 black cuticular spots basally (northern Mexico) .... *Stethophotopsis* (2.3, p. 45)

Color not as above, with metasoma entirely dark, with setae denser, or T2 with continuous black band basally **16**

16 (15) Nocturnal species with generally dull-brown coloration, at least with mesosoma reddish .... **various nocturnal genera** (see chapter 6, p. 285)

Diurnal species with body entirely black and marked with whitish setae (tropical Central America) .... *Xystromutilla* (7.2.13)

17 (14) Mesosoma fiddle-shaped; vertex usually clearly wider than mesosoma **18**

Mesosoma pear-shaped; vertex usually not much wider than mesosoma **23**

18 (17) Gena with a sharp angular carina that extends onto the vertex **19**

Genal carina simple or absent, not extending onto vertex **20**

19 (18) Mandible narrowly acuminate apically, unidentate or bidentate (tropical Central America) .... *Calomutilla* (7.2.3)

Mandible broadly dilated apically, tridentate (tropical Central America) .... *Pertyella* (7.2.10)

20 (18) T1 shape disciform, often with dorsal tooth or spine mesally **21**

T1 shape sessile, never with dorsal tooth mesally **22**

21 (20) Mandible narrowly acuminate apically, unidentate or bidentate (tropical Mexico and Central America) .... *Lophostigma* (7.2.8)

Mandible somewhat dilated apically, tridentate (tropical Central America) .... *Silvorientilla* (7.2.11)

22 (20) T1 slender, less than half width of T2 (tropical Mexico and Central America) .... *Lophomutilla* (7.2.7)

T1 wider, greater than half width of T2 (widespread) .... *Pseudomethoca*, **in part** (2.2, p. 41)

23 (17) T1 shape sessile, T1 more than half as wide as T2 .... **24**

T1 shape petiolate or disciform, T1 less than half as wide as T2 .... **25**

24 (23) Head and mesosoma sparsely setose with areolate sculpture, areas inside areolations microreticulate (tropical Central America) .... *Nanotopsis* (7.2.12)

Head and mesosoma densely punctate and setose (widespread) .... *Protophotopsis* (2.3, p. 42)

25 (23) T1 shape disciform; body color usually dull brown with sparse setae; T2 disc lacking distinct cuticular spots (widespread) .... *Lomachaeta* (2.1, p. 39)

T1 shape petiolate; body color black or reddish with dense setae; T2 disc with 2–4 yellow cuticular spots **26**

26 (25) T2 marked with 2 yellowish cuticular spots; mesosoma with distinct large scutellar scale (tropical Central America) .... *Frigitilla* (7.2.1)

T2 marked with 4 yellowish cuticular spots; mesosoma usually without scutellar scale (widespread in Mexico and Central America) .... *Dasymutilla*, **in part** (2.1, p. 36)

Males—wings usually present, antenna with 11 flagellomeres, metasoma with 7 tergites

1 Eye with setae, vertically ovate without notch; forewing with veins distinct to outer margin (northern Mexico) .... *Myrmosula* (4.6, p. 178)

Eye bare, subcircular or ovate with notch; forewing with veins not distinct near margin **2**

2 (1) Eye vertically ovate with notch **3**

Eye subcircular, without notch **4**

3 (2) T1 shape slender cylindrical; T2 without felt line (widespread) .... *Ephuta* (4.4, p. 175)

T1 shape subsessile; T2 with felt line (widespread) .... *Timulla* (4.5, p. 175)

4 (2) Apical tarsomere with lamella covering base of tarsal claws **5**

Apical tarsomere without such a lamella, base of tarsal claws exposed **6**

5 (4) Mandible clothed with dense microsetae (tropical Central America) ...... *Pappognatha* (7.2.9)

Mandible with scattered normal setae (tropical Central America) .... *Hoplomutilla* (7.2.5)

6 (4) Axilla armed with posterior tooth, truncate lobe, or impunctate swelling; T1 shape usually petiolate or disciform **7**

Axilla unarmed; if axilla with apparent angulation, then T1 shape sessile **10**

7 (6) T1 shape disciform or subsessile; small-bodied insects with sparse setae and punction (widespread) .... *Lomachaeta* (4.1, p. 169)

T1 shape petiolate; generally larger species with dense setae and punction **8**

8 (7) S2 considerably swollen mesally with small distinct seta-filled pit mesally (tropical Central America) .... *Frigitilla* (7.2.1)

S2 with or without seta-filled pit, but not considerably swollen **9**

9 (8) Body color entirely black, marked with black and whitish setae; tibial spurs white (tropical Central America) .... *Traumatomutilla* (7.2.2)

Body color and setae variable, often partly bright red or yellowish; tibial spurs usually black or dark reddish (widespread) ...... *Dasymutilla* (4.1, p. 167)

10 (6) T2 fringe composed of plumose setae **11**

T2 fringe composed of simple or vaguely brachyplumose setae **12**

11 (10) Clypeus with two large teeth anteromesally, separated by deep sinus; diurnal species with body entirely black and wings extensively blackened (tropical Central America) .... *Xystromutilla* (7.2.13)

Clypeus not bidentate, or with teeth smaller and not separated by deep sinus; often nocturnal species with dull brown coloration and wings not darkened .... **various nocturnal genera** (see chapter 6, p. 285)

12 (10) T1 shape wide and petiolate, in lateral view with conspicuous constriction between T1 and T2 .... **13**

T1 shape not broad and petiolate, if broad, then subsessile with T2, if petiolate, then slender **14**

13 (12) Head and mesosoma sparsely setose with areolate sculpture, areas inside areolations microreticulate (tropical Central America) .... ***Nanotopsis*** (7.2.12)

Head and mesosoma densely punctate and setose (widespread) .... ***Protophotopsis*** (4.3, p. 173)

14 (12) Metasomal cuticle predominantly orange or reddish **15**

Metasomal cuticle predominantly blackish **16**

15 (14) Head dorsoventrally flattened, vertex extended with posterior transverse carina; paramere with pre-apical ventral setose lobe (tropical Mexico and Central America) .... ***Silvorientilla*** (7.2.11)

Head not flattened, lacking posterior carina; paramere without pre-apical lobe (widespread) .... ***Pseudomethoca***, in part (4.2, p. 169)

16 (14) Epaulet raised to form triangular tooth; T1 shape generally slender subsessile or subdisciform **17**

Epaulet not so raised; T1 shape usually short or widely sessile **18**

17 (16) Hypopygium with posterior margin convex or truncate (tropical Mexico and Central America) .... ***Lophomutilla*** (7.2.7)

Hypopygium with posterior margin concave (tropical Mexico and Central America) .... ***Lophostigma*** (7.2.8)

18 (16) Postgena armed with ventral tooth or lamella **19**

Postgena unarmed **20**

19 (18) Mandible vertical to apex, bidentate; postgenal armature usually consisting of flattened lamella (widespread) .... ***Pseudomethoca***, in part (4.2, p. 169)

Mandible apically oblique, tridentate; postgenal armature conical or tooth-like (tropical Mexico and Central America) .... ***Hoplognathoca*** (7.2.4)

20 (18) Paramere bifurcated apically; mesosternal area armed with smooth tubercle or tooth (tropical Central America) .... ***Horcomutilla*** (7.2.6)

Paramere not bifurcated; mesosternal area unarmed **21**

21 (20) Clypeus usually blackish, coarsely punctate, somewhat expanded anteriorly, and with teeth (when present) closely spaced and situated between antennal tubercles (widespread) .... ***Pseudomethoca***, in part (4.2, p. 169)

Clypeus usually pale yellow-brown, weakly punctate, flat without anterior expansion, and with 2 widely separated small teeth placed antolateral to antennal tubercles **22**

22 (21) Mandible with basal three-fourths broad before abruptly narrowed tridentate apical third of mandible, middle apical mandibular tooth distinct and usually equidistant between upper and lower teeth; inner surface of mandible usually with small tooth in basal half; penis valve with dorsal margin concave to straight (tropical Central America) .... ***Calomutilla*** (7.2.3)

Mandible with basal two-thirds relatively broad before abruptly narrowed tridentate apical third of mandible, middle apical mandibular tooth usually smaller and located closer to lower tooth; inner surface of mandible usually without tooth in basal half; penis valve with dorsal margin clearly convex (tropical Central America) .... ***Pertyella*** (7.2.10)

9.3.8—Subfamilies and tribes around the world

These keys are expected to work for only about 90% of specimens; many structural exceptions cannot be incorporated into such a key without expanding it by dozens of couplets and characters. Since such groups have a relatively manageable size, tribes are not keyed for subfamilies with fewer than 500 species, and subtribes are ignored entirely. The following codes are used for geographic realms: AT=Afrotropical; AU=Australasian; NA=Nearctic; NT=Neotropical; OR=Oriental; PA=Palaearctic. The following prefixes are used for subfamilies: Mut.=Mutillinae; Sph.=Sphaeropthalminae.

Females—antenna with 10 flagellomeres (usually), metasoma with 6 tergites

1	Mesosoma divided, pronotum separated as articulating plate (NA, OR, PA)	**Myrmosinae**
	Mesosoma entire, pronotum fused to remainder of mesosoma	**2**
2 (1)	Ocelli present; tarsal claws bidentate; pronotum and mesonotum separated by well-defined carina and groove (AT, OR, PA)	**Pseudophotopsidinae**
	Ocelli absent; tarsal claws simple; pronotum and mesonotum usually not strongly differentiated	**3**
3 (2)	Mesosoma, in dorsal view, more or less parallel-sided or clearly widest in propodeum	**4**
	Mesosoma, in dorsal view, with swollen or projecting mesonotum and narrowed propodeum	**9**
4 (3)	T2 lacking felt line, S2 with or without felt line (AT, OR, PA)	**Ticoplinae**
	T2 with felt line, S2 with felt line, if present, shorter than tergal felt line	**5**
5 (4)	Mesopleural lamella present and distinct; metapleural suture terminating near midpoint of mesopleuron (AT, OR, PA)	**Myrmillinae**
	Mesopleural lamella absent or indistinct; metapleural suture terminating near pronotal suture	**6 (Mutillinae)**
6 (5)	Mesosoma posterodorsally with transverse row of spines or teeth (AT, OR, PA)	**Mut.Ctenotillini**
	Mesosoma unarmed posteriorly, at most with some scattered tubercles	**7**
7 (6)	T1 broadly sessile with T2, usually with distinct dorsal and anterior faces; mesosoma without scutellar scale (AT, AU, OR, PA)	**Mut.Mutillini**
	T1 narrower; scutellar scale usually present	**8**
8 (7)	T2 disc marked with one or three pale setal spots (AT, OR, PA)	**Mut.Smicromyrmini**
	T2 disc marked with two pale setal spots side by side (AT, AU, NA, NT, OR, PA)	**Mut.Trogaspidiini**
9 (3)	T2 and S2 lacking felt lines; eye minute, with its diameter shorter than the flagellar width (AT, OR)	**Rhopalomutillinae**
	Felt line usually present on T2 and/or S2; eye generally larger, with its diameter much greater than the flagellar width	**10**
10 (9)	Species from the Old World	**11**
	Species from the New World	**14**
11 (10)	Eye generally protruding and smooth; plumose or brachyplumose setae usually present, especially on gena or tergal fringes	**12**
	Eye generally flat and distinctly faceted; plumose setae absent	**13**
12 (11)	Species from AU and southeastern OR	**Sph.Dasymutillini** (*Ephutomorpha* & kin)
	Species from PA and northeastern OR	**Sph.Sphaeropthalmini, in part** (*Cystomutilla* & kin)

13 (11) T1 shape broadly sessile/disciform, usually with distinctly separated dorsal and anterior faces (AT, AU, OR, PA) .... **Odontomutillinae**

T1 shape variable, generally petiolate or smoothly subsessile (AT, AU, OR, PA) .... **Dasylabrinae**

14 (10) Eye longitudinally ovate; T1 shape slender subcylindrical (NA, NT) .... **Sph.Ephutini**

Eye circular or transversely ovate; T1 shape variable **15**

15 (14) T2 generally with distinct fringe of plumose setae; generally nocturnal forms (NA, NT) .... **Sph.Sphaeropthalmini, in part (*Sphaeropthalma* & kin)**

T2 fringe simple or brachyplumose; generally diurnal forms **16**

16 (15) T1 generally narrow petiolate or disciform; mesosoma with lateral propodeal margin converging posterior to propodeal spiracle (NA, NT) .... **Sph.Dasymutillini, in part (*Dasymutilla* & kin)**

T1 generally broadly disciform or sessile with T2; mesosoma constricted at propodeal spiracles, lateral propodeal margin slightly diverging posterior to spiracle (NA, NT) .... **Sph.Pseudomethocini**

Males—antenna with 11 flagellomeres, metasoma with 7 tergites

1 Forewing with sclerotized veins reaching wing margin; T2 and S2 lacking felt line; eye usually setose (NA, OR, PA) .... **Myrmosinae**

Forewing with sclerotized veins not reaching wing margin; T2 and/or S2 usually having felt line; eye usually bare **2**

2 (1) Hindwing with jugal lobe present; tarsal claws bidentate (AT, OR, PA) .... **Pseudophotopsidinae**

Hindwing lacking jugal lobe; tarsal claws simple **3**

3 (2) Fully winged forms **4**

Apterous or brachypterous forms **18**

4 (3) Second submarginal cell of forewing petiolate with marginal cell or absent; eye often setose; T2 lacking felt line, S2 sometimes having felt line (AT, OR, PA) .... **Ticoplinae, in part**

Second submarginal cell of forewing sessile with marginal cell; eye never setose; T2 and/or S2 usually having felt line **5**

5 (4) Eye emarginate along inner margin; *if* weakly emarginate, *then* tegula ovate, longer than broad **6**

Eye entire along inner margin, *if* weakly emarginate, *then* tegula small, usually as wide as long **12**

6 (5) T2 lacking felt line; T1 petiolate or cylindrical **7**

T2 with felt line; T1 usually sessile or subsessile with T2 .... **8**

7 (6) T1 cylindrical; hypopygium entire (NA, NT) .... **Sph.Ephutini**

T1 elongate petiolate; hypopygium emarginate, formed of separated lateral lobes (AT, OR) .... **Rhopalomutillinae**

8 (6) Mesoscutellum armed with two posterior teeth (AT, AU, OR, PA) .... **Odontomutillinae**

Mesoscutellum unarmed, at most with single mesal swelling or tooth **9**

9 (8) T1 broadly sessile with T2, usually with distinct dorsal and anterior faces; tegula elongate, surpassing anterior scutellar margin (AT, AU, OR, PA) .... **Mut.Mutillini**

T1 narrower; tegula usually shorter **10**

10 (9)   Tergal junctions, especially T2–4, constricted in dorsal view
(AT, OR, PA) .... **Mut.Ctenotillini**

Tergal junctions not constricted **11**

11 (10, 21)   F1 usually much shorter than F2; vertex usually flattened with
longitudinal ridges or quadrate with the sides parallel behind the
eyes (AT, OR, PA) .... **Mut.Smicromyrmini**

F1 usually nearly as long as or longer than F2; vertex usually
simply rounded (AT, AU, NA, NT, OR, PA) .... **Mut.Trogaspidiini**

12 (5)   Species from the Old World **13**

Species from the New World **16**

13 (12)   Eye longitudinally ovate, flat; T1 shape short sessile (AT, OR, PA) .... **Myrmillinae, in part**

Eye usually subcircular, often protruding; T1 shape variable, often petiolate **14**

14 (13)   Eye generally flat and distinctly faceted; plumose
setae absent (AT, AU, OR, PA) .... **Dasylabrinae, in part (*Dasylabris* & kin)**

Eye generally protruding and smooth; plumose or brachyplumose
setae usually present, especially on gena or tergal fringes **15**

15 (14)   Species from AU and southeastern OR .... **Sph.Dasymutillini (*Ephutomorpha* & kin)**

Species from PA and northeastern OR .... **Sph.Sphaeropthalmini,
in part (*Cystomutilla* & kin)**

16 (12, 18)   T2 generally with distinct fringe of plumose setae;
generally nocturnal forms (NA, NT) .... **Sph.Sphaeropthalmini, in
part (*Sphaeropthalma* & kin)**

T2 fringe simple or brachyplumose; generally diurnal forms **17**

17 (16)   T1 generally narrow petiolate or disciform; mesosoma
with lateral propodeal margin converging posterior
to propodeal spiracle (NA, NT) .... **Sph.Dasymutillini, in part
(*Dasymutilla* & kin)**

T1 generally broadly disciform or sessile with T2; mesosoma
constricted at propodeal spiracles, lateral propodeal margin
slightly diverging posterior to spiracle (NA, NT) .... **Sph.Pseudomethocini**

18 (3)   Species from the New World .... **Sphaeropthalminae, back to 16**

Species from the Old World **19**

19 (18)   In dorsal view, mesonotum wider than pronotum
and propodeum, or pronotum clearly wider than
mesonotum and propodeum (AT) .... **Dasylabrinae, in part (*Brachymutilla* & kin)**

In dorsal view, pronotum and propodeum both wider
than mesonotum, propodeum clearly wider than
pronotum and mesonotum, or mesosoma parallel-sided **20**

20 (19)   Eye usually setose; T2 lacking felt line, S2 sometimes
with felt line (AT, OR, PA) .... **Ticoplinae, in part**

Eye bare; T2 with felt line, S2 lacking felt line **21**

21 (20)   Head generally quadrate, clearly wider than mesosoma
(AT, OR, PA) .... **Myrmillinae, in part**

Head generally rounded, usually not wider than mesosoma
(AT, OR) .... **Mutillinae, back to 11**

REFERENCES

André, E. 1901. Nouvelle contribution a la connaissance des Mutillides de l'Australie. *Mémoires de la Société Zoologique de France* 14(4): 467–513.

Argaman, Q. 1988. Description of the female of *Ticolpa*, with biological and taxonomic notes (Hymenoptera, Mutillidae). *Fragmenta Balcanica* 14(5): 33–46.

Bartholomay, P. R., Williams, K. A., Cambra, R. A., Oliveira, M. L. 2020. Revision of the *Traumatomutilla juvenilis* species group (Hymenoptera, Mutillidae). *Journal of Natural History* 53(43–44): 2639–2683.

Bartholomay, P. R., Williams, K. A., Cambra, R. A., Oliveira, M. L. 2021. Revision of the *Traumatomutilla gemella* species-group (Hymenoptera, Mutillidae) with the description of its hitherto unknown males. *Zoosystema* 43(1): 1–28.

Bartholomay, P. R., Williams, K. A., Cambra, R. A., Oliveira, M. L. 2022. Revision of the *Traumatomutilla indica* species-group (Hymenoptera: Mutillidae). *Zootaxa* 5108: 1–97.

Bartholomay, P. R., Williams, K. A., Lopez, V. M., Oliveira, M. L. 2019. Revision of the *Traumatomutilla americana* species group (Hymenoptera: Mutillidae). *Zootaxa* 4608(1): 1–34.

Bartholomay, P. R., Williams, K. A., Luz, D. R., Cambra, R. A., Oliveira, M. L. 2019. *Traumatomutilla* André miscellanea: Revision of the bellica, bifurca, diabolica, and vitelligera species groups, and a new group for the new species *T. pilkingtoni* Bartholomay and Williams (Hymenoptera: Mutillidae: Sphaeropthalminae: Dasymutillini). *Insecta Mundi* 0709: 1–37.

Bartholomay, P. R., Williams, K. A., Luz, D. R., Morato, E. F. 2015. *Frigitilla* gen. nov., a new genus of Amazonian Mutillidae (Hymenoptera). *Zootaxa* 3957(1): 49–58.

Bartholomay, P. R., Williams, K. A., Luz, D. R., Oliveira, M. L. 2018. New species of *Traumatomutilla* André in the *T. tabapua* and *T. integella* species-groups (Hymenoptera, Mutillidae). *Zootaxa* 4433(2): 361–385.

Bayliss, P. S., Brothers, D. J. 2001. Behavior and host relationships of *Dolichomutilla sycorax* (Smith, 1855) (Hymenoptera: Mutillidae, Specidae). *Journal of Hymenoptera Research* 10(1): 1–9.

Bischoff, H. 1920. Monographie der Mutilliden Afrikas. *Archiv für Naturgeschichte, Abteilung A* 86(1–3): 1–480.

Bischoff, H. 1921. Monographie der Mutilliden Afrikas. *Archiv für Naturgeschichte, Abteilung A* 86(4): 481–830.

Brabant, C. M., Williams, K. A., Pitts, J. P. 2010. True females of the subgenus *Photomorphina* Schuster (Hymenoptera: Mutillidae). *Zootaxa* 2559(1): 58–68.

Brach, V. 1978. *Brachynemurus nebulosus* (Neuroptera: Myrmeleontidae): a possible Batesian mimic of Florida mutillid wasps (Hymenoptera: Mutillidae). *Entomological News, Philadelphia* 89: 153–156.

Brothers, D. J. 1971. *Ascetotilla*, a new genus of Mutillidae from New Guinea (Hymenopt era). *Pacific Insects* 13(3–4): 471–485.

Brothers, D. J. 1972. Biology and immature stages of *Pseudomethoca f. frigida,* with notes on other species (Hymenoptera: Mutillidae). *The University of Kansas Science Bulletin* 50(1): 1–38.

Brothers, D. J. 1978. Biology and immature stages of *Myrmosula parvula* (Hymenoptera: Mutillidae). *Journal of the Kansas Entomological Society* 51(4): 698–710.

Brothers, D. J. 1994. A new genus and four new species of Mutillidae associated with *Brachyponera lutea* Mayer (Formicidae) in Western Australia (Hymenoptera). *Australian Journal of Entomology* 33(2): 143–152.

Brothers, D. J. 2006. Capíto 54: Familia Mutillidae. In: Fernández, F., Sharkey, M. J., eds. *Introducción a los Hymenoptera de la Región Neotropical.* Universidad Nacional de Colombia: Bogotá, pp. 577–593.

Brothers, D. J. 2015. Revision of the Rhopalomutillinae (Hymenoptera, Mutillidae): I, generic review with descriptions of three new genera. *Journal of Hymenoptera Research* 46: 1–24.

Brothers, D. J. 2022. Critical analysis of the type material of Mutillidae described from the Australasian Region (Hymenoptera). *Zootaxa* 5140(1): 1–215.

Brothers, D. J., Lelej, A. S. 2017. Phylogeny and higher classification of Mutillidae (Hymenoptera) based on morphological reanalyses. *Journal of Hymenoptera Research* 60(1): 1–97.

Cambra, R. A. 1996. Sistemática y Biología del Género Lophomutilla Mickel (Hymenoptera: Mutillidae). Thesis, Universidad de Panamá: Panama City, 99 pp.

Cambra, R. A., Quintero, A. D. 1996. The Mexican and Central American species of *Lophostigma* Mickel, including records, and taxonomic notes for the genus (Hymenoptera: Mutillidae). *Pan-Pacific Entomologist* 72(2): 92–101.

Cambra, R. A., Quintero, A. D. 1997. A revision of *Protophotopsis* Schuster (Hymenoptera: Mutillidae). *Journal of Hymenoptera Research* 6(2): 263–272.

Cambra, R. A., Quintero, A. D. 2004. New species of *Xystromutilla* Andre (Hymenoptera: Mutillidae) and the first illustrated key for the males of the genus. *Transactions of the American Entomological Society* 130(4): 463–478.

Cambra, R. A., Quintero, A. D. 2008. New species and new distribution records for *Pseudomethoca* Ashmead (Hymenoptera: Mutillidae) from Central America. *Transactions of the American Entomological Society* 134(1–2): 69–86.

Cambra, R. A., Quintero, A. D. 2010. Review of the Neotropical genus *Nanotopsis* Schuster (Hymenoptera: Mutillidae: Sphaeropthalminae). *Transactions of the American Entomological Society* 136(3–4): 289–298.

Cambra, R. A., Quintero, A. D., Brothers, D. J. 2014. A new species of *Hoplognathoca* Suárez, 1962 from Central America and description of male *Hoplognathoca costarricensis* Suárez, 1962 (Hymenoptera: Mutillidae). *Zootaxa* 3884(3): 295–300.

Cambra, R. A., Brothers, D. J., Quintero, A. D. 2020. Review of *Calomutilla* Mickel, 1952, a new species, and comparison with *Pertyella* Mickel, 1952 (Hymenoptera: Mutillidae). *Zootaxa* 4789(2): 466–480.

Cambra, R. A., Williams, K. A., Bartholomay, P. R. 2022. Two new species of *Horcomutilla* Casal, 1962 (Hymenoptera: Mutillidae), description of hitherto unknown males, and an illustrated key to all known species of the genus. *Neotropical Entomology* 51(3): 427–446.

Cameron, P. 1895. Fam. Mutillidae. In: Goodman, F. D., Salvin, O., eds. *Biologia Centrali-Americana. Insecta. Hymenoptera. (Fossores.) Vol. II.* R. H. Porter; London, pp. 1–413.

Casal, O. H. 1969. Sobre *Traumatomutilla* André (Hymenoptera, Mutillidae). *Physis* 28(77): 279–298.

Chen, C-W. 1957. A revision of the velvety ants or Mutillidae of China (Hymenoptera). *Quarterly Journal of the Taiwan Museum* 10(3–4): 135–224, 6 plates.

Christ, J. L. 1791. *Naturgeschichte, Klassification und Nomenclatur der Insekten vom Bienen, Wespen und Ameisengeschlecht; als der fünften Klasse fünfte Ordnung des Linneischen Natursystems von den Insekten: Hymenoptera. Mit häutigen Flügeln.* In der Hermannischen Buchhandlung: Frankfurt am Main, 535 pp.

Evans, H. E. 1957. Studies on the larvae of digger wasps (Hymenoptera: Sphecidae). Part III: Philanthinae, Trypoxyloninae, and Crabroninae. *Transactions of the American Entomological Society* 83: 79–117.

Evans, H. E., Yoshimoto, C. M. 1962. The ecology and nesting behavior of the Pompilidae (Hymenoptera) of the Northeastern United States. *Miscellaneous Publications of the Entomological Society of America* 3: 66–119.

Fales, H. M., Jaouni, T. M., Schmidt, J. O., Blum, M. S. 1980. Mandibular gland allomones of *Dasymutilla occidentalis* and other mutillid wasps. *Journal of Chemical Ecology* 6: 895–903.

Fritz, M. A. 1990. Notas taxonómicas sobre los géneros *Lophomutilla* y *Pertyella* (Hymenoptera: Mutillidae). I. *Revista de la Sociedad Entomológica Argentina* 48(1–4): 129–142.

Fritz, M. A., Pagliano, G. 1993. Sobre *Lophomutilla* Mickel y *Pertyella* Mickel (Hymenoptera, Mutillidae). *Bollettino della Società entomologica italiana* 124(3): 209–220.

Gall, B. G., Spivey, K. L., Chapman, T. L., Delph, R. J., Brodie, E. D., Jr., Wilson, J. S. 2018. The indestructible insect: velvet ants from across the United States avoid predation by representatives from all major tetrapod clades. *Ecology and Evolution* 8(11): 5852–5862.

Haddock, J. D. 1967. Mutillidae. In: Krombein, K. V., Burks, B. D. Aprocrita (Aculeata). Hymenoptera of America north of Mexico, Synoptic Catalog No. 2. United States Department of Agriculture: Washington, DC, pp. 326–341.

Hines, H. M., Witkowski, P., Wilson, J. S., Wakamatsu, K. 2017. Melanic variation underlies aposematic color variation in two hymenopteran mimicry systems. *PloS one* 12(7): e0182135.

Jensen, T., Walker, A. A., Nguyen, S. H., Jin, A. H., Deuis, J. R., Vetter, I., King, G. F., Schmidt, J. O., Robinson, S. D. 2021. Venom chemistry underlying the painful stings of velvet ants (Hymenoptera: Mutillidae). *Cellular and Molecular Life Sciences* 78(12): 5163–5177.

Krombein, K. V. 1940 [1939]. Studies in the Tiphiidae (Hymenoptera, Aculeata). I. V. A revision of the Myrmosinae of the New World with a discussion of the Old World species. *Transactions of the American Entomological Society* 65: 415–465, 24 plates.

Krombein, K. V. 1954. Taxonomic notes on some wasps from Florida with descriptions of new species and subspecies (Hymenoptera, Scolioidea and Sphecoidea). *Transactions of the American Entomological Society* 80(1): 1–27.

Krombein, K. V. 1971. A monograph of the Mutillidae of New Guinea, Bismarck Archipelago and Solomon Islands. Part 1: Mutillinae (Hymenoptera, Aculeata). In: Yasumatsu, K., Asahina, S., eds. *Entomological Essays to Commemorate the Retirement of Professor K. Yasumatsu.* Hokuryukan Publishing Company: Tokyo, pp. 25–60.

Krombein, K. V. 1992. Host relationships, ethology and systematics of *Pseudomethoca* Ashmead (Hymenoptera: Mutillidae, Andrenidae, Halictidae and Anthophoridae). *Proceedings of the Entomological Society of Washington* 94(1): 91–106.

Krombein, K. V., Evans, H. E. 1955. An annotated list of wasps collected in Florida, March 20 to April 3, 1954 (Hymenoptera, Aculeata). *Proceedings of the Entomological Society of Washington* 57: 223–235.

Krombein, K. V., Hurd, P. D., Smith, D. R., Burks, B. D. 1979. *Volume 2. Aprocrita (Aculeata). Catalog of Hymenoptera in America north of Mexico.* Smithsonian Institution Press: Washington, DC, pp. i–xvi, 1199–2209.

Lelej, A. S. 1980. The genus *Pseudophotopsis* André, 1896 (Hymenoptera, Mutillidae) from the USSR and neighbouring countries. *Entomologicheskoe Obozrenie* 59(3): 634–649. [In Russian]

Lelej, A. S. 1985. *The velvet ants (Hymenoptera, Mutillidae) of the USSR and neighbouring countries. Nauka, Leningrad, 268 pp. [In Russian]*

Lelej, A. S. 2002. *Catalogue of the Mutillidae (Hymenoptera) of the Palaearctic Region.* Dal'nauka: Vladivostok, 171 pp.

Lelej, A. S. 2005. *Catalogue of the Mutillidae (Hymenoptera) of the Oriental Region.* Dal'nauka: Vladivostok, 252 pp.

Lelej, A. S., Schmid-Egger, C. 2005. The velvet ants (Hymenoptera, Mutillidae) of Central Europe. *Linzer biologische Beiträge* 37(2): 1505–1543.

Linnaeus, C. 1758. Systema naturæ per regna tria naturæ, secundum classes, ordines, genera, species, cum characteribus, differentiis, synonymis, locis. Editio decima, reformata [10th revised edition]. Volume 1, Laurentius Salvius: Holmia, 824 pp.

Luz, D. R., Williams, K. A., Bartholomay, P. R. 2016. The mutillid wasps of the *Dasymutilla paradoxa* species-group (Hymenoptera, Mutillidae). *Zootaxa* 4193(2): 361.

Manley, D. G. 1999. A synonymy for *Pseudomethoca donaenae* (Cockerell & Fox) (Hymenoptera: Mutillidae). *Pan-Pacific Entomologist* 75(1): 32–34.

Manley, D. G., Neff, J. L. 1989. *Pseudomethoca ilione* (Fox), a new synonym of *P. gila* (Blake) (Hymenoptera: Mutillidae). *Pan-Pacific Entomologist* 65(1): 12–14.

Manley, D. G., Pitts, J. P. 2002. A key to genera and subgenera of Mutillidae (Hymenoptera) in America north of Mexico with description of a new genus. *Journal of Hymenoptera Research* 11(1): 72–10.

Manley, D. G., Pitts, J. P. 2007. Tropical and subtropical velvet ants of the genus *Dasymutilla* Ashmead (Hymenoptera: Mutillidae) with descritions of 45 new species. *Zootaxa* 1487: 1–128.

Manley, D. G., Radke, W. R. 2002. Synonymy of *Dasymutilla sicheliana* (Saussure) (Hymenoptera: Mutillidae). *Pan-Pacific Entomologist* 78: 230–234.

Manley, D. G., Radke, W. R. 2006. Velvet ants (Hymenoptera: Mutillidae) of the Bitter Lake National Wildlife Refuge, New Mexico, with descriptions of new species. *Southwestern Naturalist* 51(4): 536–382.

Manley, D. G., Taber, S. III. 1978. A mating aggregation of *Dasymutilla foxi* in southern Arizona (Hymenoptera: Mutillidae). *The Pan-Pacific Entomologist* 54: 231–235.

Manley, D. G., Williams, K. A., Pitts, J. P. 2020. Keys to Nearctic velvet ants of the genus *Dasymutilla* Ashmead (Hymenoptera: Mutillidae), with notes on taxonomic changes since Krombein (1979). *Proceedings of the Entomological Society of Washington* 122(2): 335–414.

Mellor, J. E. M. 1927. A note on the mutillid *Ephutomma continua* Fabr. and of *Bembex mediterranea* Hdl. in Egypt with a summary of the distribution and of some previously recorded habits of the Mutillidae. *Bulletin de la Société Entomologique d'Égypte* 20: 69–79.

Mickel, C. E. 1924. A revision the mutillid wasps of the genera *Myrmilloides* and *Pseudomethoca* occurring in America north of Mexico. *Proceedings of the United States National Museum* 64(2505): 1–52, 4 plates.

Mickel, C. E. 1928. Biological and taxonomic investigations on the mutillid wasps. *United States National Museum Bulletin* 143: 1–351.

Mickel, C. E. 1935a. Descriptions and records of Nearctic mutillid wasps of the genera *Myrmilloides* and *Pseudomethoca* (Hymenoptera: Mutillidae). *Transactions of the American Entomological Society* 61: 383–398.

Mickel, C. E. 1935b. The mutillid wasps of the islands of the Pacific Ocean (Hymenoptera: Mutillidae). *Transactions of the Royal Entomological Society of London* 83(2): 177–307.

Mickel, C. E. 1936a. Two new genera and five new species of Mutillidae. *Annals of the Entomological Society of America* 29: 289–297.

Mickel, C. E. 1936b. New species and records of nearctic mutillid wasps of the genus *Dasymutilla* (Hymenoptera). *Annals of the Entomological Society of America* 29: 29–60.

Mickel, C. E. 1937. The mutillid wasps of the genus *Timulla* which occur in North America north of Mexico. *Entomologica Americana* 17(1): 1–118.

Mickel, C. E. 1938. The neotropical mutillid wasps of the genus *Timulla* Ashmead (Hymenoptera: Mutillidae). *Transactions of the Royal Entomological Society of London* 87(2): 529–679, 9 plates.

Mickel, C. E. 1939a. A monograph of the Neotropical mutillid genus *Hoplomutilla* Ashmead (Hymenoptera: Mutillidae). *Revista de Entomologia* 10: 337–403.

Mickel, C. E. 1939b. Monograph of a new Neotropical mutillid genus, *Pappognatha. Annals of the Entomological Society of America* 32: 329–343.

Mitchell, A., Brothers, D. J. 2002. Phylogeny of genera of Ticoplinae (Hymenoptera: Mutillidae). *Journal of Hymenoptera Research* 11(2): 312–325.

Nentwig, W. 1985. A mimicry complex between mutillid wasps (Hymenoptera: Mutillidae) and spiders (Araneae). *Studies on Neotropical Fauna and Environment* 20(2): 113–116.

Nonveiller, G., Suárez, F. J., Fritz, M. A., Achterberg, C. van. 1990. *Catalogue of the Mutillidae, Myrmosidae and Bradynobaenidae of the Neotropical region including Mexico (Insecta: Hymenoptera)*. SPB Academic Publishing: The Hague, 150 pp.

Okayasu, J., Williams, K. A., Lelej, A. S. 2018. A remarkable new species of *Sinotilla* Lelej (Hymenoptera: Mutillidae: Smicromyrmini) from Taiwan and an overview of color diversity in East Asian mutillid females. *Zootaxa* 446(3): 301–324.

Pagliano, G. 2007. Contributo alla conoscenza dei Mutillidae italiani (Hymenoptera, Scolioidea). *Bollettino Museo Regionale di Scienze Naturali* 24(1): 25–110.

Pagliano, G., Brothers, D. J., Cambra, R., Lelej, A. S., Lo Cascio, P., Matteini Palmerini, M., Scaramozzino, P. L., Williams, K. A., Romano, M. 2020. Checklist of names in Mutillidae (Hymenoptera), with illustrations of selected species. *Bollettino del Museo Regionale di Scienze Naturali* 36(1–2): 5–425.

Pan, A. D., Williams, K. A., Wilson, J. S. 2017. Are diurnal iguanian lizards the evolutionary drivers of New World female velvet ant (Hymenoptera: Mutillidae) Müllerian mimicry rings? *Biological Journal of the Linnean Society* 120(2): 436–447.

Pitts, J. P. 2003. *Schusterphotopsis*, a new genus of Sphaeropthalminae (Hymenoptera: Mutillidae) from California, with notes on the closely related genera *Acrophotopsis* Schuster and *Dilophotopsis* Schuster. *Zootaxa* 333(1): 1–7.

Pitts, J. P. 2007a. Description of the female of *Sphaeropthalma galapagensis* Williams and notes on *Morsyma* Fox and *Caenotilla* Pitts and Manley (Hymenoptera: Mutillidae). *The Pan-Pacific Entomologist* 83(2): 95–100.

Pitts, J. P. 2007b. Revision of *Odontophotopsis* Viereck (Hymenoptera: Mutillidae), Part 1, with a description of a new genus *Laminatilla*. *Zootaxa* 1619(1): 1–43.

Pitts, J. P., McHugh, J. V. 2000. *Stethophotopsis*, a new genus of Sphaeropthalmini (Mutillidae: Sphaeropthalminae) with a brachypterous male from Arizona. *Journal of Hymenoptera Research* 9(1): 29–33.

Pitts, J. P., McHugh, J. V. 2002. Review of *Acrophotopsis* Schuster (Hymenoptera: Mutillidae: Sphaeropthalminae), with description of a new species from Baja California. *Journal of Hymenoptera Research* 11(2): 338–349.

Pitts, J. P., Wilson, J. S. 2009. Description of the female of *Acrophotopsis* (Hymenoptera: Mutillidae) and synonymy of *Sphaeropthalma dirce. Journal of Hymenoptera Research* 18: 205–211.

Quintero, A. D., Cambra, T. R. A. 2001a. On the identity of *Scaptopoda* F. Lynch Arribálzaga, new taxonomic changes and new distribution records for Neotropical Mutillidae (Hymenoptera), with notes on their biology. *Transactions of the American Entomological Society* 127(3): 291–304.

Quintero, A. D., Cambra, T. R. A. 2001b. On the endemic genus *Morsyma* Fox (Hymenoptera: Mutillidae: Sphaeropthalminae), with a description of the previously unknown female. *Transactions of the American Entomological Society* 127(4): 451–458.

Quintero, A. D., Cambra, T. R. A. 2005. *Pappognatha* Mickel (Hymenoptera: Mutillidae: Sphaeropthalminae): New species, sex associations, hosts, and new distribution records. *Journal of Hymenoptera Research* 14(2): 191–199.

Ricketts, T. H., Dinerstein, E., Olso, D. M., Eichbaum, W., Loucks, C. J., DellaSala, D. A., Kavanagh, K., Hedao, P., Hurley, P., Carney, K., Abell, R., Walters, S. 1999. *Terrestrial Ecoregions of North America: A Conservation Assessment*. Island Press: Washington, DC, 485 pp.

Sadler, E. A., Pitts, J. P., Wilson, J. S. 2018. Stinging wasps (Hymenoptera: Aculeata), which species have the longest sting? *PeerJ* 6: e4743.

Schmidt, J. O., Blum, M. S. 1977. Adaptations and responses of *Dasymutilla occidentalis* (Hymenoptera: Mutillidae) to predators. *Entomologia Experimentalis et Applicata* 21(2): 99–111.

Schmidt, J. O., Hook, A. W. 1979. A record population of *Pseudomethoca simillima (Hymenoptera: Mutillidae)*. *Florida Entomologist* 62: 152.

Schmidt, J. O., Schmidt, L. S., Schmidt, D. K. 2021. The paradox of the velvet-ant (Hymenoptera, Mutillidae). *Journal of Hymenoptera Research* 84: 327–337.

Schultz, T. D., Puchalski, J. 2001. Chemical defenses in the tiger beetle *Pseudoxycheila tarsalis* Bates *(Carabidae: Cicindelinae). Coleopterists Bulletin* 55: 164–166.

Schuster, R. M. 1944. Notes and records of the eastern representatives of the Photopsidine genera of Mutillidae with description of new forms. *Bulletin of the Brooklyn Entomological Society* 39: 139–155.

Schuster, R. M. 1947 [1946]. A revision of the Sphaeropthalmine Mutillidae of America north of Mexico. *Annals of the Entomological Society of America* 39: 692–703.

Schuster, R. M. 1949. Contributions toward a monograph of Mutillidae of the Neotropical Region III. A key to the subfamilies and several new genera (Hymenoptera). *Entomologica Americana* 24: 59–140.

Schuster, R. M. 1951. A revision of the genus *Ephuta* (Mutillidae) in America north of Mexico. *Journal of the New York Entomological Society* 59: 1–43.

Schuster, R. M. 1956. A revision of the genus *Ephuta* (Mutillidae) in America north of Mexico. Part 2. *Journal of the New York Entomological Society* 64: 7–84.

Schuster, R. M. 1958. A revision of the Sphaeropthalmine Mutillidae of America north of Mexico, II. *Entomologica Americana, New Series* 37: 1–130.

Smith, G. B., Mitchell, A. 2021. Are these the world's most colourful silverfish? Possible mutillid mimics from Western Australia (Zygentoma: Lepismatidae). *Records of the Western Australian Museum* 36: 13–32.

Suárez, F. J. 1963. Datos sobre mutílidos neotropicales V. Sinópsis del género *Hoplognathoca* Suárez (Hymenoptera-Mutillidae). *Archivos del Instituto de Aclimatación* 12: 55–66, 2 plates.

Tanner, D. A., Boehme, N. F., Pitts, J. P. 2009. Review of *Acanthopsis* Schuster (Hymenoptera: Mutillidae). *Journal of Hymenoptera Research* 18(2): 192–204.

Torrico-Bazoberry, D., Muñoz, M. I. 2019. High-frequency components in the distress stridulation of Chilean endemic velvet ants (Hymenoptera: Mutillidae). *Revista Chilena de Entomología* 45(1): 5–13.

Turner, R. E. 1914. Notes on fossorial Hymenoptera. XIV. On the Mutillidae of Western Australia and Tasmania. *Annals and Magazine of Natural History*, Series 8, 14: 429–450.

Turrisi, G. F., Palmerini, M. M., Brothers, D. J. 2015. Systematic revision and phylogeny of the genera *Blakeius* Ashmead, 1903 and *Liomutilla* André, 1907, with description of two new genera (Hymenoptera: Mutillidae, Myrmillinae). *Zootaxa* 4010(1): 1–78.

Velásquez, A., Toledo, V. M., Luna, I. 2000. Mexican Temperate Vegetation. In: Barbour, M. G., Billings, W. D., eds. *North American Terrestrial Vegetation*, 2nd edition. Cambridge University Press: Cambridge, pp. 573–592.

Waldren, G. C. 2021. The velvet ants (Hymenoptera: Mutillidae): Systematics, biology, and biogeography of a little-known family. PhD dissertation, Utah State University: Logan, 336 pp.

Waldren, G. C., Sadler, E. A., Murray, E. A., Bossert, S., Danforth, B. N., Pitts, J. P. 2023. Phylogenomic Inference of the Higher Classification of Velvet Ants (Hymenoptera: Mutillidae). *Systematic Entomology* 2023: 1–25.

Waldren, G. C., Williams, K. A., Cambra, R. T., Pitts, J. P. 2020. Systematic revision of the North American velvet ant genus *Invreiella* Suárez (Hymenoptera: Mutillidae) with description of eleven new species. *Zootaxa* 4894(2): 151–205.

Wasbauer, M. 1973. Some new taxa in the Myrmosinae with keys to the females in North America (Hymenoptera, Tiphiidae). *Pan-Pacific Entomologist* 49(4): 325–337.

Williams, K. A., Bartholomay, P. R., Oliveira, M. L. 2017. Species groups of *Traumatomutilla* André (Hymenoptera: Mutillidae). *Insecta Mundi* 0533: 1–38.

Williams, K. A., Cambra, R. A., Waldren, G. C., Quintero, A. D., Pitts, J. P. 2019. *Silvorientilla*, gen. nov. A new neotropical velvet ant genus (Hymenoptera: Mutillidae). *Folia Entomológica Mexicana* 4(3): 91–109.

Williams, K. A., Cambra, R. A., Bartholomay, P. R., Luz, D. R., Quintero, A. D., Pitts, J. P. 2019. Review of the genus *Lomachaeta* Mickel, 1936 (Hymenoptera: Mutillidae) with new species and sex associations. *Zootaxa* 4564(1): 101–136.

Williams, K. A., Manley, D. G., Pilgrim, E. M., von Dohlen, C. D., Pitts, J. P. 2011. Multifaceted assessment of species validity in the *Dasymutilla bioculata* species group (Hymenoptera: Mutillidae). *Systematic Entomology* 36(1): 180–191.

Wilson, J. S., Pitts, J. P. 2008. Revision of velvet ant genus *Dilophotopsis* Schuster (Hymenoptera: Mutillidae) by using molecular and morphological data, with implications for desert biogeography. *Annals of the Entomological Society of America* 101(3): 514–524.

Wilson, J. S., Jahner, J. P., Forister, M. L., Sheehan, E. S., Williams, K. A., Pitts, J. P. 2015. North American velvet ants form one of the world's largest known Müllerian mimicry complexes. *Current Biology* 25(15): R704–R706.

Wilson, J. S., Pan, A. D., Limb, E. S., Williams, K. A. 2018. Comparison of African and North American velvet ant mimicry complexes: Another example of Africa as the "odd man out." *PLOS One* 13(1): e0189482. doi: 10.1371/journal.pone.0189482.

Wilson, J. S., Williams, K. A., Forister, M. L., von Dohlen, C. D., Pitts, J. P. 2012. Repeated evolution in overlapping mimicry rings among North American velvet ants. *Nature Communication* 3: 1272. Doi:10.1038/ncomms2275.

FORMAL ACKNOWLEDGMENTS

Most photographs were taken by the authors, especially the photographs of museum pinned specimens. We are grateful to Efram Goldberg, Daniel Roueche, Jillian Cowles, Alexander Wild, and Clarence Holmes for providing some excellent photographs of live specimens, which are published here. We also thank George Waldren and David Wahl for providing some museum specimen photographs. Regarding the museum specimens, the authors examined and photographed specimens from dozens of museums. We especially thank Christine LeBeau and James Carpenter at the American Museum of Natural History, New York, USA; Gavin Broad at the British Natural History Museum, London, UK; Stuart Fullerton and Sandor "Shawn" Kelly at the Bug Closet collection, Orlando, USA; Robert Zuparko and Vince Lee at the California Academy of Sciences Collection in San Francisco, USA; Jacqueline Airoso and Martin Hauser at the California State Collection of Arthropods, Sacramento, USA; John Rawlins and Bob Androw at the Carnegie Museum of Natural History, Pittsburgh, USA; Matthias Buck at the Royal Alberta Museum, Calgary, Canada; Andrew Bennett, John Huber, and Lubomir Masner at the Canadian National Collection of Insects, Ottawa, Canada; Boris Kondratieff at Colorado State University, Fort Collins, USA; James Liebherr at Cornell University Insect Collection, Ithaca, USA; Gabriel Melo at Colecao Entomological Pe. Jesus Santiago Moure, Curitiba, Brazil; Cheryl Barr and Robert Zuparko at Essig Museum Entomology Collection, Berkeley, USA; James Pitts and David Wahl at the Entomological Museum of Utah State University, Logan, USA; Jim Wiley, Kyle Schnepp, and Elijah Talamas at the Florida State Collection of Arthropods, Gainesville, USA; Chris Grinter at the Illinois Natural History Survey, Champagne, USA; Pedro Bartholomay at Instituto Nacional das Pesquisas Amazonas, Manaus, Brazil; Stefan Cover at MCZC; Felipe Vivallo at Museu Nacional de Rio de Janeiro, Brazil; Frederique Bakker at the Naturalis Biodiversity Museum, Royal Museum of Natural History, Leiden, the Netherlands; Zack Falin and Michael Engel at Snow Entomological Museum Collection, Lawrence, USA; Ed Riley and Karen Wright at the Texas A & M University Insect Collection, College Station, USA; Sangmi Lee at Arizona State University, Tempe, USA; Gene Hall at University of Arizona Insect Collection, Tucson, USA; Lynn Kimsey and Steve Heydon at the Bohart Museum, University of California at Davis, USA; Doug Yanega at University of California at Riverside, USA; Mark O'Brien at University of Michigan Museum of Zoology, Ann Arbor, USA; Robin Thomson and Philip Clausen at University of Minnesota Insect Collection, St. Paul, USA; and Brian Harris at the Smithsonian Institution, Washington, USA. We also thank the following individuals who allowed us to study their collected material: Donald Manley, Denis Brothers, Christian Schmid-Egger, Frank Parker, Mike Irwin, and Bill Warner. Many of the datapoints used to construct our maps came from the Symbiota Collections of Arthropods Network (scan-bugs.org). Many useful velvet ant observations came from the community at inaturalist.org; we are very grateful to everyone who photographed and posted their velvet ant observations there. We thank all the authors of previous velvet ant literature, dating back to Carl Linnaeus in 1758; many of their works are cited above, and many others were consulted over the decades that it took to research this book.

Kevin is especially thankful to James Pitts, Don Manley, and Justin Schmidt, who taught him so much about velvet ants in the early stages and put up with so many questions from an undergraduate amateur. Thanks to James especially for being Kevin's PhD program advisor, friend, and mentor; he had the single biggest role in Kevin's professional development. Gratitude is also in order to his various coauthors on dozens of papers about velvet ants, especially Arkady Lelej, Denis Brothers, Pedro Bartholomay, Roberto Cambra, George Waldren, David Luz, Narit Thaochan, Juriya Okayasu, Vinicius Lopez, Yostin Anino, Guido Pagliano, Ben Parslow, Craig Brabant, Celso Azevedo, Carol von Dohlen, Erik Pilgrim, Nicole Boehme, Mark Deyrup, and especially Joe Wilson and Aaron Pan, fellow authors of the book you are reading now. Kevin thanks his past and present employers and colleagues at the California Department of Food and Agriculture and Florida Division of Plant Industry.

CHECKLIST OF DIURNAL VELVET ANTS IN THE USA

Subfamily MYRMOSINAE Fox, 1894
Genus *Leiomyrmosa* Wasbauer, 1973
☐ *Leiomyrmosa spilota* Wasbauer, 1973

Genus *Myrmosa* Latreille, 1797
☐ *Myrmosa blakei* Bradley, 1917
☐ *Myrmosa bradleyi* Roberts, 1929
☐ *Myrmosa nocturna nocturna* Krombein, 1940
☐ *Myrmosa nocturna rufigastra* Krombein, 1940
☐ *Myrmosa peculiaris* Krombein, 1940
☐ *Myrmosa texana* Krombein, 1940
☐ *Myrmosa unicolor* Say, 1824

Genus *Myrmosula* Bradley, 1917
☐ *Myrmosula boharti* Wasbauer, 1973
☐ *Myrmosula exaggerata* (Krombein, 1940)
☐ *Myrmosula nasuta* Wasbauer, 1973
☐ *Myrmosula parvula* (Fox, 1893)
☐ *Myrmosula peregrinatrix* (Krombein, 1946)
☐ *Myrmosula rufiventris* (Blake, 1879)
☐ *Myrmosula rutilans* (Blake, 1879)

Subfamily MUTILLINAE Latreille, 1802
Tribe TROGASPIDIINI Bischoff, 1920
Genus *Timulla* Ashmead, 1899
☐ *Timulla barbata* (Fox, 1899)
☐ *Timulla barbigera* (Bradley, 1916)
☐ *Timulla compressicornis* Mickel, 1937
☐ *Timulla contigua* Mickel, 1937
☐ *Timulla dubitata* (Smith, 1855)
☐ *Timulla dubitatiformis* Mickel, 1937
☐ *Timulla euphrosyne* Mickel, 1937
☐ *Timulla euterpe* (Blake, 1879)
☐ *Timulla ferrugata* (Fabricius, 1804)
☐ *Timulla floridensis* (Blake, 1879)
☐ *Timulla grotei* (Blake, 1871)
☐ *Timulla hollensis* (Melander, 1903)
☐ *Timulla kansana* Mickel, 1937
☐ *Timulla leona* (Blake, 1871)
☐ *Timulla navasota navasota* (Bradley, 1916)
☐ *Timulla navasota cohahuila* Krombein, 1951
☐ *Timulla neobule* Mickel, 1937
☐ *Timulla nicholi* Mickel, 1937
☐ *Timulla oajaca* (Blake, 1871)
☐ *Timulla ocellaria* Mickel, 1937
☐ *Timulla ornatipennis* (Bradley, 1916)
☐ *Timulla rufosignata* (Bradley, 1916)

☐ *Timulla sayi* (Blake, 1871)
☐ *Timulla subhyalina* Mickel, 1937
☐ *Timulla suspensa* (Gerstaecker, 1874)
☐ *Timulla tolerata* Mickel, 1937
☐ *Timulla vagans* (Fabricius, 1798)
☐ *Timulla wileyae* Mickel, 1937

Subfamily SPHAEROPTHALMINAE Schuster, 1949
Tribe DASYMUTILLINI Brothers & Lelej, 2017
Genus *Dasymutilla* Ashmead, 1899
☐ *Dasymutilla albiceris* Mickel, 1936
☐ *Dasymutilla angulata* Krombein, 1951
☐ *Dasymutilla apicalata* (Blake, 1871)
☐ *Dasymutilla arachnoides* (Smith, 1855)
☐ *Dasymutilla archboldi* Schmidt & Mickel, 1979
☐ *Dasymutilla arenivaga* Mickel, 1928
☐ *Dasymutilla arenerronea* Bradley, 1916
☐ *Dasymutilla asopus* (Cresson, 1865)
☐ *Dasymutilla asteria* Mickel, 1936
☐ *Dasymutilla atricauda* Mickel, 1936
☐ *Dasymutilla aureola* (Cresson, 1865)
☐ *Dasymutilla bioculata* (Cresson, 1865)
☐ *Dasymutilla birkmani* (Melander, 1903)
☐ *Dasymutilla bonita* Mickel, 1928
☐ *Dasymutilla californica* (Radoszkowski, 1861)
☐ *Dasymutilla calorata* Mickel, 1928
☐ *Dasymutilla campanula* Mickel, 1928
☐ *Dasymutilla canella* (Blake, 1871)
☐ *Dasymutilla cirrhomeris* Manley & Pitts, 2007
☐ *Dasymutilla clotho* (Blake, 1872)
☐ *Dasymutilla coccineohirta* (Blake, 1871)
☐ *Dasymutilla connectens* (Cameron, 1895)
☐ *Dasymutilla creon* (Blake, 1872)
☐ *Dasymutilla curialis* Mickel, 1928
☐ *Dasymutilla curticeps* Mickel, 1928
☐ *Dasymutilla dammersi* Mickel, 1936
☐ *Dasymutilla digressa* Mickel, 1928
☐ *Dasymutilla dilucida* Mickel, 1928
☐ *Dasymutilla dionysia* Mickel, 1928
☐ *Dasymutilla eminentia* Mickel, 1928
☐ *Dasymutilla erythrina* (Say, 1836)
☐ *Dasymutilla eurynome* Mickel, 1928
☐ *Dasymutilla fasciventris* Mickel, 1938
☐ *Dasymutilla ferruginea* (Smith, 1879)
☐ *Dasymutilla flammifera* Mickel, 1928
☐ *Dasymutilla foxi* (Cockerell, 1894)
☐ *Dasymutilla furina* Mickel, 1928

❏ *Dasymutilla gentilis* Mickel, 1928
❏ *Dasymutilla gibbosa* (Say, 1836)
❏ *Dasymutilla gloriosa* (Saussure, 1868)
❏ *Dasymutilla gorgon* (Blake, 1871)
❏ *Dasymutilla hector* (Blake, 1871)
❏ *Dasymutilla heliophila* (Cockerell, 1900)
❏ *Dasymutilla imperialis* Manley & Pitts, 2004
❏ *Dasymutilla iztapa* (Blake, 1871)
❏ *Dasymutilla klugii* (Gray, 1832)
❏ *Dasymutilla klugioides* Mickel, 1936
❏ *Dasymutilla leda* (Blake, 1872)
❏ *Dasymutilla macilenta* (Blake, 1871)
❏ *Dasymutilla macra* (Cresson, 1865)
❏ *Dasymutilla magna* (Cresson, 1865)
❏ *Dasymutilla magnifica* Mickel, 1928
❏ *Dasymutilla meracula* Mickel, 1928
❏ *Dasymutilla monticola* (Cresson, 1865)
❏ *Dasymutilla montivagoides* (Viereck, 1906)
❏ *Dasymutilla myrice* Mickel, 1928
❏ *Dasymutilla neomexicana* Manley, 2006
❏ *Dasymutilla nigra* Manley & Pitts, 2004
❏ *Dasymutilla nigricauda* (Viereck, 1906)
❏ *Dasymutilla nigripes* (Fabricius, 1787)
❏ *Dasymutilla nocturna* Mickel, 1928
❏ *Dasymutilla nogalensis* Mickel, 1928
❏ *Dasymutilla nupera* Mickel, 1928
❏ *Dasymutilla occidentalis* (Linnaeus, 1758)
❏ *Dasymutilla parksi* Mickel, 1936
❏ *Dasymutilla pseudopappus* (Cockerell, 1895)
❏ *Dasymutilla pulchra* (Smith, 1855)
❏ *Dasymutilla quadriguttata* (Say, 1823)
❏ *Dasymutilla radkei* Manley, 2006
❏ *Dasymutilla rubicunda* Bradley, 1916
❏ *Dasymutilla sackenii* (Cresson, 1865)
❏ *Dasymutilla saetigera* Mickel, 1928
❏ *Dasymutilla satanas* Mickel, 1928
❏ *Dasymutilla scaevola* (Blake, 1871)
❏ *Dasymutilla scitula* Mickel, 1928
❏ *Dasymutilla serenitas* Mickel, 1928
❏ *Dasymutilla sicheliana* (Saussure, 1868)
❏ *Dasymutilla snoworum* (Cockerell, 1897)
❏ *Dasymutilla sophrona* Mickel, 1928
❏ *Dasymutilla stevensi* Mickel, 1928
❏ *Dasymutilla testaceiventris* (Fox, 1899)
❏ *Dasymutilla texanella* Mickel, 1928
❏ *Dasymutilla thetis* (Blake, 1886)
❏ *Dasymutilla toluca* (Blake, 1871)
❏ *Dasymutilla uniguttata* Gillaspy & Lara, 1979
❏ *Dasymutilla ursus* (Fabricius, 1793)
❏ *Dasymutilla vestita* (Lepeletier, 1845)

❏ *Dasymutilla waco* (Blake, 1871)
❏ *Dasymutilla wileyae* Mickel, 1928

Genus *Lomachaeta* Mickel, 1936
❏ *Lomachaeta argenta* Pitts & Manley, 2004
❏ *Lomachaeta beadugrimi* (Pitts & Manley, 2004)
❏ *Lomachaeta calamondin* Williams in Williams *et al.*, 2019
❏ *Lomachaeta cirrhomeris* Pitts & Manley, 2004
❏ *Lomachaeta crocopinna* Pitts & Manley, 2004
❏ *Lomachaeta hicksi* Mickel, 1936
❏ *Lomachaeta ilex* Williams & Pitts, 2009
❏ *Lomachaeta litosisyra* Williams & Pitts, 2009
❏ *Lomachaeta osita* Williams in Williams *et al.*, 2019
❏ *Lomachaeta polemomechana* Williams & Pitts, 2009
❏ *Lomachaeta powelli* (Mickel, 1964)
❏ *Lomachaeta ptilohyalus* Pitts & Manley, 2004
❏ *Lomachaeta snellingella* Williams & Pitts, 2009
❏ *Lomachaeta vacamuerta* Williams & Pitts, 2009
❏ *Lomachaeta warneri* Williams in Williams *et al.*, 2019

Tribe EPHUTINI Ashmead, 1903
Genus *Ephuta* Say, 1836
❏ *Ephuta albiceps* Schuster, 1951
❏ *Ephuta argenticeps* Schuster, 1951
❏ *Ephuta auricapitis* Schuster, 1951
❏ *Ephuta battlei* Bradley, 1916
❏ *Ephuta cephalotes* Schuster, 1951
❏ *Ephuta coloradella* Schuster, 1951
❏ *Ephuta conchate* Mickel, 1928
❏ *Ephuta copano* (Blake, 1871)
❏ *Ephuta ecarinata* Schuster, 1951
❏ *Ephuta eurygnathus* Schuster, 1951
❏ *Ephuta floridana* Schuster, 1951
❏ *Ephuta grisea* Bradley, 1916
❏ *Ephuta krombeini* Schuster, 1957
❏ *Ephuta margueritae* Schuster, 1951
❏ *Ephuta minuta* Schuster, 1951
❏ *Ephuta pauxilla* Bradley, 1916
❏ *Ephuta psephenophila* Schuster, 1951
❏ *Ephuta puteola* (Blake, 1879)
❏ *Ephuta rufisquamis* (André, 1905)
❏ *Ephuta sabaliana* Schuster, 1951
❏ *Ephuta scrupea* (Say, 1836)
❏ *Ephuta slossonae* (Fox, 1899)
❏ *Ephuta spinifera* Schuster, 1951
❏ *Ephuta stenognatha* Schuster, 1951
❏ *Ephuta sudatrix* (Melander, 1903)
❏ *Ephuta tegulica* Bradley, 1916
❏ *Ephuta tumacacori* Schuster, 1951

Tribe PSEUDOMETHOCINI Brothers, 1975
Genus *Invreiella* Suárez, 1966
❏ *Invreiella cephalargia* (Mickel, 1924)
❏ *Invreiella manleyi* Waldren in Waldren et al., 2020

Genus *Myrmilloides* André, 1902
❏ *Myrmilloides grandiceps* (Blake, 1872)

Genus *Pseudomethoca* Ashmead, 1896
❏ *Pseudomethoca ajattara* Williams & Pitts, 2008
❏ *Pseudomethoca anthracina* (Fox, 1892)
❏ *Pseudomethoca athamas* (Fox, 1899)
❏ *Pseudomethoca aureovestita* Bradley, 1924
❏ *Pseudomethoca bequaerti* Mickel, 1924
❏ *Pseudomethoca bethae* Krombein, 1991
❏ *Pseudomethoca brazoria* (Blake, 1871)
❏ *Pseudomethoca carbonaria* Mickel, 1924
❏ *Pseudomethoca connectens* (Cresson, 1865)
❏ *Pseudomethoca contumax* (Cresson, 1865)
❏ *Pseudomethoca dentigula* Mickel, 1935
❏ *Pseudomethoca donaeanae* (Cockerell, 1897)
❏ *Pseudomethoca flammigera* Mickel, 1924
❏ *Pseudomethoca flavida* (Blake, 1871)
❏ *Pseudomethoca frigida* (Smith, 1855)
❏ *Pseudomethoca gila* (Blake, 1871)
❏ *Pseudomethoca klotsi* Mickel, 1935
❏ *Pseudomethoca meritoria* Mickel, 1924
❏ *Pseudomethoca mulaiki* Mickel, 1938
❏ *Pseudomethoca nephele* (Fox, 1899)
❏ *Pseudomethoca nigricula* Mickel, 1924
❏ *Pseudomethoca nudula* Mickel, 1924
❏ *Pseudomethoca oceola* (Blake, 1871)
❏ *Pseudomethoca oculata* (Banks, 1921)
❏ *Pseudomethoca paludata* Mickel, 1924
❏ *Pseudomethoca perditrix* Krombein, 1991
❏ *Pseudomethoca peremptrix* Williams, 2023
❏ *Pseudomethoca pigmentata* Mickel, 1924
❏ *Pseudomethoca praeclara* (Blake, 1886)
❏ *Pseudomethoca propinqua* (Cresson, 1865)

❏ *Pseudomethoca quadrinotata* Mickel, 1938
❏ *Pseudomethoca russeola* Mickel, 1924
❏ *Pseudomethoca sanbornii* (Blake, 1871)
❏ *Pseudomethoca scaevolella* (Cockerell & Casad, 1895)
❏ *Pseudomethoca simillima* (Smith, 1855)
❏ *Pseudomethoca sonorae* Williams, 2023
❏ *Pseudomethoca torrida* Krombein, 1954
❏ *Pseudomethoca toumeyi* (Fox, 1894)
❏ *Pseudomethoca wickhami* (Cockerell & Casad, 1895)

Tribe SPHAEROPTHALMINI Schuster, 1949
Genus *Morsyma* Fox, 1899
❏ *Morsyma ashmeadii* Fox, 1899

Genus *Photomorphus* Viereck, 1903
❏ *Photomorphus alogus* Viereck, 1903
❏ *Photomorphus archboldi* Manley & Deyrup, 1987
❏ *Photomorphus auriventris* Schuster, 1958
❏ *Photomorphus banksi* (Bradley, 1916)
❏ *Photomorphus impar* (Melander, 1903)
❏ *Photomorphus myrmicoides* (Cockerell, 1895)
❏ *Photomorphus paulus* (Bradley, 1916)
❏ *Photomorphus quintilis* (Viereck, 1906)
❏ *Photomorphus spinci* (Bradley, 1916)

Genus *Protophotopsis* Schuster, 1947
❏ *Protophotopsis venenaria* (Melander, 1903)

Genus *Sphaeropthalma* Blake, 1871
❏ *Sphaeropthalma auripilis* (Blake, 1871)
❏ *Sphaeropthalma boweri* Schuster, 1945
❏ *Sphaeropthalma contracta* (Blake, 1879)
❏ *Sphaeropthalma edwardsii* (Cresson, 1875)
❏ *Sphaeropthalma luiseno* Schuster, 1958
❏ *Sphaeropthalma marpesia* (Blake, 1879)
❏ *Sphaeropthalma pensylvanica* (Lepeletier, 1845)
❏ *Sphaeropthalma unicolor* (Cresson, 1865)

Genus *Stethophotopsis* Pitts in Pitts & McHugh, 2000
❏ *Stethophotopsis maculata* Pitts, 2000

INDEX